AUTONOMIC NEUROIMMUNOLOGY

The Autonomic Nervous System
A series of books discussing all aspects of the autonomic nervous system.
Edited by Geoffrey Burnstock, Autonomic Neuroscience Institute, Royal
Free Hospital School of Medicine, London, UK.

This book is part of a series. The publisher will accept continuation orders which may be cancelled
at any time and which provide for automatic billing and shipping of each title in the series upon
publication. Please write for details.

AUTONOMIC NEUROIMMUNOLOGY

Edited by

John Bienenstock, Edward J. Goetzl

and

Michael G. Blennerhassett

Taylor & Francis
Taylor & Francis Group

LONDON AND NEW YORK

First published 2003
by Taylor & Francis
11 New Fetter Lane, London EC4P 4EE

Simultaneously published in the USA and Canada
by Taylor & Francis Inc,
29 West 35th Street, New York, NY 10001

Taylor & Francis is an imprint of the Taylor & Francis Group

© 2003 Taylor & Francis

Typeset in Times New Roman by
Newgen Imaging Systems (P) Ltd, Chennai, India
Printed and bound in Great Britain by
TJ International Ltd, Padstow, Cornwall

British Library Cataloguing in Publication Data
A catalogue record for this book is available from the British Library

Library of Congress Cataloging in Publication Data
Autonomic neuroimmunology / edited by John Bienenstock, Edward J. Goetzl & Michael G. Blennerhassett.
p.; cm.
Includes bibliographical references and index.
1. Neuroimmunology. 2. Autonomic nervous system – Immunology.
3. Viscera – Innervation. [DNLM: 1. Autonomic Nervous System – immunology.
2. Neuroimmunomodulation. WL600 A93964 2003] I. Bienenstock, John.
II. Goetzl, Edward J. III. Blennerhassett, Michael G.

QP356.47 .A986 2003
616.8'8079–dc21   2002151761

ISBN 0–415–30658–2

Contents

Preface to the Series – Historical and Conceptual Perspective of the Autonomic Nervous System Book Series

The pioneering studies of Gaskell (1886), Bayliss and Starling (1899), and Langley and Anderson (*see* Langley, 1921) formed the basis of the earlier and, to a large extent, current concepts of the structure and function of the autonomic nervous system; the major division of the autonomic nervous system into sympathetic, parasympathetic and enteric subdivisions still holds. The pharmacology of autonomic neuroeffector transmission was dominated by the brilliant studies of Elliott (1905), Loewi (1921), von Euler and Gaddum (1931), and Dale (1935), and for over 50 years the idea of antagonistic parasympathetic cholinergic and sympathetic adrenergic control of most organs in visceral and cardiovascular systems formed the working basis of all studies. However, major advances have been made since the early 1960s that make it necessary to revise our thinking about the mechanisms of autonomic transmission, and that have significant implications for our understanding of diseases involving the autonomic nervous system and their treatment. These advances include:

(1) Recognition that the autonomic neuromuscular junction is not a 'synapse' in the usual sense of the term where there is a fixed junction with both pre- and post-junctional specialization, but rather the transmitter is released from mobile varicosities in extensive terminal branching fibres at variable distances from effector cells or bundles of smooth muscle cells which are in electrical contact with each other and which have a diffuse distribution of receptors (see Hillarp, 1959; Burnstock, 1986a).

(2) The discovery of non-adrenergic, non-cholinergic nerves and the later recognition of a multiplicity of neurotransmitter substances in autonomic nerves, including monoamines, purines, amino acids, a variety of different peptides and nitric oxide (Burnstock *et al.*, 1964, 1986b, 1997; Rand, 1992; Milner and Burnstock, 1995; Lincoln *et al.*, 1995; Zhang and Snyder, 1995; Burnstock and Milner, 1999).

(3) The concept of neuromodulation, where locally released agents can alter neurotransmission either by prejunctional modulation of the amount of transmitter released or by postjunctional modulation of the time-course or intensity of action of the transmitter

(Marrazzi, 1939; Brown and Gillespie, 1957; Vizi, 1979; Fuder and Muscholl, 1995; MacDermott *et al.*, 1999).

(4) The concept of cotransmission that proposes that most, if not all, nerves release more than one transmitter (Burnstock, 1976; Hökfelt, Fuxe and Pernow, 1986; Burnstock, 1990a; Burnstock and Ralevic, 1996) and the important follow-up of this concept, termed 'chemical coding', in which the combinations of neurotransmitters contained in individual neurones are established, and whose projections and central connections are identified (Furness and Costa, 1987).

(5) Recognition of the importance of 'sensory-motor' nerve regulation of activity in many organs, including gut, lungs, heart and ganglia, as well as in many blood vessels (Maggi, 1991; Burnstock, 1993), although the concept of antidromic impulses in sensory nerve collaterals forming part of 'axon reflex' vasodilatation of skin vessels was described many years ago (Lewis, 1927).

(6) Recognition that many intrinsic ganglia (e.g., those in the heart, airways and bladder) contain interactive circuits that are capable of sustaining and modulating sophisticated local activities (Saffrey *et al.*, 1992; Ardell, 1994). Although the ability of the enteric nervous system to sustain local reflex activity independent of the central nervous system has been recognized for many years (Kosterlitz, 1968), it has been generally assumed that the intrinsic ganglia in peripheral organs consist of parasympathetic neurones that provided simple nicotinic relay stations.

(7) The major subclasses of receptors to acetylcholine and noradrenaline have been recognized for many years (Dale, 1914; Ahlquist, 1948), but in recent years it has become evident that there is an astonishing variety of receptor subtypes for autonomic transmitters (see *Pharmacol. Rev.*, **46**, 1994). Their molecular properties and transduction mechanisms have been characterised (see *IUPHAR Compendium of Receptor Characterisation and Classification* 2000). These advances offer the possibility of more selective drug therapy.

(8) Recognition of the plasticity of the autonomic nervous system, not only in the changes that occur during development and ageing, but also in the changes in expression of transmitter and receptors that occur in fully mature adults under the influence of hormones and growth factors following trauma and surgery, and in a variety of disease situations (Burnstock, 1990b; Saffrey and Burnstock, 1994; Milner and Burnstock, 1995; Milner *et al.*, 1999).

(9) Advances in the understanding of 'vasomotor' centres in the central nervous system. For example, the traditional concept of control being exerted by discrete centres such as the vasomotor centre (Bayliss, 1923) has been supplanted by the belief that control involves the action of longitudinally arranged parallel pathways involving the forebrain, brain stem and spinal cord (Loewy and Spyer, 1990; Jänig and Häbler, 1995).

In addition to these major new concepts concerning autonomic function, the discovery by Furchgott that substances released from endothelial cells play an important role in addition to autonomic nerves, in local control of blood flow, has made a significant impact on our analysis and understanding of cardiovascular function (Furchgott and Zawadski, 1980; Burnstock and Ralevic, 1994). The later identification of nitric oxide as the major endothelium-derived relaxing factor (Palmer *et al.*, 1988; see Moncada *et al.*, 1991) (confirming the independent suggestion by Ignarro and by Furchgott) and endothelin as an

G. Burnstock – Editor of The Autonomic Nervous System Book Series

endothelium-derived constricting factor (Yanagisawa *et al.*, 1988; see Rubanyi and Polokoff, 1994) have also had a major impact in this area.

In broad terms, these new concepts shift the earlier emphasis on central control mechanisms towards greater consideration of the sophisticated local peripheral control mechanisms.

Although these new concepts should have a profound influence on our considerations of the autonomic control of cardiovascular, urogenital, gastrointestinal and reproductive systems and other organs like the skin and eye in both normal and disease situations, few of the current textbooks take them into account. This is largely because revision of our understanding of all these different specialised areas in one volume by one author is a near impossibility. Thus, this Book Series of 14 volumes is designed to try to overcome this dilemma by dealing in depth with each major area in separate volumes and by calling upon the knowledge and expertise of leading figures in the field. Volume I, deals with the basic mechanisms of *Autonomic Neuroeffector Mechanisms* which sets the stage for later volumes devoted to autonomic nervous control of particular organ systems, including *Heart, Blood Vessels, Respiratory System, Urogenital Organs, Gastrointestinal Tract, Eye Function, Autonomic Ganglia, Autonomic-Endocrine Interactions, Development, Regeneration and Plasticity and Comparative Physiology and Evolution of the Autonomic Nervous System.*

Abnormal as well as normal mechanisms will be covered to a variable extent in all these volumes depending on the topic and the particular wishes of the Volume Editor, but one volume edited by Robertson and Biaggioni, 1995, has been specifically devoted to *Disorders of the Autonomic Nervous System* (see also Mathias and Bannister, 1999).

A general philosophy followed in the design of this book series has been to encourage individual expression by Volume Editors and Chapter Contributors in the presentation of the separate topics within the general framework of the series. This was demanded by the different ways that the various fields have developed historically and the differing styles of the individuals who have made the most impact in each area. Hopefully, this deliberate lack of uniformity will add to, rather than detract from, the appeal of these books.

G. Burnstock
Series Editor

REFERENCES

Ahlquist, R.P. (1948). A study of the adrenotropic receptors. *Am. J. Physiol.*, **153**, 586–600.
Ardell, J.L. (1994). Structure and function of mammalian intrinsic cardiac neurons. In *Neurocardiology*, edited by J.A. Armour and J.L. Ardell, pp. 95–114. Oxford: Oxford University Press.
Bayliss, W.B. (1923). *The Vasomotor System*. Longman: London.
Bayliss, W.M. and Starling, E.H. (1899). The movements and innervation of the small intestine. *J. Physiol. (Lond.)*, **24**, 99–143.
Brown, G.L. and Gillespie, J.S. (1957). The output of sympathetic transmitter from the spleen of a cat. *J. Physiol. (Lond.)*, **138**, 81–102.
Burnstock, G. (1976). Do some nerve cells release more than one transmitter? *Neuroscience*, **1**, 239–248.
Burnstock, G. (1986a). Autonomic neuromuscular junctions: Current developments and future directions. *J. Anat.*, **146**, 1–30.
Burnstock, G. (1986b). The non-adrenergic non-cholinergic nervous system. *Arch. Int. Pharmacodyn. Ther.*, **280**(suppl.), 1–15.

Burnstock, G. (1990a). Co-transmission. The fifth heymans lecture – Ghent, February 17, 1990. *Arch. Int. Pharmacodyn. Ther.*, **304**, 7–33.

Burnstock, G. (1990b). Changes in expression of autonomic nerves in aging and disease. *J. Auton. Neiv. Syst.*, **30**, 525–534.

Burnstock, G. (1993). Introduction: changing face of autonomic and sensory nerves in the circulation. In *Vascular Innervation and Receptor Mechanisms: New Perspectives*, edited by L. Edvinsson and R. Uddman, pp. 1–22. San Diego: Academic Press Inc.

Burnstock, G. (1997). The past present and future of purine nucleotides as signalling molecules. *Neuropharmacology*, **36**, 1127–1139.

Burnstock, G., Campbell, G., Bennett, M. and Holman. M.E. (1964). Innervation of the guinea-pig taenia coli: are there intrinsic inhibitory nerves which are distinct from sympathetic nerves? *Int. J. Neuropharmacol.*, **3**, 163–166.

Burnstock, G. and Milner, P. (1999). Structural and chemical organisation of the autonomic nervous system with special reference to non-adrenergic, non-cholinergic transmission. In *Autonomic Failure: A Textbook of Clinical Disorders of the Autonomic Nervous System*. 4th edn, edited by C.J. Mathias and R. Bannister, pp. 63–71, Oxford: Oxford University Press.

Burnstock, G. and Ralevic, V. (1994). New insights into the local regulation of blood flow by perivascular nerves and endothelium. *Br. J. Plast. Surg.*, **47**, 527–543.

Burnstock, G. and Ralevic, V. (1996). Cotransmission. In *The Pharmacology of Smooth Muscle*, edited by C.J. Garland and J. Angus., Oxford: Oxford University Press.

Dale, H. (1914). The action of certain esters and ethers of choline and their reaction to muscarine. *J. Pharmacol. Exp. Ther.*, **6**, 147–190.

Dale, H. (1935). Pharmacology and nerve endings. *Proc. Roy. Soc. Med.*, **28**, 319–332.

Elliott, T.R. (1905). The action of adrenalin. *J. Physiol. (Lond.)*, **32**, 401–467.

Fuder, H. and Muscholl, E. (1995). Heteroceptor-mediated modulation of noradrenaline and acetylcholine release from peripheral nerves. *Rev. Physiol. Biochem. Physiol.*, **126**, 265–412.

Furchgott, R.F. and Zawadski, J.V. (1980). The obligatory role of endothelial cells in the relaxation of arterial smooth muscle by acetylcholine. *Nature*, **288**, 373–376.

Furness, J.B. and Costa, M. (1987). *The Enteric Nervous System*. Edinburgh: Churchill Livingstone.

Gaskell, W.H. (1886). On the structure, distribution and function of the nerves which innervate the visceral and vascular systems. *J. Physiol. (Lond.)*, **7**, 1–80.

Hillarp, N.-Å. (1959). The construction and functional organisation of the autonomic innervation apparatus. *Acta Physiol. Scand.*, **46** (suppl. 157), 1–38.

Hökfelt, T., Fuxe, K. and Pernow, B. (Eds.) (1986). Coexistence of neuronal messengers: a new principle in chemical transmission. In *Progress in Brain Research*, Vol. 68. Amsterdam: Elsevier.

IUPHAR Compendium of Receptor Characterisation and Classification 2000, IUPHAR Media Ltd (London), UK.

Jänig, W. and Häbler, H.-J. (1995). Visceral-autonomic integration. In *Visceral Pain, Progress in Pain Research and Management*, edited by G.F. Gebhart, Vol. 5, pp. 311–348. Seattle: IASP Press.

Kosterlitz, H.W. (1968). The alimentary canal. In *Handbook of Physiology*, edited by C.F. Code, Vol. IV, pp. 2147–2172. Washington, DC: American Physiological Society.

Langley, J.N. (1921). *The Autonomic Nervous System*, part 1. Cambridge: W. Heffer.

Lewis, T. (1927). *The Blood Vessels of the Human Skin and Their Responses*. Shaw & Sons: London.

Lincoln, J., Hoyle, C.H.V. and Burnstock, G. (1995). Transmission: nitric oxide. In *The Autonomic Nervous System*, Vol. 1 (reprinted): *Autonomic Neuroeffector Mechanisms*, edited by G. Burnstock and C.H.V. Hoyle, pp. 509–539. The Netherlands: Harwood Academic Publishers.

Loewi, O. (1921). Über humorale Übertrangbarkeit der Herznervenwirkung. XI. Mitteilung. *Pflügers Arch. Gesamte Physiol.*, **189**, 239–242.

Loewy, A.D. and Spyer, K.M. (1990). *Central Regulations of Autonomic Functions*. New York: Oxford University Press.

MacDermott, A.B., Role, L.W. and Siegelbaum, S.A. (1999). Presynaptic iootropic receptors and the control of transmitter release. *Ann. Rev. Neurosci.*, **22**, 443–485.

Maggi, C.A. (1991). The pharmacology of the efferent function on sensory nerves. *J. Auton. Pharmacol.*, **11**, 173–208.

Marrazzi, A.S. (1939). Electrical studies on the pharmacology of autonomic synapses. II. The action of a sympathomimetic drug (epinephrine) on sympathetic ganglia. *J. Pharmacol. Exp. Ther.*, **65**, 395–404.

Mathias, C.J. and Bannister, R. (eds.) (1999). *Autonomic Failure*, 4th edn., Oxford: Oxford University Press.

Milner, P. and Burnstock, G. (1995). Neurotransmitters in the autonomic nervous system. In *Handbook of Autonomic Nervous Dysfunction*, edited by A.D. Korczyn, pp. 5–32. New York: Marcel Dekker.

Milner, P., Lincoln, J. and Burnstock, G. (1999). The neurochemical organisation of the autonomic nervous system. In *Handbook of Clinical Neurology Vol 74(30): The autonomic nervous system – Part 1 – Normal Functions*, edited by O. Appezeller, pp. 87–134, Amsterdam: Elsevier Science.

Moncada, S., Palmer, R.M.J. and Higgs, E.A. (1991). Nitric oxide: physiology, pathophysiology, and pharmacology. *Pharmacol. Rev.*, **43**, 109–142.

Palmer, R.M.J., Rees, D.D., Ashton, D.S. and Moncada, S. (1988). Arginine is the physiological precursor for the formation of nitric oxide in endothelium-dependent relaxation. *Biochem. Biophys. Res. Commun.*, **153**, 1251–1256.

Rand, M.J. (1992). Nitrergic transmission: nitric oxide as a mediator of non-adrenergic, non-cholinergic neuro-effector transmission. *Clin. Exp. Pharmacol. Physiol.*, **19**, 147–169.

Rubanyi, G.M. and Polokoff, M.A. (1994). Endothelins: molecular biology, biochemistry, pharmacology, physiology, and pathophysiology. *Pharmacol. Rev.*, **46**, 328–415.

Saffrey, M.J. and Burnstock, G. (1994). Growth factors and the development and plasticity of the enteric nervous system. *J. Auton. Nerv. Syst.*, **49**, 183–196.

Saffrey, M.J., Hassall, C.J.S., Allen, T.G.J. and Burnstock, G. (1992). Ganglia within the gut, heart, urinary bladder and airways: studies in tissue culture. *Int. Rev. Cytol.*, **136**, 93–144.

Vizi, E.S. (1979). Prejunctional modulation of neurochemical transmission. *Prog. Neurobiol.*, **12**, 181–290.

von Euler, U.S. and Gaddum, J.H. (1931). An unidentified depressor substance in certain tissue extracts. *J. Physiol.*, **72**, 74–87.

Yanagisawa, M., Kurihara, H., Kimura, S., Tomobe, Y., Kobayashi, M., Mitsui, Y., Yazaki, Y., Goto, K. and Masaki, T. (1988). A novel potent vasoconstrictor peptide produced by vascular endothelial cells. *Nature*, **332**, 411–415.

Zhang, J. and Snyder, S.H. (1995). Nitric oxide in the nervous system. *Annu. Rev. Pharmacol. Toxicol.*, **35**, 213–233.

Preface

Many seminal observations have established the evolution and ontogenetic development of anatomic connections between elements of the nervous system and cells of adaptive immunity. On this morphologic foundation, diverse findings of early functional studies documented the capacity of neuromediators to evoke and control immune responses, and of immune cytokines to alter neural activities. Contributions in the present volume describe in rich scientific detail our current understanding of these vital neuroimmunological interactions and their implications for normal physiology and disease states.

The immune system sends messages to neurons and glia other neural cells in many forms, including histamine and proteinases from mast cells, eicosanoids from macrophages and mast cells, and an array of cytokines from macrophages and T cells. The chapters by Drs Befus, Bunnett, Pothoulakis, Theoharides, and Undem and Weinreich describe the many aspects of neural development and function which are influenced by immune factors.

Neuroendocrine factors, encompassing hormones of the hypothalamic–pituitary axis with systemic activities and various neuropeptides with principally compartmental functions, are major stimuli or inhibitors of immune cell mobilization and activation. Neuropeptide activation of mast cells was one of the first observations in the field and the section by Drs Blennerhassett and Bienenstock documents the cell biological mechanisms for these events. That one or more adrenergic and peptidergic messengers modulate immunity is analyzed by Drs Cooke, Felten and Nance, whereas Dr. Tracey elucidates the pathways by which cholinergic factors may suppress immunological inflammation. Prime examples of how a single neuropeptide may regulate the nature and intensity of T cell and B cell involvement in compartmental immune responses with a high degree of specificity are provided by Drs Goetzl, McGillis and Weinstock. Their findings are supported not only by comprehensive biochemical and pharmacological data, but also by genetic manipulations of expression of neuropeptides and neuropeptide receptors in dedicated lines of mice with clear neuroimmunological phenotypes. Potential implications of organ system-selective neuroimmunology for host defense and diseases are elegantly elucidated by Dr Luger for the skin, Drs Perdue and Pothoulakis for the intestines, Dr Theoharides for the bladder, and Drs Renz, and Undem and Weinreich for the lungs.

The clear message of the volume is that modern neuroimmunology is firmly established as a scientific discipline, an avenue for approaching distinctive mechanisms of mammalian biology, and a pathway to novel diagnostic techniques and therapies for diseases of the neural, endocrine and immune systems.

Contributors

A. Dean Befus
Pulmonary Research Group
Department of Medicine
Faculty of Medicine
University of Alberta
Edmonton, Alberta
Canada, T6G 2S2

John Bienenstock
Department of Pathology and Molecular
  Medicine
McMaster University
1200 Main Street West,
Hamilton, Ontario L8N 3Z5
Canada

Michael G. Blennerhassett
Gastrointestinal Diseases
  Research Unit
Department of Medicine
Queen's University
Kingston, Ontario,
Canada

Armin Braun
Fraunhofer Institute of Toxicology and
  Aerosol Research
Drug Research and Clinical Inhibition
30625 Hannover
Germany

Nigel W. Bunnett
Departments of Surgery and Physiology
University of California
Box 0660, Room C317
521 Parnassus Avenue
San Francisco, CA 94143-0660
USA

Ignazio Castagliuolo
Institute of Microbiology
University of Padua, Padua
Italy

Robert C. Chan
Departments of Medicine and
  Microbiology–Immunology
University of California Medical Center
UB8B, Box 0711
San Francisco, CA 94143-0711
USA

Fievos L. Christofi
Department of Anesthesiology
The Ohio State University
Columbus, Ohio 43210
USA

Helen J. Cooke
Department of Neuroscience
The Ohio State University
Columbus, Ohio 43210
USA

Christopher J. Czura
Laboratory of Biomedical Science
North Shore-LIJ Research Institute
350 Community Drive
Manhasset, NY 11030
USA

Joseph S. Davison
Department of Physiology and Biophysics
Faculty of Medicine
University of Calgary
Calgary, Alberta
Canada, T2N 4N1

Rene E. Déry
Pulmonary Research Group
Department of Medicine
Faculty of Medicine
University of Alberta
Edmonton, Alberta
Canada, T6G 2S2

Glenn Dorsam
Departments of Medicine and
  Microbiology–Immunology
University of California Medical Center
UB8B, Box 0711
San Francisco, CA 94143-0711
USA

David L. Felten
Susan Samueli Center for Complementary
  and Alternative Medicine and Department
  of Anatomy and Neurobiology
University of California
Irvine College of Medicine
Irvine, CA 92612-5850
USA

Stefan Fernandez
Department of Microbiology, Immunology
  and Molecular Genetics
University of Kentucky College of
  Medicine
MS 415, 800 Rose Street
Lexington, KY 40536-0084
USA

Paul Forsythe
Pulmonary Research Group
Department of Medicine
Faculty of Medicine
University of Alberta
Edmonton, Alberta
Canada, T6G 2S2

Pierangelo Geppetti
Department of Experimental and
  Clinical Medicine
Section of Pharmacology
University of Ferrara
Via Fossato di Mortara 19,
44100 Ferrara
Italy

Edward J. Goetzl
Departments of Medicine and
  Microbiology–Immunology
University of California Medical Center
UB8B, Box 0711
San Francisco, CA 94143-0711
USA

Najma Javed
Department of Physiology and
  Health Science
Ball State University
Muncie, IN
USA

Sheila P. Kelley
Center for Psychoneuroimmunology
  Research
University of Rochester Medical Center
300 Girtendenn Boulevard, Rochester
  NY 14642
USA

Thomas A. Luger
Department of Dermatology and
  Boltzmann Institute for Cell- and
  Immunobiology of the Skin
University of Münster
Von-Esmarch-Str. 58
48219 Münster
Germany

Ronald Mathison
Department of Physiology and Biophysics
Faculty of Medicine
University of Calgary
Calgary, Alberta
Canada, T2N 4N1

Joseph P. McGillis
Department of Microbiology, Immunology
  and Molecular Genetics
University of Kentucky College of Medicine
MS415, 800 Rose Street
Lexington, KY 40536-0084
USA

Jonathan C. Meltzer
Departments of Pathology and Anatomy
University of Manitoba
Winnipeg
Manitoba R3E 0W3
Canada

Dwight M. Nance
Susan Samueli Centre for Complementary
  and Alternative Medicine
UCI College of Medicine
Orange, CA 92868
USA

Mary H. Perdue
Intestinal Disease Research Program
HSC3N5E
McMaster University
Hamilton, Ontario
Canada

Charalabos Pothoulakis
Division of Gastroenterology
Beth Israel Deaconess Medical Center
Harvard Medical School
Boston, 02115 Massachusetts
USA

Harald Renz
Department of Clinical Chemistry
  and Molecular Diagnostics
Philipps-University Marburg

35033 Marburg, Baldingerstr.
Germany

Grannum R. Sant
Department of Urology
Tufts University
School of Medicine and New England
  Medical Center
Boston, MA 02111
USA

Sonja Ständer
Department of Dermatology and
  Boltzmann Institute for Cell- and
  Immunobiology of the Skin
University of Münster
Von-Esmarch-Str. 58
48129 Münster
Germany

Martin Steinhoff
Department of Dermatology and
  Boltzmann Institute for Cell- and
  Immunobiology of the Skin
University of Münster
Von-Esmarch-Str. 58
48129 Münster
Germany

Johan D. Söderholm
Division of Surgery
Department of Biomedicine and Surgery
Linköping University
Sweden

Theoharides C. Theoharides
Department of Pharmacology and
  Experimental Therapeutics
Tufts University School of Medicine
136 Harrison Avenue
Boston, MA 02111
USA

Kevin J. Tracey
Laboratory of Biomedical Science
North Shore-LIJ Research Institute
350 Community Drive
Manhasset, NY 11030
USA

Bradley J. Undem
Department of Medicine
Johns Hopkins School of Medicine
Baltimore, MD 21205
USA

Daniel Weinreich
Department of Pharmacology and
  Experimental Therapeutics
University of Maryland School of Medicine
655 West Baltimore Street
Room 4-002
Baltimore, MD 21201
USA

Joel V. Weinstock
Division of Gastroenterology–Hepatology
Department of Medicine
University of Iowa
Rm 4607, JCP
Iowa City, IA 52242-1009
USA

Hanneke P.M. van der Kleij
Department of Pharmacology and
  Pathophysiology
Utrecht University
Utrecht
The Netherlands

1 Sympathetic Innervation of the Immune System and its Role in Modulation of Immune Responses, Immune-Related Diseases Models, and Integrative Medicine

David L. Felten[1] and Sheila P. Kelley[2]

[1]Susan Samueli Center for Complementary and Alternative Medicine and Department of Anatomy and Neurobiology, University of California, Irvine College of Medicine, Irvine, CA 92612-5850, USA
[2]Center for Psychoneuroimmunology Research, University of Rochester Medical Center, 300 Girtendenn Boulevard, Rochester, NY 14642, USA

Sympathetic noradrenergic nerve fibres innervate the vasculature and parenchyma of primary lymphoid organs (bone marrow, thymus), secondary lymphoid organs (spleen, lymph nodes), and other lymphoid tissue. Many lymphoid cells express adrenoceptors that are responsive to circulating and nerve-derived catecholamines; these receptors are differentially expressed and can be highly regulated. β-Receptor stimulation can alter functional activities of neutrophils, antigen-presenting cells, T and B lymphocytes, and NK cells. Noradrenergic sympathetic denervation of secondary lymphoid organs generally results in diminished cell-mediated immune responses and enhanced antibody responses, with a cytokine shift towards Th1. Circulating catecholamines markedly suppress mature effector cells, including NK cells. Catecholamines and sympathetic nerves significantly modulate the severity and expression of viral infections, wound healing, autoimmune reactivity, mammary tumour growth, pro-liferation, and spread, and immune senescence. Chronic stressors are accompanied by a general pattern of immune changes that include diminished cell-mediated and NK cell functions, and activation of latent viruses. Some complementary medical interventions, such as aerobic exercise, humour and laughter, and music interven-tions, can evoke changes that are directionally opposite to those of chronic stressors. We propose that comple-mentary interventions utilize neuroendocrine and sympathetic neural signalling of immune responses to achieve beneficial effects for wellness and chronic disease interventions.

KEY WORDS: catecholamines; sympathetic nerves; cell-mediated immunity; natural killer cell activity; cancer; autoimmune disease; complementary medicine; immune senescence.

INTRODUCTION TO NEURAL MODULATION OF IMMUNE RESPONSES

Over the past three decades, evidence pointing to bi-directional communication between the nervous system and the immune system has been clearly established. Several lines of research support this contention, including the following evidence: (1) lesions in the central nervous system (CNS) lead to altered immune responses (Katayama *et al.*, 1987; Felten *et al.*, 1991); (2) immune responses can be classically conditioned (Ader and Cohen, 1975, 2001), involving a CNS learning paradigm; (3) primary and secondary lymphoid organs are innervated with noradrenergic (NA) sympathetic nerve fibres and other peptidergic nerve fibres (Felten and Felten, 1991; Bellinger *et al.*, 2001); (4) cells of the immune system possess receptors for a host of hormones and neurotransmitters, including receptors for catecholamines, ACTH, opioid peptides, substance P, and vasoactive intestinal peptide (VIP), to name a few (Plaut, 1987; Carr and Blalock, 1991; Madden *et al.*, 1995); (5) functions of these same immune system cells that possess receptors are altered by the appropriate hormones and neurotransmitters (see Madden and Felten, 1995); and (6) cytokines produced in the periphery can affect the brain, producing illness behaviour and activation of neuroendocrine and autonomic stress-related circuits (Hori *et al.*, 2001; Maier *et al.*, 2001; Rivier, 2001).

The two major routes by which neural-immune communication occurs include the hypo-thalamo-pituitary-endocrine organ axis (neuroendocrine signalling) and direct neural connections with cells in lymphoid organs. There are extensive reviews of neuroendocrine influences on immune responses (e.g. Madden and Felten, 1995), a topic that will not be pursued further in this chapter. The direct neural connections are via a two-neuron chain from the thoraco-lumbar spinal cord, through the sympathetic chain ganglia or collateral ganglia, to primary and secondary lymphoid organs and other lymphoid tissue. The main neurotransmitter for sympathetic innervation is norepinephrine, although co-localized neuropeptides such as neuropeptide Y are abundant, but not invariable. Although extensive peptidergic innervation of lymphoid organs has been documented (Bellinger *et al.*, 1990), we still do not know the origin for most of these peptidergic fibres. While some may be co-localized with norepinephrine (NE) in sympathetic nerve fibres, others may be independently present in autonomic or primary sensory nerves. The routes of communication by which cytokines and other immune-derived products can signal the brain are varied and complex (Maier *et al.*, 2001), and include direct crossing into the brain, stimulation of the vagus nerve and other peripheral nerves, action on cells in the circumventricular organs, and stimulation of small molecules such as nitric oxide (NO) and prostaglandin E2 (PGE2).

IDENTIFICATION OF ADRENOCEPTORS ON CELLS OF THE IMMUNE SYSTEM

Receptors responsive to NE and epinephrine are classified as α- and β-adrenoceptors. The β-adrenoceptors are generally coupled intracellularly to the Gs protein of the adenylate cyclase complex, leading to the generation of intracellular cAMP. The $\alpha1$ receptor is linked with increased phosphatidylinositol turnover and a rise in intracellular calcium, while the $\alpha2$ receptor is linked with the Gi subunit of the adenylate cyclase complex. High affinity β-adrenoceptors, mainly of the $\beta2$ class, are found on lymphocytes (Brodde *et al.*, 1981; Landmann *et al.*, 1981, 1985). $\alpha2$-Adrenoceptors have been identified on human lymphocytes (Titinchi and Clark, 1984; Goin *et al.*, 1991). Activated macrophages express both $\alpha2$- and β-adrenoceptors (Abrass *et al.*, 1985; Spengler *et al.*, 1990). Other inflammatory cells such as neutrophils, basophils, and eosinophils also express β-adrenoceptors (Plaut, 1987; Yukawa *et al.*, 1990).

β-Adrenoceptor density appears to vary according to cell type and the stage of development of that cell (Fuchs *et al.*, 1988). B lymphocytes possess the highest density of β-adrenoceptors (perhaps reflecting upregulation from their distance from NA innervation). T lymphocytes have a lower density of β-adrenoceptors, with cytotoxic T cells possessing a higher density than helper T cells. In addition, the density of β-adrenoceptors increases on thymocytes as they mature. Activation of lymphocytes in the spleen led to a lower density of β-adrenoceptor expression (Fuchs *et al.*, 1988), while activation of lymphocytes in lymph nodes led to a higher density of β-adrenoceptor expression (Madden *et al.*, 1989).

Sanders and colleagues (Sanders *et al.*, 1994, 1997) demonstrated the expression of β-adrenoceptors on resting Th1 cells (cytokines favouring cell-mediated immunity), but not on resting Th2 cells (cytokines favouring humoural immunity). Thus, the expression of adrenoceptors appears to be highly regulated, and may vary according to strain, cell type, state of activation, level of maturation, and the local microenvironment of signals that can influence the expression of these receptors. An additional feature of adrenoceptor activation is the phenomenon of dual signalling, whereby activation of two or more receptors can influence the intracellular effects of that signalling; a clear example is T cell receptor stimulation and β-adrenoceptor stimulation exerting synergistic effects on cAMP production (Carlson *et al.*, 1989).

AUTONOMIC INNERVATION OF ORGANS OF THE IMMUNE SYSTEM

Both primary and secondary lymphoid organs of the immune system are abundantly innervated by postganglionic NA sympathetic nerve fibres (see Felten and Felten, 1991; Madden and Felten, 1995; Bellinger *et al.*, 2001 for detailed reviews). NA nerve fibres supply both the arteries and the parenchyma of the bone marrow (Felten *et al.*, 1996). NA nerve fibres travel along the capsule and trabeculae of the thymus, with some fibres extending into the cortex and medullary regions (Felten *et al.*, 1985). As the thymus involutes with age, more nerve fibres appear in the parenchyma of the thymus.

Secondary lymphoid organs also receive abundant NA sympathetic innervation (Williams and Felten, 1981; Williams *et al.*, 1981). NA nerve fibres enter the spleen and distribute along the central artery and its branches, and also extend into the periarteriolar lymphatic sheath (PALS) (Ackerman *et al.*, 1987). In the PALS, nerve terminals end in close apposition to T lymphocytes of both CD4 and CD8 subsets (Felten and Olschowka, 1987). NA nerve fibres also distribute along the marginal sinus, and extend into the marginal zone. In early development, NA innervation appears in the parenchyma of the PALS several days before NA nerve fibres extend along the vasculature (Ackerman *et al.*, 1991), suggesting that separate systems of NA nerves may innervate these two compartments. NA innervation of lymph nodes enters the node with the vasculature, extends through the medullary cords, and arborizes extensively in the paracortex and cortex, among T lymphocytes. As a similar feature in spleen and lymph nodes, NA innervation supplies T lymphocyte and macrophage/antigen presenting cell zones, but does not abundantly supply the follicles/germinal centres. However, it still is possible that with high sympathetic nerve activity, released NE can diffuse through the parenchyma in a paracrine-like fashion, and bind to adrenoceptors on B lymphocytes. This relative location of NA nerve terminals with regard to the presence of potential target cells with adrenoceptors may constitute an important regulatory feature of neural–immune signalling. Cells close to NA nerve terminals may see a constant presence of NE, while cells distant to NA nerve terminals may only have NE present in conditions of high sympathetic nerve activity.

SYMPATHETIC NERVES MODULATE INNATE AND ACQUIRED IMMUNE RESPONSES

One straightforward approach to looking at ligand–receptor interactions of catecholamines with lymphoid cells is the use of specific catecholamines or their agonists and antagonists in dose–response effects on specific cell types of the immune system. These studies, while giving little information about catecholamine influences on complex immune responses that involve cooperativity of multiple cell types, nonetheless provide some insight into individual cellular responses to adrenoceptor stimulation.

Several studies of neutrophils revealed that β-adrenoceptor stimulation inhibited respiratory burst activity (Nielson, 1987; Gibson-Berry et al., 1993) as well as spontaneous activity and chemotaxis (Rivkin et al., 1975). The influence of catecholamine stimulation of lymphocytes is complex. β-Adrenoceptor stimulation of B lymphocytes either enhanced or inhibited proliferation, depending upon which stimulator of B cell activity was used (see Madden et al., 1995 for discussion). Effects of catecholamines on T cell proliferation are mainly inhibitory, but are not uniformly so, depending again upon other molecules present. NK cell activity (Katz et al., 1982; Hellstrand et al., 1985) and lytic activity of cytotoxic T lymphocytes (Strom et al., 1973) were inhibited by β-adrenoceptor agonists.

Another approach to studying sympathetic neural influences on immune responses is chemical sympathectomy with 6-hydroxydopamine, an oxidative catecholamine-related neurotoxin that selectively destroys peripheral NA nerve terminals. With this technique, more than 90% of the NA nerve terminals in the spleen and lymph nodes can be selectively destroyed (Williams et al., 1981; Livnat et al., 1985; Madden et al., 1994a). With use of chemical sympathectomy, our laboratory has found that primary antibody responses are generally increased (Williams et al., 1981; Livnat et al., 1985; Madden et al., 1995), cytotoxic T lymphocyte responses and delayed-type hypersensitivity responses (cell-mediated immunity) are generally diminished (Livnat et al., 1985; Madden et al., 1989, 1994a,b). These responses generally were obtained when the chemical sympathectomy was initiated prior to immune challenge; thus, these effects reflect events occurring at the initiative phase of an immune response. In addition to these alterations in acquired immune responses, there were alterations in conA-induced T lymphocyte proliferation in vivo (generally decreased), altered cellularity (increased B cells, decreased T cells), an immunoglobulin isotype switch from IgM to IgG, and altered lymphocyte trafficking in lymph nodes of non-immune mice. Indeed, the effects of sympathectomy are highly dependent upon whether or not the cells in question have been challenged by antigen.

The influence of sympathetic nerve fibres on immune responses, as noted above, is not simple or unidirectional. For example Kruszewska *et al.* (1995, 1998) found that in a Th1 dominant strain of mice (C57Bl/6) acutely or chronically chemically sympathectomized, antibody responses to keyhole limpet hemocyanin (KLH) were increased (both IgG and IgM), and KLH-specific IL-2 and IL-4 production *in vitro* were enhanced. However, in a Th2 dominant strain (Balb/c), antibody responses were not elevated, although both IL-2 and IL-4 production *in vitro* were increased. Interestingly, these effects were not altered by the presence of RU-486, a glucocorticoid receptor antagonist, suggesting that these observed effects were not due to a secondary alteration of glucocorticoid secretion from altered hypothalamo-pituitary-adrenal (HPA) activity, but were due to the actual removal of the sympathetic nerves themselves.

Before concluding the section on catecholamine and sympathetic nerve effects on immune responses, an important distinction should be drawn. From the collective literature, it appears that the effects of β-adrenoceptor stimulation during the activational phase of the immune response results in enhanced responses, but β-adreneceptor stimulation late in an immune response, especially the effects on mature effector cell function, is markedly inhibitory (see Madden *et al.*, 1995 for discussion). Thus, it appears that the timing of the catecholamine with regard to the antigen challenge is critical. In addition, the sympathetic nerve fibres supplying the spleen and lymph nodes, when activated during the initiative phase of an immune response appear to be immuno-enhancing, while circulating catecholamine effects on mature effector cells, particularly those in the periphery responding to infectious challenge, are markedly inhibiting.

SYMPATHETIC NERVES MODULATE IMMUNE REACTIVITY IN DISEASE MODELS, AND AFFECT THE OUTCOME OF DISEASE

INFECTIOUS DISEASE AND WOUND HEALING

The process of wound healing, although complex and involving many cell types and mediators, appears to be highly sensitive to stressors, such as examination stress in medical students (Marucha *et al.*, 1998). Some aspects of impairment of wound healing appear to be attributable to catecholamine influences. In a related vein, studies from Sheridan's laboratory (Sheridan *et al.*, 1991, 1994; Sheridan, 1998; Bonneau *et al.*, 2001) have investigated viral infections in experimental animals, including influenza and herpes simplex infections. They have described impaired cell-mediated immune responses in animals exposed to restraint stress. Some of the effects of stress-induced modulation of immune responses are due to corticosterone, and can be reversed by corticosterone blockade, while other effects are due to catecholamine secretion, and can be reversed by β-adrenoceptor blockade (Moynihan and Stevens, 2001). Rice *et al.* (2001) demonstrated that chemical sympathectomy increases the innate immune response, and decreases specific acquired immune response in the spleen to infectious challenge with Listeria monocytogenes. Thus, not all responses to catecholamine alteration follow a unidimensional pattern, and the site (spleen and lymph nodes versus periphery) where measurements may determine the specific response to catecholamine challenge or removal.

In a retroviral infectious model in C57Bl/6 mice, a challenge with LP-BM5 mixture (murine AIDS, or MAIDS), a retroviral mixture that damages both T and B lymphocytes and has many characteristics of human AIDS infection, chemical sympathectomy was carried out prior to challenge. Chemical sympathectomy had no effect at all on the severity, course, or cytokine expression in murine AIDS, nor did use of implanted nadolol β-adrenoceptor blockade (Kelley *et al*., 2001). In view of the marked changes, both *in vitro* and *in vivo*, induced by chemical sympathectomy and β-adrenoceptor blockade in innate and acquired immune responses, and in most animal models of immune-related disease, this finding was surprising. However, an explanation became apparent in the control group (non-sympathectomized) infected with murine AIDS. The viral infection itself markedly depleted norepinephrine levels in the spleen and produced a functional sympathectomy. We do not yet know the mechanism by which the LP-BM5 retroviral mixture produces sympathectomy; there is no evidence for direct viral invasion of sympathetic NA neurons. Therefore, another indirect mechanism must be identified.

AUTOIMMUNE DISORDERS

In several models of autoimmune disease, evidence is pointing towards both the HPA axis and sympathetic catecholamines as important modulators of the severity and expression of autoimmunity. Sternberg's laboratory (Sternberg *et al*., 1989, 1992; Sternberg, 1997) has demonstrated that strains of rats (Lewis/N) susceptible to induced autoimmunity, show diminished activation of the HPA axis, and thus are unable to generate sufficient glucocorticoid responsiveness to prevent autoimmunity. This finding was demonstrated in streptococcal cell-wall induced rheumatoid arthritis. They also found that normally non-susceptible rats (Fischer 344) could be made susceptible to induced rheumatoid arthritis if the HPA axis were inhibited.

Felten and colleagues (Felten *et al*., 1992a,b; Lorton *et al*., 1996) demonstrated that chemical sympathectomy resulted in earlier onset and greater severity of adjuvant-induced rheumatoid arthritis. Studies by Lorton *et al*. (1996, 1999) used very specific local NA denervation of the popliteal and inguinal lymph nodes with 6-hydroxydopamine to avoid systemic effects of this neurotoxin. This selective procedure was sufficient to markedly exacerbate the severity of induced rheumatoid arthritis. Systemic catecholamine depletion with guanethidine had the opposite effect, and reduced the severity of induced rheumatoid arthritis, probably due to its effects on NA innervation of the joint and subsequent inflammatory responses at that site. Interestingly, when substance P nerve fibres were denervated in these same lymph nodes using capsaicin, the opposite effect was achieved, that of delayed onset and reduced severity of induced rheumatoid arthritis (Lorton *et al*., 2000). These studies are consistent with earlier findings using systemic β-adrenoceptor blockade (Levine *et al*., 1988). Systemic β-blockade blocked inflammatory catecholamine responses in the joints in rheumatoid arthritis, thus superceding whatever other effects may have occurred in the lymph nodes. Thus, it appears that separate noradrenergic mechanisms are at play in the lymph nodes and the peripheral target tissue, the joints.

Chelmicka-Schorr *et al*. (1988, 1989) demonstrated that chemical sympathectomy augments the severity of experimental allergic encephalomyelitis (a model for central demyelinating disease), while a β-agonist, isoproterenol, suppresses the expression of this induced autoimmune demyelinating disease. These findings are consistent with the local denervation studies in experimental rheumatoid arthritis. A further intriguing finding is found in the genetically autoimmune strain of mice, the MRL lpr/lpr strain, that dies of a lupus-like autoimmune disease (Brenneman *et al.*, 1993). These mice demonstrate a decrease in the innervation of the spleen that coincides with the onset of their autoimmune pathology. Thus, it appears that the sympathetic NA nerve fibres are an important contributory system to the regulation of autoimmune reactivity, both in genetic and induced autoimmune models.

Breast cancer models

Several studies from our laboratory have focused on carcinogenically induced (DMBA) mammary tumours and spontaneously occurring mammary tumours in Sprague–Dawley rats. We have used deprenyl, an MAO-B inhibitor that has additional antioxidant and anti-angiogenic properties, as well as the ability to stimulate the sprouting of sympathetic nerve fibres. We found that low-dose deprenyl could stimulate the regrowth of sympathetic NA nerve fibres into the spleen in young sympathectomized rats or old rats with natural age-related depletion of sympathetic NA nerve fibres (ThyagaRajan *et al.*, 1998a). Deprenyl-stimulated regrowth of nerve fibres into the spleen of old rats was accompanied by increased IL-2 production, increases T cell proliferation, and increased NK cell activity (ThyagaRajan *et al.*, 1998c). When given to young female Sprague–Dawley rats with carcinogenically induced mammary tumours, deprenyl prevented the growth, increased number, and spread of tumours (ThyagaRajan *et al.*, 1998b) by a process that involved effects on both central and peripheral catecholamine neurotransmission (ThyagaRajan *et al.*, 1999). Deprenyl administration is equally effective at stopping the growth and spread of spontaneously occurring mammary tumours in old Sprague–Dawley rats (ThyagaRajan *et al.*, 2000). These findings point to NK cell activity as a possible immunological component of deprenyl treatment that helps to hold mammary tumour growth and spread in check.

Work by Ben-Eliyahu's laboratory has demonstrated that NK cell activity is important in regulating metastatic spread of mammary tumours, and that catecholamines exert an inhibitory effect on NK cell activity (Ben-Eliyahu and Shakhar, 2001). This work has important implications for stress-induced damage to NK cell activity at the time of surgery, or during treatment for breast cancer in women, at which time metastatic spread may be most likely to occur (Ben-Eliyahu *et al.*, 1999). Both pharmacological and behavioural approaches should be considered for preventing catecholamine-induced inhibition of NK cell activity, especially in light of evidence that catecholamines and not glucocorticoids modulate NK cell activity (Bodner *et al.*, 1998).

IMMUNE SENESCENCE

Studies in aging Fischer 344 rats demonstrate an age-related loss of sympathetic NA innervation of spleen and lymph nodes (Felten *et al.*, 1986; Bellinger *et al.*, 1987, 1992). Interestingly, these old rats demonstrate similar immune deficits (diminished cell-mediated immune responses, diminished T cell proliferation) as young rats with chemical sympathectomy. This led to the hypothesis that some of the cell-mediated immune deficits in immune senescence are neural in origin. An important component to testing this hypothesis was the demonstration that stimulation of regrowth and sprouting of the sympathetic NA nerves back into the spleen in old immunodeficient rodents led to the restoration of sympathetic NA innervation, and enhanced IL-2 production, T cell proliferation, and NK activity (ThyagaRajan *et al.*, 1998a,c). Our current hypothesis regarding the mechanism by which the site-specific sympathetic NA denervation occurs, only in secondary lymphoid organs, is based upon the oxidative hypothesis of toxic free-radical derivatives of catecholamines that occur during times of high turnover and release of catecholamines, followed by reuptake into the nerve terminals by the high-affinity uptake carrier, resulting in destruction of those nerve terminals (Felten *et al.*, 1992b). We have hypothesized that due to surges of NE release during an immune response at the site of challenge, an age-related denervation occurs; thus, the price an animal pays for a robust enhancement of the initiative phase of an immune response in the spleen or lymph nodes is the age-related destruction of the NA terminals that are involved.

CHRONIC STRESS, IMMUNE SENESCENCE, AND A PATTERN OF IMMUNE CHANGES

A general pattern of immunological changes is emerging as part of chronic stressors. As a generalized observation in animal models of stress (Dhabhar and McEwen, 2001), acute stressors appear to enhance, and chronic stressors appear to diminish, cell-mediated immune responses and other immune reactivity. These findings are consistent with the extensive work of Kiecolt-Glaser and Glaser (review in Kiecolt-Glaser, 1999), demonstrating a general pattern of immunological deficits that accompanies chronic stressors in humans. They have investigated such chronic stressors as examination stress in medical students, accompanying loneliness in these students, marital discord, and caregiving to a loved one with dementia. The generalized pattern of deficits seen with these chronic stressors includes: (1) diminished lymphocyte proliferation; (2) diminished production of cell-mediated immune cytokines such as IL-2 and IFN-γ; (3) diminished NK cell activity; (4) diminished vaccine responses; and (5) increased antibody titers to specific epitopes on latent viruses, indicating that these latent viruses are becoming activated. These immunological alterations often were accompanied by evidence of elevated production of stress mediators, such as epinephrine and cortisol. These altered measures of immune reactivity are implicated in health consequences, particularly related to anti-viral responses, anti-tumour immunity, and vaccine protection from infection.

It is this pattern of findings that has led us to a heightened interest in complementary and integrative medical interventions that appear to produce immunological effects directionally opposite to those seen in chronic stressors, with some effects persisting long beyond the interval of the intervention.

SYMPATHETIC NERVES AND CATECHOLAMINES IN COMPLEMENTARY AND INTEGRATIVE MEDICINE

Complementary and integrative approaches to health care and wellness have become the subject of increasing numbers of studies, based on increasing interests by the general public and increasing scientific interest at the NIH with the establishment of the National Center for Complementary and Alternative Medicine. Evidence-based studies have fallen into two general categories: (1) life style interventions, such as exercise, dietary and nutritional interventions (including supplements), habits related to substance abuse and smoking; and (2) psychosocial and mind/body interventions, such as music therapy, humour and laughter, guided imagery, meditation, counselling and peer support, Qi Gong, acupuncture, to name a few. We have proposed the scientific foundations for evidence-based studies of these interventions (Felten, 2000).

Our laboratories have undertaken some investigations of some of these approaches. Aerobic exercise in moderation (up to 70–80% capacity) results in enhanced NK cell function (Berk et al., 1990), enhanced T helper/T suppressor cell ratio, enhanced blastogenesis, and enhanced total lymphocyte count (Nieman et al., 1989), while continuing the exercise to exhaustion in conditioned long-distance marathoners results in suppression of these measures, with recovery occurring the next day.

Extensive investigations have been conducted with interventions of watching a humour video of the individual's choice for 30–60 min (Berk et al., 1989, 2001a). The humour intervention resulted in significant decreases in cortisol and epinephrine, and an elevation in growth hormone. The immunological changes included enhanced NK cell activity, increased markers for activated T cells and active cytotoxic T lymphocytes, increased T helper/T suppressor ratio, elevated IFN-γ production, and increased IgG production. Many of these changes persisted into the next day.

An interesting follow up preliminary investigation was undertaken with random assignment into the control or experimental humour groups two days prior to the intervention. In this case, those assigned to the humour intervention group arrived at the intervention site with a marked alteration in the profile of mood states instrument (POMS), demonstrating enhanced positive mood and diminished negative mood (Berk et al., 2001b). This finding suggests that the anticipation of a positive intervention can produce its own CNS alterations, with implications for anticipatory-induced changes in neuroendocrine and immune reactivity.

A recurring question with complementary interventions such as humour or music therapy is the duration and biological significance of these approaches. To evaluate this issue in humour interventions, a randomized study of diabetic patients with a first heart attack was undertaken (Tan et al., 1997). The controls received conventional cardiac rehabilitation alone, while the experimental group received 30 min of viewing of a humour video as part of their cardiac rehabilitation. They were assessed blindly for 1 year. Daily humour for 1 year was accompanied by a 50% decrease in arrhythmias, a 50% decrease in use of nitroglycerin for angina, a 15+ mm Hg decline in systolic blood pressure, and an 80% decrease in recurrent myocardial infarcts, all highly significant statistically. In addition, both epinephrine secretion and NE secretion were diminished by approximately 65% based on both 24-h urine collections and plasma levels. Epinephrine secretion dropped within a month of the intervention, while NE levels came down over 6+ months at a more gradual rate. This decline in catecholamine secretion probably accounts for the remarkable clinical results. This preliminary study demonstrates that a simple intervention such as humour, included as part of cardiac rehabilitation, can exert profound and lasting effects of stress mediator secretion, with remarkable clinical benefits.

We also conducted a music therapy intervention in normal subject using a variety of drum circle interventions (Bittman *et al.*, 2001). This intervention was accompanied by increased DHEA/cortisol ratios, increased NK cell activity, and increased lymphokine-activated killer cell activity.

Clearly, we are just beginning to explore the neuroendocrine, immunological, and inflammatory mediators and responses involved in complementary and integrative interventions that can be incorporated as adjunctive therapy with conventional medicine. However, both life style interventions and mind/body interventions appear to generally result in diminished secretion of stress-related mediators such as cortisol and catecholamines (especially epinephrine), and enhanced NK cell activity and T cell functions, often including cell-mediated cytokines. These responses, while transient, can be cumulative when an intervention such as humour video viewing is used on a daily basis. These responses appear to be directionally opposite to those induced by chronic stressors. It is likely that each specific life style and mind/body intervention will have a characteristic pattern of alterations in neuroendocrine mediators, cytokines, inflammatory mediators, and immune responses, both in magnitude and temporal sequence, and that individual differences such as age, gender, and other factors will play a role in the final effectiveness of these interventions. However, this approach permits the beginning of an evidence-based approach to the study of complementary and integrative interventions, and permits us to apply our knowledge of neural–immune signalling to the treatment of human subjects as a component of wellness programmes, and as adjunctive therapies in chronic diseases.

REFERENCES

Abrass, C.K., O'Connor, S.W., Scarpace, P.J., and Abrass, I.B. (1985) Characterization of the beta-adrenergic receptor of the rat peritoneal macrophage. *J. Immunol.*, **135**, 1338–1341.

Ackerman, K.D., Bellinger, D.L., Felten, S.Y., and Felten, D.L. (1991) Ontogeny and senescence of noradrenergic innervetion of the rodent thymus and spleen. In *Psychoneuroimmunology*, 2nd edn, edited by R. Ader, D.L. Felten, and N. Cohen, pp. 70–125. San Diego, CA: Academic Press.

Ackerman, K.D., Felten, S.Y., Bellinger, D.L., Livnat, S., and Felten, D.L. (1987) Noradrenergic sympathetic innervation of spleen and lymph nodes in relation to specific cellular compartments. *Prog. Immunol.*, **6**, 588–600.

Ader, R. and Cohen, N. (1975) Behaviorally conditioned immunosuppression. *Psychosom. Med.*, **37**, 333–340.

Ader, R. and Cohen, N. (2001) Conditioning and immunity. In *Psychoneuroimmunology*, 3rd edn, vol. 2, edited by R. Ader, D.L. Felten, and N. Cohen, pp. 3–34. San Diego, CA: Academic Press.

Bellinger, D.L., Ackerman, K.D., Felten, S.Y., Pulera, M., and Felten, D.L. (1992) A longitudinal study of age-related loss of noradrenergic nerves and lymphoid cells in the rat spleen. *Exp. Neurol.*, **116**, 295–311.

Bellinger, D.L., Felten, S.Y., Collier, T.J., and Felten, D.L. (1987) Noradrenergic sympathetic innervation of the spleen: IV. Morphometric evidence for age-related loss of noradrenergic fibers in the spleen. *J. Neurosci. Res.*, **18**, 55–63, 126–129.

Bellinger, D.L., Lorton, D., Lubahn, C., and Felten, D.L. (2001) Innervation of lymphoid organs – association of nerves with cells of the immune system and their implications in disease. In *Psychoneuroimmunology*, 3rd edn, vol. 1, edited by R. Ader, D.L. Felten, and N. Cohen, pp. 55–111. San Diego, CA: Academic Press.

Bellinger, D.L., Lorton, D., Romano, T.D., Olschowka, J.A., Felten, S.Y., and Felten, D.L. (1990) Neuropeptide innervation of lymphoid organs. In *Neuropeptides and Immunopeptides: Messengers in a Neuroimmune Axis*, edited by M.S. L-Dorisio and A. Panerai. *Ann. N.Y. Acad. Sci.*, **594**, 17–33.

Ben-Eliyahu, S., Page, G.G., Yirmiya, R., and Shakhar, G. (1999) Evidence that stress and surgical interventions promote tumor development by suppressing natural killer cell activity. *Int. J. Cancer*, **80**, 880–888.

Ben-Eliyahu, S. and Shakhar, G. (2001) The impact of stress, catecholamines, and the menstrual cycle on NK activity and tumour development: from in vitro studies to biological significance. In *Psychoneuroimmunology*, 3rd edn, vol. 2, edited by R. Ader, D.L. Felten, and N. Cohen, pp. 545–563. San Diego, CA: Academic Press.

Berk, L.S., Felten, D.L., Tan, S.A., Bittman, B.B., and Westengard, J. (2001a) Modulation of neuroimmune parameters during the eustress of humor-associated mirthful laughter. *Altern. Ther. Health Med.*, **7**, 62–76.

Berk, L.S., Felten, D.L., Tan, S.A., and Westengard, J. (2001b) The anticipation of a humor eustress event modulates mood states prior to the actual experience. *Brain Behav. Immun.*, **15**, 137.

Berk, L.S., Nieman, D.C., Youngberg, W.S., Arabatzis, K., Simpson-Westerbergy, M., Lee, J.W., Tan, S.A., and Eby, W.C. (1990) The effect of long endurance running on natural killer cells in marathoners. *Med. Sci. Sports Exerc.*, **22**, 207–212.

Berk, L.S., Tan, S.A., Fry, W.F., Napier, B.J., Lee, J.W., Hubbard, R.W., Lewis, J.E., and Eby, W.C. (1989) Neuroendocrine and stress hormone changes during mirthful laughter. *Am. J. Med. Sci.*, **298**, 390–396.

Bittman, B.B., Berk, L.S., Felten, D.L., Westengard, J., Simonton, O.C., Pappas, J., and Ninehouser, M. (2001) Composite effects of group drumming music therapy on modulation of neuroendocrine-immune parameters in normal subjects. *Altern. Ther. Health Med.*, **7**, 38–47.

Bodner, G., Ho, A., and Kreek, M.J. (1998) Effect of endogenous cortisol levels on natural killer activity in healthy humans. *Brain Behav. Immun.*, **12**, 285–296.

Bonneau, R.H., Padgett, D.A., and Sheridan, J.F. (2001) Psychoneuroimmune interactions in infectious disease: studies in animals. In *Psychoneuroimmunology*, 3rd edn, vol. 2, edited by R. Ader, D.L. Felten, and N. Cohen, pp. 483–497. San Diego, CA: Academic Press.

Brenneman, S.M., Moynihan, J.A., Grota, L.J., Felten, D.L., and Felten, S.Y. (1993) Splenic norepinephrine is decreased in MRL-lpr/lpr mice. *Brain Behav. Immun.*, **7**, 135–143.

Brodde, O.E., Engel, G., Hoyer, D., Block, K.D., and Weber, F. (1981) The beta-adrenergic receptor in human lymphocytes – sub-classification by the use of a new radioligand (±) [125 Iodo]cyanopindolol. *Life Sci.*, **29**, 2189–2198.

Carlson, S.L., Brooks, W.H., and Roszman, T.L. (1989) Neurotransmitter-lymphocyte interactions: dual receptor modulation of lymphocyte proliferation and cAMP production. *J. Neuroimmunol.*, **24**, 155–162.

Carr, D.J.J. and Blalock, J.E. (1991) Neuropeptide hormones and receptors common to the immune and neuroendocrine systems: bi-directional pathways of intersystem communication. In *Psychoneuroimmunology*, 2nd edn, edited by R. Ader, D.L. Felten, and N. Cohen, pp. 573–588. San Diego, CA: Academic Press.

Chelmicka-Schorr, E., Checinski, M., and Arnason, B.G.W. (1988) Chemical sympathectomy augments the severity of experimental allergic encephalomyelitis. *J. Neuroimmunol.*, **17**, 347–350.

Chelmicka-Schorr, E., Kwasniewski, M.N., Thomas, B.E., and Arnason, B.G.W. (1989) The β-adrenergic agonist isoproterenol suppresses experimental allergic encephalomyelitis in Lewis rats. *J. Neuroimmunol.*, **25**, 203–207.

Dhabhar, F.S. and McEwen, B.S. (2001) Bidirectional effects of stress and glucocorticoid hormones on immune function: possible explanations for paradoxical observations. In *Psychoneuroimmunology*, 3rd edn, vol. 1, edited by R. Ader, D.L. Felten, and N. Cohen, pp. 301–338. San Diego, CA: Academic Press.

Felten, D.L. (2000) Neural influence on immune responses: underlying suppositions and basic principles of neural-immune signaling. *Prog. Brain Res.*, **122**, 381–389.

Felten, D.L., Cohen, N., Ader, R., Felten, S.Y., Carlson, S.L., and Roszman, T.L. (1991) Central neural circuits involved in neural-immune interactions. In *Psychoneuroimmunology*, 2nd edn, edited by R. Ader, D.L. Felten, and N. Cohen, pp. 3–25. San Diego, CA: Academic Press.

Felten, D.L., Felten, S.Y., Carlson, S.L., Olschowka, J.A., and Livnat, S. (1985) Noradrenergic and peptidergic innervation of lymphoid tissue. *J. Immunol.*, **135**, 755s–765s.

Felten, D.L., Felten, S.Y., Bellinger, D.L., and Lorton, D. (1992a) Noradrenergic and peptidergic innervation of secondary lymphoid organs: role in experimental rheumatoid arthritis. *Eur. J. Clin. Invest. Suppl. 1*, **22**, 37–41.

Felten, D.L., Felten, S.Y., Steece-Collier, K., Date, I., and Clemens, J.A. (1992b) Age-related decline in the dopaminergic nigrostriatal system: the oxidative hypothesis and protective strategies. *Ann. Neurol.*, **32**, 133s–136s.

Felten, D.L., Gibson-Berry, K., and Wu, J.H.D. (1996) Innervation of bone marrow by tyrosine hydroxylase-immunoreactive nerve fibers and hemopoiesis-modulating activity of a β-adrenergic agonist in mouse. *Mol. Biol. Hematopoiesis*, **5**, 627–636.

Felten, S.Y., Bellinger, D.L., Collier, T.J., Coleman, P.D., and Felten, D.L. (1986) Decreased sympathetic innervation of spleen in aged Fischer 344 rats. *Neurobiol. Aging*, **8**, 159–165.

Felten, S.Y. and Felten, D.L. (1991) The innervation of lymphoid tissue. In *Psychoneuroimmunology*, 2nd edn, edited by R. Ader, D.L. Felten, and N. Cohen, pp. 27–69. San Diego, CA: Academic Press.

Felten, S.Y. and Olschowka, S.Y. (1987) Noradrenergic sympathetic innervation of the spleen. II. Tyrosine hydroxylase (TH)-positive nerve terminals form synaptic-like contacts on lymphocytes in the splenic white pulp. *J. Neurosci. Res.*, **18**, 37–48.

Fuchs, B.A., Albright, J.W., and Albright, J.F. (1988) β-adrenergic receptors on murine lymphocytes: density varies with cell maturity and lymphocyte subtype and is decreased after antigen administration. *Cell. Immunol.*, **114**, 231–245.

Gibson-Berry, K.L., Whitin, J.C., and Cohen, H.J. (1993) Modulation of the respiratory burst in human neutrophils by isoproterenol and dibutyryl cyclic AMP. *J. Neuroimmunol.*, **43**, 59–68.

Goin, J.C., Sterin-Borda, L., Borda, E.S., Finiasz, M., Fernandez, J., and de Bracco, M.M.E. (1991) Active alpha$_2$ and beta adrenoceptors in lymphocytes from patients with chronic lymphocytic leukemia. *Int. J. Cancer*, **49**, 178–181.

Hellstrand, K., Hermodsson, S., and Strannegard, O. (1985) Evidence for a β-adrenoceptor-mediated regulation of human natural killer cells. *J. Immunol.*, **134**, 4095–4099.

Hori, T., Katafuchi, T., and Oka, T. (2001) Central cytokines: effects on peripheral immunity, inflammation, and nociception. In *Psychoneuroimmunology*, 3rd edn, vol. 1, edited by R. Ader, D.L. Felten, and N. Cohen, pp. 517–545. San Diego, CA: Academic Press.

Katayama, M., Kobayashi, S., Kuramoto, N., and Yokoyama, M.M. (1987) Effects of hypothalamic lesions on lymphocyte subsets in mice. *Ann. N. Y. Acad. Sci.*, **496**, 366–376.

Katz, P., Zaytoun, A.M., and Fauci, A.S. (1982) Mechanisms of human cell-mediated cytotoxicity. I. Modulation of natural killer cell activity by cyclic nucleotides. *J. Immunol.*, **129**, 287–296.

Kelley, S.P., Moynihan, J.A., Stevens, S.Y., Grota, L.J., and Felten, D.L. (2001) Chemical sympathectomy has no effect on the severity of murine AIDS: murine AIDS alone depletes norepinephrine levels in infected spleen. *Brain Behav. Immun.*, doi: 10.1006/brbi.2001.0627.

Kiecolt-Glaser, J.K. (1999) Stress, personal relationships, and immune function: health implications. *Brain Behav. Immun.*, **13**, 61–72.

Landmann, R., Bittiger, H., and Buhler, F.R. (1981) High affinity beta-2 adrenergic receptors in mononuclear leucocytes: similar density in young and old subjects. *Life Sci.*, **29**, 1761–1771.

Landmann, R., Burgisser, E., West, M., and Buhler, F.R. (1985) Beta adrenergic receptors are different in subpopulations of human circulating lymphocytes. *J. Recept. Res.*, **4**, 37–50.

Levine, J.D., Coderre, T.J., Helms, C., and Basbaum, A.I. (1988) β-2 adrenergic mechanisms in experimental arthritis. *Proc. Natl. Acad. Sci. USA*, **85**, 4553–4556.

Livnat, S., Felten, S.Y., Carlson, S.L., Bellinger, D.L., and Felten, D.L. (1985) Involvement of peripheral and central catecholamine systems in neural–immune interactions. *J. Neuroimmunol.*, **10**, 5–30.

Lorton, D., Bellinger, D.L., Duclos, M., Felten, S.Y., and Felten, D.L. (1996) Application of 6-hydroxydopamine into the fat pad surrounding the draining lymph nodes exacerbates the expression of adjuvant-induced arthritis. *J. Neuroimmunol.*, **64**, 103–113.

Lorton, D., Lubahn, C., Engan, C., Schaller, J., Felten, D.L., and Bellinger, D.L. (2000) Local application of capsaicin into the draining lymph nodes attenuates expression of adjuvant-induced arthritis. *Neuroimmunomodulation*, **7**, 115–125.

Lorton, D., Lubahn, C., Klein, N., Schaller, J., and Bellinger, D.L. (1999) Dual role for noradrenergic innervation of lymphoid tissue and arthritic joints in adjuvant-induced arthritis. *Brain Behav. Immun.*, **13**, 315–334.

Madden, K.S. and Felten, D.L. (1995) Experimental basis for neural-immune interactions. *Physiol. Rev.*, **75**, 77–106.

Madden, K.S. and Livnat, S. (1991) Catecholamine action and immunologic reactivity. In *Psychoneuroimmunology*, 2nd edn, edited by R. Ader, D.L. Felten, and N. Cohen, pp. 283–310. San Diego, CA: Academic Press.

Madden, K.S., Felten, S.Y., Felten, D.L., Sundaresan, P.V. and Livnat, S. (1989) Sympathetic neural modulation of the immune system: I. Depression of T cell immunity in vivo and in vitro following chemical sympathectomy. *Brain Behav. Immun.*, **3**, 72–89.

Madden, K.S., Felten, S.Y., Felten, D.L., Hardy, C.A., and Livnat, S. (1994a) Sympathetic nervous system modulation of the immune system: II. Induction of lymphocyte proliferation and migration in vivo by chemical sympathectomy. *J. Neuroimmunol.*, **49**, 67–75.

Madden, K.S., Moynihan, J.A., Brenner, G.J., Felten, S.Y., Felten, D.L., and Livnat, S. (1994b) Sympathetic nervous system modulation of the immune system: III. Alterations in T and B cell proliferation and differentiation in vitro following chemical sympathectomy. *J. Neuroimmunol.*, **49**, 77–87.

Madden, K.S., Sanders, V.M., and Felten, D.L. (1995) Catecholamine influences and sympathetic neural modulation of immune responsiveness. *Annu. Rev. Pharmacol. Toxicol.*, **35**, 417–448.

Maier, S.F., Watkins, L.R., and Nance, D.M. (2001) Multiple routes of action of interleukin 1 on the nervous system. In *Psychoneuroimmunology*, 3rd edn, vol. 1, edited by R. Ader, D.L. Felten, and N. Cohen, pp. 563–583. San Diego, CA: Academic Press.

Marucha, P.T., Kiecolt-Glaser, J.K., and Favagehi, M. (1998) Mucosal wound healing is impaired by examination stress. *Psychosom. Med.*, **60**, 362–365.

Moynihan, J.A. and Stevens, S.Y. (2001) Mechanisms of stress-induced modulation of immunity in animals. In *Psychoneuroimmunology*, 3rd edn, vol. 2, edited by R. Ader, D.L. Felten, and N. Cohen, pp. 227–249. San Diego, CA: Academic Press.

Nielson, C.P. (1987) β-adrenergic modulation of the polymorphonuclear leukocyte respiratory burst is dependent upon the mechanism of cell activation. *J. Immunol.*, **139**, 2392–2397.

Nieman, D.C., Berk, L.S., Simpson-Westerberg, M., Arabatzis, K., Youngberg, S., Tan, S.A., Lee, J.W., and Eby, W.C. (1989) Effects of long-endurance running on immune system parameters and lymphocyte function in experienced marathoners. *Int. J. Sports Med.*, **10**, 317–323.

Plaut, M. (1987) Lymphocyte hormone receptors. *Annu. Rev. Immunol.*, **5**, 621–669.

Rice, P.A., Boehm, G.W., Moynihan, J.A., Bellinger, D.L., and Stevens, S.Y. (2001) Chemical sympathectomy increases the innate immune response and decreases the specific immune response in the spleen to infection with Listeria monocytogenes. *J. Neuroimmunol.*, **114**, 19–27.

Rivier, C. (2001) The hypothalamo-pituitary-adrenal axis response to immune signals. In *Psychoneuroimmunology*, 3rd edn, vol. 1, edited by R. Ader, D.L. Felten, and N. Cohen, pp. 633–648. San Diego, CA: Academic Press.

Rivkin, I., Rosenblatt, J., and Becker, E.L. (1975) The role of cyclic AMP in the chemotactic responsiveness and spontaneous motility of rabbit peritoneal neutrophils. The inhibition of neutrophil movement and the elevation of cyclic AMP levels by catecholamines, prostaglandins, theophylline, and cholera toxin. *J. Immunol.*, **115**, 1126–1134.

Sheridan, J.F., Feng, N., Bonneau, R.H., Allen, C.M., Huneycutt, B.S., and Glaser, R. (1991) Restraint stress differentially affects anti-viral cellular and humoral immune responses in mice. *J. Neuroimmunol.*, **31**, 245–255.

Sheridan, J.F., Dobbs, C., Brown, D., and Zwilling, B. (1994) Psychoneuroimmunology: stress effects on pathogenesis and immunity during infection. *Clin. Microbiol. Rev.*, **7**, 200–212.

Sheridan, J. (1998) Stress-induced modulation of anti-viral immunity. *Brain Behav. Immun.*, **12**, 1–6.

Spengler, R.N., Allen, R.M., Remick, D.G., Strieter, R.M., and Kunkel, S.L. (1990) Stimulation of alpha-adrenergic receptor augments the production of macrophage-derived tumor necrosis factor. *J. Immunol.*, **145**, 1430–1434.

Sternberg, E.M., Hill, J.M., Chrousos, G.P., Kamilario, T., Listwak, S.J., Gold, P.W., and Wilder, R.L. (1989) Inflammatory mediator-induced hypothalamic–pituitary–adrenal axis activation is defective in streptococcal cell wall arthritis-susceptible rats. *Proc. Natl. Acad. Sci.*, **86**, 2374–2378.

Sternberg, E.M., Chrousos, G.P., Wilder, R.L., and Gold, P.W. (1992) The stress response and the regulation of inflammatory disease. *Ann. Intern. Med.*, **117**, 854–866.

Sternberg, E.M. (1997) Neural-immune interactions in health and disease. *J. Clin. Invest.*, **100**, 2641–2647.

Strom, T.B., Carpenter, C.B., Garovoy, M.R., Austen, K.F., Merrill, J.P., and Kaliner, M. (1973) The modulating influence of cyclic nucleotides upon lymphocyte-0mediated cytotoxicity. *J. Exp. Med.*, **138**, 381–393.

Tan, S.A., Tan, L.G., Berk. L.S., Lukman, S.T., and Lukman, L.F. (1997) Mirthful laughter, an effective adjunct in cardiac rehabilitation. *Can. J. Cardiol.*, **13**, Suppl. B, 190B.

ThyagaRajan, S., Felten, S.Y., and Felten, D.L. (1998a) Restoration of sympathetic noradrenergic nerve fibers in the spleen by low doses of l-deprenyl treatment in young sympathectomized and old Fischer 344 rats. *J. Neuroimmunol.*, **81**, 144–157.

ThyagaRajan, S., Felten, S.Y., and Felten, D.L. (1998b) Anti-tumor effect of deprenyl in rats with carcinogen-induced mammary tumors. *Cancer Lett.*, **123**, 177–183.

ThyagaRajan, S., Madden, K.S., Kalvass, J.C., Dimitrova, S.S., Felten, S.Y., and Felten, D.L. (1998c) L-deprenyl-induced increase in IL-2 and NK cell activity accompanies restoration of noradrenergic nerve fibers in the spleens of old F344 rats. *J. Neuroimmunol.*, **92**, 9–21.

ThyagaRajan, S., Madden, K.S., Stevens, S.Y., and Felten, D.L. (1999) Anti-tumor effect of L-deprenyl is associated with enhanced central and peripheral neurotransmission and immune reactivity in rats with carcinogen-induced mammary tumors. *J. Neuroimmunol.*, **109**, 95–104.

ThyagaRajan, S., Madden, K.S., Stevens, S.Y., and Felten, D.L. (2000) Inhibition of tumor growth by L-deprenyl involves neural-immune interactions in rats with spontaneously developing mammary tumors. *Anti-Cancer Res.*, **19**, 5023–5028.

Titinchi, S. and Clark, B. (1984) Alpha$_2$-adrenoceptors in human lymphocytes: direct characterization by [^{3}H]yohimbine binding. *Biochem. Biophys. Res. Commun.*, **121**, 1–7.

Williams, J.M. and Felten, D.L. (1981) Sympathetic innervation of murine thymus and spleen: a comparative histofluorescence study. *Anat. Rec.*, **199**, 531–542.

Williams, J.M., Peterson, R.G., Shea, P.A., Schmedtje, J.F., Bauer, D.C., and Felten, D.L. (1981) Sympathetic innervation of murine thymus and spleen: evidence for a functional link between the nervous and immune systems. *Brain Res. Bull.*, **6**, 83–94.

Yukawa, T., Ukena, D., Kroegel, C., Chanez, P., Dent, G., *et al.* (1990) Beta2-adrenergic receptors on eosinophils. *Am. Rev. Respir. Dis.*, **141**, 1446–1452.

2 Interactions between the Adrenergic and Immune Systems

Dwight M. Nance[1] and Jonathan C. Meltzer[2]

[1]*Susan Samueli Center for Complementary and Alternative Medicine, UCI College of Medicine, Orange, CA 92868, USA*
[2]*Departments of Pathology and Anatomy, University of Manitoba, Winnipeg, Manitoba R3E 0W3, Canada*

The concept of a central neuroimmune regulatory network that is responsive to signals produced by the immune system and that is capable of generating endocrine and autonomic responses that modify immune function continues to be supported. However, the view that this regulatory system operates by means of a counter-regulatory feedback network requires revision. If the central regulatory network is first activated by central inflammatory stimuli or stress, then subsequent immune responses to a variety of challenges are modified via endocrine and sympathetic nervous system mechanisms. However, if the neuroimmune regulatory system is not activated *prior* to exposure to an immune challenge, counter-regulatory responses produced by this regulatory system may have minimal impact on an ongoing immune response. Although an immune challenge simultaneously signals the central regulatory network and initiates peripheral immune responses, the counter-regulatory responses produced by the central nervous system appear to occur too late to modify ongoing immune reactions generated by that same immune challenge. Thus, the conceptual model of a negative feedback system, at least with regards to immune activation, must be modified to account for the fact that the timing of endocrine and autonomic responses are too late to impact on an ongoing immune response. Although the outputs of the neuroimmune regulatory system can have a counter-regulatory action, this action must be occurring at a later time period during the course of an immune response. Thus, the neuroimmune regulatory system may delimit the length of time an immune reaction is allowed to proceed and to limit the number of immune challenges that can be processed at one time. A feed-forward model system may be required too incorporate the likely actions of the endocrine and sympathetic nervous system on immune reactions occurring later during an immune response. Finally, further analysis of the potential role of the sympathetic nervous system in mediating conditioned immune responses would appear to be worthy of vigorous investigation and may provide a unique model system to examine further the interactions between the adrenergic and immune systems.

KEY WORDS: neuroimmune regulatory system; paraventricular nucleus; counter-regulatory responses.

INTRODUCTION

The research field of psychoneuroimmunology, or neuroimmunology, has experienced rapid growth and development during the past 25 years. The idea of bi-directional communication between the brain and the immune system, as originally proposed by Besedovsky *et al.* (1979a,b, 1983, 1984, 1986), continues to provide a primary conceptual framework for analysing the complex interactions among the three fundamental regulatory systems of the body, the central nervous system, the endocrine system and the immune system. Multidisciplinary approaches and techniques have characterized this research field and have provided the basis for advancements in our understanding of this fundamental and complex regulatory network. Despite the intrinsic difficulties presented by attempts to integrate these diverse regulatory systems into a unitary theoretical framework, there are several constituent elements that provide the foundation for any conceptual model of the neuroimmune regulatory system. Similar to other homeostatic physiological processes, the hypothalamus and its connections comprise the central integrative and regulatory elements for this proposed regulatory system. The dual efferent arms of this central regulatory system are composed of the endocrine and autonomic nervous systems. Thus, any actions by the nervous system on any immunological processes, by necessity, must be mediated by neuroendocrine responses and/or changes in sympathetic nerve activity. Within the hypothalamus, the paraventricular nucleus (PVN) has been proposed as a nodal region for the regulation of brain-immune interactions (Nance and MacNeil, 2001). Focus on the PVN is due to the fact that the cell bodies that produce the neuroendocrine releasing factor CRF are primarily localized within this nucleus and that the immunomodulatory actions of adrenal steroids are well established (Selye, 1950; Rivier, 2001). In addition, the PVN has efferent connections with brain stem and spinal cord autonomic nuclei indicating that this same hypothalamic region can directly modulate the activity of both the sympathetic and parasympathetic nervous systems (Swanson, 1985). Thus, both neuroanatomical and functional evidence suggest that the PVN and it's connections are strategically positioned to simultaneously regulate the two efferent arms of the neuroimmune regulatory system.

HOW DOES THE IMMUNE SYSTEM SIGNAL THE PVN?

Changes in neurotransmitter levels and turnover, electrical activity, and induction of the activity dependent cellular marker c-fos have all been utilized to determine the activational effects of immune-related stimuli on the hypothalamus and its connections (Saphier, 1989; Wan *et al.*, 1993a, 1994; Zalcman *et al.*, 1994). Central inflammation associated with microbial infections of the brain, neurological injury and neurodegenerative process all produce potent and direct action on the central nervous system. Microglial cells, the central representatives of the immune system and the same as shown by their peripheral counterparts, rapidly produce inflammatory cytokines in response to microbial products and tissue injury. The action of these immune dependent signals on the PVN are illustrated by the activational effects of central injections of endotoxin (lipopolysaccharide, LPS), proinflammatory cytokines such as Interleukin-1 (IL-1), and immune-related intermediaries associated with the inflammatory cascade, such as prostaglandin (PGE2) and nitric oxide (NO).

Central injections of LPS (Wan *et al.*, 1993a), IL-1 (Rivest *et al.*, 1992), PGE2 (Lacroix *et al.*, 1996; Jackson, 1999; Nance and MacNeil, 2001) and NO (Lee *et al.*, 1999) have all been shown to induce the rapid induction of c-fos mRNA and protein primarily within the PVN, supraoptic and arcuate nuclei. Additional brain areas activated by these central injections typically include brain stem areas providing afferent input to these hypothalamic nuclei and limbic structures such as the bed nucleus of the stria terminalis (BNST) and the central nucleus of the amygdala which project to the hypothalamus and brain stem auto-nomic nuclei. Central injections of PGD2 have been shown to produce an increase in the turnover rate of norepinephrine (NE) in the hypothalamus (Terao *et al.*, 1995). Central injections of IL-1 increase the multi-unit electrical activity in the PVN, whereas central injections of interferon-alpha inhibit multi-unit activity in the PVN and preoptic area of the hypothalamus (Saphier, 1989).

How do immune stimuli that originate outside of the central nervous system signal the PVN? Systemic injections of LPS and IL-1 induce c-fos mRNA and protein in the PVN and related brain areas in a pattern similar to that generated by central injections of these same microbial and immune-dependent stimuli. Also, systemic injections of IL-1 produce an increase in the turnover rate of NE in the hypothalamus (Zalcman *et al.*, 1994) and immunization with sheep red blood cells (SRBCs) increase electrical activity in the preop-tic area and PVN 5 and 6 days after immunization, respectively (Saphier *et al.*, 1990). Given these immune-related changes in the PVN, what are the afferent pathways that medi-ate the activation of the hypothalamus by systemic immune-related stimuli? Sensory nerves are capable of transmitting immune signals generated in the periphery to visceral sensory nuclei in the brain stem which are relayed to the hypothalamus. Abdominal vagal afferents have been shown to mediate the central activational effects of i.p. injections of LPS and IL-1 on the induction of c-fos protein in the PVN, whereas subdiaphragmatic vagotomy has only a minor effect on these same immune stimuli following an i.v. injection (Wan *et al.*, 1994). Cutaneous nerves can also signal the hypothalamus following inflam-mation of the skin and muscle as demonstrated for cutaneous or muscle injections of the general inflammatory reagent, turpentine (Turnbull and Rivier, 1996).

How do immune stimuli that reach the systemic circulation (i.v.) activate the hypothala-mus? Circumventricular organs (CVOs) are unique structures within the central nervous system that do not possess a blood–brain barrier and have been long implicated as chemoreceptive organelles (Scammell *et al.*, 1996). The molecular size of LPS, as well as cytokines, precludes their penetration of the blood–brain barrier; however, CVOs con-stitute potential sites for these large immune-related molecules to gain access to central neuroregulatory systems. For example, the OVLT, located at the rostral end of the third ventricle, has been implicated in the generation of fever in response to pyrogenic immune stimuli (Scammell *et al.*, 1996). With regards to the neuroimmune regulatory system, the area postrema (AP), located in the fourth ventricle adjacent to the primary visceral sensory nucleus in the brain stem, the nucleus tractus solitaius (NTS), may mediate the central acti-vational effects of blood-borne immune stimuli (Lee *et al.*, 1998). Both LPS and IL-1 induce c-fos protein in cells located in the AP, and lesions of the AP are reported to attenuate the central activational effects of i.v. injections of IL-1. However, others have reported that AP lesions, some of which extended to include portions of the NTS, failed to block the central activational effects of i.v. immune stimuli (Ericsson *et al.*, 1994). Despite these divergent results, it remains likely that the AP may constitute one of several redundant afferent pathways through which products of the immune system can signal central neuroregulatory systems.

Finally, cerebral microvascular endothelial cells distributed throughout the brain appear to posses the necessary sensory transduction machinery for microbial and immune-related products to access central neuroimmunoregulatory centres. In this regard, the PVN and SON are unique brain nuclei in that these brain regions are highly vascularized, relative to all other brain areas (Gross *et al.*, 1986). Thus, this large endothelial interface between the systemic circulation and these important integrative and regulatory nuclei would appear to provide ample opportunity for immune related stimuli to activate neurons within these nuclei if, in fact, endothelial cells were targets for immune stimuli. Of course, the actions of the immune system on the vasculature and endothelial cells are well established (Abbas *et al.*, 1997). Consistent with this, following an i.v. injection of LPS or IL-1, the cerebral microvasculatur endothelial cells has been shown to be the initial site of induction of inflammatory cytokines, such as IL-1, as well as the site of production for prostaglandins as indexed by the induction of inducible cyclooxygenase-2 (Ericsson *et al.*, 1995; Matsumura *et al.*, 1998; Quan *et al.*, 1998). Thus, cerebral endothelial cells provide a critical interface between systemic immune responses and related stimuli and the brain.

In summary, it is likely that there are multiple and redundant afferent pathways by which the immune system gains access to the neuroimmune regulatory network. Thus, the immune system may be regarded as an additional and diffuse sense organ capable of detecting microbial invasion and tissue injury in any region or organ of the body and by means of multiple pathways signal this information to central homeostatic regulatory regions which in turn initiate appropriate biochemical, physiological and behavioural responses. Despite this widely distributed afferent network for the neuroimmune regulatory system, these signals show great convergence onto very specific and neurochemically identified neuroanatomical nuclei and pathways that we now identify as the central components of the neuroimmune regulatory system.

NEUROANATOMICAL PATHWAYS AND CONNECTIONS OF THE NEUROIMMUNE REGULATORY SYSTEM

The numerous studies which have examined the central induction of c-fos mRNA and protein following numerous immunological challenges have repeatedly shown the PVN and its afferent and efferent connections with the brain stem and spinal cord are consistently activated by immune related stimuli. Examination of this pattern of central activation in the brain and spinal cord reveals the neuroanatomical substrates for both the neuroendocrine system and the autonomic nervous system. The detailed work of Swanson (1985), in particular, has delineated the neuroanatomical pathways and connections of the PVN, and foremost among the inputs to the hypothalamus and PVN arise from catecholamine cell bodies located in the dorsal and ventrolateral medulla and the pons. The central ascending input to the PVN from NE, as well as epinephrine (E) and serotonin (5HT) neurons in the ventromedial medulla constitute primary and critical neuroanatomical pathways for the transmission of immune-related stimuli to the PVN. In support of this, we have shown that systemic injection of IL-1 produces a dramatic increase in the turnover of NE in the hypothalamus (Zalcman *et al.*, 1994). However, the most direct and powerful demonstration of the critical role of ascending catecholamine inputs to the PVN from the brain stem and the requirement of this pathway in order for immune-related signals to access the PVN has been provided by stereotaxic knife cuts and deafferentation experiments (Li *et al.*, 1996; Jackson, 1999; Nance and MacNeil, 2001). We and others have shown that knife cuts located in the brain stem that sever part of the ascending catecholamine inputs to the PVN produced a dramatic decrease in catecholamine fibres in the PVN. Importantly, these cuts also dramatically reduce the induction of c-fos protein in the PVN following an i.v. injection of LPS or IL-1. In additional studies, we showed that posterolateral deafferentation of the hypothalamus, which eliminated all posterior, dorsal and lateral connections of the PVN, completely eliminated all NE-containing fibres in the PVN and also completely eliminated the ability of i.v. LPS to activate the PVN (Jackson, 1999; Nance and MacNeil, 2001). Finally, this ascending catecholamine pathway was shown further to be specific to immune-related signals and not the result of some generalized response to stress. The effects of footshock on the induction c-fos protein in the PVN was unaltered by the brain stem knife cuts, whereas posterolateral deafferentation of the PVN only produced a partial reduction in the number of c-fos positive neurons in the PVN following footshock. Together, these results illustrate that while the immune system and psychological stress may activate a common neuroimmunoregulatory network, they access these hypothalamic regulatory nuclei via separate neuroanatomical pathways. Since only the rostral inputs to the PVN from limbic forebrain structures was spared by the posterior hypothalamic deafferentation surgeries, psychological stressors must access the PVN and the neuroimmuoregulatory system via rostral connections.

Do cutaneous inflammatory signals reach the PVN via ascending catecholamine pathways? Intramuscular injections of turpentine have been shown to stimulate activation of the hypothalamo-pituitary-adrenal (HPA) axis and corticosterone release (Turnbull and Rivier, 1996). Pretreatment with a cyclooxygenase inhibitor was subsequently shown to inhibit this hypothalamic response to a peripheral inflammatory challenge. Consistent with a role for prostaglandins in mediating this central response, i.m. injections of turpentine were shown to induce the expression of the prostaglandin synthesizing enzyme Cox-2 in cerebral vascular endothelial cells (Laflamme *et al.*, 1999). Thus, the response to inflammation produced by a peripheral injection of turpentine is similar to other immune challenges which can also be blocked by cyclooxygenase inhibitors (see below); therefore, we would predict that the ascending catecholamine pathways to the hypothalamus are critical also for central activation induced by turpentine. The physiological responses to systemic injections of turpentine have provided some of the most convincing evidence that a cytokine may mediate the central activational effects of a peripheral inflammatory challenge. Utilizing IL-1β knockout mice, Horai *et al.* (1998) demonstrated that relative to wild-type controls, the effects of turpentine injections on corticosterone release and fever was not observed in IL-1β knockout mice.

The central adrenergic system also comprises a major descending projection from brain stem autonomic nuclei to the sympathetic preganglionic neurons located in the intermediolateral cell column of the thoracolumbar spinal cord (Swanson, 1985). In collaboration with direct peptidergic projections from the PVN and other hypothalamic nuclei to the sympathetic preganglionic neurons, these brain stem and hypothalamic projections comprise the major output pathway from the neuroimmune regulatory system to sympathetic premotor neurons located in the spinal cord and motor sympathetic neurons located in prevertebal and paravertebral sympathetic ganglia.

BIOCHEMICAL MEDIATORS OF THE ACTIONS OF IMMUNE STIMULI ON THE CENTRAL NERVOUS SYSTEM

As already alluded to above, pretreatment with cyclooxygenase inhibitors, such as indomethacin, blocks the majority of the physiological effects of LPS and related inflammatory cytokines including the central induction of c-fos protein, fever and increased NE turnover in the brain (Masana *et al.*, 1990; Wan *et al.*, 1994). Thus, the production of prostaglandins, such as PGE2, represent a primary mediator of the central and peripheral actions of immune stimuli. Acting perhaps in the periphery, we have shown that pretreatment with a HI, but not a H2, histamine antagonist would also attenuate the activation of c-fos protein in the PVN following i.p. injections of LPS. Lastly, pretreatment with the NMDA glutamate receptor antagonist, MK801, could also prevent the central induction of c-fos protein in the brain following both i.p. and i.v. injections of LPS (Wan *et al.*, 1994). Together, these results indicate that prostaglandins, histamine and glutamate neural transmission are all members of a cascade of biochemical and neurochemical mediators that translate the peripheral actions of microbial and immune products into neural signals capable of reaching the central neuroimmune regulatory system.

FUNCTIONAL CONSEQUENCES OF IMMUNE STIMULI ON THE NEUROIMMUNE REGULATORY SYSTEM

Activation of the HPA axis is produced by microbial products and related inflammatory cytokines (Rivier, 2001) and CRF neurons are a major target for the central effects of immune stimuli on the brain. LPS injections produce a dose-related increase in plasma corticosterone levels which is paralleled by the induction of c-fos protein in CRF neurons localized in the PVN (Wan *et al.*, 1993a). Also, other neuropeptide and transmitter specific neurons in the PVN are also activated by LPS, such as vasopressin, oxytocin and NO producing neurons (Jackson, 1999). However, the release of adrenal steroids in response to a wide range of immune-related stimuli (LPS, double stranded DNA, viruses, turpentine, etc.) as well as in response to immunization with foreign antigens, indicates that activation of the HPA axis is a generalized and common response to the majority of immune-related challenges. However, the release of pituitary hormones, such as vasopressin, oxytocin, α-MSH, prolactin, growth hormone and β-endorphin, may also impact on peripheral immune processes and may represent yet an additional avenue for modulation of immune function by the central nervous system.

Concurrent with the general activation of the HPA axis, a variety of immune stimuli activate the sympathetic nervous system. Increased turnover rate of NE in immune organs, such as the spleen, have been shown to occur in response central injections of IL-1 and prostaglandins (Vriend *et al.*, 1993; Terao *et al.*, 1995), and splenic levels of NE have been shown to vary during the course of an immune response following immunization with SRBCs (Besedovsky et al., 1984; Green-Jonhson et al., 1996). However, electrophysiological recordings of sympathetic nerves innervating the spleen have provided the most direct and unequivocal evidence that the sympathetic nervous system is a major target for the central actions of immune stimuli. We have shown that relative to sympathetic nerves supplying the kidney, LPS produces a selective and sustained increase in splenic sympathetic nerve activity (MacNeil et al., 1996). Peripheral and central injections of the cyclooxygenase inhibitor indomethacin blocked or attenuated the effects of i.v. LPS on splenic sympathetic nerve activity (MacNeil et al., 1997). Consistent with the proposed intermediary role of prostaglandins in the central activational effects of LPS, central injections of PGE2 produced an immediate increase in splenic sympathetic nerve activity. Although the HPA axis can be activated independently of the sympathetic nervous system by the judicious selection of very low doses of LPS (Meltzer, 2000), it is likely that both of these efferent arms of the neuroimmune regulatory system are engaged simultaneously in response to most immune challenges and inflammatory stimuli.

EFFECTS OF HPA-AXIS ACTIVATION ON IMMUNE FUNCTION

The immunosuppressive effects of adrenal steroids are well known and continue to be utilized clinically for treatment of a variety of inflammatory condition. Due to the potent anti-inflammatory effects of adrenal hormones, they typically exert an immunosuppressive action on the innate immune system, comprised primarily of macrophages and NK cells. However, with regards to the adaptive or acquired immune system, primarily mediated by T and B cells, adrenal steroids generally produce a bias towards humoral, or TH2-type immune responses, in contrast to TH1-type, or cellular immune responses. The immunological consequences of alterations in the HPA axis are illustrated by the reported differences in the functional immune responses of different strains of rats. In comparison to Fisher rats, which demonstrate a 'normal' HPA axis, Lewis rats have been shown to have a hyporesponsive HPA axis which is associated with a defect in CRF neurons in the PVN (Sternberg et al., 1989). One functional consequence of this altered HPA axis in Lewis rats is their susceptibility to a number of experimentally induced autoimmune diseases, such as adjuvant induced arthritis and experimental allergic encephalitis. Lastly, experimental removal of the adrenal glands, or adrenal insufficiency, is generally associated with the overproduction of inflammatory cytokines and sepsis in response to bacterial infections or endotoxin challenges (Nagy and Berczi, 1978; Ramachandra et al., 1992) thereby illustrating the general restraint exerted by the HPA axis on the immune system.

EFFECTS OF SYMPATHETIC NERVOUS SYSTEM ACTIVATION ON IMMUNE FUNCTION

We have identified two experimental paradigms for which we have been able to demonstrate a functional role for adrenergic nerves on immune function, central inflammatory stimuli and stress. Both innate and acquired immune responses are modulated by the sympathetic nervous system. As summarized above, central inflammatory stimuli, such as intracranial injections of IL-1 or PGE2, activate the sympathetic nervous system. These central inflammatory signals inhibit innate immune responses and produce an immune suppression in splenic NK cell activity and splenic macrophage cytokine production (Sundar *et al.*, 1990; Brown *et al.*, 1991; Nance and MacNeil, 2001). Importantly, injections of β-adrenergic blocking agents or surgically cutting the splenic nerve have been shown to block or attenuate the immunosuppressive effects of central inflammatory stimuli on splenic NK cell activity and cytokine production by splenic macrophage.

The sympathetic nervous system is activated by stress, such as intermittent footshock and physical restraint. These 'stressors' typically produce a rapid induction of c-fos protein in hypothalamic and brain stem nuclei identified as the neuroanatomical substrates for the autonomic nervous system as well as the central neural elements of the HPA axis (Wan *et al.*, 1994). Most of the hypothalamic and brain stem regions activated by stress are the same brain areas activated by LPS or IL-1 (Wan *et al.*, 1993a,b, 1994); however, stress also induces additional and more widespread central activation of limbic forebrain structures than that observed with immune challenges and includes brain areas such as the lateral septal region, amygdala, BNST and medial frontal cortex. Based upon our hypothalamic knife cut experiments reviewed earlier, we believe that it is the rostral inputs from these limbic forebrain structures that mediate the effects of stress on the hypothalamic neuroimmune regulatory system. Although the immunological consequences have yet to be clarified, it has recently been shown that stress produces a dramatic degranulation of mast cells in the skin (Singh *et al.*, 1999). The mediation of this effect by cutaneous sensory nerves and/or sympathetic nerve fibres would appear to be worthy of further analysis. The effects of stress on splenic NK cell activity are well documented and represents one of the most reliable immunological changes induced by stress (Shimizu *et al.*, 1996). NK cells possess an abundance of β-adrenergic receptors and it appears likely that NE from sympathetic nerve terminals mediate the stress-induced suppression of splenic NK cell activity. As shown by Shimizu *et al.* (1996) stress-induced suppression of splenic NK cell activity was attenuated by cutting the splenic nerve. We have examined in some detail the effects of intermittent footshock on splenic macrophage function (Meltzer, 2000). We have developed an *in vivo* procedure for assessing splenic cytokine production in response to stress. Based upon dose–response experiments, we have identified a low dose of i.v. LPS (0.1 μg) that produces *intermediate* levels of inflammatory cytokine mRNA and protein production in the rat. Lower doses produced minimal or undetectable levels of cytokine production, and a dose of 1.0 μg and higher produced maximal levels of cytokine mRNA in the spleen. Thus, by utilizing this 0.1 μg dose of LPS/rat we have been able to detect both inhibition and facilitation of splenic cytokine production. Finally, double labelling immunocytochemistry verified that splenic macrophages were the primary, if not exclusive, source of inflammatory cytokine production following i.v. LPS injections (Meltzer *et al.*, 1997; Meltzer, 2000). We have found that as little as 15 min of intermittent footshock administered either immediately before or immediately following an i.v. injection of 0.1 μg LPS produced a dramatic suppression in splenic TNF-α mRNA and protein when assessed 1 h after the LPS injection. Splenic levels of IL-1β mRNA and protein followed a similar pattern, but were less dramatic than the changes observed for TNF-α. Also, the effects of footshock on plasma levels of TNF-α and IL-1β protein followed the same pattern observed for the spleen. To determine the contribution of the adrenal gland and sympathetic nervous system to this stress-induced suppression of *in vivo* splenic cytokine production, we examined the effects of splenic nerve cuts alone, adrenalectomy alone, and the effects of combining both adrenalectomy and nerve cuts. In contrast to some of our earlier *in vitro* assessments of the effects of stress on splenic immune function (see Wan *et al.*, 1993b), we found that cutting the splenic nerve had no detectable effect of the immunosuppressive effects of stress on *in vivo* splenic cytokine production in rats with intact adrenal glands. Likewise, we found that adrenalectomy had no effect on the magnitude of the stress-induced suppression of *in vivo* splenic cytokine production in animals with an intact sympathetic nerve supply to the spleen. Together, these results demonstrated that the adrenal steroids could mediate the immediate immunosuppressive effects of stress on *in vivo* splenic cytokine production in the absence of a sympathetic nerve supply to the spleen. Significantly, stress was equally effective at suppressing splenic cytokine production in adrenalectomized animals, thereby leaving only the sympathetic nerve supply to the spleen to mediate this immunosuppression. Finally, we directly tested the effects of splenic nerve cuts in animals that were adrenalectomized and stressed. Results of this study verified that it was the splenic nerve, that is, the sympathetic nervous system, that mediated the inhibition of splenic macrophage cytokine production induced by stress in adrenalectomized rats. Thus, under these experimental conditions, the adrenal gland and sympathetic appear to be equal partners with each being entirely capable of independently suppressing splenic macrophage function in response to stress. What has yet to be established is whether it is the adrenal cortex and/or the adrenal medulla that is primarily responsible for the stress-induced suppression of splenic macrophage function. However, given the effectiveness of the splenic nerve in the total absence of the adrenal gland, and the fact that the adrenal medulla is essentially a specialized sympathetic ganglion, it appears likely that the sympathetic nervous system is primarily responsible for stress-induced suppression of *in vivo* splenic macrophage function. Finally, it could be argued that the reason that treatment with β-adrenergic blockers are so effective at reducing the effects of stress on immune function, even in adrenal intact animals, is that the physiological effects of both sympathetic nerve terminals and epinephrine released from the adrenal medulla would be inhibited.

ADRENERGIC REGULATION OF ADAPTIVE IMMUNITY

The adaptive immune system can be divided into humoral immunity, mediated by antibody producing B cells, and cell-mediated immunity, mediated by T lymphocytes. Corresponding to this division in the adaptive immune system are two types of T-helper cells, designated TH1 and TH2, which direct adaptive immune responses in the direction of cellular or humoral immune responses, respectively. These divergent functional categories of T cells are based upon the specific cytokines produced by these cell types. Interleukin-2 (IL-2), interferon gamma (IFN-γ) and lymphotoxin (TNF-β) are cytokines produced by TH1 cells and generate cellular immune responses. IL-4, -5, -6, -9, -10 and -13 are produced by TH2 cells and promote humoral immune responses. The actions of TH1- and

TH2-type immune responses are mutually inhibitory such that one or the other of these adaptive responses will dominate during a specific immune response. For example, TH1 responses predominate in response to viral infections whereas TH2 responses are effective against parasites. The bias of adaptive immunity towards a humoral or cell-mediated response is directly linked with the actions and signal molecules (cytokines) generated by innate immune responses. The demonstration that TH1, but not TH2 cells, possess adrenergic receptors suggest that catecholamines may play an important role in regulating adaptive immune responses (Sanders et al., 1997). Also, adrenergic receptors on B cells suggest that the sympathetic nervous system may also regulate antibody production. However, due to the complexity of cellular interactions involved in the regulation of adaptive immunity, our understanding of the exact role of the sympathetic nervous in regulating this aspect of host defense is incomplete. Contributing to the complexity of analysing this system is the bi-directional interactions between the innate and adaptive immune systems. For example, actions of the sympathetic nervous system on cellular responses of the innate immune system, such as inflammation, antigen processing and presentation, will impact on subsequent antigen-specific adaptive immune responses. Plus, it is highly likely that the sympathetic nervous system can impact further upon the adaptive immune system at various stages which typically occur across several days, in contrast to innate immune responses which generally occur within a few hours. Also, many of the anti-microbial and host defense functions of the adaptive immune system are carried out by cells of the innate immune system in that signals generated by T cells have a dramatic effect on the anti-microbial capacity of macrophages and NK cells. Finally, since the adaptive immune system is 'antigen-specific', there are substantial strain, species and individual differences in the responses to specific protein antigens. With these considerations in mind, we will summarize some of our work and others on this aspect of immune regulation.

Utilizing footshock as a means of activating the sympathetic nervous system, we have examined the effects of cutting the splenic nerve on a humoral immune response (Wan et al., 1993b). Prior studies showed that the effects of stress on the PFC response following immunization with SRBCs was critically related to the timing of the stressor in relation to the antigenic challenge (Zalcman et al., 1988). Stress was found to be immunosuppressive only when administered on day 3 following immunization, with the PFC response being assessed on day 4. Related to this is the observation that splenic NE levels decrease on day 3 following immunization with SRBCs (Green-Johnson et al., 1996), which corresponds to the period just prior to when antigen-specific B cells and IgM antibodies first begin to appear, with high levels of antigen-specific IgG antibodies appearing a few days later. We found that animals stressed on day 3 after immunization with SRBCs showed a 50% inhibition in the number of PFCs when assessed on day 4 (Wan et al., 1993b). Cutting the splenic nerve prior to immunization produced a small and nonsignificant increase in the number of PFCs, but importantly, splenic nerve cuts completely abrogated the inhibitory effects of stress on the PFC response. In support of an inhibitory role for splenic NE on humoral immunity, we utilized a substrain of BALB/c mice which had been selectively bred for susceptibility to audiogenic seizures (epilepsy prone, EP), and were compared to epilepsy resistant (ER) mice (Green-Johnson et al., 1996). Following immunization with SRBCs, we found that the EP mice showed significantly lower SRBC-specific IgG PFCs and antibody titres. Examination of splenic levels of NE following immunization indicated that while both strains showed a significant drop in NE levels on day 3 after immunization, the EP strain maintained significantly higher splenic NE levels that the ER mice for all time periods tested. *In vitro* examination of T-cell function (proliferative responses and cytokine production) in the two strains revealed no differences in T-cell responses between the two strains, indicating no intrinsic defect in T-cell function and suggesting further that the *in vivo* environment and possibly splenic NE were mediating the strain differences in humoral immunity. Consistent with this, we found that treatment of the ER strain with an adrenergic receptor agonist significantly reduced the IgG PFC response and IgG antibody titres following immunization with SRBC, relative to vehicle injected controls. Somewhat analogous data are provided by comparisons between two strains of mice well character-ized for demonstrating predominately TH1 or TH2-type immune responses (Kruszewska *et al.*, 1995). When challenged with a variety of pathogens, C57BL/6J mice show a strong cell-mediated response and produce TH1 cytokines (IL-2 and IFN-γ). In contrast BALB/cj mice generate a humoral immune response and produce TH2 cytokines (IL-4 and IL-10). Additional data provided by Kruszewska *et al.* (1995) indicated that splenic NE content in the BALB/cj strain was approximately twice the level of NE in the C57BL/6j strain. These results are compatible with the evidence that only TH1 cells possess adrenergic receptors and therefore the relative bias of the BALB/cj mice towards TH2-type immune responses, in comparison to C57BL/cj mice, may be in part related to their chronically higher levels of splenic NE. They also found that chemical sympathectomy induced by 6OHDA injec-tions enhanced humoral immune responses in both strains (Kruszewska *et al.*, 1995). Thus, while there is evidence that the sympathetic nervous system is inhibitory on the adaptive immune system, the studies reviewed above do not exclude the possibility that the actual inhibitory action of NE on adaptive immunity is due to NE induced alterations in cells of the innate immune system, specifically macrophages and antigen presenting cells. In this regard, there are also experiments that clearly indicate that NE and the sympathetic ner-vous system facilitates adaptive immunity, a few of which will be considered next.

IL-2 is a cytokine produced by activated T cells and stimulates T-cell prolifera-tion and potentiates B-cell antigen-specific antibody production. We have examined the immunoenhancing effects of IL-2 on the PFC response to SRBC immunization and tested the role of NE and the sympathetic nervous system on this immune response (Zalcman *et al.*, 1994). First, we found that IL-2 injections potentiated the splenic PFC response of both mice and rats, but only if IL-2 was injected in close proximity to SRBC immuniza-tion. IL-2 injections 2 days after SRBC administration had no effect on the PFC response. In mice, the immunostimulatory effects of IL-2 on the PFC response could be blocked by administration of a β-adrenergic antagonist whereas an α-adrenergic antagonist had no effect. Finally, we found that cutting the splenic nerve in rats blocked the immunostimula-tory effects of IL-2 on the antigen-specific PFC response. These results show that enhance-ment of humoral immunity by IL-2 is mediated via β-adrenergic receptors and requires an intact splenic nerve.

The work of Sanders and colleagues has provided some of the most compelling evidence that NE facilitates humoral immunity. They demonstrated that NE acts upon both T cells and B cells and found that when NE was added at the time of *in vitro* immunization the PFC response to SRBC was increased (Sanders and Munson, 1984a). This effect was blocked by a β-adrenergic receptor antagonist and reproduced by a β2-adrenergic receptor agonist (Sanders and Munson, 1984b). We reviewed earlier Sanders' observation that only TH1 cells have β2-adrenergic receptors (Ramer-Quinn *et al.*, 1997; Sanders *et al.*, 1997). These binding studies were supported further by the fact that a β2-adrenergic agonist decreased IL-2 and IFN-γ expression in TH1 clones but had no effect on cytokine production in TH2 cells. Since the T cells were exposed to NE prior to reconstitution with B cells, this verified the effects of the adrenergic agonist were on T cells. Importantly, these results have recently been verified *in vivo*. SCID (severe combined immunodeficiency) mice were reconstituted with antigen-specific T and B cells. Since the T cells were TH2 clones, only the B cells had adrenergic receptors (Kohm and Sanders, 1999). Mice that were given a chemical sympathectomy with 6OHDA prior to reconstitution demonstrated a decreased primary IgM response and decreased primary and secondary IgG responses. Decreased antibody responses could also be produced in animals treated with a β-adrenergic receptor antagonist and partially restored with a β2-adrenergic agonist. They also found that the important T-cell costimulatory molecule B7–2 produced by antigen presenting cells, was increased in B cells treated with NE. These studies clearly show that NE and the sympathetic nervous system can regulate both B- and T-cell function.

Finally, sympathetic fibres and NE have also been implicated in another immunological phenomenon, the Shwartzman reaction. First described in rabbits, Shwartzman demonstrated that if a sublethal i.v. dose of LPS was injected systemically, and then followed 24 h later by a second i.v. injection of LPS, the animals showed a widespread intravascular thrombus formation and disseminated intravascular coagulation (Shwartzman, 1928). Interestingly, if the first injection of LPS was administered intradermal and the i.v. injection given 24 h later, haemorrhagic necrosis occurred exclusively at the intradermal injection site. The latter paradigm was referred to as the localized Shwartzman reaction. Although regarded by some as not representing a true immunological response (lack of antigen specificity), other studies indicate a similarity to DTH reactions, and antigen-specific (T-cell dependent) Shwartzman reactions have been demonstrated (De Weck *et al.*, 1968). Nonetheless, it is a powerful and potentially lethal immunological response and the fact that inflammatory cytokines (TNF-α and IL-1) can for the most part be substituted for the LPS injections indicates its similarity to septic shock and the clinical relevance of this phenomenon (Movat *et al.*, 1987). Likewise, LPS induced IL-12 and IFN-γ have been shown to be critical cytokines for the induction of the Shwartzman reaction that dramatically potentiate macrophage TNF-α production (Hermans *et al.*, 1990; Ozmen *et al.*, 1994). TNF-α was initially identified by it's ability to produce haemorrhagic necrosis of tumours which may be analogous to a localized Shwartzman reaction. Some of the early studies of the Shwartzman reaction indicated a primary role for NE and the sympathetic nervous system in this phenomenon (Collins *et al.*, 1972; Shapiro *et al.*, 1974). Most dramatic was an examination of the effects of cutting the sympathetic nerve supply to one kidney in an animal prior to inducing a generalized Shwartzman reaction (Palmerio *et al.*, 1962). The kidneys are a primary target for haemorrhagic necrosis during a Shwartzman reaction. Although the generalized Shwartzman reaction was observed in the animal, the sympathectomized kidney was entirely spared from damage. Other studies have shown that pretreatment with α-adrenergic antagonist can block the generalized Shwartzman reaction (Latour and Leger-Gauthier, 1987). Likewise, in appropriately primed animals, localized injections of NE can reproduce the localized Shwartzman reaction (Selye and Tuchweber,

1966). Thus, further examination of the role of the sympathetic nervous system in this phenomenon would appear worthy of consideration.

DOES IT REALLY WORK LIKE THAT?

Besedovsky's feedback model of the neuroimmune regulatory system has provided an important conceptual framework for assessing the counterregulatory role of the HPA axis and sympathetic nervous system in the regulation of immune responses. Basic elements of this model system have been supported by an expanding volume of literature. That is, products generated by immune cells and immune responses can signal the brain via identifiable pathways and specific brain areas and chemically defined populations of neurons are functional targets for these immune related stimuli. Likewise, the dual output pathways for this regulatory system, the HPA axis and sympathetic nervous system, have clearly been shown to modulate numerous immune responses. Despite the general support for this model system, what remains to be established is whether this proposed counter-regulatory feedback model system actually applies to *in vivo* immune responses. A persistent and unresolved problem has been the timing of the afferent signals produced by *in vivo* immune responses with regards to the activation of the output pathways and their subsequent actual impact on an ongoing immune reaction. Thus far, convincing demonstrations of an effect of both the HPA axis and the sympathetic nervous system on immune responses have required that the neuroimmune regulatory system be engaged, or activated, *prior* to the initiation of an immune challenge. This is exactly the case with most experiments involving stress, as well as our own work involving the central injections of cytokines or prostaglandin. Thus, while it is amply clear that activation of these counter-regulatory systems *before* an immune reaction becomes established can have profound effects on the magnitude of that immune response, there is an absence of evidence that this counterregulatory system is functioning during an *in vivo* immunological challenge. Specifically, although an immune challenge produces counter-regulatory responses from the HPA axis and sympathetic nervous system, these responses appear to be too late to impact upon ongoing immune reactions initiated by the same immune challenge that activates these counter-regulatory responses. This is most apparent with an endotoxin challenge, but may be applicable to real infections with live microbes. The continued reliance upon *in vitro* assessments of immune function following various experimental manipulations have done little to clarify this issue. As an example, we have found that cutting the splenic nerve prior to *in vitro* assessment of splenic macrophage cytokine production in response to LPS results in increased cytokine production by macrophages (see Brown *et al.*, 1991). However, we have found that if the same experiment is performed *in vivo*, that is, the splenic nerve is cut and then the animal treated with LPS i.v., there is no detectable effect of the nerve cut on splenic levels of cytokine mRNA or protein. Likewise with adrenalectomy and adrenalectomy plus a splenic nerve cut (Meltzer, 2000). We previously showed that *in vitro*, the combination of adrenalectomy plus splenic nerve cuts produced a profound potentiation in macrophage cytokine production in response to LPS (Brown *et al.*, 1991 unpublished data); however, a similar experimental manipulation *in vivo* had no effect on the production of splenic cytokines in animals given an i.v. injection of LPS.

WHEN DO THE ADRENAL GLAND AND SYMPATHETIC NERVOUS SYSTEM EXERT THEIR COUNTER-REGULATORY EFFECTS?

As reviewed earlier, engagement of the neuroimmune regulatory system *prior* to the activation of immune cells by microbial stimuli or antigen typically produces robust alterations in subsequent immune responses. However, if the counter-regulatory responses are dependent upon the same stimuli that initiate the immune response, the counter-regulatory responses may be too late to modulate ongoing immune reactions. This is very likely the case with the innate immune system, macrophage cell function, and acute inflammatory reactions. Under these more physiological circumstances, it may be that the neuroimmune regulatory system primarily impacts upon secondary responses following the inital microbial insult, such as antigen presentation and the adaptive/specific immune system. Thus, under normal-basal conditions (absence of stress or central infection), the early immunological events generated by bacteria and viruses may proceed without significant modulation by the counter-regulatory neuroimmune regulatory system despite its full engagement by the immune stimuli. Conceptually, this would allow the full extent and magnitude of the initial immune challenge or injury to be expressed and assessed by the central regulatory system such that counter-regulatory reactions are matched appropriately to the severity of the inital insult. For example, shutting down TNF-α production early on in an immune reaction following a severe microbial infection could be disasterous to an organism. Yet, unrestrained inflammatory cytokine production and release possesses an equal or greater risk to an organism. When viewed in this context, counter-regulatory reactions may do little to restrain the initial reaction to a microbial challenge, but rather, serve to impose time limitions on immune reactions which are tailored to the nature and severity of the initial threat to the organism. Also, *in vivo* counter-regulatory responses may reduce or delay immune reactions to a second immune challenge that occurs before the inital microbial threat has been resolved. This latter process may account for why experimentally activating the neuroimmune-regulatory system by central inflammatory stimuli or stress *prior* to an immune challenge is so effective. It is as if this immune challenge is processed as if it were the second challenge to the organism, with central inflammation or stress acting like the first. Representing a relatively simple model of this proposed process is the so-called counter-irritant effect which is illustrated by the 'anti-inflammatory effects of turpentine'. If turpentine is injected s.c. in the abdominal wall of rats and then carrageenan injected into the footpad 24 h later, assessment of foot oedema at 2 and 5 h after carrageenan injections show a dramatic reduction in oedema for animals pretreated with turpentine, relative to untreated controls (Damas and Deflandre, 1984). Although adrenalectomy was shown to reduce the amount of carrageenan-induced oedema in the foot, the antiinflammatory effect of pretreatment with turpentine was still observed in adrenalectomized rats and the magnitude of the decrease was comparable to animals with intact adrenal glands. More recently, Levine and associates (Green *et al.*, 1997; Miao *et al.*, 2000) described a similar negative feedback effect on plasma extravasation of the knee joint produced by prior intradermal injections of capsaicin or electrical stimulation of the rat paw. Both the adrenal medualla and sympathetic postganglionic neurons mediated the inhibition of inflammation in the knee joint. Since the counter-irritation effect can also be demonstrated for systemic treatments, implications for the neuroimmune regulatory system are not limited to cutaneous immune responses and may reflect a fundamental process whereby the immune system can remain focused on a specific pathogen. Finally, interactions between sequential microbial infections, such as a viral infection followed by a bacterial challenge, have been noted (Doughty *et al.*, 2001; Nansen and Thomsen, 2001). Except with this sequence of treatments, viral infection produced a rapid sensitization to the effects of endotoxin with lethal consequences, an effect that is reminiscent of the generalized Shwartzman reaction. In all of these paradigms, there is a two-step process wherein the inflammatory or immunological consequences of the initial stimulus exerts a powerful modulatory effect on subsequent challenges. Again, this is analogous to the modulatory effects of stress or inflammation in the brain (step 1) on the functional response to a subsequent immunological challenge (step 2) that was reviewed earlier. The potential role of the neuroimmune regulatory system in mediating these interactions has not been fully explored.

What about conditioned immune responses? Conditioned immune responses would appear to present a situation during which the sympathetic nervous system might be capable of exerting a counter-regulatory effect. First, conditioned physiological responses are usually generated rapidly following exposure to the conditioned stimulus (CS) and therefore might be capable of exerting modulatory control of an ongoing immune response. We have utilized the well-established classically conditioned taste aversion paradigm to test whether the neuroendocrine, autonomic and immunological effects of LPS could be conditioned in rats (Janz *et al.*, 1996). First, we determined the unconditioned response (UCR) to an i.p. injection of 50 μg LPS. The UCR included a dramatic increase in serum corticosterone, a suppression in the *in vitro* production of IL-2 by T cells, and a small and nonsiginificant decrease in splenic NE levels. Subsequently, animals were classically conditioned by pairing a saccharin solution (conditioned stimulus), with LPS injection (unconditioned stimulus). Various conditioning control groups, such as animals given saccharin and LPS in an unpaired manner were included. Relative to control animals, conditioned rats that were reexposed to the CS showed a modest, but significant increase in serum corticosterone, a significant suppression in splenic T-cell production of IL-2, and importantly, a significant decrease in splenic NE levels. Thus, the conditioning procedure reproduced the immunological effects of LPS on T-cell function, whereas the conditioned effects on corticosterone levels were modest in comparison to the potent unconditioned effects of LPS on the HPA axis. But importantly, the conditioning procedure produced a significant decrease in splenic NE levels by over 50%, an effect which was dramatically greater than the unconditioned response to LPS. These results are consistent with the possibility that the sympathetic nervous system may be a primary mediator of the effects of conditioning on the immune system. In direct support of this proposal, Exton *et al.* (1999) utilized a similar conditioned taste aversion paradigm and demonstrated that the conditioned immunosuppressive effects of cyclosporin on heart allograft rejection could be completely blocked by prior cutting of the splenic nerve. Thus, further analysis of the effects of behavioral conditioning on functional *in vivo* immune responses may provide a unique avenue to analyze further the function role of the sympathetic nervous system in the regulation of the immune system and host defense.

SUMMARY AND CONSIDERATION OF A NEW MODEL

The concept of a central neuroimmune regulatory network that is responsive to signals produced by the immune system and that is capable of generating endocrine and autonomic responses that modify immune function continues to be supported by an expanding volume of data. However, the view that this neuroimmune regulatory system operates by means of a counter-regulatory feedback network requires revision. If the central regulatory network is first activated by central inflammatory stimuli or stress, then subsequent immune responses to a variety of challenges are modified via endocrine and sympathetic nervous system mechanisms. However, if the neuroimmune regulatory system is not activated *prior* to exposure to an immune challenge, counter-regulatory responses produced by this regulatory system may have minimal impact on an ongoing immune response. Although an immune challenge simultaneously signals the central regulatory network and initiates peripheral immune responses, the counter-regulatory responses produced by the central nervous system appear to occur too late to modify ongoing immune reactions generated by that same immune challenge. This is most applicable to the early cascade of immunological reactions initated by an immune challenge which can proceed unabated despite the subsequent full engagement of endocrine and sympathetic responses. Thus, the conceptual model of a negative feedback counter-regulatory system, at least with regards to the initial stages of immune activation, must be modified to account for the fact that the timing of endocrine and autonomic responses are typically too late to impact on an ongoing immune response. Although the outputs of the neuroimmune regulatory system can no doubt have a counter-regulatory action, by necessity, this action must be occurring at a later time period during the course of an immune response. Thus, the counter-regulatory aspect of the neuroimmune regulatory system may be to delimit the length of time an immune reaction is allowed to proceed and to limit the number of immune challenges that can be processed at one time. Thus, a feedforward model system may be required to incorporate the likely actions of the endocrine and sympathetic nervous system on immune reactions occurring later during an immune response, such as dendritic cell maturation, antigen processing and presentation, and subsequent regulation of T- and B-cell function and the adaptive immune system. Finally, further analysis of the potential role of the sympathetic nervous system in mediating conditioned immune responses would appear to be worthy of vigorous investigation and may provide a unique model system to examine further the interactions between the adrenergic and immune systems.

ACKNOWLEDGEMENTS

This research was supported by NIH grant # MH 43778 and the Canadian Institutes of Health Research.

REFERENCES

Abbas, F., Amin, Z., Burk, R.M., Krauss, A.H., Marshall, K., Senior, J. and Woodward, D.F. (1997) A comparative study of thromboxane (TP) receptor mimetics and antagonists on isolated human umbilical artery and myometrium. *Adv. Exp. Med. Biol.*, **407**, 219–230.

Besedovsky, H., del Rey, A., Da Prada, M. and Keller, H.H. (1979a) Immunoregulation mediated by the sympathetic nervous system. *Cell. Immunol.*, **48**, 346–355.

Besedovsky, H., Sorkin, E., Felix, D. and Haas, H. (1979b) Hypothalamic changes during the immune response. *Eur. J. Immunol.*, **7**, 323–325.

Besedovsky, H., del Rey, A., Sorkin, E., Da Prada, M., Burri, R. and Honegger, C. (1983) The immune response evokes changes in brain non-adrenergic neurons. *Science*, **221**, 564–566.

Besedovsky, H., del Rey, A. and Sorkin, E. (1984). Immunoregulation by neuroendocrine mechanisms. In P. Behah and F. Spreafico (eds). *Neuroimmunology*, New York: Raven Press, pp. 445–450.

Besedovsky, H., del Rey, A., Sorkin, E. and Dinarello, C.A. (1986) Immunoregulatory feedback between interleukin-1 and glucocorticoid hormones. *Science*, **233**, 652–654.

Brown, R., Zuo, L., Vriend, C., Nirula, R., Janz, L., Falk, J., Nance, D., Dyck, D. and Greenberg, A.H. (1991) Suppression of splenic macrophage IL-1 secretion following intracerebroventricular injection of interleukin-1: Evidence for pituitary-adrenal and sympathetic control. *Cell. Immunol.*, **132**, 84–93.

Collins, A.D., Edmonds, M.S., Henson, C., Izard, S.R. and Brunson, J.G. (1972) Norepinephrine, endotoxin shock, and the generalized Shwartzman reaction. *Arch. Pathol.*, **93**, 82–88.

Damas, J. and Deflandre, E. (1984) The mechanism of the anti-inflammatory effect of turpentine in the rat. *Arch. Pharmacol.*, **327**, 143–147.

De Weck, A.L., Frey, J.R. and Geleick, H. (1968) Immunologic specificity of the localized Shwartzman phenomenon induced in guinea pigs by simple chemical haptens. *J. Immunol.*, **100**, 1–6.

Doughty, L.A., Nguyen, K.B., Durbin, J.E. and Biron, C.A. (2001) A role for IFN-$\alpha\beta$ in virus infection-induced sensitization to endotoxin. *J. Immunol.*, **166**, 2658–2664.

Ericsson, A., Kovacs, K.J. and Sawchenko, P.E. (1994) A functional anatomical analysis of central pathways subserving the effects of interleukin-1 on stress-related neuroendocrine neurons. *J. Neurosci.*, **14**, 897–913.

Ericsson, A., Liu, C., Hart, R.P. and Sawchenko, P.E. (1995) Type 1 interleukin-1 receptor in the rat brain: distribution, regulation, and relationship to sites of IL-1-induced cellular activation. *J. Comp. Neurol.*, **361**, 681–698.

Exton, M.S., Schult, M., Donath, S., Strubel, T., Bode, U., Del Rey, A. *et al.* (1999) Conditioned immunosuppression makes subtherapeutic cyclosporin effective via splenic innervation. *Am. J. Physiol.*, **45**, R1710–R1717.

Green, P.G., Janig, W. and Levine, J.D. (1997) Negative feedback neuroendocrine control of inflammatory response in the rat is dependent on the sympathetic postganglionic neuron. *J. Neurosci.*, **17**, 3234–3238.

Green-Johnson, J.M., Zalcman, S., Vriend, C.Y., Nance, D.M. and Greenberg, A.H. (1996) Role of norepinephrine in suppressed IgG production by eplipsy-prone mice. *Life Sci.*, **59**, 1121–1132.

Gross, P.M., Sposito, N.M., Pettersen, S.E. and Fenstermacher, J.D. (1986) Differences in function and structure of the capillary endothelium in the supraoptic nucleus and pituitary neural lobe of rats. Evidence for the supraoptic nucleus as an osmometer. *Neuroendocrinology*, **44**, 401–407.

Hermans, H., Van Damme, J., Dillen, C., Dijkmans, R. and Billiau A. (1990) Interferon γ, a mediator of lethal lipopolysaccharide-induced Shwartzman-like shock ractions in mice. *J. Exp. Med.*, **171**, 1853–1869.

Horai, R., Asano, M., Sudo, K., Kanuka, H., Suzuki, M., Nishihara, M., Takahashi, M. and Iwakura, Y. (1998) Production of mice deficient in genes for interleukin (IL)-1alpha, IL-1beta, IL-1alpha/beta, and Il-1 receptor antagonist shows that IL-1beta is crucial in turpentine-induced fever development and glucocorticoid secretion. *J. Exp. Med.*, **187**, 1463–1475.

Jackson, A.T.K. (1999) Chemical specificity of endotoxin-induced c-fos expressing neurons in the rat hypothalamus, *MSc Thesis*, University of Manitoba, Winnipeg, Canada.

Janz, L.J., Green-Johnson, J., Murray, L., Vriend, C.Y., Nance, D.M., Greenberg, A.H. and Dyck, D.G. (1996) Pavlovian conditioning of LPS-induced responses: effects on corticosterone, splenic NE and IL-2 production. *Physiol. Behav.*, **59**, 1103–1109.

Kohm, A.P. and Sanders, V.M. (1999) Suppression of antigen-specific Th2 cell-dependent IgM and IgG1 production following norepinephrine depletion in vivo. *J. Immunol.*, **162**, 5299–5308.

Kruszewska, B., Felten, S.Y. and Moynihan, J.A. (1995) Alterations in cytokine and antibody production following chemical sympathectomy in two strains of mice. *J. Immunol.*, **155**, 4613–4620.

Lacroix, S., Vallieres, L. and Rivest, S. (1996) C-fos mRNA pattern and corticotropin-releasing factor neuronal activity throughout the brain of rats injected centrally with a prostaglandin of E2 type. *J. Neuroinmumol.*, **70**, 163–179.

Laflamme, N., Lacroix, S. and Rivest, S. (1999) An essential role of interleukin-1β in mediating NF-KB activity and Cox-2 transcription in cells of the blood-brain barrier in response to a systemic and localized inflammation but not during endotoxemia. *J. Neurosci.*, **19**, 10923–10930.

Latour, J.-G. and Leger-Gauthier, C. (1987) Vasoactive agents and production of thrombosis during intravascular coagulation. *Am. J. Pathol.*, **126**, 569–580.

Lee, H.Y., Whiteside, M.B. and Herkenham, M. (1998) Area postrema removal abolishes stimulatory effects of intravenous interleukin-1beta on hypothalamic-pituitary-adrenal axis activity and c-fos mRNA in the hypothalamic paraventricular nucleus. *Brain Res. Bull.*, **46**, 495–503.

Lee, S., Kim, K. and Rivier, C. (1999) Nitric oxide stimulates ACTH secretion and the transcription of the genes encoding for NGFI-B, corticotropin-releasing factor, corticotropin-releasing factor receptor type 1, and vasopressin in the hypothalamus of the intact rat. *J. Neurosci.*, **19**, 7640–7647.

Li, H.Y., Ericsson, A. and Sawchenko, P.E. (1996) Distinct mechanisms underlie activation of hypothalamic neurosecretory neurons and their medullary catecholaminergic afferents in categorically different stress paradigms. *Proc. Natl. Acad. Sci. USA*, **93**, 2359–2364.

MacNeil, B.J., Jansen, A.H., Greenberg, A.H. and Nance, D.M. (1996) Activation and selectivity of splenic sympathetic nerve electrical activity response to bacterial endotoxin. *Am. J. Physiol.*, **270**, R264–R270.

MacNeil, B.J., Jansen, A.H., Janz, L.J., Greenberg, A.H. and Nance, D.M. (1997) Peripherial endotoxin increases splenic sympathetic nerve activity via central prostaglandin synthesis. *Am. J. Physiol.*, **273**, R609–R614.

Masana, M.I., Heyes, M.P. and Mefford, I.N. (1990) Indomethacin prevents increased catecholamine turnover in rat brain following systemic endotoxin challenge. *Prog. Neuropsychopharmacol. Biol. Psychiatry*, **14**, 609–621.

Matsumura, K., Cao, C., Ozaki, M., Morii, H., Nakadate, K. and Watanabe, Y. (1998) Brain endothelial cells express cyclooxygenase-2 during lipopolysaccharide-induced fever: light and electron microscopic immunocytochemical studies. *J. Neurosci.*, **18**, 6279–6289.

Meltzer, J.C. (2000) Neural and endocrine regulation of in vivo splenic immune function in the rat. *PhD thesis*, University of Manitoba, Winnipeg, Canada.

Meltzer, J.C., Grimm, P.C., Greenberg, A.H. and Nance, D.M. (1997) Enhanced immunohistochemical detection of autonomic nerve fibers, cytokines and inducible nitric oxide synthase by light and fluorescent microscopy in rat spleen. *J. Histochem. Cytochem.*, **45**, 599–610.

Miao, F.J., Janig, W. and Levine, J.D. (2000) Nociceptive neuroendocrine negative feedback control of neurogenic inflammation activated by capsaicin in the rat paw: role of the adrenal medulla. *J. Physiol.*, **527**, 601–610.

Movat, H.Z., Burrowes, C.E., Cybulsky, M.I. and Dinarello, C.A. (1987) Acute inflammation and a Shwartzman-like reaction induced by interleukin-1 and tumor necrosis factor. *Am. J. Pathol.*, **129**, 463–476.

Nagy, E. and Berczi, I. (1978) Immunodeficiency in hypophysectomized rats. *Acta Endocrinol. Copenh.*, **89**, 530–537.

Nance, D.M. and MacNeil, B.J. (2001) Immunoregulation by the sympathetic nervous system. In I. Berczi and R.M. Gorcynski (eds). *New Foundation of Biology*, Amsterdam: Elsevier Science, pp. 121–139.

Nansen, A. and Thomsen, A.R. (2001) Viral infection causes rapid sensitization to lipopolysaccharide: central role of IFN-$\alpha\beta$. *J. Immunol.*, **166**, 982–988.

Ozmen, L., Pericin, M., Hakimi, J., Chizzonite, R.A., Wysocka, M., Trinchieri, G. *et al.* (1994) Interleukin 12, interferon γ, and tumor necrosis factor α are the key cytokines of the generalized Shwartzman reaction. *J. Exp. Med.*, **180**, 907–915.

Palmerio, R., Ming, S.C., Frank, E. and Fine, J. (1962) The role of the sympathetic nervous system in the generalized Shwartzman reaction. *J. Exp. Med.*, **115**, 609–615.

Quan, N., Whiteside, M. and Herkenham, M. (1998) Time course and localization patterns of interleukin-1beta messenger RNA expression in brain and pituitary after peripheral administration of lipopolysaccharide. *Neuroscience*, **83**, 281–293.

Ramachandra, R.N., Sehon, A.H. and Berczi, I. (1992) Neuro-hormonal host defence in endotoxin shock. *Brain Behav. Immun.*, **6**, 157–169.

Ramer-Quinn, D.S., Baker, R.A. and Sanders, V.M. (1997) Activated T helper 1 and T helper 2 cells differentially express the beta-2-adrenergic receptor: a mechanism for selective modulation of T helper 1 cell cytokine production. *J. Immunol.*, **159**, 4857–4867.

Rivest, S., Torres, G. and Rivier, C. (1992) Differential effects of central and peripheral injection interleukin-1β on brain c-fos expression and neuroendocrine functions. *Brain Res.*, **587**, 13–23.

Rivier, C. (2001) The hypothalamo-pituitary-adrenal axis response to immune signals. In R. Ader, D.L. Felten and N. Cohen (eds). *Psychoneuroimmunology*, 3rd edition, New York: Academic Press, pp. 633–648.

Sanders, V.M., Baker, R.A., Ramer-Quinn, D.S., Kasprowicz, D.J., Fuchs, B.A. and Street, N.E. (1997) Differential expression of the beta2-adrenergic receptor by Th1 and Th2 clones: implications for cytokine production and B cell help. *J. Immunol.*, **158**, 4200–4210.

Sanders, V.M. and Munson, A.E. (1984a) Beta adrenoceptor mediation of the enhancing effect of norepinephrine on the murine primary antibody response in vitro. *J. Pharmacol. Exp. Ther.*, **230**, 183–192.

Sanders, V.M. and Munson, A.E. (1984b) Kinetics of the enhancing effect produced by norepinephrine and terbutaline on the murine primary antibody response in vitro. *J. Pharmacol. Exp. Ther.*, **231**, 527–531.

Saphier, D. (1989) Neurophysiological and endocrine consequences of immune activity. *Psychoneuroendocrinology*, **14**, 63–87.

Saphier, D., Ovadia, H. and Abramsky, O. (1990) Neural responses to antigenic challenges and immunomodulatory factors. *Yale J. Biol. Med.*, **63**, 109–119.

Scammell, T.E., Elmquist, J.K., Griffin, J.D. and Saper, C.B. (1996) Ventromedial preoptic prostaglandin E2 activates fever-producing autonomic pathways. *J. Neurosci.*, **16**, 6246–6254.

Selye, H. and Tuchweber, B. (1966) Cutaneous zones of thrombohemorrhagic reactivity to noradrenaline. *Acta Anat.*, **63**, 1–7.

Selye, H. (1950) *Stress.* Montreal: Acta Inc.

Shapiro, L. Cuevas, P., Stallard, R.E. and Ruben, M.P (1974) Absence of the localized Schwartzman reaction following 6-OH dopamine sympathectomy. *J. Periodontal Res.*, **9**, 207–210.

Shimizu, N., Kaizuka, Y., Hori, T. and Nakane, H. (1996) Immobilization increases norepinephrine release and reduces NK cytotoxicity in spleen of conscious rat. *Am. J. Physiol.*, **271**, R537–R544.

Shwartzman, G. (1928) A new phenomenon of local skin reactivity to *B. typhosus* culture filtrate. *Proc. Soc. Exp. Biol. Med.*, **25**, 560–561.

Singh, L.K., Pang, X., Alexacos, N., Letourmeau, R. and Theoharides, T.C. (1999) Acute immobilization stress triggers skin mast cell degranulation via corticotropin releasing hormone, neurotensin, and substance P: A link to neurogenic skin disorders. *Brain Behav. Immun.*, **13**, 225–239.

Sternberg, E.M., Young, W.S., Bernardini, R., Calogero, A.E., Chrousos, G.P., Gold, P.W. and Wilder, R.L. (1989) A central nervous system defect in biosynthesis of corticotropin-releasing hormone is associated with susceptibility to streptococcal cell wall-induced arthritis in Lewis rats. *Proc. Natl. Acad. Sci. USA*, **86**, 4771–4775.

Sundar, S.K., Cierpial, M.A., Kilts, C., Ritchie, J.C. and Weiss, J.M. (1990) Brain IL-1-induced immunosuppression occurs through activation of both pituitary-adrenal axis and sympathetic nervous system by corticotropin-releasing factor. *J. Neurosci.*, **10**, 3701–3706.

Swanson, L.W. (1985) The hypothalamus. In A. Bjorklund, T. Hokfelt and L.W. Swanson (eds). *Handbook of Chemical Neuroanatomy*, Vol. 5, Amsterdam: Elsevier, pp. 1–124.

Terao, A., Kitamura, H., Asano, A., Kobayashi, M. and Saito, M. (1995) Role of prostaglandins D2 and E2 in interleukin-1-induced activation of norepinephrine turnover in the brain and peripheral organs of rats. *J. Neurochem.*, **65**, 2742–2747.

Turnbull, A.V. and Rivier, C. (1996) Corticotropin-releasing factor, vasopressin and prostaglandins mediate, and nitric oxide restrains, the hypothalamic-pituitary-adrenal response to acute local inflamation in the rat. *Endocrinology*, **137**, 455–463.

Vriend, C.Y., Zuo, L., Dyck, D., Nance, D.M. and Greenberg, A.R. (1993) Central administration of Interleukin-1b increases norepinephrine turnover in the spleen. *Brain Res. Bull.*, **31**, 39–41.

Wan, W., Janz, L., Vriend, C.Y., Sorensen, C.M., Greenberg, A.H. and Nance, D.M. (1993a) Differential induction of c-fos immunoreactivity in hypothalamus and brain stem nuclei following central and peripheral administration of endotoxin. *Brain Res. Bull.*, **32**, 581–587.

Wan, W., Vriend, C.Y., Wetmore, L., Gartner, J.G., Greenberg, A.H. and Nance, D.M (1993b) The effects of stress on splenic immune function are mediated by the splenic nerve. *Brain Res. Bull.*, **30**, 101–105.

Wan, W., Wetmore, L., Sorensen, C.M., Greenberg, A.H. and Nance, D.M. (1994) Neural and biochemical mediators of endotoxin and stress-induced c-fos expression in the rat brain. *Brain Res. Bull.*, **34**, 7–14.

Zalcman, S., Minkiewicz-Janda, A., Richter, M. and Anisman, H. (1988) Critical periods associated with stressor effects on antibody titers and on the plaque-forming cell response to sheep red blood cells. *Brain Behav. Immun.*, **2**, 254–266.

Zalcman, S., Green-Johnson, J.M., Murray, L., Nance, D.M., Dyck, D., Anisman, H. and Greenberg, A.H. (1994) Cytokine-specific central monoamine alterations induced by interleukin (IL)-1, IL-2 and IL-6. *Brain Res.*, **643**, 40–49.

3 Enteric Neural Reflexes and Secretion

Helen J. Cooke[1], Najma Javed[2] and Fievos L. Christofi[3]

[1]*Department of Neuroscience, The Ohio State University, Columbus, OH, USA*
[2]*Department of Physiology and Health Science, Ball State University, Muncie, IN, USA*
[3]*Department of Anesthesiology, The Ohio State University, Columbus, OH, USA*

Chloride secretion, accompanied by obligate transport of water, is important for lubricating the intestinal lining or for flushing out unwanted microbes or noxious chemicals. The rate of secretion is dependent on the enteric nervous system. Sensory cells which detect chemical or mechanical signals release mediators such as 5-hydroxytryptamine (5-HT) which triggers afferent neurons in the submucosal plexus. Electrical signals are transmitted to secretomotor neurons or to interneurons synaptically coupled to secretomotor neurons and chloride secretion occurs. Input from myenteric neurons to submucosal neurons modulates secretion rates. The models that are beginning to emerge show considerable complexity in the neurochemical codes and microcircuits that control intestinal secretion. Purines are emerging as fundamental regulators of enteric neural reflexes.

KEY WORDS: chloride secretion; submucosal plexus; enteric reflexes.

Motility and secretory disorders of the gastrointestinal tract are very common. These disorders often occur in patients with enteric infections, or with inflammatory bowel disease such as Crohn's disease or ulcerative colitis (Gay *et al.*, 1999; Palmer and Greenwood-Vanmeerveld, 2001; Schneider *et al.*, 2001). The hallmarks of enteric infections or inflammatory bowel disease are abdominal malaise and diarrhoea. Structural changes in the enteric nervous system may be the basis for the pathogenesis of disturbances in gut function (Schnieder *et al.*, 2001). During inflammation, microbial penetration or intestinal allergic responses, chloride secretion and fluid volume are amplified and motility patterns are activated, resulting in flushing out of the luminal contents, that is, diarrhoea (Eklund *et al.*, 1985; Palmer and Greenwood-Vanmeerveld, 2001). Neural reflex pathways within the enteric nervous system are responsible for these motility and secretory patterns. While reflex regulation of motility or secretion has been studied individually, little is known about the neurons that link the two and how these pathways coordinate such distinct functions.

An understanding of how the neural circuits function in health and disease will provide insights into therapeutic interventions for gastrointestinal motility and secretory disorders.

Chloride secretion in the intestine is essential for providing a driving force for fluid movement into the lumen where it hydrates sticky mucins. This is important for providing lubrication for movement of digested products along the length of the intestine. Chloride secretion is usually low, but can reach extraordinary rates when challenged with enterotoxins or secretagogues. Regulation of chloride secretion to meet the needs of the moment is provided by the enteric nervous system. Neural reflex programmes are called up to make adjustments in chloride secretion. This chapter will focus on the components of neural reflex pathways that regulate intestinal chloride secretion. It will address chemo- and mechanotransduction in relation to gut reflexes. It will examine several of the established neurotransmitters such as substance P and acetylcholine involved in synaptic transmission as well as present some 'new kids on the block' who come from a large family of nucleosides and nucleotides. Finally these components will be used to illustrate the integrated response to stimulation.

The guinea pig has been the animal model of organization of neural circuits in most studies; however, certain aspects of organization of neurons differ among species. Substantial differences in neurochemical codes can be found across species as well as across regions of gut in the same species. The reader is referred to recent reviews for further details (Furness *et al.*, 1999; Neunlist *et al.*, 1999a,b; Costa *et al.*, 2000; Lomax and Furness, 2000; Brookes, 2001; Timmermans *et al.*, 2001).

The enteric nervous system is essential for regulating motility and secretion appropriate for a particular digestive state. The enteric nervous system is organized into ganglionated plexuses embedded in the wall of the intestine. It consists of submucosal ganglia in the connective tissue of the submucosa and myenteric ganglia distributed between the longitudinal and circular muscles. While the submucosal plexus is the predominant player in secretory reflexes, in some instances secretion may require pathways within the myenteric plexus (Jodal *et al.*, 1993). In small animals the submucosal plexus is often arranged as a single layer; however, in large animals such as pig and human, the submucosal plexus contains at least two layers that differ in the neurochemical codes and their projections (Hens *et al.*, 2001; Timmermans *et al.*, 2001). The ganglia contain neurons, nerve fibres and varicosities, as well as glia and other supporting structures. Based on immunohistochemical staining techniques, selective lesioning, retrograde or anterograde transport of neuronal tracers, electrophysiological recordings from enteric neurons and functional studies, neurons within submucosal ganglia can be classified as intrinsic primary afferent neurons (IPANs), putative interneurons, secretomotor and vasodilator neurons (Moore and Vanner 1998; Furness *et al.*, 1999; Hens *et al.*, 2000, 2001; Lomax and Furness, 2000; Pan and Gershon, 2000; Timmermans *et al.*, 2001). These types of neurons (Figure 3.1) comprise the framework necessary for evoking neural reflexes that regulate secretion of chloride and fluid as well as blood flow. A similar classification holds for neurons in myenteric ganglia in that both IPANs and interneurons are present; however, they also contain motor neurons, which innervate the longitudinal, and circular muscle layers, and secretomotor neurons, which project to the mucosa (Kunze and Furness, 1999). In addition, subsets of myenteric interneurons, which project aborally synapse with submucosal neurons and modify their

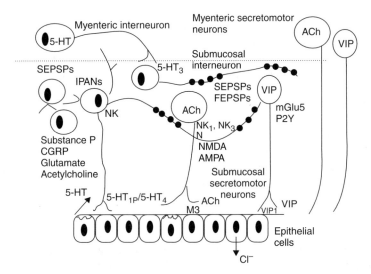

Figure 3.1   Diagram of the components of a neural reflex regulating intestinal secretion. Mechanical stimulation releases 5-HT from enterochromaffin cells. 5-HT binds to 5-HT_{1P} (guinea pig) or 5-HT_4 (humans) receptors on IPANs and causes excitation. IPANs release transmitters, substance P, CGRP, glutamate and/or acetylcholine (ACh) at synapses with downstream neurons, most likely secretomotor neurons. Activation of a subset of submucosal cholinergic (ACh) secretomotor neurons coded by the presence of NPY or another cholinergic subset coded by calretinin (not shown) release transmitters at neuroepithelial junctions. VIP neurons can also be activated and can release VIP at synapses with epithelial cells. ACh at M3 receptors causes a transient secretion and VIP evokes a sustained secretion. Secretomotor neurons receive excitatory or inhibitory input from myenteric neurons indirectly when they synapse with submucosal interneurons. Ganglionic transmission may involve NK_1, NK_3, nicotinic (N), NMDA and AMPA receptors on secretomotor neurons or mGlu5 and P2Y receptors on VIP secretomotor neurons. Projections of myenteric secretomotor neurons indicate secretion can occur through the myenteric plexus as well. Other inputs from extrinsic nerves can further alter reflex evoked secretion.

activity. With these neural components the enteric nervous system can illicit neural reflexes that control and coordinate motility, secretion and blood flow independently from any input from the central nervous system.

COMPONENTS OF ENTERIC MICROCIRCUITS

SENSORY CELLS

There are specialized cells, which can convert a chemical or a mechanical force to a biological response. Neurons, enteroendocrine cells and sensitized mast cells are some of the sensory transducers in the gut. As an illustration, IPANs and enterochromaffin cells will be discussed.

Chemical stimuli

Food intake and glucose homeostasis involves a complex regulatory system that includes hypothalamic neurons, intestinal vagal and spinal afferent neurons and the pancreatic β-cell, which releases insulin. It is now clear from the work of Liu *et al.* (1999b) that a subset of enteric neurons in the guinea pig is reported to detect glucose. This subset of myenteric neurons is characterized as AH/type 2. They have a long after hyperpolarization following an action potential; they receive slow excitatory post-synaptic potentials (slow EPSPs); they project circumferentially to synapse with each other; they costore choline acetyltransferase (ChAT, a synthetic enzyme for acetylcholine) with substance P, glutamate and/or calbindin, a chemical code characteristic of IPANs; they have K_{ATP} channels associated with sulfonylurea receptors and respond to leptin. Like certain neurons in the hypothalamus, they can be divided into two classes: (1) glucoresponsive neurons which track extracellular glucose levels. They are excited by increases in glucose greater than 5 mM and are hyperpolarized when extracellular glucose is removed. (2) Glucosensitive neurons have a different behavioural pattern in that they depolarize in response to a decrease in glucose concentration (Liu *et al.*, 1999b).

The mechanisms by which enteric neural circuits respond to glucose may be similar to those occurring in the pancreatic β-cell because these neurons have K_{ATP} channels and cellular elements necessary for glucose detection. The properties of glucoresponsive neurons suggest that when extracellular glucose levels fall, a decrease in intracellular ATP opens K_{ATP} channels and hyperpolarizes the cell (Liu *et al.*, 1999b). It becomes less excitable and release of transmitter is curtailed. The absence of glucose has the opposite effect. Similar studies as those done in myenteric glucosensitive neurons are needed for submucosal neurons. Submucosal IPANs also have K_{ATP} channels but the presence of K_{ATP} channels alone is not necessarily a signature of glucosensitive neurons. The presence of two populations of glucose detecting IPANs with opposite effects suggests that extracellular glucose is closely monitored and this will impact on gastrointestinal reflexes controlling motility and possibly secretion.

K_{ATP} channels are composed of two subunits, an inwardly rectifying K channel 6.2 (Kir6.2) and the sulfonylurea receptor, SUR1. K_{ATP} channels are regulated by the ob gene product, leptin. Leptin has anti-obesity properties by virtue of its ability to activate K_{ATP} channels (Liu *et al.*, 1999b). Leptin causes a hyperpolarization in a subset of neurons that contain leptin immunoreactivity. These are identified as ChAT/NPY secretomotor neurons and IPANs. Subunits of the K_{ATP} channel and SUR1, are of interest clinically because a mutation in SUR1 results in loss of K_{ATP} channels that causes neurons and β-cells to be chronically depolarized, that is, excitable. Associated with this mutation is hyperinsulinaemia, hypoglycaemia which can be complicated by gastrointestinal problems of unknown origin. If mechanisms in myenteric neurons reflect those in submucosal neurons, the gastrointestinal symptoms are likely to result from abnormal secretory and motility reflexes.

D-glucose placed in the duodenum inhibits gastric emptying and stimulates pancreatic secretion by activation of 5-HT$_3$ receptors on extrinsic vagal and spinal afferents (Raybould, 2001). Because 5-HT$_3$ receptors are present in submucosal neurons, as well as in extrinsic afferents, 5-HT may play an important role in glucose-evoked secretory reflexes in the intestine as well as in glucose-induced delays in gastric emptying. Because the 5-HT secreting cell, the enterochromaffin cell, is sparsely distributed in the intestine, it has been difficult to examine the mechanisms and signalling pathways at the cellular level. Recently BON cells have been used as a model of enterochromaffin cells (Kim *et al.*, 2001a).

They were derived from a metastasis of a pancreatic carcinoid tumour of enterochromaffin cell origin. Therefore some caution must be exercised in extrapolation to non-transformed enterochromaffin cells (Racke and Schworer, 1991; Kim *et al.*, 2001a). In BON cells, 5-HT release is triggered by D-glucose or galactose, two hexoses that are transported by the energy dependent sodium-glucose-like transporter-1 (SGLT1) into intestinal epithelial cells (Wright *et al.*, 1997). Non-metabolizable hexoses such as α-methyl-D-glucopyranoside also trigger 5-HT release. Phloridzin, which inhibits the binding of D-glucose to SGLT1 in intestinal epithelial cells, also attenuates 5-HT release in BON cells. It is unclear whether D-glucose must be transported into the enterochromaffin cell in order for 5-HT to be released or whether binding to the transporter is sufficient to trigger 5-HT release.

A 55–60 kDa protein with homology to SGLT-1, was identified in BON cells by Western blotting (Kim *et al.*, 2001a). The SGLT1-like protein has similarities and differences to SGLT1. It is smaller in size and has a high threshold for detection of hexoses in the concentration range of 50–100 mM. This is a much higher threshold for 5-HT release than the 5 mM threshold for depolarization of myenteric neurons in response to increasing extracellular glucose (Liu *et al.*, 1999b). At 50–100 mM glucose the release of 5-HT is not due to osmotic changes, because there was no effect of comparable concentrations of mannitol on 5-HT release. Another important difference from glucose activated IPANs is that non-metabolizable hexoses were just as effective in releasing 5-HT as was the metabolizable D-glucose. Although the signalling pathways have not been completely delineated, it is clear that only hexose substrates for SGLT1 in epithelial cells, namely glucose and galactose, can trigger 5-HT release from BON cells. Fructose, which is a substrate in epithelial cells for the facilitative transporter, GLUT5, does not release 5-HT from BON cells. These results imply that hexose-induced release of 5-HT may be an important determinant of postprandial gastrointestinal function.

Mechanical stimuli

Mechanical forces generated by mucosal stroking, pressure, touch or stretch are detected by enterochromaffin cells, causing release of 5-HT. Until recently, little was known about the intracellular signalling pathways involved. Activation of enteric neurons through release of 5-HT during mucosal stroking is a common pathway in both guinea pig and human (Cooke *et al.*, 1997a,b; Kellum *et al.*, 1999). Species differences are apparent by examining the subsets of 5-HT receptors mediating secretion which include 5-HT$_4$ for human and 5-HT$_{1P}$ in guinea pig. If carcinoid BON cells are representative of enterochromaffin cells, then it would appear that G-protein-coupled receptors and G proteins are involved in releasing 5-HT. Kim *et al.* (2001b) showed that mechanical stimulation of BON cells activates the α subunit of Gq. To demonstrate that a receptor couples to Gαq, they treated BON cells with an inhibitory peptide, which competes with a domain in the C-terminus of Gαq known to interact with G-protein-coupled receptors. Activation of Gαq triggers downstream signalling events, which include activation of phospholipase Cβ (PLCβ), formation of inositol 1,4,5-triphosphate (IP$_3$) and diacylglycerol followed by mobilization of calcium from intracellular stores and 5-HT release (Kim *et al.*, 2001b). Although this peptide can bind to G-protein-coupled receptors, it cannot release 5-HT because it does not have a sequence necessary for activation of PLCβ in the signalling pathway (Kim *et al.*, 2001b). These studies demonstrate that a Gq-coupled receptor must be activated to cause mechanically-evoked 5-HT release from BON cells. One of the possibilities is a purinoceptor whose ligand is a nucleotide. This possibility is likely since mechanical stimulation of airway epithelia released ATP and UTP which act as autocrine mediators (Lazarowski and Boucher, 2001). Kim *et al.* (2001b) provide convincing evidence that the downstream signalling events in the Gαq pathway occur in BON cells. They show that mobilization of calcium from intracellular stores and not influx of extracellular calcium is associated with 5-HT release. Further studies are necessary to identify how a mechanical stimulus evokes nucleotide release in the enterochromaffin cell model.

Although mechanical stimulation and activation of a purinoceptor is dependent on the mobilization of intracellular calcium other receptors coupled to 5-HT release may be dependent on influx of calcium through voltage-regulated calcium channels with L-type characteristics. Enterochromaffin cells have a plethora of cell surface receptors that are either inhibitory or excitatory with respect to 5-HT release (Racke and Schworer, 1991). The presence of multiple receptors suggests that there is considerable complexity in the regulation of 5-HT release that ranges from autocrine, paracrine, hormonal and neural control mechanisms. Enterochromaffin cells play a key paracrine role in intestinal secretory reflexes by virtue of release of 5-HT in proximity to $5\text{-}HT_{1P}/5\text{-}HT_4$ receptors on IPANs.

IPANs

IPANs in the guinea pig ileum are morphologically multipolar Dogiel type II neurons with two or more processes, one of which projects to the mucosa directly beneath (Furness *et al.*, 1998). These are called AH/type 2 neurons based on their long lasting after-hyperpolarizing potential which limits the frequency for firing action potentials repetitively. It is well accepted today that IPANs exist with cell bodies in the submucosal ganglia and in myenteric ganglia. Submucosal AH/type 2 neurons and 82–85% of myenteric AH/type 2 neurons are immunoreactive for calbindin, one of the markers for IPANs (Nuenlist and Schemann, 1997; Kunze and Furness, 1999; Quinson *et al.*, 2001).

AH/type 2 neurons in the distal colon of the guinea pig project up and down the gut for several millimetres to the region which they innervate. In the submucosal plexus of the guinea pig IPANs are chemically coded as follows: (1) ChAT, substance P, calbindin and dynorphin (1–8) accounting for 15% of submucosal neurons in guinea pig small intestine; (2) one subset of these cholinergic neurons has substance P and glutamate; and (3) another has ChAT/CGRP immunoreactivity.

In the myenteric ganglia, approximately 32% of myenteric neurons are AH/type 2 neurons with one process projecting to the mucosa (Quinson *et al.*, 2001). In the small intestine, overlapping fields of IPANs supply the villi and provide a dense innervation in the human and guinea pig small intestine (Hens *et al.*, 2001). Considerably fewer IPANs per cubic millimetres are found for the submucosal plexus compared to a greater density in the myenteric plexus. The presence of IPANs in each of the plexuses raises the question whether many are specialized to transduce different stimuli. The lower number of IPANs in the submucosal plexus may reflect a greater sensitivity to chemical or mechanical stimuli. For example, submucosal IPANs respond to mechanical stimulation such as movement of the villi, touch, distortion or shaking whereas myenteric IPANs may have unique properties enabling them to respond to stretch or tension and chemicals such as inorganic acids and fatty acids (Kunze and Furness, 1999).

IPANs form intertwining networks by synapsing with each other and releasing a transmitter that causes slow EPSPs. This intertwining configuration accounts for spread of activity around the circumference of the intestine ensuring that a ring of contraction and secretion will occur. The neurotransmitters, which mediate slow synaptic transmission among these IPAN networks, are not completely known, although evidence for CGRP is strong. CGRP is necessary for spread of excitation within the submucosal plexus as well (Pan and Gershon, 2000). It is likely that substance P is also a mediator of slow EPSPs for IPANs, because NK_1 or in some species NK_3 receptors are present on IPANs.

Acetylcholine seems to be a transmitter in IPANs because the vesicular acetylcholine transporter is present (Li and Furness, 1998). Since all IPANs are cholinergic, putative transmitters for the slow EPSP (CGRP, substance P and glutamate) and acetylcholine may be co-released when the reflex is activated by applying 5-HT. Evidence points to the existence of monosynaptic pathways between IPANs and secretomotor neurons. Consistent with monosynaptic transmission is the short stimulus–response delay of 7 ms in neural transmission from neurons in the mucosa to the impaled second-order neuron. There is evidence that fast EPSPs evoked by mechanical stimulation of the mucosa can be blocked with nicotinic cholinergic receptor blockade, but not by low calcium/high magnesium containing medium that blocks polysynaptic pathways (Pan and Gershon, 2000). Thus mechanically evoked secretion is a consequence of release of transmitter at synapses between IPANs and secretomotor neurons. If this is the case, the appropriate receptor for the transmitter should be present on cell somas of secretomotor neurons. Substance P is released from IPANs during mechanical stimulation and its targets are NK_1 and NK_3 receptors on secretomotor neurons in the guinea pig colon (Frieling et al., 1999). If this secretory reflex pathway is monosynaptic, then IPANs may play a unique role as an afferent, which releases the transmitter for slow synaptic transmission and acetylcholine which mediates fast synaptic transmission.

INTERNEURONS

Submucosal interneurons

There are several types of interneurons that are involved in the regulation of secretion and its coordination with muscle contraction. In guinea pig ileum VIP interneurons in the submucosal plexus that project orally to myenteric neurons without branching to other submucosal neurons have been identified (Song et al., 1992, 1998; Li and Furness, 1998). These neurons undoubtedly play a role in coordination of secretion and motility.

Submucosal interneurons containing ChAT that synapse with other submucosal neurons have been proposed. These are distinguished from other ChAT neurons by failure to express markers found for the other subclasses of neurons. Projections from ChAT/NPY secretomotor neurons and substance P IPANs form axonal tufts around other submucosal neurons suggesting that they make functional connections. This anatomical arrangement suggests that other classes of neurons (afferents or secretomotor neurons) may function as interneurons. The concept of monosynaptic pathways that eliminate the need for interneurons to communicate with secretomotor neurons in some reflexes may apply to the mechanically evoked secretory reflexes through the submucosal plexus. When this reflex is activated secretion over a short distance would be anticipated.

Myenteric inputs

Myenteric interneurons provide synaptic input to submucosal S/type 1 neurons which are either interneurons or secretomotor neurons in the guinea pig ileum (Moore and Vanner, 1998, 2000). Most S cells receive fast EPSPs and more than 30% receive slow EPSPs. Synaptic potentials could be recorded as far as 25 mm aborally from the stimulus site. Of the fast EPSPs, all were nicotinic cholinergic and were blocked with hexamethonium. Thus aborally projecting myenteric interneurons provide synaptic inputs to submucosal S/type 1 cells that stimulate secretion (Moore and Vanner, 2000). Myenteric ChAT/5-HT neurons (2% of myenteric neurons) project aborally and synapse with other myenteric 5-HT interneurons and with myenteric and submucousal non-5-HT containing neurons via 5-HT$_3$ receptors (Wardell *et al.*, 1994; Kuwahara *et al.*, 1998; Meedeniya *et al.*, 1998; Costa *et al.*, 2000; Brooks, 2001). Activation of these receptors causes colonic secretion that is reduced by nicotinic cholinergic blockers, atropine and nitric oxide synthase inhibitors (Frieling *et al.*, 1991; Wang *et al.*, 1991; Kuwahara *et al.*, 1998). The noncholinergic component of 5-HT evoked secretory response is mediated in part by nitric oxide generating neurons. The overall secretory response would be a composite of secretion controlled by reflexes in the submucosal plexus and of reflexes through the myenteric plexus that provide excitatory (5-HT) or inhibitory (somatostatin) inputs to submucosal interneurons or secretomotor neurons. Myenteric somatostatin interneurons project aborally and may project to the submucosal neurons (Song *et al.*, 1998). Somatostatin interneurons have filamentous characteristics and appear to be a hydrid between AH/type 2 and S/type 1 neurons. This mechanism may be separate from myenteric reflexes that control myenteric secretomotor neurons that project to the epithelial cells. The latter pathway may be an example of long reflexes that coordinate secretion and motility over relatively longer distances. Although this discussion summarizes excitatory myenteric inputs to submucosal interneurons, inhibitory inputs from somatostatin and enkephalin containing myenteric neurons occurs as well and these will impact on secretion as well (Cooke and Reddix, 1994).

SECRETOMOTOR NEURONS

Secretomotor neurons are S/type 1 cells that are unipolar in the guinea pig ileum and colon and are classified as myenteric and submucosal. In the submucosal class, there are two types in the small intestine, which include the following (Song *et al.*, 1992; Lomax and Furness, 2000; Brookes, 2001): (1) VIP, dynorphin, galanin comprising 41–43% of the population; (2) ChAT, NPY, cholecystokinin, CGRP, dynorphin (1–8), somatostatin +/− enkephalin which account for 26–33% of the neurons. In the guinea pig colon, there are three types of submucosal secretomotor neurons: VIP with NOS, ChAT with NPY, and ChAT with calretinin. In the guinea pig proximal and distal colon ChAT/NPY neurons projected orally and VIP neurons aborally. However, this polarization was not seen for the small intestine of guinea pigs and humans (Neunlist and Schemann, 1998; Brookes, 2001; Hens *et al.*, 2001). The myenteric plexus also makes a contribution to secretion by way of cholinergic and VIPergic secretomotor neurons, which project to the mucosa.

5-HT pulsed onto a submucosal ganglion activates $5\text{-}HT_3$ receptors on S/type 1 ChAT neurons and evokes chloride secretion (Frieling *et al.*, 1991). The secretory response is reduced by nicotinic and muscarinic receptor blockade suggesting the involvement of cholinergic interneurons (Frieling, 1991; Wang *et al.*, 1991). In this case, there may be submucosal ChAT interneurons that relay information from myenteric neurons to other submucosal neurons. The projections of submucosal 'interneurons' are short neural elements confined to the submucosa where they may control secretory and vasomotor function by providing a very localized response (Cooke *et al.*, 1997a,b; Vanner, 2000). Submucosal interneurons may provide the interface between myenteric interneurons and submucosal secretomotor neurons and may function to coordinate secretion through the myenteric plexus.

INTEGRATIVE FUNCTION

What initiates enteric reflexes? A majority of them occur when 5-HT is released from enterochromaffin cells. Secretory and motility reflexes are activated by chemical and mechanical stimuli that are transduced to a biological response. Enteric toxins, inflammatory mediators, inorganic acids, short chain fatty acids, D-glucose and 5-HT are chemicals, which can initiate intestinal reflex activity (Eklund *et al.*, 1989; Frieling *et al.*, 1999; Furness *et al.*, 1998, 1999). Mechanical stimuli such as mucosal stroking, touch, shear stress, pressure, stretch, agitation of the buffer and nitrogen puffs can activate neural reflexes (Cooke *et al.*, 1997a,b) (Figure 3.1). 5-HT is a key player when released into the interstitial space from enterochromaffin cells. In the guinea pig another mediator released by mechanical stimulation is prostaglandin. The influence of this mediator can be eliminated by blocking cyclooxygenase pathways, or bypassing the enterochromaffin cell by pulsing 5-HT on to the mucosa. Thus, 5-HT acts as a paracrine mediator, which binds to $5\text{-}HT_{1P}/5\text{-}HT_4$ receptors on IPANs in the guinea pig colon and human jejunum and evokes slow depolarizing responses (Cooke *et al.*, 1997a,b; Kellum *et al.*, 1999). This sets up a cascade of events that includes release of substance P, probably from IPANs, and activation of submucosal ChAT/NPY and VIP secretomotor neurons in the guinea pig colon. Reflex circuits can be long through the myenteric plexus and submucosal plexus or short through the submucosal plexus only. During mucosal stroking in the human jejunum, 5-HT activates $5\text{-}HT_4$ receptors which stimulate secretion via activation of nicotinic cholinergic receptors and not muscarinic as reported for the guinea pig colon (Cooke *et al.*, 1997a,b; Kellum *et al.*, 1999). Intestinal secretion is mediated in part by input from extrinsic primary afferent neurons; the latter is abolished with capsaicin, which first releases and then prevents substance P and CGRP release. In the small intestine, conduction from IPANs to secretomotor neurons is monosynaptic (Pan and Gershon, 2000). Therefore, IPANs release putative transmitters (substance P, CGRP or glutamate) which illicit slow EPSPs in secreto-, vasomotor neurons or release acetylcholine, which causes fast EPSPs.

Submucosal IPANs do not appear to have 5-HT$_3$ receptors on their terminals although they can be found on cell bodies of myenteric IPANs. The 5-HT receptors that cause reflex chloride secretion through the submucosal plexus are 5-HT$_{1P}$/5HT$_4$ receptors in the human small intestine and guinea pig colon. 5-HT$_3$ receptors are not present on submucosal IPANs. Thus, 5-HT$_3$ receptor blockade has little effect on reflex driven chloride secretion through the submucosal plexus; yet 5-HT$_3$ blockers are effective in inhibiting propulsion, because they are present on myenteric IPANs. This receptor does not appear to play a role in secretory reflexes that occur entirely through the submucosal plexus; however, there are myenteric 5-HT neurons that project to and synapse with submucosal neurons containing 5-HT$_3$ receptors. These neurons may play a role in secretory reflexes that are conducted through myenteric ganglia. From previous work done in the guinea pig colon and ileum, it is well documented that activation of 5-HT$_3$ receptors on submucosal neurons leads to secretion due to stimulation of submucosal ChAT neurons, release of acetylcholine at nicotinic and muscarinic synapses (Wang et al., 1991; Kellum et al., 1999). Why are some secretory reflexes through the submucosal plexus where others are through the myenteric plexus? Although not proven, it is likely the myenteric plexus is called in when secretion over longer distances is required. An example is secretion through the myenteric plexus when exposed to cholera toxin.

SUBSTANCE P IN INTESTINAL REFLEXES

Mechanical stimulation of submucosa/mucosa preparations of guinea pig colon results in tetrodotoxin-sensitive chloride secretion. This response is due to release of substance P from intrinsic neurons, because NK$_1$ receptor antagonists attenuate the response even when extrinsic afferents, which also contain substance P, have been eliminated by capsaicin treatment (Cooke et al., 1997c). The observation that substance P is found predominantly in submucosal IPANs and that NK$_1$ receptors are on epithelial cells suggests that all the appropriate elements are present for an axon reflex to account for part of the mechanically evoked response (Cooke et al., 1997c). Axon reflexes occur when an action potential is propagated antidromically to collaterals that are in proximity to a responding cell such as another neuron or epithelial cells. Antidromic release of substance P from IPANs cannot be excluded as possible means for stimulating chloride secretion, since NK$_1$ receptors are present on epithelial cells (Cooke et al., 1997c). Demonstration of this axon reflex might be dependent on the intensity of the stimulus as well as the number of epithelial receptors and their special arrangement relative to the nerve fibres.

While it is unclear what contribution axon reflexes within the submucosal plexus make to the overall secretory response, axon reflexes involving extrinsic afferents are reported to stimulate intestinal secretion mediated by NK$_1$ receptors in guinea pig small intestine and colon (MacNaughton et al., 1997; Moriarty et al., 2001). Sensory fibre stimulation of rat colon released substance P from mast cells and caused secretion, which was reduced by NK$_1$ receptor antagonist. The authors suggested that NK1 receptors mediate secretory effects of mast cells during stimulation of extrinsic afferents. In the human jejunum, capsaicin reduces mechanically evoked secretion without affecting 5-HT release. These observations suggest that axon reflexes operate through extrinsic primary afferents (MacNaughton et al., 1997; Moriarty et al., 2001).

Another pathway for substance P to stimulate secretion is by IPANs releasing substance P directly at synapses with NK_1 and NK_3 receptors on ChAT/NPY, ChAT/Calretinin and VIP secretomotor neurons in the guinea pig colon (Lomax *et al.*, 1998; Frieling *et al.*, 1999). NK_1 and NK_3 receptor agonists give slow excitatory responses equally in both types of secretomotor neurons. This distribution differs somewhat in the guinea pig ileum where NK_1 and NK_3 receptor agonists depolarized, respectively, 100% and 50% of submucosal neurons studied (MacNaughton *et al.*, 1997; Lomax *et al.*, 1998). NK_3 receptors were on 81% of ChAT/NPY neurons, only 2% of VIP neurons and 65% of calretinin-immunoreactive secretomotor/vasodilator neurons in the submucosal plexus and 75% of myenteric ChAT/NPY secretomotor neurons (Jenkinson *et al.*, 1999).

CGRP IN INTESTINAL REFLEXES

Mucosal stroking evokes both a secretory response and the peristaltic reflex (Cooke, 1992; Cooke *et al.*, 1993; Foxx-Orenstein *et al.*, 1996; Wang and Cooke, 1999). Mucosal stroking releases 5-HT which acts on CGRP afferents to cause secretion and muscle contraction. The response to stroking was decreased by $5\text{-}HT_{1P}/5\text{-}HT_4$ receptor antagonist in the guinea pig. 5-HT applied to the mucosal surface evokes slow depolarizing responses (slow EPSPs) which are attenuated by the CGRP antagonist, CGRP (8–37) and rarely by cholinergic muscarinic blockade. 5-HT added to the mucosal side also causes fast EPSPs, which are blocked by hexamethonium. Considerable evidence suggests that conduction from IPANs appear to be monosynaptic without a connecting interneuron between the afferent and the secretomotor neuron via nicotinic, CGRP and rarely muscarinic synapses (Pan and Gershon, 2000). While these studies cannot ascertain whether these reflexes involve regulation of submucosal blood vessels or epithelial cells, other studies in the guinea pig colon suggest that the CGRP is a neurotransmitter for secretory reflexes through the submucosal plexus (Wang and Cooke, 1999). This conclusion is based on the observation that the CGRP antagonist (8–37) also attenuates the mechanically evoked secretory responses in the guinea pig colon. Since secretion does not occur in guinea pig colon unless the submucosal and myenteric plexuses are intact, CGRP containing IPANs may be the pathway activated when myenteric secretory reflexes are involved.

GLUTAMATE IN INTESTINAL REFLEXES

Glutamate is concentrated in varicosities of ChAT neurons and co-localizes with the neuronal glutamate transporter, EAAC1, one of the proteins necessary for uptake of glutamate. The two taken together support the concept that glutamate is a neurotransmitter in the submucosal plexus (Liu and Kirchgessner, 2000). Glutamate acts at ionotrophic (iGlu) or metabotrophic (mGlu) receptors, which are glutamate-gated ion channels or GTP-binding proteins. The iGlu are divided into *N*-methyl-D-aspartate receptors (NMDA) and non-NMDA receptors for kinate and α-amino-3-hydroxyl-5-methyl-4-isoxazole propionic acid (AMPA). Nearly, all submucosal neurons express NR1 and NR2A/B subunits of NMDA receptors (Kirchgessner *et al.*, 1997). The widespread pattern of expression predicts that most enteric neurons are susceptible to NMDA excitotoxicity; however only a small subset of neurons display this characteristic. Excitotoxicity occurs when high concentrations of glutamate cause intracellular calcium to rise too high causing necrosis and apoptosis of glutamate neurons (Kirchgessner *et al.*, 1997). An unsuspecting diner who consumes a plate of mussels contaminated with domoic acid which binds to NMDA receptors is likely to experience excitotoxicity first hand when nausea, and vomiting begin. The clue to the puzzle of why only a small subset of neurons have this characteristic may be due to the combination of subunits that determine the channel's activation state.

Most neurons are immunoreactive for the iGlu2/3 subunit of the AMPA receptor and most also appear to have NMDA receptors (Liu and Kirchgessner, 2000). In the rat and guinea pig intestine, messenger RNA for NMDA1 is present in VIP neurons, which are mostly secreto-vasomotor neurons. Thus, glutamate might play a role in intestinal reflexes to modulate secretion (Burns and Stephens, 1995; Kirchgessner *et al.*, 1997). While glutamate acting at AMPA receptors is thought to be a transmitter for fast EPSPs in myenteric neurons, this is less clear for submucosal neurons, because no fast depolarizing responses were recorded in submucosal neurons even though they have been shown to express AMPA receptors as well as NMDA receptors (Ren *et al.*, 1999). How can these two views be reconciled? Glutamate once released is rapidly taken up by the EAAC1 transporter and this may mask the ability to act at cell surface receptors. Lack of detection of a fast response could reflect failure to implement the appropriate experimental conditions and may not necessarily exclude its existence.

The mGlu include three groups based on sequence similarity, pharmacology and signal transduction mechanisms: group I includes mGlu1, mGlu5, which are coupled to phospho-inositide hydrolysis; group II (mGlu2, mGlu3) and group III (mGlu4, mGluR6, mGlu7 and mGlu8), which inhibits cyclic AMP (cAMP).

Glutamate, acting via group I, a metabotrophic receptor, mGLu5, causes a slow depolarization of submucosal S/type 1 neurons with uniaxonal morphology (Liu and Kirchgessner, 2000). Glutamate receptor mGlu5 is found in enteric secretomotor neurons containing VIP and not in IPANs containing CGRP or substance P. VIP secretomotor neurons express functional group I, mGlu receptors that mediate slow depolarizing response to glutamate. Other mGlu receptors may be involved such as Group I, mGlu1 receptor because it is expressed on guinea pig submucosal neurons (Liu and Kirchgessner, 2000). Glutamate is important in intestinal reflexes governing secretion. Distention of the gut with an intraluminal balloon evokes slow EPSPs in enteric neurons. Glutamate receptor antagonists (group I) suppresses stimulus evoked slow EPSPs and increases the amplitude of inhibitory post-synaptic potentials (IPSPs) in submucosal neurons of the guinea pig small intestine (Ren *et al.*, 1999). Glutamate effects were mimicked by agonists of group I metabotropic receptors but not by group II or III or iGlu5 receptors. Slow depolorizations evoked by 5-HT or substance P also were suppressed whereas the slow inhibition by norepinephrine was potentiated. These observations point to glutamate's role as an excitatory neurotransmitter in secretory reflexes via its stimulatory action on VIP secretomotor neurons. In group II, mGlu2/3 receptor is present in submucosal cell bodies of the rat small intestine and this may have implications for gut function (Larzabal *et al.*, 1999).

During mechanical stimulation of the villi by agitating the fluid in the bath, or by distention of small intestinal segments of guinea pig ileum mGlu5 receptor internalizes, like some of the other G-protein-coupled receptors (Liu and Kirchgessner, 2000). The stimuli were sufficient to release endogenous glutamate from IPANs found in guinea pig. One of the downstream events in the signalling pathway is the phosphorylation of the cAMP response element binding protein, CREB, (pCREB). These studies imply that mechanically evoked reflexes involve release of glutamate in myenteric neurons probably from IPANs at mGlu5 receptor synapses with VIP secretomotor neurons (Liu and Kirchgessner, 2000).

Glutamate also suppresses fast excitatory potentials in S/type 1 submucosal neurons without any effect on the cell's resting membrane potential. This response is mimicked by agonists of group II and group III mGlu receptors, and is due to presynaptic inhibition of transmitter release. Thus the presence of glutamate in IPANS along with expression of ionotropic and metabotropic glutamate receptors at pre- and postsynaptic sites on neurons in the submucous plexus adds to the complexity of the neural circuits that control intestinal secretion and its coordination with motility.

NUCLEOSIDES AND NUCLEOTIDES

Adenosine modulation of intestinal reflexes

Recent data is accumulating that adenosine is an important endogenous modulator of inflammatory processes, exerting anti- or pro-inflammatory effects, depending on the specific adenosine receptor subtype that is involved (Bouma et al., 1997; Roman and Fitz, 1999). Adenosine has affects that may be of therapeutic potential against ischemia reperfusion and inflammatory bowel disease (IBD). Recent efforts have focused on interventions that elevate endogenous adenosine levels and enhance its protecting actions at its local cellular site of formation, while reducing or eliminating systemic side effects. One such intervention is the purine nucleoside acadesine (riboside 5-amino-1β-D-ribofuranosyl-imidazole-4-carboxamide, AICA) that is protective in ischemic injury and is promising in treating IBD (Schoenberg et al., 1995). After it is taken up by cells, it becomes incorporated in the de novo purine biosynthetic pathway by phosphorylation to 5-aminoimidazole-4-carboxamide ribonucleotide (AICA ribotide or AICAR); it can further be metabolized to inosine monophosphate (IMP). The adenosine enhancing activity and protective actions of acadesine are believed to be due to the combined weak inhibitory action of AICAR on enzymes that are reciprocally involved in the IMP to AMP conversion, and of adenosine kinase and adenosine deaminase. It has recently been recognized that the anti-inflammatory effects of methotrexate and sulfasalazine used in the treatment of IBD and rheumatoid arthritis, is a result of their ability to induce the accumulation of AICAR and thereby increase local adenosine release at sites of inflammation (Cronstein et al., 1993; Gadangi et al., 1996). The proposed cytoprotective effect of adenosine in IBD may also involve its antioxidant properties to both inhibit oxidant production, as well as stimulate the activity of antioxidant enzymes superoxide dismutase, glutathione peroxidase, catalase via adenosine A3 receptor activation (Maggirwar et al., 1994).

After haemorrhagic shock and resuscitation or in various animal models of sepsis, administration of $ATP\text{-}MgCl_2$ enhances the recovery of intestinal, hepatic and other organ functions, as well as improved survival. The beneficial effects of $ATP\text{-}MgCl_2$ are believed to be due to various actions of adenosine to improve tissue energy stores as ATP, improve microcirculatory blood flow, and suppress pro-inflammatory cytokines IL-6 and TNF-α, inhibit oxidant production and stimulate antioxidant enzymes (Wang, P. *et al.*, 1991, 1992; Harkema and Chaudry, 1992; Kerner *et al.*, 1995).

Adenosine has diverse actions in the nervous, cardiopulmonary, renal and gastrointestinal systems where it exerts its actions by binding to A1, A2a, A2b or A3 receptors (Galligan and Bertrand, 1994; Christofi and Cook, 1997; Fredholm *et al.*, 2000; Guzman *et al.*, 2000; Christofi, 2001; Sharp *et al.*, 2001). A1 and A3 receptors are coupled to Gi and inhibition of adenylyl cyclase whereas A2a and A2b are positively coupled to Gs. Gene expression and distribution of adenosine receptors in human intestine are shown in Tables 3.1

TABLE 3.1
RT-PCR to detect human adenosine receptor in human intestine.

Adenosine receptors	A1	A2a	A2b	A3
Jejunum				
Whole thickness	−(−)	(−)+?	++	++
Mucosa/submucosa	−(−)	++	++	+
Mucosa	−(−)	+	++	+
Mucosa	−(+)?	++	++	++
Submucous plexus	−(+)	−(+)	+	+
Submucous plexus	−(−)	−(+)	+	+
Longitudinal/circular	+(++)	++	++	+
Ileum				
Whole thickness	+(+)	++	+	+
Mucosa/submucosa	nd	nd	nd	nd
Mucosa	−(+)	++	+	+
Submucous plexus	nd	nd	nd	nd
Cecum				
Whole thickness	−(−)	+	++	+
Mucosa/submucosa	−(+)	++	+	+
Mucosa	−(−)	+	++	+
Submucous plexus	−(+)	−(−)	+	−(+)
Colon				
Whole thickness	−(−)	−(−)	+	−(+)?
Mucosa/submucosa	−(−)	−(−)	++	+
Mucosa	nd	nd	nd	nd
Submucous plexus	−(+)	−(+)?	++	+
Human Cell Lines				
HT-29 (colonic epithelium)	++	+++	+++	−(−)?
T-84 (colonic epithelium)	−(+)	+	+++	−(−)?
T98G (glioblastoma)	++	+++	+++	−(+)

nd, not done; +, − indicate receptor mRNA detected or not detected, respectively. Failure to detect adenosine receptor mRNA after the first round of PCR reflects either its absence or low expression. If the results of RT-PCR were negative, an aliquot of first PCR reaction was amplified a second time and the result is indicated in parentheses. Failure to detect adenosine receptor mRNA after the second round of PCR is likely to be due to its absence. A question mark (?) indicates a possible faint expression of receptor mRNA.

TABLE 3.2
Cellular localization of adenosine A1, A2a, A2b, and A3 receptor
immunoactivities in human small and large intestine.

Human intestinal region/cell type	Adenosine receptor subtypes			
	A1	A2a	A2b	A3
Jejunum				+
Longitudinal muscle	±	±	−	
Myenteric plexus neurons	+	+	+	
Glia	−	−	+++	
Circular muscle	+	−	−	
Submucous plexus neurons	−	+++	+++	+++
Nerve fibres/neurites	−	+	+	+
Epithelia	−	+	+	+
Colon				+
Longitudinal muscle	±	−	−	
Myenteric plexus neurons	−	+	+	
Circular muscle	+	+	+	
Submucous plexus neurons	+++	+	+	
Gilia	−	−	+	
Epithelia	+			
T98G				+
U373				+
BON Cells				+

'−', absent; '+', present (or present ≤ 2 neurons); '±', marginally detectable; '++', 3–6 neurons; '+++', > 6 neurons.

and 3.2 (Linden *et al.*, 1993; Handcock and Coupar, 1995; Christofi *et al.*, 2001). Adenosine A1, A2a, A2b or A3 receptor mRNAs were differentially expressed in neural and non-neural layers of the jejunum, ileum, colon and cecum, in human epithelial cells (HT-29, T-84), glial cells (T98G) or enterochromaffin cells (BON cells) (Table 3.1). In human intestine, expression of the adenosine A1, A2a, A2b and A3 receptors differs from that in rat intestine (Dixon *et al.*, l996; Yu *et al.*, 2000). The differences imply additional functions for adenosine receptors in the human.

In the submucous plexus, adenosine A2b receptor immunoreactivity is expressed exclusively in 50% of VIP neurons that may represent interneurons, secretomotor or motor neurons. Adenosine A2a receptor is also expressed in other neurons. Adenosine A3 receptors are expressed in 57% of substance P-positive/IPANs in jejunal submucosal plexus and less than 10% of VIP neurons. In submucosal neurons of the human intestine about 38% of VIP neurons innervate the mucosa, 45% are interneurons in the submucosal plexus, and 12% are circular muscle motor neurons (Porter *et al.*, 1999). Adenosine A2b receptors may be expressed in one or more of these functional subsets of VIP-ergic neurons, whereas, A3 receptors may be restricted to a subset of VIP neurons that projects to the circular muscle. Retrograde dye labelling of submucosal neurons can give the identity of neurons expressing adenosine receptors.

A1 receptors are expressed in jejunal myenteric neurons and colonic submucosal neurons. A2a and A2b receptors are co-expressed in enteric neurons and epithelial cells (Barrett *et al.*, 1990; Puffinbarger *et al.*, 1995; Lelievre *et al.*, 1998; Christofi *et al.*, 2000). Since adenosine A2a and A2b receptors have high and low affinities, respectively, for endogenous adenosine, A2a or A2b receptors may be preferentially activated in the human submucosal plexus depending on the levels. Furthermore in cells that co-express A2a and A2b receptors, it is unknown whether each receptor will convey a distinct response (Feoktistov and Biaggioni, 1995). In general, A2b receptor immunoreactivity was more prominent than A2a receptor immunoreactivity in myenteric neurons, glia or nerve fibres. Adenosine A2a receptors and A2b receptors are expressed on the cell somas, neurites and varicose nerve fibres of submucosal neurons. The location of these receptors supports a role for the A2 receptors in both the pre- and post-synaptic neuromodulatory actions of adenosine. Functional evidence exists only for post-synaptic excitatory A2a receptors on submucosal S/type 1 neurons of the guinea pig small bowel (Barajas-Lopez *et al.*, 1991).

During mechanical stimulation, 5-HT and prostaglandins are released and act on IPANS and secretomotor neurons. Activaton of this reflex involves release of substance P from intrinsic sources (Cooke *et al.*, 1997a–c). Downstream events include stimulation of cholinergic and VIP-containing secretomotor neurons and these are all potential targets for endogenous adenosine's action. Mechanically evoked chloride secretion is modulated by adenosine acting at A1 receptors (Cooke *et al.*, 1999). The A1 receptor agonist, 8-cyclopentyltheophylline, enhances reflex evoked secretion suggesting that endogenous adenosine suppressed the reflex. Manipulating the endogenous adenosine levels by adding adenosine deaminase and nucleoside transport inhibitors enhanced or reduced secretion. Addition of an A1 receptor agonist attenuated the reflex secretory response and this was prevented by the presence of A1 receptor antagonists (Cooke *et al.*, 1999). Adenosine modulates both the 5-HT and prostaglandin limbs of the mucosal reflex in the guinea pig distal colon. These observations all point to adenosine's role as a physiological brake to suppress mechanicallly evoked secretion in the intestine. The effects of adenosine are anticipated if adenosine accumulates in physiological (i.e. intense contracture) or pathological states (i.e. during injury from abdominal surgery, ischaemia or hypoxia).

Electrophysiological studies on gut neurons showed that adenosine interacts with pre- and post-synaptic A1 receptors to inhibit slow synaptic transmission and at pre-synaptic A1 receptors to inhibit fast synaptic transmission (Barajas-Lopez *et al.*, 1991, 1995; Christofi and Wood, 1993, 1994; Christofi *et al.*, 1994). In myenteric AH/type 2 neurons, the main postsynaptic action of adenosine is A1 receptor-mediated suppression of neuronal excitability, in association with a decrease in cell input resistance and a sustained membrane hyperpolarization (Christofi and Wood, 1994; Christofi *et al.*, 1994). In a significant proportion of submucosal S/type 1 neurons, adenosine elevates excitability and causes a slow EPSP-like effect (Barajas-Lopez *et al.*, 1991). Extracellular adenosine is sufficient to suppress neuronal excitability, fast EPSPs, and slow EPSPs, and enhance inhibitory slow synaptic transmission. These effects serve to complement the ability of adenosine to shut down excitatory neural activity in the gut *microcircuits* through its dual pre- and post-synaptic actions (Christofi and Cook, 1986, 1987; Broad *et al.*, 1992; Christofi and Wood, 1993).

NUCLEOTIDES

Nucleotides in intestinal reflexes

Nucleotides like ATP exert their actions by binding to P2X and P2Y purinoceptors families (Burnstock, 2001). The P2X receptor family represents at least six ligand-gated ionotropic receptors. The P2Y receptor family represents six metabotrophic G-protein-coupled receptors. These include P2Y1, P2Y2, P2Y4, P2Y6, P2Y11 and P2Y12. The endogenous ligands at P2Y receptors are ADP (P2Y1), UTP, ATP (P2Y2), UTP (P2Y4), UDP (P2Y6), ATP (P2Y11). All of these receptors can couple to PLC and in the case of P2Y11 or P2Y12 to Gs or Gi and adenylyl cyclase as well.

Nucleosides and nucleotides are beginning to be implicated in mucosal reflexes regulating chloride secretion (Cooke et al., 1999). Mechanical stimulation releases 5-HT from enterochromaffin cells and initiates secretory reflexes. Mechanical stimulation is reported to release nucleotides as well (Lazarowski and Boucher, 2001). An early indication that nucleotides were involved in this secretory reflex comes from studies with apyrase. This enzyme that hydrolyses nucleotides decreased mechanically evoked secretion. Apyrase breaks down 5'-nucleotides to products that are inactive at nucleotide P2Y receptors. On the other hand, inhibitors of membrane ecto-ATPases prevent breakdown of ATP (Chen and Lin, 1997) and enhance secretion.

To explore more completely the possibility that ATP or other nucleotides participate in secretory reflexes through the submucosal plexuses, various inhibitors of purinoceptors have been used. In the distal colon, recent studies suggest that the secretory response due to mucosal stroking involves ATP or other nucleotides. This is evidenced by a reduction in secretion by P2 receptor antagonists PPADS, suramin, reactive blue 2 or MRS 2179 (Cooke et al., 2000). The P2 receptor antagonist PPADS is known to block actions of nucleotides at P2X and receptors as well as P2Y receptors. In submucosal neurons, PPADS blocks the fast (P2X) depolarization response elicited by ATP, but does not affect the slow (P2Y) depolarization response in the same subset or different subsets of S/type 1 neurons. Suramin also has no effect on the slow depolarization whereas it auguments the fast depolarization. Therefore the inhibitory effect of PPADS and suramin on reflex evoked chloride secretion cannot be explained by an action at P2X2,3,5 receptors, because suramin should have augmented the response (Barajas-Lopez et al., 1996, 2000). Instead it inhibited secretion. This coupled with the finding that a P2Y1 receptor antagonist reduced the reflex response argues for the involvement of a P2Y1 receptor and possibly others.

The identity of the P2Y receptor subtypes on the cell somas of S/type 1 neurons of the submucosal S/type 1 plexus remains unknown. Clearly though, these receptors are linked to the rise in intracellular calcium in the neurons, are not coupled to elevations in intracellular cAMP levels (Barajas-Lopez et al., 2000). P2Y11 and P2Y12 receptors are linked to adenylyl cyclase and elevations or reduction in cAMP. Therefore, the P2Y receptor on submucosal neurons linked to slow depolarization cannot be P2Y11 or P2Y12 receptors; however, adenosine is a breakdown product of ATP, and can act at A2a receptors to elicit a slow depolarization in the S/type 1 neurons that is mediated through an AC/cAMP pathway (Liu et al., 1999a).

Submucosal S/type 1 neurons in the ileum can be divided into subsets based on their response to ATP: those with only fast P2X depolarizations (10% of neurons), those with only slow P2Y depolarizations (35%), or those with both types of responses (55%). In separate experiments, parallel findings were obtained with ATP on Ca^{2+} transients in submucosal neurons (Barajas-Lopez et al., 2000). These S/type 1 neurons likely represent VIP, ChAT and NPY types of secretomotor neurons, and perhaps some submucosal VIP interneurons projecting to myenteric plexus (Li and Furness, 1998; Porter et al., 1999). In rodents, cell somas of a subset of submucosal neurons with VIP, CHAT, NPY, nNOS or calretinin express P2Y1 receptors. Cell somas of substance P or calbindin immunoreactive neurons do not express P2Y1 receptor immunoreactivity. Therefore, submucosal IPANs that are identified by their substance P or calbindin immunoreactivity, do not express P2Y1Rs on their cell somas. Therefore, it is likely that nucleotides released during mucosal stroking activate P2Y1Rs on secretomotor or interneurons. ATP does activates a cholinergic secretomotor-pathway since the muscarinic antagonist atropine partially blocks the neural Isc response to exogenous ATP (Cooke et al., 2000).

COORDINATION OF SECRETION AND MOTILITY

Mucosal stroking causes secretion and associated with this event is muscle contraction. The amplitude of secretion correlates with the amplitude of contraction (Cooke et al., 1993). Details on innervation of the two plexes and mechanisms of coordination are not known. In both guinea pig and rat tissues, the majority of the secretory and contractile responses were sensitive to the neural blocking agent, tetrodotoxin and therefore involved enteric nerves. In the guinea pig distal colon, the coordinated reflex response was attenuated by PPADS or MRS 2179 and this supports the possible involvement of P2X and P2Y receptors. Direct effects on muscle and epithelial cells in the presence of neural blockade reveals the distinct effects of nucleotides on effectors when contraction and secretion are uncoupled by interrupting the enteric nervous system. Spencer et al. (2000) provided evidence for ATP/or a related nucleotide involvement in the mucosal stimulation of excitatory reflex responses in circular and longitudinal muscle of the guinea pig ileum. In these studies, they concluded that PPADS or suramin-sensitive neurally mediated nucleotide responses occurred in smooth muscles in both ascending and descending excitatory motor pathways.

It is clear that nucleotides are involved in mucosal reflex responses in functionally different regions of the gut and in different species. Furthermore, nucleotides are involved in the reflex-evoked coordination of contraction/motility and secretion. Whether different mechanical stresses evoke a different combination of mediators and therefore activate different receptors is an unanswered question.

ENTEROCHROMAFFIN CELL MODEL – BON CELLS

Mechanical forces generated by mucosal stroking, pressure, touch or stretch, are detected by enterochromaffin cells, causing release of 5-HT (Racke and Schworer, 1991). BON cells serve as a model of human enterochromaffin cells, (Evers et al., 1994) (Figure 3.2). Mechanical stimulation of carcinoid BON cells by rotational shaking elicits 5-HT release through activation of the Gq-PLCβ and AC/cAMP signalling pathways via calcium as

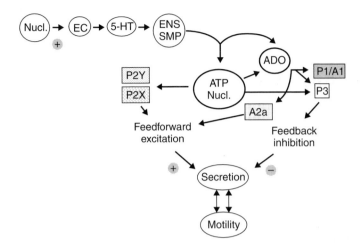

Figure 3.2   Autocrine and neurocrine regulation of mucosal reflexes by nucleotides. Nucelotides (Nucl.) such as ATP, UTP, ADP, UDP or others, are released by mechanical activation of enterochromaffin cells. Nucleotides stimulate enterochromaffin cells in an autocrine fashion to release 5-HT that initiates the mucosal reflex by activation of axon collaterals on IPANs in the submucous plexus (SMP). Release of ATP, other nucleotides or adenosine provides a dual effect in the enteric nervous system. Nucleotides activate excitatory post-synaptic P2X (fast EPSPs) or P2Y receptors (slow EPSPs) on S/type 1/putative secretomotor neurons to cause feedforward excitation resulting in stimulation of chloride secretion and a coordinated motility response. Pre-synaptic P2Y receptors are also involved in the net excitatory response. Adenosine release or formed by breakdown of ATP activates pre-synaptic P1/A1 or P3 inhibitory receptors to cause feedback inhibition and termination or attenuation of the reflex response.

described earlier (Liu *et al.*, 1999b; Kim *et al.*, 2000a,b and unpublished observations). The identity of the mechanosensitive receptor however remains elusive. Evidence is accumulating that nucleosides and nucleotides may act in an autocrine/paracrine manner via P2 receptors in a variety of cell types. Mechanically evoked 5-HT release is nearly abolished by apyrase as is ATP-evoked 5-HT release. RT-PCR analysis has identified mRNAs for P2Y receptors in BON cells. These observations are beginning to implicate P2Y receptors in mediating 5-HT release from enterochromaffin cells. The downstream signalling events linked to P2Y receptors include activation of the Gq/PLCβ signalling cascade, generation of IP_3 and mobilization of intracellular calcium and 5-HT release (Kim *et al.*, 2001b). Further studies are necessary to identify how a mechanical stimulus results in nucleotide release in the enterochromaffin cell model.

Nucleotides like ATP are high-energy phosphate compounds involved in intermediary metabolism. Any process that depletes energy supply leading to elevated adenosine levels creates an imbalance between energy demand and availability, as well as an imbalance between nucleotides and nucleoside/adenosine levels (Deshpande *et al.*, 1999) (Figure 3.3). Such an imbalance will have fundamental influence on enteric neural reflexes that can be predicted by the sites of action of purines at epithelial, enterochromaffin and neural levels. An imbalance is likely to occur in such diverse disease states of the gut as inflammation, hyper-motility, ischaemia, irritable bowel syndrome, prolonged diarrhoea or

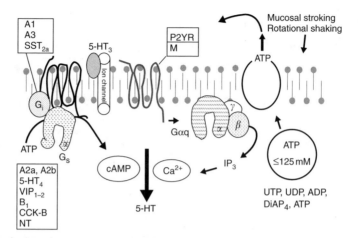

Figure 3.3   Receptor regulation of 5-HT release from enterochromaffin cells. 5-HT release occurs via activation of several converging pathways that include the AC/cAMP and $G\alpha q$/IP3-Ca^{2+} intracellular signalling pathways, as well as activation of ion channels. On the one hand, activation of adenosine A1 or A3 receptors and somatostatin SST_{2a} coupled to Gi protein inhibits the AC/cAMP pathway resulting in suppression of 5-HT release. On the other hand, activation of adenosine A2a or A2b receptors, 5-HT4 receptors, VIP receptors, adrenergic β1 receptors, gastrin/CCK-B receptors or neurotensin (NT) receptors activates Gs protein and stimulates cAMP production and 5-HT release. 5-HT activation of 5-HT_3 receptors linked to an ion channel also leads to 5-HT release. Autocrine release of ATP or other nucleotides activates P2Y receptors (P2YR) linked to $G\alpha q$ leading to IP3 formation and stimulation of intracellular free Ca^{2+} levels. A rise in free intracellular Ca^{2+} or cAMP inside the enterochromaffin cells leads to stimulation of 5-HT release to activate the mucosal reflex muscarinic receptor (M).

cancer. Endogenous adenosine, ATP, UTP, ADP, UDP and possibly others, likely play a funda-mental role in the neuroregulation of intestinal motility and secretion under normal and patho-physiological conditions.

SUMMARY

Neural reflexes through the submucosal plexus are important for maintaining chloride and water secretion at optimal levels for the appropriate digestive state in health and disease. The myenteric plexus also contributes to secretion by providing excitatory and inhibitory inputs onto submucosal interneurons or secretomotor neurons. Alternatively, secretion could occur entirely through the myenteric plexus since myenteric secretomotor neurons are found projecting to the mucosa. Transmitters that may be involved in regulation of secretion include acetylcholine, substance P, CGRP, glutamate at ganglionic synapses with acetylcholine and VIP acting at neuroepithelial junctions. Evidence is accumulating that nucleotides and nucleosides interact with many different kinds of receptors on neural and non-neural cells and are candidate transmitters and neuromodulators in the intestine. The picture that emerges is one of considerable complexity in the enteric neural circuits that control intestinal secretion.

REFERENCES

Barajas-Lopez, C., Espinosa-Luna, R. and Christofi, F.L. (2000) Changes in intracellular Ca2+ by activation of P2 receptors in submucosal neurons in short-term cultures. *Eur. J. Pharmacol.*, **409**, 243–257.

Barajas-Lopez, C., Huizinga, J.D., Collins, S.M., Gerzanich, V., Espinosa-Luna, R. and Peres, A.L. (1996) P2x-purinoreceptors of myenteric neurons from the guinea-pig ileum and their unusual pharmacological properties. *Br. J. Pharmacol.*, **119**, 1541–1548.

Barajas-Lopez, C., Muller, M.J., Prieto-Gomez, B. and Espinosa-Luna, R. (1995) ATP inhibits the synaptic release of acetylcholine in submucosal neurons. *J. Pharmacol. Exp. Ther.*, **274**, 1238–1245.

Barajas-Lopez, C., Surprenant, A. and North, R.A. (1991) Adenosine A1 and A2 receptors mediate presynaptic inhibition and postsynaptic excitation in guinea pig submucosal neurons. *J. Pharmacol. Exp. Ther.*, **258**, 490–495.

Barrett, K.D., Cohn, J.A., Huott, P.A., Wasserman, S.I. and Dharmsathaphorn, K. (1990) Immune-related intestinal chloride secretion. II. Effect of adenosine on T84 cell line. *Am. J. Physiol.*, **258**, C902–C912.

Bouma, M.G., van den Wildenberg, F.A.J.M. and Buurman, W.A. (1997) The anti-inflammatory potential of adenosine in ischemia-reperfusion injury: established and putative beneficial actions of a retaliatory metabolite. *SHOCK*, **8**, 313–320.

Broad, R.M., McDonald, T.J., Brodin, E., Cook, M.A. (1992) Adenosine A1 receptors mediate inhibition of tachykinin release from perifused enteric nerve endings. *Am. J. Physiol.*, **262**, G525–G531.

Brookes, S.J. (2001) Retrograde tracing of enteric neuronal pathways. *Neurogastroenterol. Motil.*, **13**, 1–18.

Burns, G.A. and Stephens, K.E. (1995) Expression of mRNA for the *N*-methyl-D-aspartate (NMDAR1) receptor and vasoactive intestinal polypeptide (VIP) co-exist in enteric neurons of the rat. *J. Auton. Nervous Syst.*, **55**, 207–210.

Burnstock, G. (2001) Purinergic signaling in gut. Purinergic and pyrimidinergic signaling II, cardiovascular, respiratory, immune, metabolic and gastrointestinal tract function. In *Handbook of Experimental Pharmacology*, vol. 151/II, M.P. Abbracchio and M. Williams (eds), pp. 141–238.

Chen, B.C. and Lin, W.W. (1997) Inhibition of ecto-ATPase by the P2 purinergic agonists, ATP gamma S, alpha, beta-methylene-ATP, and AMP-PNP, in endothelial cells. *Biochem. Biophys. Res. Commun.*, **233**, 442–446.

Christofi, F.L. (2001) Unlocking the mysteries of gut sensory transmission. Is adenosine the key? *News Physiol. Sci., NIPS*, **16**, 201–207.

Christofi, F.L., Baidan, L.D., Fertel, R.L. and Wood, J.D. (1994) Adenosine A₂ receptor-mediated excitation of a subset of AH/Type II neurons and elevation of cAMP levels in myenteric ganglia of guinea-pig small bowel. *Neurogastroenterol. Motil.*, **6**, 67–78.

Christofi, F.L. and Cook, M.A. (1986) The affinity of various purine nucleosides for adenosine receptors on purified enteric varicosities compared with their efficacy as presynaptic inhibitors of acetylcholine release. *J. Pharmacol. Exp. Ther.*, **237**, 305–311.

Christofi, F.L. and Cook, M.A. (1987) Possible heterogeneity of adenosine receptors on myenteric nerve endings. *J. Pharmacol. Exp. Ther.*, **243**, 302–309.

Christofi, F.L., Cook, M.A. (1997) Purinergic modulation of gastrointestinal function. In *Purinergic Approaches in Experimental Therapeutics*, K. Jacobson and M. Williams (eds), New York: Wiley Press. pp. 261–282.

Christofi, F.L., Wood, J.D. (1993) Effects of pituitary adenylate cyclase activating peptide on morphologically identified myenteric neurons in guinea-pig small bowel. *Am. J. Physiol. Gastrointest. Liver Physiol.*, **264**, G414–G421.

Christofi, F.L., Wood, J.D. (1994) Electrophysiological subtypes of inhibitory P₁ purinoceptors on myenteric neurons of guinea-pig small bowel. *Br. J. Pharmacol.*, **113**, 703–710.

Christofi, F.L., Yu, J.G., Xue, J., Wang, Y.Z., Kim, M. and Cooke, H.J. (2000) Adenosine A3 receptors are involved in the activation of an inhibitory neural motor pathway in rat colon. *Drug. Dev. Res.*, **50**, 80 (abstract).

Christofi, F.L., Zhang, H., Yu, J.G., Guzman, J., Xue, J., Kim, M., Wang, Y.-Z. and Cooke, H.J. (2001) Differential gene expression of adenosine A1, A2a, A2b, and A3 receptors in the human enteric nervous system. *J. Comp. Neurol.*, **439**, 46–64.

Cooke, H.J. (1992) Calcitonin gene-related peptides: influence on intestinal ion transport. In *Calcitonin Gene-Related Peptide: The First Decade of a Novel Pleiotropic Neuropeptide*, Y. Tache, P. Holzer and M.G. Rosenfeld (eds), New York: New York Academy of Sciences, pp. 364–372.

Cooke, H.J. and Reddix, R.A. (1994) Neural regulation of intestinal electrolyte transport. In *Physiology of the Gastrointestinal Tract*, 3rd edition, L.R. Johnson (ed.), New York: Raven Press, pp. 2083–2132.

Cooke, H.J., Sidhu, M. and Wang, Y.-Z. (1997a) 5-HT activates neural reflexes regulating secretion in the guinea pig colon. *Neurogastroenterol. Motil.*, **9**, 181–186.

Cooke, H.J., Sidhu, M. and Wang, Y.-Z. (1997b) Activation of 5-HT$_{1P}$ receptors on submucosal afferents subsequently triggers VIP neurons and chloride secretion in the guinea-pig colon. *J. Autonom. Nervous Syst.*, **66**, 105–110.

Cooke, H.J., Sidhu, M., Fox, P., Wang, Y.-Z. and Zimmermann, E.M. (1997c) Substance P as a mediator of colonic secretory reflexes. *Am. J. Physiol. Gastrointest. Liver Physiol.*, **272**, G238–G245.

Cooke, H.J., Wang, Y.-Z. and Rogers, R. (1993) Coordination of Cl$^-$ secretion and contraction by a histamine H$_2$-receptor agonist in guinea pig distal colon. *Am. J. Physiol. Gastrointest. Liver Physiol.*, **265**, G973–G978.

Cooke, H.J., Wang, Y.-Z., Liu, C.Y., Zhang, H. and Christofi, F.L. (1999) Activation of neuronal adenosine A$_1$ receptors suppresses secretory reflexes in the guinea pig colon. *Am. J. Physiol. Gastrointest. Liver Physiol.*, **276**, G451–G462.

Cooke, H.J., Wang, Y-Z., Xue, J., Yu, J., Christofi, F. (2000) P2 receptors mediate chloride secretory reflexes in guinea pig distal colon. *Drug Dev. Res.*, **50**, 53.

Costa, M., Brookes, S.J.H. and Hennig, G.W. (2000) Anatomy and physiology of the enteric nervous system. *Gut*, **47** (Suppl. IV), iv15–iv19.

Cronstein, B.N., Naime, D. and Ostad, E. (1993) The anti-inflammatory mechanism of methotrexate: increased adenosine release at inflamed sites diminishes leukocyte accumulation in an *in vivo* model of inflammation. *J. Clin. Invest.*, **92**, 2675–2682.

Deshpande, N.A., McDonald, T.J. and Cook, M.A. (1999) Endogenous interstitial adenosine in isolated myenteric neural networks varies inversely with prevailing PO$_2$. *Am. J. Physiol. Gastrointest. Liver Physiol.*, **276**, G875–G885.

Dixon, A.K., Gubitz, A.K., Richardson, P.J., Freeman, T.C. (1996) Tissue distrubution of adenosine receptor mRNA in the rat. *Br. J. Pharm.*, **118**, 146–148.

Eklund, S., Jodal, M. and Lundgren, O. (1985) The enteric nervous system participates in the secretory response to the heat stable enterotoxins of *Escherichia coli* in the rats and cats. *Neuroscience*, **14**, 673–681.

Eklund, S., Karlstrom, L., Rokaeus, A., Theodorsson, E., Jodal, M. and Lundgren, O. (1989) Effects of cholera toxin, *Escherichia coli* heat stable toxin and sodium deoxycholate on neurotensin release from the ileum *in vivo*. *Regul. Pept.*, **26**(3), 241–252.

Evers, B.M., Ishizuka, J., Townsend, C.M. Jr. and Thompson, J.C. (1994) The human carcinoid cell line, BON. A model system for the study of carcinoid tumors. *Ann. N.Y. Acad. Sci.*, **733**, 393–406.

Feoktistov, I. and Biaggioni, I. (1995) Adenosine A2b receptors evoke interleukin-8 secretion in human mast cells: an enprofylline-sensitive mechanism with implications for asthma. *J. Clin. Invest.*, **96**, 1979–1986.

Foxx-Orenstein, A.E., Kuemmerle, J.F. and Grider, J.R. (1996) Distinct 5-HT receptors mediate the peristaltic reflex induced by mucosal stimuli in human and guinea pig intestine. *Gastroenterology*, **111**, 1281–1290.

Fredholm, B.B., Arslan, G., Halldner, L., Kull, B., Schulte, G. and Wasserman, W. (2000) Structure and function of adenosine receptors and their genes. *Naunyn-Schmiedeberg's Arch. Pharmacol.*, **362**, 364–374.

Frieling, T., Cooke, H.J. and Wood, J.D., (1991) Serotonin receptors on submucous neurons in the guinea pig colon. *Am. J. Physiol.*, **261**, G1017–G1023.

Frieling, T., Dobreva, G., Weber, E., Becker, K., Ruprecht, C., Neulist, M. and Schemann, M. (1999) Different tachykinin receptors mediate chloride secretion in the distal colon through activation of submucosal neurones. *Naunyn-Schmiedeberg's Arch. Pharmacol.*, **359**, 71–79.

Furness, J.B., Kunze, W.A., and Clerk, N. (1999) Nutrient tasting and signaling mechanisms in the gut. II. The intestine as a sensory organ: neural, endocrine, and immune responses. *Am. J. Physiol. Gastrointest. Liver Physiol.*, **277**, G922–G928.

Furness, J.B., Kunze, W.A., Bertrand, P.P., Clerc, N. and Bornstein, J.C. (1998) Intrinsic primary afferent neurons of the intestine. *Prog. Neurobiol.*, **54**, 1–18.

Gadangi, P., Longaker, M. and Naime, D. *et al.* (1996) The anti-inflammatory mechanism of sulfasalazine is related to adenosine release at inflamed sites. *J. Immunol.*, **156**, 1937–1941.

Galligan, J.J. and Bertrand, P.P. (1994) ATP mediates fast synaptic potentials in enteric neurons. *J. Neurosci.*, **14**, 7563–7571.

Gay, J., Fioraqmonti, J., Garcia-Villar, R., Emonds-Alt, X. and Bueno, L. (1999) Involvement of tachykinin receptors in sensitization to cow's milk proteins in guinea pigs. *Gut*, **44**, 497–503.

Guzman, J.E., Yu, J.-G., Xue, J., Wang, Y.-Z., Kim, M., Cooke, H.J. and Christofi, F.L. (2000) Adenosine A3 receptor gene expression in the intestinal tract of humans, rats and guinea-pigs. *Drug Dev. Res.*, **50**, 79 (abstract).

Handcock, D.L. and Coupar, I.M. (1995) Functional characterization of the adenosine receptor mediating inhibition of intestinal secretion. *Br. J. Pharmacol.* **114**, 52–56.

Harkema, J.M. and Chaudry, I.H. (1992) Magnesium-adenosine triphosphate in the treatment of shock, ischemia and sepsis. *Crit. Care Med.*, **20**, 263–275.

Hens, J., Schrodl F., Frehmer A., Adriaensen Dd., Neuhuber W., Scheuermann D.W., Schemann M. and Timmermans J.P. (2000). Mucosal projections of enteric neurons in the porcine small intestine. *J. Comp. Neurol.*, **421**, 429–436.

Hens, J., Vanderwinden, J.M., De Laet, M.H. Scheurmann, D.W. and Timmermans, J.P. (2001) Morphological and neurochemical identification of enteric neurons with mucosal projections in the human small intestine. *J. Neurochem.*, **76**, 464–471.

Jenkinson, K.M., Morgan, J.M., Furness, J.B. and Southwell, B.R. (1999) Neurons bearing NK₃ tachykinin receptors in the guinea-pig ileum revealed by specific binding of flourescently labeled agonists. *Histochem. Cell Biol.*, **112**, 233–246.

Jodal, M., Holmgren, S., Lundgren, O. and Sjoqvist, A. (1993) Involvement of the myenteric plexus in the cholera toxin-induced net fluid secretion in the rat small intestine. *Gastroenterology*, **105**, 1286–1293.

Kellum, J.M., Albuquereque, F.C., Stoner, M.C. and Harris, R.P. (1999) Stroking human jejunal mucosa induces 5-HT release and Cl-secretion via afferent neurons and 5-HT4 receptors. *Am. J. Physiol. Gastrointest. Liver Physiol.*, **277**, G515–G520.

Kerner, J., Wang, P. and Chaudry, I.H. (1995) Impaired gut lipid absorptive capacity after trauma-hemorrhage and resuscitation. *Am. J. Physiol.*, **269**, R869–R873.

Kim, M., Cooke, H.J., Javed, N.H., Carey, H.V., Christofi, F. and Reybould, H.E. (2001a) D-glucose releases 5-hydroxytryptamine from human BON cells as a model of enterochromaffin cells. *Gastroenterology*, **121**, 1400–1406.

Kim, M., Javed, N.H., Yu, J.G., Christofi, F. and Cooke, H.J. (2001b) Mechanical stimulation activates Gαq signaling pathways and 5-hydroxytryptamine release from human carcinoid BON cells. *J. Clin. Invest.*, **108**, 1051–1059.

Kim, M., Javed, N., Christofi, F.L, Xue, J., Raybould, H. and Cooke, H.J. (2000) Adenosine acts at A1, A2a and A2b receptors to modulate 5-hydroxytryptamine release from human BON cells. *Gastroenterology*, **118**, P-262 (abstract).

Kirchgessner, A.L., Liu, M.T. and Alcantra, F. (1997) Excitotoxicity in the enteric nervous system. *J. Neurosci.*, **17**, 8804–8816.

Kunze, W.A. and Furness, J.B. (1999) The enteric nervous system and regulation of intestinal motility. *Annu. Rev. Physiol.*, **6**, 117–142.

Kuwahara, A., Kuramoto, H. and Kadowaki, M. (1998). 5-HT activates nitric oxide-generating neurons to stimulate chloride secretion in guinea pig distal colon. *Am. J. Physiol. Gastrointest. Liver Physiol.*, **275**, G829–G834.

Kuwahara, A., Tien, X.Y., Wallace, L. and Cooke, H.J. (1987) Cholinergic receptors mediating secretion in guinea pig colon. *J. Pharmacol. Exp. Ther.*, **242**, 600–606.

Larzabal, A., Losada, J., Mateos, J.M., Benitez, R., Garmilla, I.J., Kuhn, R., Grandes, P. and Sarria, R. (1999) Distribution of the group II metabotropic glutamate receptors (mGluR2/3) in the enteric nervous system of the rat. *Neurosci. Lett.*, **276**, 91–94.

Lazarowski, E.R. and Boucher, R.C. (2001) UTP as an extracellular signaling molecule. *News Physiol. Sci.*, **16**, 1–5.

Lelievre, V., Muller, J.M., Falcon. (1998) Adenosine modulates cell proliferation in human colonic adenocarcinoma. I. Possible involvement of adenosine A1 receptor subtypes in HT29 cells. *Eur. J. Pharmacol.*, **341**, 289–297.

Li, Z.S. and Furness, J.B. (1998) Immunohistochemical localisation of cholinergic markers in putative intrinsic primary afferent neurons of the guinea pig small intestine. *Cell Tissue Res.*, **294**, 35–43.

Linden, J., Taylor, H.E., Robeva, A.S., Tucker, A.L., Stehle, J.H., Rivkees, S.A., Fink, J.S. and Reppert, S.M. (1993) Molecular cloning and functional expression of a sheep A3 adenosine receptor with widespread tissue distribution. *Mol. Pharmacol.*, **44**, 532–542.

Liu, C.Y., Jamaleddin, A.J., Zhang, H. and Christofi, F.L. (1999a) FICRhR/cyclic AMP signaling in myenteric ganglia and calbindin-D₂₈ intrinsic primary afferent neurons involves adenylyl cyclases I, III and IV. *Brain Res.*, **826**, 253–269.

Liu, M. Seino, S. and Kirchgessner, A.L. (1999b) Identification and characterization of glucoresponsive neurons in the enteric nervous system. *J. Neurosci.*, **19**, 10305–10317.

Liu, M. and Kirchgessner, A.L. (2000) Agonist- and reflex-evoked internalization of metabotropic glutamate receptor 5 in enteric neurons. *J. Neurosci.*, **20**, 3200–3205.

Lomax, A.E. and Furness, J.B. (2000) Neurochemical classification of enteric neurons in the guinea-pig distal colon. *Cell Tissue Res.*, **302**, 59–72.

Lomax, A.E., Bertrand, P.P. and Furness, J.B. (1998) Identification of the populations of enteric neurons that have NK1 tachykinin receptors in the guinea pig small intestine. *Cell Tissue Res.*, **294**, 27–33.

MacNaughton, W., Moore, B. and Vanner, S. (1997) Cellular pathways mediating tachykinin-evoked secretomotor responses in guinea-pig ileum. *Am. J. Physiol. Gastrointest. Liver Physiol.*, **273**, G1127–G1134.

Maggirwar, S.B., Dhanraj, D.N., Somani, S.M. and Ramkumar, V. (1994) Adenosine acts as an endogenous activator of the cellular antioxidant defense system. *Biochem. Biophys. Res. Commun.*, **201**, 508–515.

Meedeniya, A.C., Brookes, S.J., Hennig, G.W. and Costa, M. (1998) The projections of 5-hydroxytryptamine-accumulating neurons in the myenteric plexus of the small intestine of the guinea-pig. *Cell Tissue Res.*, **291**, 375–384.

Moore, B.A. and Vanner, S. (1998) Organization of intrinsic cholinergic neurons projecting within submucosal plexus of guinea pig ileum. *Am. J. Physiol. Gastrointest. Liver Physiol.*, **275**, G490–G497.

Moore, B.A. and Vanner, S. (2000) Properties of synaptic input from myenteric neurons innervating submucosal S neurons in guinea pig ileum. *Am. J. Physiol. Gastrointest. Liver Physiol.*, **278**, G273–G280.

Moriarty, D., Selve, N., Baird, A.W. and Goldhill, J. (2001) Potent NK1 antagonism by SR-140333 reduces rat colonic secretory response to immunocyte activation. *Am. J. Physiol. Cell Physiol.*, **280**, C852–C858.

Neunlist, M. and Schemann, M. (1998) Polarized innervation pattern of the mucosa of the guinea pig distal colon. *Neurosci. Lett.*, **246**, 161–164.

Neunlist, M. and Schemann, M. (1997) Projections and neurochemical coding of myenteric neurons innervating the mucosa of the guinea pig proximal colon. *Cell Tissue Res.*, **287**, 119–125.

Neunlist, M., Dobreva, G. and Schemann, M. (1999a) Characteristics of mucosally projecting myenteric neurons in the guinea-pig proximal colon. *J. Physiol.*, **517**, 533–546.

Neunlist, M., Reiche, D., Michel, K., Pfannkuche, H., Hoppe, S. and Schemann, M. (1999b) The enteric nervous system: region and target specific projections and neurochemical codes. *Eur. J. Morphol.*, **37**, 233–240.

Palmer, J.M. and Greenwood-Vanmeerveld, B. (2001) Integrative neuroimmunomodulation of gastrointestinal function during enteric parasitism. *J. Parasitol.*, **87**, 483–504.

Pan, H. and Gershon, M.D. (2000) Activation of intrinsic afferent pathways in submucosal ganglia of the guinea pig small intestine. *J. Neurosci.*, **20**, 3295–3309.

Porter, A.J., Wattchow, D.A., Brookes, S.J. and Costa, M. (1999) Projections of nitric oxide synthase and vasoactive intestinal polypeptide-reactive submucosal neurons in the human colon. *J. Gastroenterol. Hepatol.*, **14**, 180–187.

Puffinbarger, N.K., Hansen, K.R., Resta, R., Laurent, A.B., Knudsen, T.B., Madara, J.L. and Thompson, L.F. (1995) Production and characterization of multiple antigenic peptide antibodies to the adensoine A2b receptor. *Mol. Pharmacol.*, **47**, 1126–1132.

Quinson, N., Robbins, H.L. and Clark, M.J. (2001) Calbindin immunoreactivity of enteric neurons in the guinea pig ileum. *Cell Tissue Res.*, **305**, 3–9.

Racke, K. and Schworer, H. (1991) Regulation of serotonin release from the intestinal mucosa. *Pharm. Res.*, **23**, 13–25.

Raybould, H.E. (2001) Primary afferent response to signals in the intestinal lumen. *J. Physiol.*, **530**, 431–432.

Ren, J., Hu, H.Z., Liu, S., Xia, Y. and Wood, J.D. (1999) Glutamate modulates neurotransmission in the submucosal plexus of guinea-pig small intestine. *Neuroreport*, **10**, 3045–3048.

Roman, R.M. and Fitz, J.G. (1999) Emerging roles of purinergic signaling in gastrointestinal epithelial secretion and hepatobiliary function. *Gastroenterology*, **116**, 964–979.

Schneider, J., Jehle, E.C., Starlinger, M.J., Neunlist, M., Hoppe, S. and Schemann, M. (2001) Neurotransmitter coding of enteric neurons in the submucous plexus is changed in non-inflamed rectum of patients with Crohn's disease. *Neurogastroenterol. Motil.*, **13**, 255–264.

Schoenberg, M.H., Poch, B., Moch, D. *et al.* (1995) Effect of acadesine treatment on post-ischemic damage to small intestine. *Am. J. Physiol.*, **269**, H1752–H1759.

Sharp, S., Yu, J.G, Guzman, J., Xue, J.J., Cooke, H.J. and Christofi, F.L. (2001) Adenosine A3 receptor expression in peptidergic neural circuits of the rat distal colon. *Gastroenterology*, A177 (abstract).

Song, Z.-M., Brookes, S.J., Steele, P.A. and Costa, M.J.H. (1992) Projections and pathways of submucous neurons to the mucosa of the guinea pig small intestine. *Cell Tissue Res.*, **399**, 255–268.

Song, Z.M., Costa, M. and Brookes, S.J. (1998) Projections of submucous neurons to the myenteric plexus in the guinea pig small intestine. *J. Comp. Neurology*, **399**, 255–268.

Spencer, N.J., Walsh, M. and Smith, T.K. (2000) Purinergic and cholinergic neuro-neuronal transmission underlying reflexes activated by mucosal stimulation in the isolated guinea-pig ileum. *J. Physiol.*, **522**, 321–331.

Timmermans, J.P., Hens, J. and Adriaensen, D. (2001) Outer submucous plexus: an intrinsic nerve network involved in both secretory and motility processes in the intestine of large mammals and humans. *Anat. Rec.*, **262**, 71–78.

Vanner, S. (2000) Myenteric neurons activate submucosal vasodilator neurons in guinea pig ileum. *Am. J. Physiol.*, **279**, G380–G387.

Wang, P., Ba, Z.F., Dean, R.E. and Chaudry, I.H. (1991) ATP-MgCl$_2$ restores the depressed hepatocellular function and hepatic blood flow following hemorrhage and crystalloid resuscitation. *J. Surg. Res.*, **50**, 368–374.

Wang, P., Ba, Z.F., Morrison, M.H., Ayala, A., Dean, R.E. and Chaudry, I.H. (1992) Mechanism of the beneficial effects of ATP-MgCl$_2$ following trauma-hemorrhage and resuscitation: downregulation of inflammatory cytokine (TFN, IL-6) release. *J. Surg. Res.*, **52**, 364–371.

Wang, X.Z. and Cooke, H.J. 1999. Calcitonin gene-related peptide receptor antagonist, CGRP 8–37, reduced neurally-evoked secretion in the guinea pig colon. *Gastroenterology*, **116**, A1100.

Wang, X.Z., Frieling, T., Wood, J.D. and Cooke, H.J. (1991). Neural 5-hydroxytryptamine receptors regulate chloride secretin in the guinea pig distal colon. *Am. J. Physiol.*, **261**, G833–G840.

Wardell, C.F., Bornstein, J.C. and Furness, J.B. (1994) Projections of 5-hydroxytryptamine-immunoreactive neurons in guinea-pig distal colon. *Cell Tissue Res.*, **278**, 379–387.

Wright, E.M., Hirsch, J.R., Loo, D.D. and Zampighi, G.A. (1997) Regulation of Na$^+$/glucose cotransporters. *J. Exp. Biol.*, **200**, 287–293.

Yu, J.G., Xue, J., Zhang, H., Wang, Y.X., Kim, M., Cooke, H.J. and Christofi, F.L. (2000) Neural adenosine A3 receptors are negatively coupled to neuromuscular transmission in rat colon. *Gastroenterology*, **118**, P-260 (abstract).

4 The Cholinergic Anti-Inflammatory Pathway

Christopher J. Czura and Kevin J. Tracey

Laboratory of Biomedical Science, North Shore-LIJ Research Institute, 350 Community Drive, Manhasset, NY 11030, USA

The innate immune system rapidly responds to bacterial infection, hypotension or haemorrhage with pro-inflammatory cytokines, including TNF, IL-1 and HMGB1, that activate macrophages, monocytes, and neutrophils, initiate tissue repair, and modulate initiate specific cellular immune responses. Failure to control this immune response leads to systemic over-expression of cytokines, which induces diffuse tissue damage, organ failure, and death. Multiple counter-regulatory and anti-inflammatory mechanisms have evolved to confine and regulate innate immune interactions within the site of infection. The central nervous system has a major role in rapidly and directly modulating the immune response. Sensory fibres within the vagus nerve detect peripheral inflammation, and transmit afferent signals that stimulate the release of centrally derived anti-inflammatory agents such as ACTH and glucocorticoids, and potentiate anorexia, hyperalgesia and pyrexia. The response to peripheral inflammation also includes anti-inflammatory signals carried through motor fibres of the vagus, which terminate in most critical organs. Acetylcholine, the principle neurotransmitter of the vagus, interacts with nicotinic cholinergic receptors expressed on macrophages and other immune cells to inhibit the release of pro-inflammatory cytokines. This 'cholinergic anti-inflammatory pathway' is uniquely positioned to directly and rapidly modulate systemic inflammatory responses.

KEY WORDS: immune response; anti-inflammatory mechanism; cytokines; vagus nerve stimulation.

INTRODUCTION

The innate immune system fights infection and facilitates wound healing in part by releasing pro-inflammatory cytokines, which activate macrophages and neutrophils, and modulate specific cellular responses. In pathological conditions such as ischaemia/ reperfusion, shock and trauma, dysregulated pro-inflammatory cytokine release induces systemic cellular injury, diffuse coagulation, unregulated apoptosis, and energy deficits. These events lead to severe complications including multiple organ failure and death. Counter-regulatory and anti-inflammatory systems have evolved to maintain homeostasis and prevent cytokine-mediated tissue injury. A growing body of evidence implicates the parasympathetic nervous system as one such critical anti-inflammatory mechanism. Bacterial products and pro-inflammatory cytokines stimulate afferent sensory signals through the vagus nerve that induce fever, anorexia, sleep, hyperalgesia and other sickness behaviours. Afferent signals also stimulate the central release of anti-inflammatory mediators including steroids and α-MSH. Recent evidence indicates that the parasympathetic nervous system also directly controls immune responses. This 'cholinergic anti-inflammatory pathway' regulates inflammation by efferent activity in the vagus nerve, which releases acetylcholine in target tissues. Macrophages and other cells of the reticuloendothelial system express nicotinic-type acetylcholine receptors that specifically modulate the activation state of the innate immune system. Stimulation of the vagus nerve by electrical or pharmacological means can recapitulate the activity of the cholinergic anti-inflammatory pathway, and can prevent lethal shock in animals challenged with the bacterial product endotoxin (lipopolysaccharide, LPS), or intra-abdominal sepsis. Because electrical vagus nerve stimulators are clinically approved devices for the treatment of epilepsy, the activity of the cholinergic anti-inflammatory pathway may be harnessed for the treatment of cytokine-mediated diseases.

SYSTEMIC INFLAMMATION AS A THERAPEUTIC TARGET

Despite the considerable effort of both basic science and clinical research programmes, the treatment of severe sepsis in the critically ill, post-operative or post-trauma patient population remains a daunting problem. Overall mortality rates are near 30%, accounting for the most common cause of death in hospital non-coronary intensive care units, with annual costs of US\$ 16.7 billion nationally (Angus *et al.*, 2001). An expansive and aggressive research effort into this problem literally created a new scientific field focused on defining the biology and pathobiology of inflammatory mediators in the development of human sepsis. The basic premise, that cytokines produced during the 'normal' course of infection can directly injure the host, is just 15 years old, and although initial optimism for a modern-day 'magic bullet' for sepsis has waned somewhat, these research efforts have provided clearer insight into the biological complexities of the proinflammatory mediator cascade.

The innate immune system, especially macrophages and neutrophils, fights infection with an arsenal of pro-inflammatory mediators designed to confine and combat invading pathogens and damaged tissues within the local site of the wound. However, the innate immune system can become over-activated, resulting in the systemic release of pro-inflammatory cytokines such as tumour necrosis factor (TNF), interferon-γ (IFN-γ), interleukin 1α (IL-1α), IL-1β and IL-6 (Ayala and Chaudry, 1996). These circulating cytokines induce widespread and diffuse coagulation, tissue damage, and hypotension, all clinical signs of septic shock. In controlled animal models, inhibition of these factors suppresses key components of the pathological sequelae of shock. For example, Tracey and colleagues discovered the inflammatory role of TNF, and demonstrated that it fulfils a modification of Koch's postulates for causation as an acute mediator of lethal septic shock syndrome, because: (1) TNF is produced during septic shock (Tracey *et al.*, 1987); (2) administration of TNF to normal mammals (including man) reproduces the haemodynamic, metabolic, immunological, and pathological sequelae of septic shock syndrome (Tracey *et al.*, 1986); and (3) removing TNF from animals with septic shock syndrome by either pharmacological means or by the use of genetic 'knock-out' technology prevents the development of lethal septic shock (Tracey and Cerami, 1994). Importantly, the same proof was never obtained for sepsis syndrome, because anti-TNF antibodies given during Gram-negative peritonitis actually worsen lethality (Remick *et al.*, 1995). Although multi-centre, randomized, double-blind clinical trials of sepsis syndrome using anti-TNF anti-bodies or soluble receptor constructs did not show statistically significant improvement in patient survival, these studies led directly to the recent FDA approval of anti-TNF anti-bodies for the treatment of rheumatoid arthritis and Crohn's disease. Thus, basic scientific knowledge derived from studies of TNF as a mediator of endotoxemia and sepsis has been translated into approved therapies for other diseases mediated by TNF. It is perhaps reassuring to note that despite the difficulties inherent in studying sepsis through a fundamental 'reductionism' approach, which focused on targeting TNF in models of endotoxemia and sepsis, thousands of patients have now derived important quality of life benefit from the development of anti-TNF antibodies for use in non-infectious, inflamma-tory diseases. It is likely that other strategies for inhibiting TNF will also be translated into new therapies for clinical use.

The identification of TNF as an essential mediator of septic, ischaemic and haemor-rhagic shock focused attention on the development of therapeutic strategies that target other endogenous toxins. Most recently, high mobility group B1 (HMGB1) has been iden-tified as a critical cytokine released from macrophages late after exposure to bacterial products, which is lethal when injected into animals (Wang *et al.*, 1999). Administration of anti-HMGB1 antibodies protects animals in models of sepsis, demonstrating that HMGB1, like TNF, is both sufficient and necessary for sepsis-like shock and death (Wang *et al.*, 1999). In contrast to TNF, however, HMGB1 is released with uniquely delayed kinetics, reaching peak levels in the serum of septic or endotoxemic mice only after 18–24 h after disease onset (Wang *et al.*, 2001). The delayed activity of HMGB1 places this newly iden-tified mediator as a potentially important therapeutic target, because its late activity makes it more clinically accessible. A number of other cytokines have been identified as key com-ponents of the inflammatory cascade, including IL-1β (Dinarello *et al.*, 1991; Dinarello, 1992), leukemia inhibitory factor (LIF) (Block *et al.*, 1993; Waring *et al.*, 1994), IFN-γ (Heinzel, 1990; Doherty *et al.*, 1992), and macrophage migration inhibitory factor (MIF) (Bernhagen *et al.*, 1993; Bozza *et al.*, 1999; Calandra *et al.*, 2000). The interaction of these and potentially other mediators induce the hypotension and tissue damage characteristic of lethal shock.

MAMMALIAN EVOLUTION CONFERRED REDUNDANT STRATEGIES FOR SUPPRESSING SYSTEMIC, LETHAL INFLAMMATION BY DEACTIVATING MACROPHAGES

Teleological reasoning suggests that evolution has conferred mechanisms that 'normally' counter-regulate the pro-inflammatory mediator response in order to prevent TNF-mediated shock and tissue injury, because lethal septic shock is a relatively unusual situation even though mammals are surrounded by potentially lethal bacteria both externally and inter-nally (Tracey *et al.*, 1987). To prevent the overly robust production of pro-inflammatory pathways during a self-limited injury or infection, redundant anti-inflammatory mechanisms that are integral to the host response effectively inhibit or suppress macrophage activation. Macrophage deactivating factors accumulate at the local site of infection (e.g. prostaglandin E2 and spermine) and are released systemically (e.g. IL-10, TGFβ, glucocorticoids). Local accumulation of prostaglandin E_2 (PGE_2) inhibits synthesis of pro-inflammatory cytokines and restrains acute cytokine responses (Knudsen *et al.*, 1986). Spermine is a ubiquitous biogenic molecule that accumulates at sites of infection or injury and post-transcriptionally inhibits the activity of several cytokines, including TNF, IL-1β, and the macrophage inflammatory proteins 1α and 1β (MIP-1α, MIP-1β) in macrophages and monocytes (Zhang *et al.*, 1997, 1999, 2000; Wang *et al.*, 1998). Anti-inflammatory cytokines such as IL-10 and transforming growth factor β (TGF-β) also serve to downregulate inflammatory responses. IL-10 deactivates macrophages in culture; in trauma patients, TNF levels are higher when IL-10 levels are depressed, an indication of pending septic complications (Bogdan *et al.*, 1991; Oswald *et al.*, 1992; Hauser *et al.*, 1995). Elevated levels of TGF-β, a potent inhibitor of monocyte activation, have been observed in monocytes derived from immunosuppressed trauma patients (Tsunawaki *et al.*, 1988; Miller-Graziano *et al.*, 1991).

The complex cytokine milieu in the septic patient is characterized by an interaction between the beneficial anti-inflammatory responses and potentially injurious pro-inflammatory responses. The impact of any future cytokine-based therapeutic strategy on patient survival for sepsis may well depend upon how the therapeutic approach influences the endogenous balance between pro- and anti-inflammatory responses (Tracey and Abraham, 1999). A safe and effective treatment strategy would ideally minimize the injurious effects of pro-inflammatory mediators; preserve the beneficial activity of anti-inflammatory mediators; and not increase the risk of secondary infections from trauma-induced immunosuppression.

SENSING PERIPHERAL INFLAMMATION

The influence of the brain on the maintenance of health and the development of disease is a topic that originated in ancient times (Blalock, 2002); more recently, this subject has been studied from the perspective of neuroimmune regulation of cytokine release (Tracey, 2002). Bacterial products and pro-inflammatory cytokines stimulate afferent sensory signaling in the vagus nerve that induces fever, anorexia, sleep, hyperalgesia and other sickness behaviours. Afferent signals also stimulate the central release of anti-inflammatory mediators including ACTH, which in turn acts on the adrenal glands to increase the release of glucocortiocoids. Glucocorticoids are potent anti-inflammatory hormones that deactivate macrophages, and inhibit the synthesis of TNF, IL-1 and other pro-inflammatory mediators. Hypophysectomy renders animals exquisitely sensitive to the lethal effects of endotoxin, in part because deficiencies in the ACTH-glucocorticoid response lead to significantly higher levels of TNF (Bloom *et al.*, 1998).

IMMUNE TO BRAIN COMMUNICATION: HUMORAL SUBSTRATES

The hypothalamus–pituitary–adrenal (HPA) axis response to inflammation can be stimulated through the humoral route, because increased inflammatory mediator levels in the bloodstream stimulate ACTH release. TNF and other cytokines are relatively large proteins that are typically incapable of diffusing across the blood–brain barrier (BBB). However, TNF and IL-1β can be actively transported across the BBB (Banks and Kastin, 1991; Gutierrez et al., 1993, 1994; Pan et al., 1997), and cytokines can enter the brain through circumventricular organs such as the pineal gland, area postrema, median eminence, and the neural lobe of the pituitary (Plotkin et al., 1996), where the BBB is non-existent or discontinuous (Gross and Weindl, 1987; Johnson and Gross, 1993). IL-1α also penetrates these areas of the brain through a putative transport mechanism in the vasculature, and accumulates to concentrations 20–160 times that of albumin (Plotkin et al., 1996). Microglia and endothelial cells of the cerebral vasculature express IL-1β receptors (Cunningham et al., 1992; Wong and Licinio, 1994; Yabuuchi et al., 1994; Ericsson et al., 1995), as well as the co-receptors for endotoxin, CD14 (Lacroix and Rivest, 1998) and Toll-like receptor 4 (TLR4) (Laflamme and Rivest, 2001). Once cytokines gain access to circumventricular organs, however, these proteins are still limited to small areas of the brain, and may induce wider-reaching effects through second messengers. Both IL-1β (Cao et al., 1996) and endotoxin (Breder et al., 1992) induce the expression of the inducible form of cyclooxygenase (COX2) in cerebrovascular endothelial cells, which processes arachadonic acid to release prostaglandin E2 (PGE2). The prostaglandins are small, lipophilic molecules that can diffuse throughout the brain and bind to PG receptors. PG receptors localize to areas of the brain that play important roles in peripheral inflammatory responses such as activation of the HPA axis and fever (Ericsson et al., 1997). These studies provide a potential mechanism for high molecular weight, blood-borne mediators to directly inform the central nervous system of inflammatory conditions in the periphery.

IMMUNE TO BRAIN COMMUNICATION: AFFERENT NEURAL SIGNALS

The HPA axis can also be stimulated via a neural route, because pro-inflammatory mediators stimulate afferent vagus nerve signals that converge on the brain stem, where they are relayed to the hypothalamus to induce ACTH release in response to inflammation (Fleshner et al., 1997). Receptors for IL-1 in the substrate of the vagus nerve are activated by increased IL-1 levels in the abdominal cavity, and IL-1 receptor–ligand interaction transduces a signal that leads to afferent neural firing (Gaykema et al., 2000). Vagotomy prevents the development of fever and increased ACTH after administration of intraperitoneal IL-1, indicating that afferent vagus nerve signals are critical to the development of the fever and ACTH responses (Gaykema et al., 2000).

The brain senses the environment through peripheral organs, and the immune system may arguably act as a diffuse sensory organ for the detection of infection and injury (Blalock, 2002). The vagus nerve seems particularly well positioned to relay information between the immune and central nervous systems, because it innervates, among other organs, the liver, lungs, kidneys, digestive tract, and other visceral organs that act either as filters for pathogens and pathogen products, or as routes of entry for pathogens. Vagus afferent fibres terminate within the area postrema and the nucleus tractus solitarius, two regions within the brain that are particularly active during peripheral inflammation, as detected by c-fos immunoreactivity (Ericsson *et al.*, 1997; Lee *et al.*, 1998). Ascending vagus nerve signals excite second-order neurons within the nucleus of the tractus solitarii (NTS) via the activity of glutamate (Smith *et al.*, 1998). The NTS forms the apex of a vago-vagal control loop that modulates visceral activity, because the NTS inhibits visceral activity receiving vagal input through two distinct mechanisms. First, NTS neurons inhibit a subset of neurons within the dorsal motor nucleus of the vagus (DMV) that provide cholinergic excitatory signals to the viscera, including the digestive tract (Rogers *et al.*, 1996). The NTS also suppresses visceral activity by activating other, inhibitory DMV efferent neurons through nonadrenergic, noncholinergic (NANC) pathways (Rogers *et al.*, 1999). Importantly, both the NTS and DMV may receive blood vessels that lack a functional BBB, making these organs important circumventricular organs. This may allow both the NTS and the DMV to receive sensory input from diffusible circulating factors such as LPS, TNF, or IL-1, in addition to afferent vagus nerve signals (Broadwell and Sofroniew, 1993; Rogers *et al.*, 1996).

THE CHOLINERGIC ANTI-INFLAMMATORY PATHWAY

Recent evidence indicates that the parasympathetic nervous system directly controls immune cell activation through a descending neural substrate. The cholinergic anti-inflammatory pathway is comprised of efferent activity in the vagus nerve, which releases acetylcholine in target tissues. Macrophages and other cells of the reticuloendothelial system express nicotinic-type acetylcholine receptors that specifically modulate the activation state of the innate immune system (Borovikova *et al.*, 2000a,b). Artificial stimulation of the vagus nerve by electrical or pharmacological means recapitulates the activity of the cholinergic anti-inflammatory pathway, and prevents lethal shock in animals challenged with endotoxin. Because electrical vagus nerve stimulators are clinically approved devices for the treatment of epilepsy and depression, the activity of the cholinergic anti-inflammatory pathway may be harnessed for the treatment of cytokine-mediated diseases.

We first described CNI-1493, a tetravalent guanylhydrazone, as a research molecule to study arginine metabolism in animal models of cachexia, but we subsequently discovered that this 891 kDa molecule is a potent macrophage deactivating agent (Bianchi *et al.*, 1995, 1996). We identified the molecular basis for the cytokine-inhibiting mechanism of drug action in macrophage cultures: it inhibits phosphorylation of p38 MAP kinase, an enzyme that occupies a critical role in regulating the synthesis of TNF and other cytokines (Cohen *et al.*, 1996). CNI-1493 confers protection in lethal endotoxemia, and against the lethality of sepsis in a standardized model of cecal ligation and puncture (D'Souza *et al.*, 1999). Following a Phase I trial (Atkins *et al.*, 2001) and a small-scale Phase II clinical trial (Hommes *et al.*, 2002), a large scale, prospective, randomized clinical trial of CNI-1493 for Crohn's disease is underway. In the early trials, CNI-1493 induced significant clinical and endoscopic improvement in 80% of the patients, and inhibited the expression of TNF in the colonic mucosa (Hommes *et al.*, 2002).

While continuing to use CNI-1493 in ongoing animal studies of systemic inflammation, we made a startling and significant discovery: the fundamental physiological mechanism of action through which CNI-1493 mediates anti-inflammatory effects *in vivo* is to stimulate efferent activity in the vagus nerve. As reviewed below, it is now clear that CNI-1493 functions *in vivo* as a pharmacological vagus nerve stimulator, and that enhanced vagus nerve activity is required for the macrophage deactivating effects of CNI-1493 *in vivo* (Borovikova *et al.*, 2000a,b; Bernik *et al.*, 2002). Vagotomy inhibits the anti-inflammatory activity of CNI-1493, whether the agent is administered via either intravenous (i.v.) or intracerebroventricular (i.c.v.) routes (Bernik *et al.*, 2002). These observations revealed a previously unrecognized mechanism to rapidly deactivate macrophages and influence the development of systemic inflammation, termed the 'cholinergic anti-inflammatory pathway' because acetylcholine is the principle neurotransmitter of the vagus nerve, and this cranial nerve is widely distributed throughout the reticuloendothelial system (Borovikova *et al.*, 2000b).

BRAIN TO IMMUNE COMMUNICATION: EFFERENT NEURAL SIGNALS

The first direct experimental insight into the role of the CNS in mediating the anti-inflammatory effects of CNI-1493 came from analysis of endotoxin-induced shock in animals receiving CNI-1493 via the i.v. or i.c.v. routes. TNF and other pro-inflammatory cytokines play pathogenic roles not only in systemic diseases, but also in local inflammatory conditions such as arthritis and stroke (Meistrell *et al.*, 1997). While examining the potential for CNI-1493 as an experimental therapeutic for stroke, animals subjected to middle cerebral artery occlusion (MCAO) received CNI-1493 through either i.v. or i.c.v. routes. The controls for this study included animals that received an i.v. dose of endotoxin, as well as an i.v. or an i.c.v. dose of CNI-1493 (Borovikova *et al.*, 2000b; Bernik *et al.*, 2002). Vehicle-treated endotoxemic rats developed significant hypotension and increased serum TNF levels within 1 h after exposure to a lethal dose of LPS. Pretreatment with intravenous CNI-1493 significantly and dose-dependently inhibited serum TNF release and prevented the development of LPS-induced hypotension. The lowest i.v. CNI-1493 dose tested (100 mg/kg) failed to prevent TNF release or hypotension. Intracerebroventricular administration of a 100-fold dilution of this ineffective i.v. dose significantly attenuated serum TNF release, and protected against the development of hypotension. Surprisingly, much lower i.c.v. doses of CNI-1493 conferred significant protection against the development of endotoxin-induced shock and inhibited serum TNF. Comparison of dose–response curves constructed from data obtained after CNI-1493 was given via either i.v. or i.c.v. routes revealed that the latter route was at least 100 000-fold more effective in preventing TNF release and shock, suggesting that the CNS participates in the systemic anti-inflammatory action of CNI-1493 during endotoxemia (Bernik *et al.*, 2002).

Intravenous CNI-1493 accumulates in brain

To determine whether CNI-1493 can cross into the CNS after intravenous dosing, and to determine the tissue distribution, we measured radioactivity in rat tissues 1 h after i.v. administration of ^{14}C-labelled CNI-1493 (Bernik *et al.*, 2002). Following transcardiac perfusion to minimize non-specific binding, the highest levels of radioactivity were observed in the spleen, liver, kidney, lungs and gastrointestinal tract. Lesser accumulation was observed in brain, skin, muscle and heart. Although peripheral organs did not accumulate significant radioactivity after i.c.v. injection of ^{14}C-CNI-1493, significant radioactive uptake persisted in brain. The amount of radioactivity detected in brain after i.c.v. administration of 100 ng/kg of ^{14}C-CNI-1493 was comparable to that observed after i.v. administration of 1 mg/kg of ^{14}C-CNI-1493. Each of these dosing regimens suppressed endotoxin-induced TNF and shock, so that brain drug levels achieved after i.v. dosing with a higher dose were comparable to those achieved with direct i.c.v. application of a lesser, but pharmacologically active, dose of CNI-1493. The i.c.v. doses that effectively inhibited systemic inflammatory responses were too low to reach detectable levels in peripheral tissues, and when considered with the results of i.v. versus i.c.v. administration routes for peripheral inflammation, it appeared that the CNS might participate in mediating the systemic anti-inflammatory mechanisms of CNI-1493 (Bernik et al., 2002).

The protective effects of CNI-1493 are abolished by surgical or chemical vagotomy

One potential explanation for the anti-inflammatory effects of centrally-administered CNI-1493 is up-regulation of ACTH, α-MSH, or circulating glucocorticoids. However, careful analysis could find no evidence of this response (Borovikova et al., 2000a), raising the possibility that an alternative mechanism is involved. Because the brain communicates with the body primarily via nerves, we considered the possibility that CNI-1493 activated vagus neural signaling, because it innervates the organs of the reticuloendothelial system. To determine whether an intact vagus nerve is required for inhibition of TNF and protection from endotoxin-induced shock by i.c.v. or i.v. CNI-1493, animals were subjected to either surgical or chemical vagotomy (Borovikova et al., 2000b; Bernik et al., 2002). Surgical vagotomy eliminated the protective effects of i.c.v. CNI-1493 against LPS-induced TNF release and hypotension. Surgical vagotomy also eliminated the protective effect of i.v. CNI-1493 against endotoxin-induced shock, indicating that the protective effects of CNI-1493, whether administered into the brain or the peripheral circulation, require an intact vagus nerve (Borovikova et al., 2000b; Bernik et al., 2002). These results indicate that cholinergic vagus neural signals mediate CNI-1493-induced systemic protection against TNF and hypotension.

CNI-1493 activates neural signalling in the intact vagus nerve

These observations suggested the testable hypothesis that CNI-1493 activated neural signals in the vagus nerve. To identify the role of CNI-1493 in efferent vagus nerve signalling, CNI-1493 was administered intravenously (i.v., 5 mg/kg) in anaesthetized rats, and electrical activity in the vagus nerve recorded (Borovikova et al., 2000a). Recording the vagus nerve activity revealed an increase in discharge rate starting at 3–4 min after CNI-1493 administration (i.v., 5 mg/kg) and lasting for 10–14 min. These results identified a previously unrecognized role of CNI-1493 in functioning as a pharmacological vagus nerve stimulator, to increase efferent vagus nerve activity, which in turn mediated the inhibition of systemic TNF release and shock (Borovikova *et al.*, 2000a).

Vagus nerve stimulation protects against endotoxin-induced hypotension

Vagus nerve stimulators have been used clinically for the treatment of seizure disorders and depression, but the effects of these devices applied to intact vagus nerves in the setting of systemic inflammatory responses are unknown (DeGiorgio *et al.*, 2000). The original observations using a 'pharmacological vagus nerve stimulator' (CNI-1493) predicted that electrical stimulation of the vagus nerve might recapitulate the anti-inflammatory action of CNI-1493. We electrically stimulated intact vagus nerves of endotoxemic rats and observed that this procedure significantly prevented the development of hypotension (Bernik *et al.*, 2002). Stimulation with either 1 or 5 V impulses for 2 ms intervals (5 Hz) prevented the development of significant hypotension. Heart rate did not increase significantly in non-stimulated endotoxemic animals, despite the development of hypotension. Stimulation of the vagus nerve, however, was associated with a significant and voltage stimulus-dependent increase in heart rate. To assess whether the right and left vagus nerves contribute a differential protective effect, we measured blood pressure and heart rate in animals after separately stimulating either nerve. We observed no significant difference between right and left cervical vagus nerve stimulation (Bernik *et al.*, 2002).

Cholinergic agonists

These studies led to our discovery that acetylcholine deactivates macrophages via specific, α-bungarotoxin, nicotinic receptors expressed on macrophages, but not monocytes (Borovikova *et al.*, 2000b). Human macrophage cultures in the presence of acetylcholine, nicotine, or muscarine are deactivated, and fail to release TNF in response to endotoxin (Borovikova *et al.*, 2000b). The molecular basis for the inhibition of TNF synthesis by cholinergic agonists is post-transcriptional, because cholinergic agonists do not inhibit endotoxin-induced upregulation of mRNA levels of TNF and other cytokines, but do inhibit the expression of TNF protein on the cell surface, and in the media (Borovikova *et al.*, 2000b). We have recently determined that the macrophage acetylcholine receptor includes the α7 subunit. α7-deficient mice are more sensitive to endotoxin, and macrophages isolated from these mice produce significantly more TNF, IL-1β, and IL-6 than wild-type. Moreover, electrical vagus nerve stimulation does not protect α7-deficient mice against endotoxin-induced hypotension or TNF release (Tracey, 2002).

Vagus nerve stimulation inhibits TNF synthesis in liver and heart

Filkins and colleagues (Kumins *et al.*, 1996) suggested that most of the circulating TNF that appears in the serum is released from the liver during lethal endotoxemia, and that most of the blood-borne TNF is not the product of circulating monocytes. In agreement with this theory, when we stimulated the vagus nerve of endotoxemic animals, we observed that TNF synthesis in liver was inhibited. The inhibition of liver TNF synthesis by vagus nerve stimulation agreed closely with the suppression of serum TNF levels. Vagus nerve stimulation during lethal endotoxemia also significantly attenuated cardiac TNF levels (Bernik *et al.*, 2002). Thus, increased neural signalling in the vagus nerve, induced by application of either an electrical device or CNI-1493, can inhibit TNF synthesis in organs, and attenuate the development of endotoxin-induced systemic inflammation and shock.

PERSPECTIVES

Our recent findings that the *in vivo* mechanism of action for CNI-1493 is dependent upon the cholinergic anti-inflammatory pathway now opens a new avenue of research into the impact of this pathway on regulation of systemic and organ-specific macrophage activation *in vivo*. It is clear from our results that CNI-1493 can inhibit the systemic pro-inflammatory response to endotoxin in established animal models of endotoxemia and sepsis, and that CNI-1493 activates the cholinergic anti-inflammatory pathway. It is not clear however, whether the cholinergic anti-inflammatory pathway can be therapeutically manipulated in the altered cytokine milieu of post-traumatic sepsis. The identification of the cholinergic anti-inflammatory pathway has opened a nascent field, and there are many experimental questions that could reasonably be pursued now. These preliminary studies suggest multiple, and perhaps divergent experimental paths including the identification of the essential macrophage receptor, and the development of anti-inflammatory therapies.

REFERENCES

Angus DC, Linde-Zwirble WT, Lidicker J, Clermont G, Carcillo J, Pinsky MR. Epidemiology of severe sepsis in the United States: analysis of incidence, outcome, and associated costs of care. *Crit Care Med* 2001; 29(7):1303–10.

Atkins MB, Redman B, Mier J, Gollob J, Weber J, Sosman J, MacPherson BL, Plasse T. A phase I study of CNI-1493, an inhibitor of cytokine release, in combination with high-dose interleukin-2 in patients with renal cancer and melanoma. *Clin Cancer Res* 2001; 7(3):486–92.

Ayala A, Chaudry IH. Immune dysfunction in murine polymicrobial sepsis: mediators, macrophages, lymphocytes, and apoptosis. *Shock* 1996; 6(Suppl 1):S27–S38.

Banks WA, Kastin AJ. Blood to brain transport of interleukin links the immune and central nervous systems. *Life Sci* 1991; 48(25):PL117–21.

Bernhagen J, Calandra T, Mitchell RA, Martin SB, Tracey KJ, Voelter W, Manogue KR, Cerami A, Bucala R. MIF is a pituitary-derived cytokine that potentiates lethal endotoxaemia. *Nature* 1993; 365(6448):756–9.

Bernik TR, Friedman SG, Ochani M, DiRaimo R, Ulloa L, Yang H, Sudan S, Czura CJ, Ivanova SM, Tracey KJ. Pharmacological stimulation of the cholinergic antiinflammatory pathway. *J Exp Med* 2002; 195(6):781–8.

Bianchi M, Bloom O, Raabe T, Cohen PS, Chesney J, Sherry B, Schmidtmayerova H, Calandra T, Zhang X, Bukrinsky M, Ulrich P, Cerami A, Tracey KJ. Suppression of proinflammatory cytokines in monocytes by a tetravalent guanylhydrazone. *J Exp Med* 1996; 183(3):927–36.

Bianchi M, Ulrich P, Bloom O, Meistrell M III, Zimmerman GA, Schmidtmayerova H, Bukrinsky M, Donnelley T, Bucala R, Sherry B *et al*. An inhibitor of macrophage arginine transport and nitric oxide production (CNI-1493) prevents acute inflammation and endotoxin lethality. *Mol Med* 1995; 1(3):254–66.

Blalock JE. Harnessing a neural-immune circuit to control inflammation and shock. *J Exp Med* 2002; 195(6):F25–8.

Block MI, Berg M, McNamara MJ, Norton JA, Fraker DL, Alexander HR. Passive immunization of mice against D factor blocks lethality and cytokine release during endotoxemia. *J Exp Med* 1993; 178(3):1085–90.

Bloom O, Wang H, Ivanova S, Vishnubhakat JM, Ombrellino M, Tracey KJ. Hypophysectomy, high tumor necrosis factor levels, and hemoglobinemia in lethal endotoxemic shock. *Shock* 1998; 10(6):395–400.

Bogdan C, Vodovotz Y, Nathan C. Macrophage deactivation by interleukin 10. *J Exp Med* 1991; 174(6):1549–55.

Borovikova LV, Ivanova S, Nardi D, Zhang M, Yang H, Ombrellino M, Tracey KJ. Role of vagus nerve signaling in CNI-1493-mediated suppression of acute inflammation. *Auton Neurosci* 2000a; 85(1–3):141–7.

Borovikova LV, Ivanova S, Zhang M, Yang H, Botchkina GI, Watkins LR, Wang H, Abumrad N, Eaton JW, Tracey KJ. Vagus nerve stimulation attenuates the systemic inflammatory response to endotoxin. *Nature* 2000b; 405(6785):458–62.

Bozza M, Satoskar AR, Lin G, Lu B, Humbles AA, Gerard C, David JR. Targeted disruption of migration inhibitory factor gene reveals its critical role in sepsis. *J Exp Med* 1999; 189(2):341–6.

Breder CD, Smith WL, Raz A, Masferrer J, Seibert K, Needleman P, Saper CB. Distribution and characterization of cyclooxygenase immunoreactivity in the ovine brain. *J Comp Neurol* 1992; 322(3):409–38.

Broadwell RD, Sofroniew MV. Serum proteins bypass the blood–brain fluid barriers for extracellular entry to the central nervous system. *Exp Neurol* 1993; 120(2):245–63.

Calandra T, Echtenacher B, Roy DL, Pugin J, Metz CN, Hultner L, Heumann D, Mannel D, Bucala R, Glauser MP. Protection from septic shock by neutralization of macrophage migration inhibitory factor. *Nat Med* 2000; 6(2):164–70.

Cao C, Matsumura K, Yamagata K, Watanabe Y. Endothelial cells of the rat brain vasculature express cyclooxygenase-2 mRNA in response to systemic interleukin-1 beta: a possible site of prostaglandin synthesis responsible for fever. *Brain Res* 1996; 733(2):263–72.

Cohen PS, Nakshatri H, Dennis J, Caragine T, Bianchi M, Cerami A, Tracey KJ. CNI-1493 inhibits monocyte/macrophage tumor necrosis factor by suppression of translation efficiency. *Proc Natl Acad Sci USA* 1996; 93(9):3967–71.

Cunningham ET Jr, Wada E, Carter DB, Tracey DE, Battey JF, De Souza EB. In situ histochemical localization of type I interleukin-1 receptor messenger RNA in the central nervous system, pituitary, and adrenal gland of the mouse. *J Neurosci* 1992; 12(3):1101–14.

DeGiorgio CM, Schachter SC, Handforth A, Salinsky M, Thompson J, Uthman B, Reed R, Collins S, Tecoma E, Morris GL, Vaughn B, Naritoku DK, Henry T, Labar D, Gilmartin R, Labiner D, Osorio I, Ristanovic R, Jones J, Murphy J, Ney G, Wheless J, Lewis P, Heck C. Prospective long-term study of vagus nerve stimulation for the treatment of refractory seizures. *Epilepsia* 2000; 41(9):1195–200.

Dinarello CA. The role of interleukin-1 in host responses to infectious diseases. *Infect Agents Dis* 1992; 1:227–36.

Dinarello CA, Thompson RC. Blocking IL-1: interleukin-1 receptor antagonist in vivo and in vitro. *Immunol Today* 1991; 12:404–10.

Doherty GM, Lange JR, Langstein HN, Alexander HR, Buresh CM, Norton JA. Evidence for IFN-gamma as a mediator of the lethality of endotoxin and tumor necrosis factor-alpha. *J Immunol* 1992; 149(5):1666–70.

D'Souza MJ, Oettinger CW, Milton GV, Tracey KJ. Prevention of lethality and suppression of proinflammatory cytokines in experimental septic shock by microencapsulated CNI-1493. *J Interferon Cytokine Res* 1999; 19(10):1125–33.

Ericsson A, Arias C, Sawchenko PE. Evidence for an intramedullary prostaglandin-dependent mechanism in the activation of stress-related neuroendocrine circuitry by intravenous interleukin-1. *J Neurosci* 1997; 17(18):7166–79.

Ericsson A, Liu C, Hart RP, Sawchenko PE. Type 1 interleukin-1 receptor in the rat brain: distribution, regulation, and relationship to sites of IL-1-induced cellular activation. *J Comp Neurol* 1995; 361(4):681–98.

Fleshner M, Silbert L, Deak T, Goehler LE, Martin D, Watkins LR, Maier SF. TNF-alpha-induced corticosterone elevation but not serum protein or corticosteroid binding globulin reduction is vagally mediated. *Brain Res Bull* 1997; 44(6):701–6.

Gaykema RP, Goehler LE, Hansen MK, Maier SF, Watkins LR. Subdiaphragmatic vagotomy blocks interleukin-1beta-induced fever but does not reduce IL-1beta levels in the circulation. *Auton Neurosci* 2000; 85(1–3):72–7.

Gross PM, Weindl A. Peering through the windows of the brain. *J Cereb Blood Flow Metab* 1987; 7(6):663–72.

Gutierrez EG, Banks WA, Kastin AJ. Murine tumor necrosis factor alpha is transported from blood to brain in the mouse. *J Neuroimmunol* 1993; 47(2):169–76.

Gutierrez EG, Banks WA, Kastin AJ. Blood-borne interleukin-1 receptor antagonist crosses the blood-brain barrier. *J Neuroimmunol* 1994; 55(2):153–60.

Hauser CJ, Lagoo S, Lagoo A, Hale E, Hardy KJ, Barber WH, Bass JD, Poole GV. Tumor necrosis factor alpha gene expression in human peritoneal macrophages is suppressed by extra-abdominal trauma. *Arch Surg* 1995; 130(11):1186–91.

Heinzel FP. The role of IFN-γ in the pathology of experimental endotoxemia. *J Immunol* 1990; 145(9): 2920–4.

Hommes D, van den Blink B, Plasse T, Bartelsman J, Xu C, Macpherson B, Tytgat G, Peppelenbosch M, Van Deventer S. Inhibition of stress-activated MAP kinases induces clinical improvement in moderate to severe Crohn's disease. *Gastroenterology* 2002; 122(1):7–14.

Johnson AK, Gross PM. Sensory circumventricular organs and brain homeostatic pathways. *FASEB J* 1993; 7(8):678–86.

Knudsen PJ, Dinarello CA, Strom TB. Prostaglandins posttranscriptionally inhibit monocyte expression of inter-leukin 1 activity by increasing intracellular cyclic adenosine monophosphate. *J Immunol* 1986; 137(10):3189–94.

Kumins NH, Hunt J, Gamelli RL, Filkins JP. Partial hepatectomy reduces the endotoxin-induced peak circulating level of tumor necrosis factor in rats. *Shock* 1996; 5(5):385–8.

Lacroix S, Rivest S. Effect of acute systemic inflammatory response and cytokines on the transcription of the genes encoding cyclooxygenase enzymes (COX-1 and COX-2) in the rat brain. *J Neurochem* 1998; 70(2):452–66.

Laflamme N, Rivest S. Toll-like receptor 4: the missing link of the cerebral innate immune response triggered by circulating gram-negative bacterial cell wall components. *FASEB J* 2001; 15(1):155–63.

Lee HY, Whiteside MB, Herkenham M. Area postrema removal abolishes stimulatory effects of intravenous inter-leukin-1beta on hypothalamic–pituitary–adrenal axis activity and c-fos mRNA in the hypothalamic paraven-tricular nucleus. *Brain Res Bull* 1998; 46(6):495–503.

Meistrell ME III, Botchkina GI, Wang H, Di Santo E, Cockroft KM, Bloom O, Vishnubhakat JM, Ghezzi P, Tracey KJ. Tumor necrosis factor is a brain damaging cytokine in cerebral ischemia. *Shock* 1997; 8(5):341–8.

Miller-Graziano CL, Szabo G, Griffey K, Mehta B, Kodys K, Catalano D. Role of elevated monocyte transform-ing growth factor beta (TGF beta) production in posttrauma immunosuppression. *J Clin Immunol* 1991; 11(2):95–102.

Oswald IP, Wynn TA, Sher A, James SL. Interleukin 10 inhibits macrophage microbicidal activity by blocking the endogenous production of tumor necrosis factor alpha required as a costimulatory factor for interferon gamma-induced activation. *Proc Natl Acad Sci USA* 1992; 89(18):8676–80.

Pan W, Banks WA, Kastin AJ. Blood–brain barrier permeability to ebiratide and TNF in acute spinal cord injury. *Exp Neurol* 1997; 146(2):367–73.

Plotkin SR, Banks WA, Kastin AJ. Comparison of saturable transport and extracellular pathways in the passage of interleukin-1 alpha across the blood-brain barrier. *J Neuroimmunol* 1996; 67(1):41–7.

Remick D, Manohar P, Bolgos G, Rodriguez J, Moldawer L, Wollenberg G. Blockade of tumor necrosis factor reduces lipopolysaccharide lethality, but not the lethality of cecal ligation and puncture. *Shock* 1995; 4(2):89–95.

Rogers RC, Hermann GE, Travagli RA. Brainstem pathways responsible for oesophageal control of gastric motil-ity and tone in the rat. *J Physiol* 1999; 514 (Pt 2):369–83.

Rogers RC, McTigue DM, Hermann GE. Vagal control of digestion: modulation by central neural and peripheral endocrine factors. *Neurosci Biobehav Rev* 1996; 20(1):57–66.

Smith BN, Dou P, Barber WD, Dudek FE. Vagally evoked synaptic currents in the immature rat nucleus tractus solitarii in an intact in vitro preparation. *J Physiol* 1998; 512(Pt 1):149–62.

Tracey KJ, Abraham E. From mouse to man: or what have we learned about cytokine-based anti-inflammatory therapies? *Shock* 1999; 11(3):224–5.

Tracey, KJ. The inflammatory reflex. *Nature* 2002; December (in press).

Tracey KJ, Beutler B, Lowry SF, Merryweather J, Wolpe S, Milsark IW, Hariri RJ, Fahey TJ III, Zentella A, Albert JD. Shock and tissue injury induced by recombinant human cachectin. *Science* 1986; 234(4775):470–4.

Tracey KJ, Cerami A. Tumor necrosis factor: a pleiotropic cytokine and therapeutic target. *Annu Rev Med* 1994; 45:491–503.

Tracey KJ, Fong Y, Hesse DG, Manogue KR, Lee AT, Kuo GC, Lowry SF, Cerami A. Anti-cachectin/TNF mono-clonal antibodies prevent septic shock during lethal bacteraemia. *Nature* 1987; 330:662–4.

Tsunawaki S, Sporn M, Ding A, Nathan C. Deactivation of macrophages by transforming growth factor-beta. *Nature* 1988; 334(6179):260–2.

Wang H, Bloom O, Zhang M, Vishnubhakat JM, Ombrellino M, Che J, Frazier A, Yang H, Ivanova S, Borovikova L, Manogue KR, Faist E, Abraham E, Andersson J, Andersson U, Molina PE, Abumrad NN, Sama A, Tracey KJ. HMG-1 as a late mediator of endotoxin lethality in mice. *Science* 1999; 285(5425):248–51.

Wang H, Yang H, Czura CJ, Sama AE, Tracey KJ. HMGB1 as a late mediator of lethal systemic inflammation. *Am J Respir Crit Care Med* 2001; 164(10 Pt 1):1768–73.

Wang H, Zhang M, Bianchi M, Sherry B, Sama A, Tracey KJ. Fetuin (alpha2-HS-glycoprotein) opsonizes cationic macrophagede activating molecules. *Proc Natl Acad Sci USA* 1998; 95(24):14429–34.

Waring PM, Waring LJ, Metcalf D. Circulating leukemia inhibitory factor levels correlate with disease severity in meningococcemia. *J Infect Dis* 1994; 170(5):1224–8.

Wong ML, Licinio J. Localization of interleukin 1 type I receptor mRNA in rat brain. *Neuroimmunomodulation* 1994; 1(2):110–15.

Yabuuchi K, Minami M, Katsumata S, Satoh M. Localization of type I interleukin-1 receptor mRNA in the rat brain. *Brain Res Mol Brain Res* 1994; 27(1):27–36.

Zhang M, Borovikova LV, Wang H, Metz C, Tracey KJ. Spermine inhibition of monocyte activation and inflammation. *Mol Med* 1999; 5(9):595–605.

Zhang M, Caragine T, Wang H, Cohen PS, Botchkina G, Soda K, Bianchi M, Ulrich P, Cerami A, Sherry B, Tracey KJ. Spermine inhibits proinflammatory cytokine synthesis in human mononuclear cells: a counter-regulatory mechanism that restrains the immune response. *J Exp Med* 1997; 185(10):1759–68.

Zhang M, Wang H, Tracey KJ. Regulation of macrophage activation and inflammation by spermine: a new chapter in an old story. *Crit Care Med* 2000; 28(4 Suppl):N60–6.

5 Developmental Regulation and Functional Integration by the Vasoactive Intestinal Peptide (VIP) Neuroimmune Mediator

Glenn Dorsam, Robert C. Chan and Edward J. Goetzl

Departments of Medicine and Microbiology–Immunology, University of California Medical Center, San Francisco, CA 94143-0711, USA

Vasoactive intestinal peptide (VIP) and the related pituitary adenylyl cyclase-activating peptide (PACAP) are derived from and act on many different types of cells (Dorsam *et al.*, 2000). Our present understanding of the breadth of physiological and pathological contributions of these mediators has superceded the circumstantial identification of their original sources and biological effect, which led to assignment of the name of each peptide. VIP is generated, stored and released by numerous cellular constituents of the nervous system and by many different types of leukocytes, including T cells, mast cells, basophils, macrophages and eosinophils (Aliakbari *et al.*, 1987; Hernanz *et al.*, 1989; Goetzl *et al.*, 1998; Schmidt-Choudhury *et al.*, 1999). Neural, endocrine and, most recently, immune effects of VIP have been recognized *in vitro* and *in vivo* in many mammalian species (Figure 5.1). The diverse activities of VIP released from cholinergic, adrenergic and other nerves are observed both during development and as trophic influences in adults. Endocrine activities of VIP are vital in hypothalamic-pituitary and pancreatic functions, and may extend to reproductive functions as well (Goetzl and Sreedharan, 1992). In immunity, the results of *in vitro* studies initially directed attention to subsets of B cells and T cells, which express G-protein-coupled receptors for VIP. The first detectable consequences of VIP actions were on T cells and encompassed enhanced adhesion and migration, decreased activation-induced apoptosis, distinctively altered generation of diverse cytokines and modified regulation of T cell-dependent B cell production of several immunoglobulins (Goetzl *et al.*, 1995). The Th2 subset of murine CD4$^+$ T cells is an important principal source and target of VIP. More recently, data to be presented here show that one VIP receptor transduces a dominant immune signal capable of controlling the Th1/Th2 balance and consequently the immune phenotype of mice.

KEY WORDS: hypothalamus; nerves; neurotrophic; parasympathetic; sympathetic; transcriptional regulation; T lymphocytes.

VIP AND VIP RECEPTORS

VIP

Vasoactive intestinal peptide (VIP) is a 28 amino acid (AA) factor encoded by a single gene in the animal kingdom. The VIP gene consists of seven exons and six introns, which generate a single preproVIP molecule of 170 AA (Nussdorfer and Malendowicz, 1998).

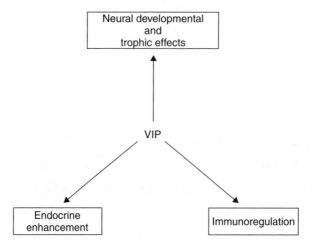

Figure 5.1   Multi-system contributions of VIP.

This protein is post-translationally processed by proteases yielding four peptides, of which three have known biological activity. Exon 5 contributes VIP and exon 4 encodes peptide histidine isoleucine (PHI) in humans and peptide histidine methionine (PHM) in rodents. In certain cells, the proteolytic cleavage of the amino end of PHI does not occur and results in a third biologically active peptide called peptide with N-terminal histidine and C-terminal valine (PHV). PHI/PHM and PHV share some biological activities with VIP. Pituitary adenylate cyclase activating polypeptide (PACAP) has a genomic organization similar to VIP with five exons and four introns generating a prepro-polypeptide that yields at least three biologically active peptides. They are PACAP 38, PACAP 27 and PACAP related peptide (PRP), all of which are processed post-translationally. The N-terminal portion of PACAP shares 68% AA sequence identity with VIP and also is C-terminally α-amidated (Nussdorfer and Malendowicz, 1998).

VIP was originally isolated from swine gastrointestinal tissue and shown to be vasoactive which led to its historical name (Said and Mutt, 1972). VIP is widespread in many organ systems and exhibits diverse biological activities including neurotrophic effects, augmented neurotransmission, growth stimulation, endocrine enhancement and immunoregulation. Thus, its limited name is more historical in nature and is not descriptive to its true biological roles. Human VIP is identical in AA sequence with that of rat and pig, and at least 82% homologous with that of mouse, cow, guinea, alligator, chicken, frog, trout, dogfish, bowfin, cod and goldfish (Nussdorfer and Malendowicz, 1998). The biological importance of VIP is suggested by its conservation throughout the vertebrate phylum. Furthermore, the VIP gene like that of other neuropeptides is reminiscent of prokaryotic poly-cistronic genes in encoding at least two specific proteins, each of which is contributed by a unique exon. In addition, PACAP, which is 68% homologous to VIP and binds some of the same receptors, is 96% identical to VIP found in the tunicate, *Chelyosoma productum*. Over 700 million years of evolution separates humans from the tunicate and further suggests an ancient origin for VIP and PACAP (Nussdorfer and Malendowicz, 1998).

VIP is widely expressed in nerves of the CNS and peripheral nervous system (PNS). VIPergic expression is plentiful in the vagus and splanchic nerves as well as the hypothalamic-pituitary axis (HP). In addition, the HP portal blood system supplying this region of the CNS has a higher VIP concentration than circulating levels elsewhere in the body (Harmar *et al.*, 1998; Nussdorfer and Malendowicz, 1998). This further supports the importance of VIP to functions of the HP. VIPergic nerves have been demonstrated to innervate many organs and tissues in several mammalian species, including salivary glands, female and male genital tracts, mucosa-associated lymphoid tissue (MALT) of the gastrointestinal and pulmonary systems, thymus, spleen, tonsils, appendix, peripheral and mesenteric lymph nodes and Peyer's patches (Dorsam *et al.*, 2000).

This wide distribution of VIPergic nerve delivery is not categorized easily into the three major subdivisions of the autonomic nervous system; sympathetic (noradrenergic), parasympathetic (cholinergic) or mesenteric (Barnes *et al.*, 1991; Jartti, 2001). The neural origins of VIP in the thymus and spleen, two organs critical to the body's ability to mount an immune response, are not known. Indeed, transection of sympathetic nerves innervating the thymus does not decrease the expression of immunoreactive VIP, suggesting a non-sympathetic, non-noradrenergic origin. However, in contrast, VIP is co-expressed with markers for sympathetic neurons including tyrosine hydroxylase and dopamine-β-hydroxylase enzymes (Barnes *et al.*, 1991). Acetylcholinesterase, a marker for parasympathetic neurons, co-localizes with VIP (Francis *et al.*, 1997). Thus, VIP expression is detected in both sympathetic and parasympathetic neurons of the autonomic nervous system.

Two new neuronal markers, vesicular monoamine transporter type 2 (VMAT2) and vesicular acetylcholine transporter (VAChT) identifying sympathetic and parasympathetic nerves, respectively, were used in a study to identify VIP's neuronal origin (Schutz *et al.*, 1998). During the development of the cholinergic and sympathetic nervous systems, both VIP and neuropeptide Y (NPY) expression overlap at first and then segregate as the rat develops into adulthood. Immunohistochemical analysis of the rat primary sympathetic chain identified early co-staining of VIP/VAChT and NPY/VMAT2. The NPY/VMAT2 co-expression was present in most cells while the VIP/VAChT co-expression was isolated to the paravertebral thoracolumbar ganglia region at E14.5. After birth, this co-expression of VIP/VAChT and NPY/VMAT2 diverged and became increasing segregated. NPY/VMAT2 maintained a ubiquitous expression whereas VIP/VAChT was further restricted to the paravertebral thoracolumbar ganglion region. Furthermore, in the sweat glands of ageing rats, VIP/VAChT co-appeared in the sudomotor fibres and became increasingly denser as the rat matured. NPY/VMAT2, in contrast, never were identified in the sudomotor fibres of the sweat gland but instead predominated in varicose fibres of the sweat glands running parallel to the vasculature. Thus in the rat VIP appears to be more prominent in cholinergic parasympathetic than noradrenergic sympathetic nerves and is consistent with the observation of low levels of VIP in sympathetic nerves of the superior cervical ganglion of adult rodents.

The expression profile of VIP in the nervous, endocrine and immune systems is directly attributed to the characteristics of a complex promoter. At least five regions of the VIP promoter monitor and reflect different intracellular signals, which describe an architecture capable of both maintaining constitutive expression and responding to diverse cellular challenges (Hahm and Eiden, 1998). These five promoter regions are termed A–E, starting

5.2 kb from the transcriptional start site (TSS), and have numerous transcriptional binding elements. Both ubiquitous and cell-specific trans-acting factors participate in transcriptional regulation of VIP. Evidence for the necessity of all five promoter regions was suggested by the altered level and distribution of expression of VIP encoded by a modified gene consisting of only 2 kb of the 5′-promoter region. Numerous signaling pathways converge at the VIP promoter, such as cAMP/PKA, PKC, RAS/MEKK/CREB kinase, JAK kinases and voltage gated Ca^{2+}/CAM kinase, which explains the responsiveness of VIP expression to many different cues. Much of the data from *in vitro* experiments in rat chromaffin cells and the pheochromocytoma cell line PC-12 have further defined the 5′-promoter regions of the VIP gene and the transcription factors known to bind to these regions. Domain A, which binds OCT1/OCT2 nuclear factors, is 435 bp in length, lies more than 5 kb from the TSS and is absolutely necessary for cell specific expression; it is referred to as tissue specific element (TSE). Region B contains E-box binding elements for tissue-specific basic helix–loop–helix (bHLH) and myocyte enhancer factor 2 (MEF-2) which are neuronal/myocyte- and neurogenic-specific transcription factors (Brand, 1997). Hence, both the TSE and region B confer cell-specific regulation of VIP. Region C constitutes the most complex region in that it recognizes multiple binding factors controlled by cytokine signalling through ciliary neurotrophic factor (CNTF), leukemia inhibitory factor (LIF), interleukin (IL)-6 and signal transducer and activator of transcription (STAT) 1 and 3 proteins via the Janus kinases (JAK). For this reason, it is referred to as the cytokine responsive element (CyRE) and also encompasses part of the upstream portion of region D, which further supports the possibility of multiple interactions between various promoter regions. PMA inducibility is thought to be mediated through this domain. Domain D binds FOS and JUN factors that make up a non-canonical AP-1 complex as determined by electrophoretic mobility shift assays. Finally, region E binds CREB and thus confers cAMP-responsiveness.

Because VIP may be expressed in many diverse tissues, it must therefore be tightly regulated to ensure precise quantitative and temporal expression. The VIP promoter interprets intracellular and intercellular physiological and pathophysiological information, including local neural and endocrine signals (Zigmond and Sun, 1997; Nussdorfer and Malendowicz, 1998; Brodski *et al.*, 2000), 'stimulus–secretion–synthesis coupling' (Waschek *et al.*, 1987), which is the feedback loop controlling the synthesis and packaging of protein after vesicle-membrane fusion, and cytokine levels (Hahm and Eiden, 1998). Consequently, the VIP promoter working through at least five distinct functionally active binding regions confers precise integrated physiological gene regulation on VIP.

VPAC-1/VPAC-2/PAC-1 RECEPTORS

Seven transmembrane G-protein-coupled receptors (GPCRs) comprise a large superfamily of plasma membrane receptors throughout eukaryotic organisms. This superfamily has evolved to recognize and respond to alkaloids, biogenic amines, glycoprotein hormones, light, odorants and peptides, such as VIP and PACAP (Lomize *et al.*, 1999; Jabrane-Ferrat *et al.*, 2000; Hamm, 2001). The VIP/PACAP neuropeptide receptors belong to a class IIA subfamily of this superfamily of G-protein coupled receptors along with receptors for glucagon, glucagon like peptide-1, secretin, growth-hormone releasing factor (GRF), parathyroid hormone (PH), calcitonin, mucin-like hormone receptor, leukocyte activation antigen (CD97) and gastric inhibitory polypeptide (Nussdorfer *et al.*, 2000). Of the three defined GPCRs in this subfamily, VPAC-1 and VPAC-2 bind VIP and PACAP, and the third receptor called PAC-1 binds preferentially to PACAP (Harmar *et al.*, 1998).

VPAC-1

VPAC-1 was first cloned by Ishihara *et al.* from rat lung in 1992 (Ishihara *et al.*, 1992). This receptor is expressed nearly ubiquitously in several species studied including humans. The tissue distribution of VPAC-1 includes the cerebral cortex, hippocampus, PNS, liver, lung, heart, kidney, skeletal muscle, smooth muscle, pancreas, spleen, eye, thyroid, thymus, submaxillary gland and gonads (De Souza *et al.*, 1985; Usdin *et al.*, 1994; Goetzl *et al.*, 1998; Harmar *et al.*, 1998; Nussdorfer and Malendowicz, 1998). Developmentally, VPAC-1 is first detected at embryonic day 7.5 (E7.5) and persists throughout gestation concentrated in neuroepithelium, foregut, heart and cells lining the embryonic vasculature by E11, and later in the CNS, spinal cord, intestines, kidney, adrenal gland, epithelial cells and liver (E18) (Gressens *et al.*, 1993; Pei, 1997; Spong *et al.*, 1999b). In adults, VPAC-1 is expressed at highest levels in lung and yet only begins to be expressed in this tissue after birth (Pei and Melmed, 1995). The human VPAC-1 gene is made up of 13 exons and 12 introns spanning 22 kb on chromosome 3q22.33-p21.31 and on chromosomes 9 and 8q32 in mouse and rat which are syntenic with human chromosome 3 (Deng *et al.*, 1994; Sreedharan *et al.*, 1995; Hashimoto *et al.*, 1999). An intriguing observation regarding the chromosomal location of VPAC-1 is that this locus in humans has been identified to contain a tumour suppressor gene (Whang-Peng *et al.*, 1982). Loss of heterozygosity resulted in 100% incidence of small cell lung carcinoma (SCLC) and 90% primary tumours of the lung (Gazdar *et al.*, 1994), making VPAC-1 a possible candidate as a tumor suppressor. Furthermore, VIP has been demonstrated to inhibit SCLC growth (Maruno *et al.*, 1998). The open reading frame for human VPAC-1 transcribes a 1395 bp mRNA message encoding a 457AA integral protein with an apparent molecular weight of 52 kDa (Sreedharan *et al.*, 1993). The primary sequence for human VPAC-1 is 83 and 89% homologous with mouse and rat VPAC-1 and 50 and 49% identical in primary sequence to human VPAC-2 and PAC-1 (Calvo *et al.*, 1996). The signalling pathways activated by VPAC-1 are predominately cAMP/PKA and to a lesser extent PLC/Ca^{2+} (Delgado *et al.*, 1996). Recently, McCulloch *et al.* demonstrated a new signalling pathway through PLD in two different cell types; albeit about one-tenth the magnitude of cAMP (McCulloch *et al.*, 2001). The rank-order of ligand binding affinities for VPAC-1 is VIP/PACAP-38/PACAP-27 (IC_{50}, 1 nM) > PHI/PHV (IC_{50}, 3 nM) $\gg$ GRF (IC_{50}, 80 nM) > secretin (IC_{50}, 300 nM). Two highly selective agonists have been documented. The VIP/GRF chimera and (Arg^{16}) chicken secretin have equivalent binding affinities as VIP (Nussdorfer and Malendowicz, 1998).

The promoter of VPAC-1 has been sequenced by our laboratory and others (Sreedharan *et al.*, 1995; Couvineau *et al.*, 2000). This promoter is similar to a metabolic 'house keeping gene' in that it is very GC-rich and lacks a functional TATA box and thus explains the widespread expression of this receptor (Koller *et al.*, 1991). It differs from a 'house keeping gene', however, in that it possesses a functional GC box and a single transcriptional start site. The minimal promoter for VPAC-1 spans approximately 250 bp from the TSS and contains an essential SP1 binding site for constitutive expression (Sreedharan *et al.*, 1995; Couvineau *et al.*, 2000). Our EMSA data have identified *in vitro* binding of Ikaros to at least one of four perfect consensus sequences found in the VPAC-1 promoter, all within 550 bp of the transcriptional start site (Dorsam and Goetzl, 2001). The physiological role of Ikaros in regulating the expression of VPAC-1 is not known and is the subject of an ongoing study. It has been recently discovered that neuronal Ikaros-like transcription factors recognize and bind to DNA sequences similar to that of Ikaros. Thus, Ikaros and Ikaros-like induced regulation of VPAC-1 may take place in T lymphocytes and neurons (Dobi *et al.*, 1997). This idea is supported by the high frequency of Ikaros core binding tetranucleotide sequences of GGGA throughout the human and rat VPAC-1 promoter and further implicate both the immune and nervous system as important targets for VIP (Dorsam and Goetzl, 2001). Furthermore, we have observed that the histone deacetylase (HDAC) inhibitor trichostatin A significantly upregulates the rate of VPAC-1 transcription as measured by reporter assays and real-time PCR in T lymphocytes (unpublished data). This effect by TSA suggests that an equilibrium of acetylation may regulate VPAC-1 expression. Because Ikaros interacts with known HDAC proteins (Koipally *et al.*, 1999; Koipally and Georgopoulos, 2000), it will be extremely interesting to investigate the extent to which these two mechanisms regulate VPAC-1 in the context of T lymphocyte development, maintenance and activation.

VPAC-2

VPAC-2 was first cloned in 1993 by Lutz *et al.* from rat olfactory bulb (Lutz *et al.*, 1993). The mRNA encodes a 457AA polypeptide with an apparent molecular weight of 52 kDa. A higher molecular weight species of VPAC-2 was observed from Tsup-1 cell line and may be due to a different glycosylation pattern in these cells. The human VPAC-2 gene spans nearly 100 kb on human chromosome 7q36.3 and on the syntenic mouse chromosome 12.F2 (Mackay *et al.*, 1996; Lutz *et al.*, 1999). The primary sequence is 50% identical to both human VPAC-1 and PAC-1 and 87% homologous to the rat and mouse VPAC-2, respectively. The hierarchy of ligand binding affinities is PACAP-38 (IC_{50}, 2 nM) $\geq$ VIP (IC_{50}, 3–4 nM) > PACAP-27 (IC_{50}, 10 nM) $\geq$ PHI/PHV (IC_{50}, 10–30 nM). To date, two selective agonists for VPAC-2 with similar affinities as PACAP and VIP are the cyclic peptides Ro25-1553 and Ro25-1392, respectively. No selective antagonists have been reported. The cAMP/PKA and PLC/Ca^{2+} signal transduction pathways are both evoked by VPAC-2 signalling as well as PLD activation. VPAC-2 activation of PLD as assessed by (^{3}H)-phosphaditylbutanol release may be greater in magnitude than VPAC-1 PLD-induced activation (McCulloch *et al.*, 2001).

The tissue distribution for VPAC-2 is strikingly complementary with that of VPAC-1 (Harmar *et al.*, 1998; Nussdorfer and Malendowicz, 1998). VPAC-2 is expressed in different tissues than VPAC-1 such as olfactory bulb, hypothalamus and suprachiasmeatic nuclei in the brain and various glands and peripheral organs such as pituitary, pineal, pancreatic islets, stomach, colon, bone marrow stromal cells, testes and ovary. The differential expression profile of these two receptors suggests different functions despite their similar VIP binding affinities and intracellular signalling mechanisms. This idea is supported by the marked dichotomy in chemotaxis effects on HUT-78 lymphoma cell lines expressing only VPAC-1, as contrasted with the Tsup-1 lymphoma cell lines expressing VPAC-2 (see text below).

The promoter region of VPAC-2 has not been widely investigated. However, since there appears to be an inverse expression profile between VPAC-1 and VPAC-2 throughout the tissues of the body and within the heterogeneous T lymphocyte cell population, it will be interesting to determine if the Ikaros family of lymphocyte-restricted transcription factors binds to the VPAC-2 promoter and mediates opposing transcriptional regulation. Recently, it was observed that IL-4 negatively regulates the expression of VPAC-2 in IL-4 knockout mice by Metwali et al. (2000) suggesting a potential STAT6 trans-repression mechanism. IL-4 is a potent inducer of the Th2 immune response and a downregulator of Th1 cytokines such as IFN-γ, IL-2 and TNF-α (Agarwal and Rao, 1998). Because VPAC-2 is induced upon CD4$^+$-T-cell activation, its specific expression profile may be higher in Th1 than Th2 CD4$^+$ T cells due to IL-4/STAT6 signalling (Metwali et al., 2000). That IL-4 negatively regulates VPAC-2 suggests VPAC-2 may be part of the Th1 CD4$^+$ repertoire potentially contributing to chemotaxis or other immune functions. Of interest, it was also noted that IL-4 repression of VPAC-2 was not seen in Th1 CD4$^+$ T cell lines suggesting a possible block in the STAT6 signalling pathway (Metwali et al., 2000).

VIP AS A NEURODEVELOPMENTAL AND NEUROTROPHIC FACTOR

VIP is considered to act as a pivotal developmental regulator based on its time of appearance in embryogenesis, the concurrent expression of VIP receptors, and its potent effects on cellular proliferation and differentiation (Waschek, 1996; Hill et al., 1999a,b). By mechanisms similar to its neuroprotective activities, VIP stimulates adjacent glial cells to secrete numerous soluble growth-factor proteins to aid in the proper closure of the neural tube during embryogenesis (DiCicco-Bloom, 1996; Hill et al., 1999a). Thus, VIP-induced neurotropic activities can be thought of as a carry-over from its developmental functions in utero. In rodents, E8–10 is the most crucial window for VIP-mediated developmental regulation. Many important developmental events are taking place in addition to neural tube closure, such as switching from yolk sac to placenta nutrition, organogenesis and an infiltration of T lymphocytes to the ducidua/trophoblast (pre-placenta region) (Hill et al., 1996; Colas and Schoenwolf, 2001).

Recently, it has been verified that the VIP source for embryonic development during E8–10 is maternal (Hill et al., 1996). VIP message is not detectable in the embryo at E8–10, but only later at E11 (Spong et al., 1999a,b). That VIP is necessary for proper development of the neural tube during a time when VIP is not synthesized by the embryo, prompted research to determine if a maternal source was possible. Radio-labelled VIP intravenously injected into pregnant mothers (E8) rapidly left the vasculature and concentrated in the embryo (Hill et al., 1996; Spong et al., 1999a). In addition, VIP mRNA and VIP antigen was enriched in the ducidua/troblast area during day E8–10. VIP antigen co-stained with CD3 and delta T cell receptor implicating gamma-delta ($\gamma\delta$) T cells were being recruited to this region, which is proximal to the developing embryo. In the periphery, (γδ) T cells constitute a small percentage of T lymphocytes (~1%) and its concentrated number in the pre-placenta region has suggested that a potential source for maternal VIP was, in part, supplied by these largely uncharacterized T cells (Hill *et al.*, 1996; Spong *et al.*, 1999a; MacDonald *et al.*, 2001). Perhaps also, maternal VIP concentrating within the foetus is preferentially chemotactic for γδ T cells and may explain the high numbers of γδ T cells proximal to the developing foetus.

A large number of VIP binding sites (see text below) have been observed in the neurotube during post-implantation E6–10 and diminish to undetectable levels at E11 and later (Hill *et al.*, 1994, 1999a). The neural tube spans the entire length of the CNS of the developing rodent and is the basis for how VIP can mediate its embryonic development and cause proper closure of the neural tube. Indeed, introduction of a VIP neutralizing antibody into pregnant mice results in severe neural tube defects in developing progeny and may be the most likely reason for embryonic lethality in VIP$^{-/-}$ mice (personal communication). The cells of the floor plate of the developing neural tube are similar in phenotype to glial cells. VIP binding to these neural floor plate glial-like cells promotes the synthesis and release of growth factors similar to those seem in the adult. One of the most potent growth regulators is ADNF, which mediates the growth and proper closure of the neural tube (Hill *et al.*, 1999a; Steingart *et al.*, 2000). In addition, VIP also shortens the cell cycle in the floor plate neuro-epithelial cells and increases cyclins A, B, D and E (Servoss *et al.*, 1996; Gressens *et al.*, 1998).

In summary, VIP acts as an important coordinator of neural tube development and closure by binding to specialized neuro-epithelial cells located on the floor plate. The developing neuro-epithelium responds by shortening its cell cycle and secretes growth regulating proteins such as ADNF allowing for the continued growth and development of the CNS. Since VIP is crucial to the proper development of the CNS, and since VIP knockouts are embryonically lethal, it is curious as to why the embryo would not synthesize and secrete this peptide internally, that is, being self-sufficient. Therefore, it is enticing to speculate that there has evolved a maternal-embryo check-point where specialized immune cells, γδ T cells, are recruited to and oversee the proper implantation and generation of placental connection. Without such orchestrated events, the development of the embryo would be spontaneously aborted in a protective measure for the female rodent. Human CD4$^+$ and CD8$^+$ T cells apparently do not synthesize VIP based on real-time fluorescent PCR amplification (Lara-Marquez *et al.*, 2001), while rodent CD4$^+$ and CD8$^+$ T cells do express VIP (Delgado and Ganea, 2001). Therefore, this dichotomy between rodents and humans in the expression of VIP suggests species-specific physiological mechanisms. Whether γδ T cells express VIP and share in its supply to the developing human embryo will require further research.

VIP is a potent sympathetic neurotrophic factor (Brenneman *et al.*, 1999a). VIP's neurotrophic actions are mediated indirectly by stimulating non-neuronal glial cells to generate numerous soluble growth-regulating proteins, which act directly on neurons to support their survival. Examples of VIP-evoked cytokine proteins are: activity-dependent neurotrophic factor (ADNF), IL-1α, IL-6, protease nexin I (PNI), macrophage inflammatory protein-1α (MIP-1) and regulated upon activation of normal T cells expressed and secreted (RANTES), brain-derived neurotrophic factor (BDNF), and heat shock protein 60 (HSP-60) (Brenneman *et al.*, 1999a).

Upon VIP binding to astrocytes, several proteins are generated and secreted; some of which have potent growth factor activities (Brenneman *et al.*, 1990; Gozes and Brenneman, 1993). One of these proteins called ADNF is a potent growth factor with an effective biological activity in the femtamolar range (10^{-13-15} M) (Brenneman and Gozes, 1996; Gozes *et al.*, 1999). ADNF is a 92 kDa secreted protein and is proteolytically processed to reveal a biologically active 14 amino acid peptide capable of acting as a neuronal growth factor (Spong *et al.*, 2001). Specific antibodies to ADNF-14 can suppress the neurotrophic activity induced by VIP and anti-VIP can block the synthesis and secretion of ADNF-14 and other growth-sustaining proteins (Brenneman *et al.*, 1999a). Taken together, these data further substantiate an indirect role for the neuroprotective activity of VIP on sympathetic neurons through astrocytes. The drug, bafilomycin A1 (BFA), is a specific inhibitor of vacuolar ATPases needed for endocytosis (Bowman *et al.*, 1988). BFA can significantly suppress the neurotrophic effects of ADNF on neurons indicating an endocytosis-mediated mechanism for removal of ADNF (Brenneman *et al.*, 1999a). Not surprisingly, other growth factors such as insulin-like growth factor (Damke *et al.*, 1991) and epidermal growth factor (Wu *et al.*, 2001) are catabolized via a similar endocytic mechanism. Acting as a secretagogue, VIP causes astrocytes to generate numerous soluble proteins, which prevent neurons from entering apoptotic cellular programmes and instead induce survival signals (Spong *et al.*, 2001). This idea is supported by the observation that anti-IL-1 α indeed suppresses VIP's neurotrophic activity. Considering that IL-1α is not normally thought of as a 'growth factor', it appears that all of these VIP- induced proteins secreted by astrocytes are acting in concert to potentate the survival of the neuron (Brenneman *et al.*, 1999a).

VIP improvement of neuronal survival despite the presence of several toxic substances and/or toxic environments is well established. These toxic substances include: *N*-methyl-D aspartate (NMDA), electrical blockade by tetrodotoxin (TXX), β-amyloid, reactive oxygen species (ROS), gp-120, cholinergic deficiencies, developmental retardation and learning impairments (Gozes *et al.*, 1996, 1997, 1999). The mechanism by which VIP protects neurons is most likely through the participation of astrocytes. Evidence for direct neuroprotective action of VIP on neurons is inconclusive. Last decade, Brenneman *et al.* showed that the HIV viral membrane glycoprotein gp-120 induces hippocampal neuronal cell death (Brenneman *et al.*, 1988). The exogenous addition of VIP to neurons in culture significantly attenuated HIV-induced neuronal cell death. The mechanism for this neuroprotective action of VIP was unknown at the time of this study. The authors did hypothesize, however, that because anti-CD4 antibodies were also protective against gp-120-induced neuronal toxicity, and gp-120 and VIP share sequence homology, it might be possible that VIP competes for gp-120 binding sites. Further experimentation by Brenneman's group has recently revealed that VIP inhibits gp-120 neuronal cell death by the glial-mediated upregulation of the chemokines macrophage inflammatory protein-1α (MIP-1α) and RANTES (Brenneman *et al.*, 1999b), which upregulate chemokine receptors CXCR4, CCR3 and/or CCR5 via an autocrine mechanism. These chemokine receptors can bind gp-120 and inhibit cellular entry. Thus, the ability for VIP to block gp-120 neuronal toxicity is not via a direct effect, such as competitive binding, but through the recruitment of the glial cells by VIP to attain secretion of MIP-1α and RANTES and subsequent upregulation of antagonistic chemokine receptors. VIP thus signals astrocytes to enter into a protein synthesizing and secretory cellular programme, which provides survival signals directly to the sympathetic neuron (Brenneman *et al.*, 2000).

The neurotrophic activities of VIP have been demonstrated *in vivo* as well. The white-matter lesions in the developing brain caused during periventricular leukomalacia are potentially lethal in premature babies (Gressens, 1999). Intracerebral administration of the drug ibotenate to neonatal mice can mimic the white-matter lesions seen in periventricular leukomalacia. VIP, co-injected with ibotenate can significantly reduce the size of the exci-totoxic white-matter cysts (Gressens, 1999). Furthermore, the intracellular signalling path-ways responsible for VIP-induced neuroprotection of neurons by activating proximal astrocytes have been studied in the excitotoxic white-matter mouse model. VIP elicits many intracellular signals in astrocytes, such as cAMP, PKA, intracellular calcium, inosi-tol phosphate and the translocation of PKC to the nuclear membrane. Pharmalogical inhibitor studies on the relevant pathways suggests that VIP binding to receptors on astro-cytes mediates its neuroprotective effect via a PKC-dependent pathway. Inhibitors of PKA, calmodulin-dependent PK inhibitor and phosphatidylinositol-3 OH kinase all had little affect on ibotenate-induced development of white matter cysts. Only the PKC/MAP kinase inhibitors, bisindolylmaleimide and PD98059, respectively, dose-dependently inhibited VIP-induced neuroprotection. Therefore, the secretagogue programme that VIP initiates in the astrocyte cell is most likely mediated through a PKC/MAPK-dependent pathway (Gressens, 1999).

Sympathetic neurons have the ability to alter their neuropeptide gene expression profile depending on their environment (Zigmond, 2000). Such plasticity has been observed when superior cervical ganglion (SCG) cells have been placed in culture (Mohney and Zigmond, 1998). Immunoreactive VIP and NPY steady-state levels were elevated up to 31-fold after 48 h in organ culture compared to no change in tyrosine hydroxylase, the rate-limiting enzyme in the biosynthesis of catecholamines (Shadiack *et al.*, 2001). VIP mRNA levels are exceedingly low *in vivo* at the time of organ explant but increase over time. This VIP upregulation was dependent on transcription, since actinomycin D ablated the elevation of VIP. One explanation for this observation is that VIP is negatively regulated in sympathetic neurons and only when removed from target-derived factor(s) is VIP upregulated (Zigmond and Sun, 1997). Another possibility is that the stress induced by organ culture evokes VIP expression. The former idea was tested by Shadiack *et al.* (2001) by adding exogenous nerve growth factor (NGF) neutralizing antibody to SCN organ cultures. Anti-NGF addition showed an increase in VIP message and protein elicited by axotomy and explantation. It will be interesting to identify the spectrum of proteins upregulated in astrocytes by VIP. Gene array technology is now becoming a more widely used biological procedure and will be invaluable for this purpose.

IMMUNE CELL VIP RECEPTORS

Immune cells of all mammalian species examined express VPAC-1, which contributes to VIP modulation of immunity (Wenger *et al.*, 1990; Goetzl *et al.*, 1998; Delgado *et al.*, 1999c; Dorsam *et al.*, 2000). Humans express VPAC-1 in peripheral blood mixed mononu-clear cells (PBC) and T lymphocytes. VPAC-1 is expressed in unstimulated human CD4[+]

and CD8$^+$ cells with a copy number of 1451±493 and 154±51 per 100 pg RNA, respectively, based on fluorescent real-time PCR (Lara-Marquez et al., 2001). Resting monocytes have 2914 ± 940 per 100 pg RNA (Lara-Marquez et al., 2001). Rats seem to express VPAC-1 on B cells, but there is little evidence for VPAC-1 in human or murine B cells, except for one cultured line of the Raji B cell lymphoma (Robichon et al., 1993). T cells activated with PMA and PMA plus anti-CD3 downregulated VPAC-1 copy number by 71% within 10 h of activation, which inversely correlated with IL-2 production and proliferation (Lara-Marquez et al., 2001). Thus, VPAC-1 levels on T cells may be indicative of the activation status of the T cell.

Developing T lymphocytes express VPAC-1 differentially at various stages of development. In mouse, CD4$^-$CD8$^-$(DN) thymocytes and CD8$^+$T cells demonstrated highest expression while CD4$^+$T cells and CD4$^+$CD8$^+$ (DP) thymocytes were lower by RT-PCR analysis (Delgado et al., 1996; Pankhaniya et al., 1998). Splenocytes showed a different pattern of VPAC-1 expression where CD4$^+$ T cells expressed more VPAC-1 than CD8$^+$ T cells, which supports recent real-time PCR data conducted with human peripheral T cells (Lara-Marquez et al., 2001). What role VPAC-1/VPAC-2 plays on developing T cells in the thymus is not well understood (Delgado et al., 1999a). It is possible that VPAC-1 aids in the proper maturation and eventual oegress of the T cell from the thymus, or in the proper trafficking within the thymus as the T cell matures. Other possibilities are that VPAC-1 aids in maintaining a non-activated state similar to its role in peripheral blood T cells or guides homing of pluripotent stem cells to the thymus.

The expression of VPAC-2 in the immune system is limited to peripheral CD4$^+$ and CD8$^+$ T cells and developing DP and CD4$^+$ thymocytes (Delgado et al., 1996; Pankhaniya et al., 1998). It is weakly expressed in resting human peripheral blood T cells with a copy number of 62 ± 18 and 66 ± 31 in CD4$^+$ and CD8$^+$ cells, respectively, but is upregulated two-fold by PMA and/or PMA plus anti-CD3 activation (Lara-Marquez et al., 2000, 2001). This small elevation in VPAC-2 was reported as not statistically significant. However, considering the dramatic downregulation of VPAC-1 and the striking increase in the ratio of VPAC-2/VPAC-1 in activated CD4$^+$ T cells, it is enticing to speculate that VPAC-2 is part of the activated CD4$^+$ T cell phenotype and contributes in chemotaxis, for example. Similar observations have been reported in rodents based solely on RT-PCR data, which is inherently poor at quantifying RNA differences and will need to be evaluated by similar real-time PCR analysis as was done in the human (Delgado et al., 1996).

Developing rodent T lymphocytes differentially express VPAC-1 and VPAC-2 receptors with rat DP and CD4$^+$ cells predominately VPAC-2, while DN and CD8$^+$ cells are predominately VPAC-1 (Delgado et al., 1996). Again, a distinct pattern of expression is observed for these two receptors in developing T cells as is seen throughout the tissues of the body. That VIP expression is most prominent in the thymus suggests a developmental role mediated by each receptor at different stages of development. Pankhaniya et al. (1998) observed that VIP treatment of developing DP DPK cells increased their differentiation rate into CD4$^+$ cells, while selective agonists for VPAC-1 had little effect. Since DP T cells are predominately VPAC-2, it was concluded that VIP contributes to single positive T cells by skewing towards a CD4$^+$ phenotype in VPAC-2 cAMP/PKA dependent mechanism.

VIP EFFECTS ON LYMPHOCYTES

TRAFFICKING, ADHESION AND MIGRATION

Over that past 20 years, many investigators have observed that neuropeptides, including VIP, modulate lymphocyte trafficking and homing to specific organs and tissues (Pankhaniya et al., 1998). In 1984, Ottaway demonstrated that a decrease in VIP binding sites (75%) on ^{51}Cr-labelled mesenteric lymph node (MLN) T cells decreased their homing to MLN and Peyer's patches (PP) of passive host mice by 30% at 1 h after transfer, as assessed by localization of radioactivity. T cell homing was not affected for other tissues and organs in this study suggesting a role in MLN and PP homing for VPAC receptors (Ottaway, 1984). It is curious that no compensatory increase in ^{51}Cr-labelled T cells was detected at other sites secondary to the decreased homing to MLN and PP, which may simply reflect limited sensitivity of the assay method. Restored homing of T cells to these nodes was observed after 18 h.

VIPergic innervation of the gastrointestinal MALT has been documented and may serve as an important chemoattractant for unstimulated T cell homing and/or trafficking (Ottaway and Husband, 1992). Several mechanisms maintain the number of cells in the vasculature, such as the total number of cells, blood flow and cell/endothelial interactions (Butcher and Picker, 1996). Lymphatics and specialized high endothelial venular cells (HEV), which border the postcapillary venules of MLN and PP, are the site of T cell entry into lymph nodes, and are known to be densely supplied by VIPergic nerves (Ottaway, 1984). This hypothesis is supported by a number of more recent in vitro reports demonstrating a VIP-induced chemotaxis of unstimulated T cells and monocytes but not PMNs through a Boyden chamber micropore filter coated with extracellular matrix protein such as Matrigel, fibronectin or type IV collagen (Johnston et al., 1994). A pulse of VIP (50 µg over 5 min) injected into the afferent popliteal lymph nodes in sheep nearly ablated all lymphocyte oegress, except for a small subpopulation of CD4$^+$ cells (Moore et al., 1988). Taken together these findings show that, VIP may contribute to the normal trafficking and homing of T cells through its chemotactic activities.

TCR engagement of T cells results in less responsiveness to VIP-induced chemotaxis, but not chemokines such as RANTES and MIP-1α (Johnston et al., 1994; Xia et al., 1996a). It is well established that VPAC-1 is significantly down-regulated upon TCR mediated activation of T cells and most likely results in the blunted chemotactic response (Johnston et al., 1994; Xia et al., 1996a). In addition, chemokine receptors are upregulated upon T cell activation (Brenneman et al., 2000). Therefore, the activated T cell becomes less sensitive to VIP-mediated signals and a heightened sensitivity to chemokine proteins such as RANTES and MIP-1α. Such a change in surface receptor phenotype may liberate the activated T cell from its normal 'unstimulated VIP mediated' trafficking to a more unrestrained movement in an attempt to seek out the inflammatory foci. Such unrestricted T cell mobility may be mediated by an array of proteolytic enzymes, which would allow for a T cell to traverse numerous extracellular matrixes. Specific inhibitors to metalloproteinases MMP2 and MMP9 such as GM6001 inhibited VIP, IL-2 and IL-4 induced migration through type IV collagen but not migration mediated by RANTES or MIP-1α (Xia et al., 1996b). This suggests that T cells utilize a broader array of proteolytic enzymes to migrate when activated compared to non-activated trafficking.

VIP can inhibit PMA-activated T cell chemotaxis by a cAMP-dependent pathway. T cell chemotaxis is mediated at least in part by PKA, and substances that elevate intracellular levels of cAMP can attenuate such activities. VIP at 10^{-9} M increases cAMP in T cells within 30 s and can therefore modulate T cell activated chemotaxis (Johnston et al., 1994). Since VPAC-1 is the predominate receptor on unstimulated T cells, chemotaxis induced by the activation of the T cell is most likely mediated by this receptor (Xia et al., 1996a). Data collected by Xia et al. (1996a) further supports this claim in that HUT-78 cells, which only express VPAC-1, dramatically suppressed IL-4 and TNF-α induced chemotaxis. Thus, if VIP is presented to T cells concomitantly to an activation signal it can dramatically alter its response to the activating signals such as chemotaxis and therefore act in limiting non-specific activation of T cells. Lastly, it stands to reason that VPAC-1, a 'non-activated' T cell receptor is downregulated upon proper T cell activation in an attempt to mount a specific and controlled immune response.

VIP and other neuropeptides such as calcitonin gene-related peptide, secretin and secroneurin have recently been observed to modulate the mobility and maturation of blood-dervived dendritic cells (Dunzendorfer et al., 2001). Dendritic cells are very mobile and move to the lymph system in order to concentrate antigens. VIP was demonstrated to induce immature dendritic cells (CD1a^{high}, CD83^{low}) to chemotax similar to its effect on T lymphocytes (Dunzendorfer et al., 2001). This chemotaxis was suppressed by rolipram a phosphodiesterase blocker, wortmannin, a PI3 kinase inhibitor and tyrphostin, a tyrosine kinase inhibitor. However, mature dendritic cells (CD1a^{low}, CD83^{high}) which are attacted by MIP-3b and 6Ckine were attenuated in a concentration dependent fashion by VIP and maximally at 10^{-10} M (Dunzendorfer et al., 2001). This implicates that neuropeptides such as VIP attract immature dendritic cells to VIPergic nerve endings and aid in their proper maturation. This idea agrees well with the observation that VIP and TNF-α can synergize to induce maturation of dendritic cells (Dunzendorfer et al., 2001). Subsequent chemotactic inhibition by VIP might ensure that the antigen-presenting dendritic cells are mobilized at the imflammatory site. For example, aerosol antigen presentation into the lung promotes dendritic cells to accumulate around peptidergic secreting nerves (Ganea and Delgado, 2001).

REGULATION OF B CELL DEVELOPMENT AND THYMOCYTE DIFFERENTIATION

Few or no PAC$_1$, VPAC$_1$ or VPAC$_2$ receptors have been identified in B cells at maturity or in their early development stages. The results of one study demonstrating VIP suppression of B cell differentiation were attributable to intermediary participation of macrophages rather than a direct action of VIP on the B cell precursors (Shimozato and Kincade, 1997); reminiscent of the neurotrophic glial-neuron mechanism of VIP (see text above). VIP significantly prevented IL-7-driven clonal proliferation of mouse pre-B cells significantly at 10^{-10} M, approximately 50% at 10^{-8} M, and more than 95% at 10^{-6} M. The inhibitory effect of VIP was dependent on the presence of macrophages or medium from VIP-treated macrophages, in large part as a result of macrophages derived IFN-α. Although reversal of the VIP effect by a peptide antagonist of VIP supported a receptor-dependent mechanism, the specific receptor(s) involved were not defined. In contrast, VIP enhances development of T cells from thymocytes by mechanism delineated in a thymic lymphoma line of

$CD4^+8^+$ cells from a transgenic mouse that express high levels of VIP receptors (Pankhaniya et al., 1998). Differentiation of mouse cultured DPK thymocyte-like cells by antigen-presenting cells resulted in conversion of the predominant type of VIP receptor from $VPAC_1$ to $VPAC_2$, without detectable PAC_1 receptors. At 1–100 nM, VIP enhanced conversion of $CD4^+8^+$ to $CD4^+8^-$ T cells with a maximal effect after 3–4 days, which was attributable entirely to $VPAC_2$ receptors and a cyclic AMP-protein kinase A signalling pathway (Pankhaniya et al., 1998). That VIP-mediated conversion of thymocytes to helper T cells was principally differentiation was confirmed by the lack of significant differences in proliferation, viability or apoptosis between $CD4^+8^+$ and $CD4^+8^-$ cells after exposure to VIP. Thus, the major effect of VIP on immune cell development so far identified is stimulation of antigen-induced generation of helper T cells.

MODULATION OF CYTOKINE PRODUCTION

Numerous immune functions of T cells are elicited, altered or suppressed by relevant concentrations of VIP and PACAP (Sreedharan et al., 1990; Merrill and Jonakait, 1995; Bellinger et al., 1996). At 10^{-10}–10^{-7} M, VIP suppresses rodent T cell proliferative responses to mitogens and anti-CD3 monoclonal antibody, and decreases IL-2 production evoked by the same stimuli. VIP also suppresses antigen-evoked proliferation of murine T cells in two different systems, including a mixed lymphocyte reaction model and Schistosome soluble egg antigen stimulation, where the VIP effect is attributable to inhibition of secretion of IL-2. In contrast, human peripheral blood mixed mononuclear leukocyte and T cell proliferation and IL-2 generation respond variably to VIP, presumably as a function of composition and isolation conditions. In some human T lymphoblast lines, nanomolar levels of VIP suppress proliferation responses, but in other the predominant effect of physiological levels of VIP is enhancement of antigen-elicited secretion of IL-2.

Of all of the regulatory effects of VIP and PACAP on immune cells, control of cytokine production has been studied most comprehensively (Sun and Ganea, 1993; Delgado et al., 1999b). The results reflect differences in the subsets of T cells, the type of stimulation of T cells and macrophages, and the duration as well as intensity of exposure(s) of immune cells to either neuropeptide. Quantification of cytokine secretion by rodent thymocytes or $CD4^+$ splenic T cells incubated with a mitogen, anti-CD3 antibody or a combination of anti-CD3 antibody with PMA showed that 10^{-11}–10^{-7} M VIP or PACAP suppressed significantly the generation of IL-2, IL-4 and IL-10 (Sun and Ganea, 1993). In contrast, rodent hepatic granuloma-derived $CD4^+$ T cells, T cell hybridomas, and helper T cell clones stimulated with known specific antigens generate IL-2, IL-5 and IFN-γ at levels that are increased strikingly by 10^{-9}–10^{-6} M VIP or PACAP (Sun and Ganea, 1993; Jabrane-Ferrat et al., 2000). The enhancing effects of VIP and PACAP are greatest after at least 24 h with the T cells, when neuropeptides are added twice daily, and at lower than maximally active concentrations of antigen.

The effects of VIP and PACAP on cytokine production by macrophages and other mononuclear leukocytes differ markedly from those identified in T lymphocytes (Delgado et al., 1999b). The production of IL-6 by rodent peritoneal macrophages without activation or after incubation with very low levels of lipopolysaccharide (LPS) is enhanced by 10^{-12}–10^{-6} M VIP or PACAP. At higher levels of LPS, however, the same concentrations of VIP or PACAP suppress production of IL-6. The generation of IL-10 by rodent macrophages incubated with LPS is significantly augmented by the same concentrations of VIP and PACAP that alter IL-2 secretion, but neither VIP nor PACAP alone has any effect in unstimulated macrophages. Production of TNF-α by rodent peritoneal macrophages and cultured mononuclear leukocytes activated with LPS or IL-1β is inhibited by 10^{-11}–10^{-7} M VIP or PACAP. Thus, the effects of VIP and PACAP on most major immune cell functions are more complex than the usual tentative postulate that these peptides are T cell and macrophage downregulatory factors.

IMMUNE CONSEQUENCES OF GENETIC MODIFICATIONS OF VPAC RECEPTORS IN T CELLS

Genetic methods have been used to investigate the separate capabilities of each of the VPAC receptors in immunity, beginning with VPAC2. VPAC2 was expressed constitutively and selectively in CD4$^+$ T cells (helper-inducer T cells) of transgenic (TG) C57BL/6 mice, directed by the LCK tyrosine kinase promoter (Voice *et al.*, 2001). CD4$^+$ T cell levels of VPAC2 similar to those in activated CD4$^+$ T cells evoked production of more Th2-type interleukins 4 and 5, and less Th1-type IFN-γ after TCR activation. VPAC2-TG mice consequently have significant elevations of blood IgE, IgG1 and eosinophils. VPAC2-TG mice also show increased antigen-specific IgE antibody responses, which mediated heightened cutaneous allergic reaction in footpad swelling and classical active cutaneous anaphylaxis models. VPAC2-TG mice also have depressed delayed-type hypersensitivity. Neuropeptide enhancement of the ratio of Th2 cell to Th1 cell cytokines thus evokes an allergic state in normally non-allergic mice, which suggests the possibility of neural contributions

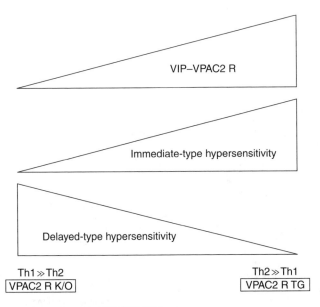

Figure 5.2   Influence on immunity of the VIP–VPAC2 R axis.

to immune phenotypic alterations in human hypersensitivity diseases. In contrast, VPAC2-null C57BL/6 mice have diminished IgE responses and allergic reactions, but significantly enhanced delayed-type hypersensitivity. TCR activation of CD4$^+$ T cells from VPAC2-null mice resulted in higher than normal IL-2 and IFN-α, but lower IL-4. Altered expression of a CD4$^+$ T cell receptor for one endogenous neuropeptide thus completely modifies the ratio of Th2/Th1 cell activity and the resultant immune phenotype (Figure 5.2).

CONCLUDING REMARKS

The quantitative prominence of VIP in immune organs and the T cell subset-specificity of expression of VPAC receptors originally suggested that this neuroendocrine anix would have important regulatory functions in immunity. Potent and selective effects of VIP on T cell development, migration, proliferation, apoptosis, interactions with B cells and cytokine generation supported the possibility that VIP and VPAC receptors mediate communication between the nervous and immune systems. That immune and hematopoietic-specific transcription factors, such as Ikaros, and immune cytokines regulate T cell expression of VPAC1 and VPAC2 increase confidence in this hypothesis. Genetic manipulation of VPAC2 alone alters the effective Th1/Th2 ratio and consequent immune phenotype in mice. Constitutive expression of the immune-inducible VPAC2 receptor selectively in Th cells of transgenic mice decreases Th1/Th2 and induces an allergic state. In contrast, VPAC2 receptor-null mice exhibit increased Th1/Th2 and enhanced delayed-type hypersensitivity. Thus, altered T cell perception of one endogenous neuropeptide evokes major changes in the immune phenotype. Future studies will be designed to determine if these neurally mediated immunophenotypic modifications alter T cell host defense against infectious challenges of specific target organs.

REFERENCES

Agarwal, S. and Rao, A. (1998) Modulation of chromatin structure regulates cytokine gene expression during T cell differentiation. *Immunity*, **9**, 765–75.

Aliakbari, J., Sreedharan, S. P., Turck, C. W. and Goetzl, E. J. (1987) Selective localization of vasoactive intestinal peptide and substance P in human eosinophils. *Biochem Biophys Res Commun*, **148**, 1440–5.

Barnes, P. J., Baraniuk, J. N. and Belvisi, M. G. (1991) Neuropeptides in the respiratory tract. Part II. *Am Rev Respir Dis*, **144**, 1391–9.

Bellinger, D. L., Lorton, D., Brouxhon, S., Felten, S. and Felten, D. L. (1996) The significance of vasoactive intestinal polypeptide (VIP) in immunomodulation. *Adv Neuroimmunol*, **6**, 5–27.

Bowman, E. J., Siebers, A. and Altendorf, K. (1988) Bafilomycins: A class of inhibitors of membrane atpases from microorganisms, animal cells, and plant cells. *Proc Natl Acad Sci USA*, **85**, 7972–6.

Brand, N. J. (1997) Myocyte enhancer factor 2 (mef 2). *Int J Biochem Cell Biol*, **29**, 1467–70.

Brenneman, D. E. and Gozes, I. (1996) A femtomolar-acting neuroprotective peptide. *J Clin Invest*, **97**, 2299–307.

Brenneman, D. E., Hauser, J., Phillips, T. M., Davidson, A., Bassan, M. and Gozes, I. (1999a) Vasoactive intestinal peptide. Link between electrical activity and glia-mediated neurotrophism. *Ann N Y Acad Sci*, **897**, 17–26.

Brenneman, D. E., Hauser, J., Spong, C. Y., Phillips, T. M., Pert, C. B. and Ruff, M. (1999b) Vip and d-ala-peptide t-amide release chemokines which prevent HIV-1 gp120-induced neuronal death. *Brain Res*, **838**, 27–36.

Brenneman, D. E., Hauser, J., Spong, C. Y. and Phillips, T. M. (2000) Chemokines released from astroglia by vasoactive intestinal peptide. Mechanism of neuroprotection from hiv envelope protein toxicity. *Ann N Y Acad Sci*, **921**, 109–14.

Brenneman, D. E., Nicol, T., Warren, D. and Bowers, L. M. (1990) Vasoactive intestinal peptide: a neurotrophic releasing agent and an astroglial mitogen. *J Neurosci Res*, **25**, 386–94.
Brenneman, D. E., Westbrook, G. L., Fitzgerald, S. P., Ennist, D. L., Elkins, K. L., Ruff, M. R. *et al.* (1988) Neuronal cell killing by the envelope protein of HIV and its prevention by vasoactive intestinal peptide. *Nature*, **335**, 639–42.
Brodski, C., Schnurch, H. and Dechant, G. (2000) Neurotrophin-3 promotes the cholinergic differentiation of sympathetic neurons. *Proc Natl Acad Sci USA*, **97**, 9683–8.
Butcher, E. C. and Picker, L. J. (1996) Lymphocyte homing and homeostasis. *Science*, **272**, 60–6.
Calvo, J. R., Pozo, D. and Guerrero, J. M. (1996) Functional and molecular characterization of VIP receptors and signal transduction in human and rodent immune systems. *Adv Neuroimmunol*, **6**, 39–47.
Colas, J. F. and Schoenwolf, G. C. (2001) Towards a cellular and molecular understanding of neurulation. *Dev Dyn*, **221**, 117–45.
Couvineau, A., Maoret, J. J., Rouyer-Fessard, C., Carrero, I. and Laburthe, M. (2000) The human vasoactive intestinal peptide/pituitary adenylate cyclase-activating peptide receptor 1 (VPAC1) promoter: characterization and role in receptor expression during enterocytic differentiation of the colon cancer cell line CACO-2CL.20. *Biochem J*, **347** (Pt 3), 623–32.
Damke, H., Klumperman, J., von Figura, K. and Braulke, T. (1991) Effects of brefeldin a on the endocytic route. Redistribution of mannose 6-phosphate/insulin-like growth factor ii receptors to the cell surface. *J Biol Chem*, **266**, 24829–33.
Delgado, M. and Ganea, D. (2001) Cutting edge: Is vasoactive intestinal peptide a type 2 cytokine? *J Immunol*, **166**, 2907–12.
Delgado, M., Martinez, C., Johnson, M. C., Gomariz, R. P. and Ganea, D. (1996) Differential expression of vasoactive intestinal peptide receptors 1 and 2 (VIP-R1 and VIP-R2) mrna in murine lymphocytes. *J Neuroimmunol*, **68**, 27–38.
Delgado, M., Martinez, C., Leceta, J. and Gomariz, R. P. (1999a) Vasoactive intestinal peptide in thymus: synthesis, receptors and biological actions. *Neuroimmunomodulation*, **6**, 97–107.
Delgado, M., Munoz-Elias, E. J., Gomariz, R. P. and Ganea, D. (1999b) Vasoactive intestinal peptide and pituitary adenylate cyclase-activating polypeptide enhance IL-10 production by murine macrophages: in vitro and in vivo studies. *J Immunol*, **162**, 1707–16.
Delgado, M., Munoz-Elias, E. J., Martinez, C., Gomariz, R. P. and Ganea, D. (1999c) Vip and pacap38 modulate cytokine and nitric oxide production in peritoneal macrophages and macrophage cell lines. *Ann N Y Acad Sci*, **897**, 401–14.
Deng, A. Y., Gu, L., Rapp, J. P., Szpirer, C. and Szpirer, J. (1994) Chromosomal assignment of 11 loci in the rat by mouse-rat somatic hybrids and linkage. *Mamm Genome*, **5**, 712–16.
De Souza, E. B., Seifert, H. and Kuhar, M. J. (1985) Vasoactive intestinal peptide receptor localization in rat forebrain by autoradiography. *Neurosci Lett*, **56**, 113–20.
DiCicco-Bloom, E. (1996) Region-specific regulation of neurogenesis by VIP and PACAP: direct and indirect modes of action. *Ann N Y Acad Sci*, **805**, 244–56; discussion 256–8.
Dobi, A., Palkovits, M., Ring, M. A., Eitel, A., Palkovits, C. G., Lim, F. *et al.* (1997) Sample and probe: a novel approach for identifying development-specific cis-elements of the enkephalin gene. *Brain Res Mol Brain Res*, **52**, 98–111.
Dorsam, G. and Goetzl, E. J. (2001) Vasoactive intestinal peptide receptor - 1 (VPAC-1) is a novel gene target of the hemolymphopoietic transcription factor ikaros. *J Biol Chem*, **277**, 13488–93.
Dorsam, G., Voice, J., Kong, Y. and Goetzl, E. J. (2000) Vasoactive intestinal peptide mediation of development and functions of T lymphocytes. *Ann N Y Acad Sci*, **921**, 79–91.
Dunzendorfer, S., Kaser, A., Meierhofer, C., Tilg, H. and Wiedermann, C. J. (2001) Cutting edge: peripheral neuropeptides attract immature and arrest mature blood-derived dendritic cells. *J Immunol*, **166**, 2167–72.
Francis, N. J., Asmus, S. E. and Landis, S. C. (1997) CNTF and LIF are not required for the target-directed acquisition of cholinergic and peptidergic properties by sympathetic neurons in vivo. *Dev Biol*, **182**, 76–87.
Ganea, D. and Delgado, M. (2001) Neuropeptides as modulators of macrophage functions. Regulation of cytokine production and antigen presentation by VIP and PACAP. *Arch Immunol Ther Exp (Warsz)*, **49**, 101–10.
Gazdar, A. F., Bader, S., Hung, J., Kishimoto, Y., Sekido, Y., Sugio, K. *et al.* (1994) Molecular genetic changes found in human lung cancer and its precursor lesions. *Cold Spring Harb Symp Quant Biol*, **59**, 565–72.
Goetzl, E. J., Pankhaniya, R. R., Gaufo, G. O., Mu, Y., Xia, M. and Sreedharan, S. P. (1998) Selectivity of effects of vasoactive intestinal peptide on macrophages and lymphocytes in compartmental immune responses. *Ann N Y Acad Sci*, **840**, 540–50.
Goetzl, E. J. and Sreedharan, S. P. (1992) Mediators of communication and adaptation in the neuroendocrine and immune systems. *FASEB J*, **6**, 2646–52.

Goetzl, E. J., Xia, M., Ingram, D. A., Kishiyama, J. L., Kaltreider, H. B., Byrd, P. K. *et al.* (1995) Neuropeptide signaling of lymphocytes in immunological responses. *Int Arch Allergy Immunol*, **107**, 202–4.

Gozes, I., Bachar, M., Bardea, A., Davidson, A., Rubinraut, S., Fridkin, M. *et al.* (1997) Protection against developmental retardation in apolipoprotein e-deficient mice by a fatty neuropeptide: Implications for early treatment of alzheimer's disease. *J Neurobiol*, **33**, 329–42.

Gozes, I., Bardea, A., Reshef, A., Zamostiano, R., Zhukovsky, S., Rubinraut, S. *et al.* (1996) Neuroprotective strategy for Alzheimer disease: intranasal administration of a fatty neuropeptide. *Proc Natl Acad Sci USA*, **93**, 427–32.

Gozes, I., Bassan, M., Zamostiano, R., Pinhasov, A., Davidson, A., Giladi, E. *et al.* (1999) A novel signaling molecule for neuropeptide action: activity-dependent neuroprotective protein. *Ann N Y Acad Sci*, **897**, 125–35.

Gozes, I. and Brenneman, D. E. (1993) Neuropeptides as growth and differentiation factors in general and VIP in particular. *J Mol Neurosci*, **4**, 1–9.

Gressens, P. (1999) VIP neuroprotection against excitotoxic lesions of the developing mouse brain. *Ann N Y Acad Sci*, **897**, 109–24.

Gressens, P., Hill, J. M., Gozes, I., Fridkin, M. and Brenneman, D. E. (1993) Growth factor function of vasoactive intestinal peptide in whole cultured mouse embryos. *Nature*, **362**, 155–8.

Gressens, P., Paindaveine, B., Hill, J. M., Evrard, P. and Brenneman, D. E. (1998) Vasoactive intestinal peptide shortens both G1 and S phases of neural cell cycle in whole postimplantation cultured mouse embryos. *Eur J Neurosci*, **10**, 1734–42.

Hahm, S. H. and Eiden, L. E. (1998) Cis-regulatory elements controlling basal and inducible vip gene transcription. *Ann N Y Acad Sci*, **865**, 10–26.

Hamm, H. E. (2001) How activated receptors couple to G proteins. *Proc Natl Acad Sci USA*, **98**, 4819–21.

Harmar, A. J., Arimura, A., Gozes, I., Journot, L., Laburthe, M., Pisegna, J. R. *et al.* (1998) International union of pharmacology. XVIII. Nomenclature of receptors for vasoactive intestinal peptide and pituitary adenylate cyclase-activating polypeptide. *Pharmacol Rev*, **50**, 265–70.

Hashimoto, H., Nishino, A., Shintani, N., Hagihara, N., Copeland, N. G., Jenkins, N. A. *et al.* (1999) Genomic organization and chromosomal location of the mouse vasoactive intestinal polypeptide 1 (vpac1) receptor. *Genomics*, **58**, 90–3.

Hernanz, A., Muelas, G. and Borbujo, J. (1989) Plasma neuropeptide pattern in acute idiopathic urticaria. *Int Arch Allergy Appl Immunol*, **90**, 198–200.

Hill, J. M., Agoston, D. V., Gressens, P. and McCune, S. K. (1994) Distribution of VIP mRNA and two distinct VIP binding sites in the developing rat brain: relation to ontogenic events. *J Comp Neurol*, **342**, 186–205.

Hill, J. M., Glazner, G. W., Lee, S. J., Gozes, I., Gressens, P. and Brenneman, D. E. (1999a) Vasoactive intestinal peptide regulates embryonic growth through the action of activity-dependent neurotrophic factor. *Ann N Y Acad Sci*, **897**, 92–100.

Hill, J. M., Lee, S. J., Dibbern, D. A., Jr., Fridkin, M., Gozes, I. and Brenneman, D. E. (1999b) Pharmacologically distinct vasoactive intestinal peptide binding sites: CNS localization and role in embryonic growth. *Neuroscience*, **93**, 783–91.

Hill, J. M., McCune, S. K., Alvero, R. J., Glazner, G. W., Henins, K. A., Stanziale, S. F. *et al.* (1996) Maternal vasoactive intestinal peptide and the regulation of embryonic growth in the rodent. *J Clin Invest*, **97**, 202–8.

Ishihara, T., Shigemoto, R., Mori, K., Takahashi, K. and Nagata, S. (1992) Functional expression and tissue distribution of a novel receptor for vasoactive intestinal polypeptide. *Neuron*, **8**, 811–19.

Jabrane-Ferrat, N., Pollock, A. S. and Goetzl, E. J. (2000) Inhibition of expression of the type I G protein-coupled receptor for vasoactive intestinal peptide (VPAC1) by hammerhead ribozymes. *Biochemistry*, **39**, 9771–7.

Jartti, T. (2001) Asthma, asthma medication and autonomic nervous system dysfunction. *Clin Physiol*, **21**, 260–9.

Johnston, J. A., Taub, D. D., Lloyd, A. R., Conlon, K., Oppenheim, J. J. and Kevlin, D. J. (1994) Human t lymphocyte chemotaxis and adhesion induced by vasoactive intestinal peptide. *J Immunol*, **153**, 1762–8.

Koipally, J. and Georgopoulos, K. (2000) Ikaros interactions with CtBP reveal a repression mechanism that is independent of histone deacetylase activity. *J Biol Chem*, **275**, 19594–602.

Koipally, J., Renold, A., Kim, J. and Georgopoulos, K. (1999) Repression by Ikaros and Aiolos is mediated through histone deacetylase complexes. *EMBO J*, **18**, 3090–100.

Koller, E., Hayman, A. R. and Trueb, B. (1991) The promoter of the chicken alpha 2(VI) collagen gene has features characteristic of house-keeping genes and of proto-oncogenes. *Nucleic Acids Res*, **19**, 485–91.

Lara-Marquez, M., O'Dorisio, M., O'Dorisio, T., Shah, M. and Karacay, B. (2001) Selective gene expression and activation-dependent regulation of vasoactive intestinal peptide receptor type 1 and type 2 in human T cells. *J Immunol*, **166**, 2522–30.

Lara-Marquez, M. L., O'Dorisio, M. S. and Karacay, B. (2000) Vasoactive intestinal peptide (VIP) receptor type 2 (VPAC2) is the predominant receptor expressed in human thymocytes. *Ann N Y Acad Sci*, **921**, 45–54.

Lomize, A. L., Pogozheva, I. D. and Mosberg, H. I. (1999) Structural organization of g-protein-coupled receptors. *J Comput Aided Mol Des*, **13**, 325–53.

Lutz, E. M., Shen, S., Mackay, M., West, K. and Harmar, A. J. (1999) Structure of the human VIPR2 gene for vasoactive intestinal peptide receptor type 2. *FEBS Lett*, **458**, 197–203.

Lutz, E. M., Sheward, W. J., West, K. M., Morrow, J. A., Fink, G. and Harmar, A. J. (1993) The VIP2 receptor: molecular characterisation of a cdna encoding a novel receptor for vasoactive intestinal peptide. *FEBS Lett*, **334**, 3–8.

MacDonald, H. R., Radtke, F. and Wilson, A. (2001) T cell fate specification and alphabeta/gammadelta lineage commitment. *Curr Opin Immunol*, **13**, 219–24.

Mackay, M., Fantes, J., Scherer, S., Boyle, S., West, K., Tsui, L. C. *et al.* (1996) Chromosomal localization in mouse and human of the vasoactive intestinal peptide receptor type 2 gene: a possible contributor to the holoprosencephaly 3 phenotype. *Genomics*, **37**, 345–53.

Maruno, K., Absood, A. and Said, S. I. (1998) Vasoactive intestinal peptide inhibits human small-cell lung cancer proliferation in vitro and in vivo. *Proc Natl Acad Sci USA*, **95**, 14373–8.

McCulloch, D. A., Lutz, E. M., Johnson, M. S., Robertson, D. N., MacKenzie, C. J., Holland, P. J. *et al.* (2001) Adp-ribosylation factor-dependent phospholipase D activation by VPAC receptors and a PAC(1) receptor splice variant. *Mol Pharmacol*, **59**, 1523–32.

Merrill, J. E. and Jonakait, G. M. (1995) Interactions of the nervous and immune systems in development, normal brain homeostasis, and disease. *FASEB J*, **9**, 611–18.

Metwali, A., Blum, A. M., Li, J., Elliott, D. E. and Weinstock, J. V. (2000) Il-4 regulates VIP receptor subtype 2 mRNA (VPAC2) expression in t cells in murine schistosomiasis. *FASEB J*, **14**, 948–54.

Mohney, R. P. and Zigmond, R. E. (1998) Vasoactive intestinal peptide enhances its own expression in sympathetic neurons after injury. *J Neurosci*, **18**, 5285–93.

Moore, T. C., Spruck, C. H. and Said, S. I. (1988) Depression of lymphocyte traffic in sheep by vasoactive intestinal peptide (vip). *Immunology*, **64**, 475–8.

Nussdorfer, G. G., Bahce4ioglu, M., Neri, G. and Malendowicz, L. K. (2000) Secretin, glucagon, gastric inhibitory polypeptide, parathyroid hormone, and related peptides in the regulation of the hypothalamus–pituitary–adrenal axis. *Peptides*, **21**, 309–24.

Nussdorfer, G. G. and Malendowicz, L. K. (1998) Role of VIP, PACAP, and related peptides in the regulation of the hypothalamo–pituitary–adrenal axis. *Peptides*, **19**, 1443–67.

Ottaway, C. A. (1984) In vitro alteration of receptors for vasoactive intestinal peptide changes the in vivo localization of mouse t cells. *J Exp Med*, **160**, 1054–69.

Ottaway, C. A. and Husband, A. J. (1992) Central nervous system influences on lymphocyte migration. *Brain Behav Immun*, **6**, 97–116.

Pankhaniya, R., Jabrane-Ferrat, N., Gaufo, G. O., Sreedharan, S. P., Dazin, P., Kaye, J. *et al.* (1998) Vasoactive intestinal peptide enhancement of antigen-induced differentiation of a cultured line of mouse thymocytes. *FASEB J*, **12**, 119–27.

Pei, L. (1997) Genomic structure and embryonic expression of the rat type 1 vasoactive intestinal polypeptide receptor gene. *Regul Pept*, **71**, 153–61.

Pei, L. and Melmed, S. (1995) Characterization of the rat vasoactive intestinal polypeptide receptor gene 5′ region. *Biochem J*, **308** (Pt 3), 719–23.

Robichon, A., Sreedharan, S. P., Yang, J., Shames, R. S., Gronroos, E. C., Cheng, P. P. *et al.* (1993) Induction of aggregation of raji human b-lymphoblastic cells by vasoactive intestinal peptide. *Immunology*, **79**, 574–9.

Said, S. I. and Mutt, V. (1972) Isolation from porcine-intestinal wall of a vasoactive octacosapeptide related to secretin and to glucagon. *Eur J Biochem*, **28**, 199–204.

Schmidt-Choudhury, A., Meissner, J., Seebeck, J., Goetzl, E. J., Xia, M., Galli, S. J. *et al.* (1999) Stem cell factor influences neuro-immune interactions: the response of mast cells to pituitary adenylate cyclase activating polypeptide is altered by stem cell factor. *Regul Pept*, **83**, 73–80.

Schutz, B., Schafer, M. K., Eiden, L. E. and Weihe, E. (1998) VIP and NPY expression during differentiation of cholinergic and noradrenergic sympathetic neurons. *Ann N Y Acad Sci*, **865**, 537–41.

Servoss, S., Glazner, G., Lee, S., Gibney, G., JY, W., Brenneman, D. *et al.* (1996) Mechanism of vasoactive inestinal peptide-stimulated embryonic growth: IGF-1, cyclins and Fos. *Soc Neurocsi Abst*, **22**, 550.

Shadiack, A. M., Sun, Y. and Zigmond, R. E. (2001) Nerve growth factor antiserum induces axotomy-like changes in neuropeptide expression in intact sympathetic and sensory neurons. *J Neurosci*, **21**, 363–71.

Shimozato, T. and Kincade, P. W. (1997) Indirect suppression of il-7-responsive b cell precursors by vasoactive intestinal peptide. *J Immunol*, **158**, 5178–84.

Spong, C. Y., Abebe, D. T., Gozes, I., Brenneman, D. E. and Hill, J. M. (2001) Prevention of fetal demise and growth restriction in a mouse model of fetal alcohol syndrome. *J Pharmacol Exp Ther*, **297**, 774–9.

Spong, C. Y., Lee, S. J., McCune, S. K., Gibney, G., Abebe, D. T., Alvero, R. *et al.* (1999a) Maternal regulation of embryonic growth: the role of vasoactive intestinal peptide. *Endocrinology*, **140**, 917–24.

Spong, C. Y., Lee, S. J., McCune, S. K., Gibney, G., Abebe, D. T., Brenneman, D. E. *et al.* (1999b) Regulation of postimplantation mouse embryonic growth by maternal vasoactive intestinal peptide. *Ann N Y Acad Sci*, **897**, 101–8.

Sreedharan, S. P., Goetzl, E. J. and Malfroy, B. (1990) Elevated synovial tissue concentration of the common acute lymphoblastic leukaemia antigen (CALLA)-associated neutral endopeptidase (3.4.24.11) in human chronic arthritis. *Immunology*, **71**, 142–4.

Sreedharan, S. P., Huang, J. X., Cheung, M. C. and Goetzl, E. J. (1995) Structure, expression, and chromosomal localization of the type I human vasoactive intestinal peptide receptor gene. *Proc Natl Acad Sci USA*, **92**, 2939–43.

Sreedharan, S. P., Patel, D. R., Huang, J. X. and Goetzl, E. J. (1993) Cloning and functional expression of a human neuroendocrine vasoactive intestinal peptide receptor. *Biochem Biophys Res Commun*, **193**, 546–53.

Steingart, R. A., Solomon, B., Brenneman, D. E., Fridkin, M. and Gozes, I. (2000) Vip and peptides related to activity-dependent neurotrophic factor protect PC12 cells against oxidative stress. *J Mol Neurosci*, **15**, 137–45.

Sun, L. and Ganea, D. (1993) Vasoactive intestinal peptide inhibits interleukin (il)-2 and il-4 production through different molecular mechanisms in t cells activated via the t cell receptor/cd3 complex. *J Neuroimmunol*, **48**, 59–69.

Usdin, T. B., Bonner, T. I. and Mezey, E. (1994) Two receptors for vasoactive intestinal polypeptide with similar specificity and complementary distributions. *Endocrinology*, **135**, 2662–80.

Voice, J. K., Dorsam, G., Lee, H., Kong, Y. and Goetzl, E. J. (2001) Allergic diathesis in transgenic mice with constitutive T cell expression of inducible vasoactive intestinal peptide receptor. *FASEB J*, **15**, 2489–96.

Waschek, J. A. (1996) VIP and PACAP receptor-mediated actions on cell proliferation and survival. *Ann N Y Acad Sci*, **805**, 290–300; discussion 300–1.

Waschek, J. A., Pruss, R. M., Siegel, R. E., Eiden, L. E., Bader, M. F. and Aunis, D. (1987) Regulation of enkephalin, VIP, and chromogranin biosynthesis in actively secreting chromaffin cells. Multiple strategies for multiple peptides. *Ann N Y Acad Sci*, **493**, 308–23.

Wenger, G. D., O'Dorisio, M. S. and Goetzl, E. J. (1990) Vasoactive intestinal peptide. Messenger in a neuroimmune axis. *Ann N Y Acad Sci*, **594**, 104–19.

Whang-Peng, J., Bunn, P. A., Jr., Kao-Shan, C. S., Lee, E. C., Carney, D. N., Gazdar, A. *et al.* (1982) A nonrandom chromosomal abnormality, del 3p(14–23), in human small cell lung cancer (sclc). *Cancer Genet Cytogenet*, **6**, 119–34.

Wu, C., Lai, C. F. and Mobley, W. C. (2001) Nerve growth factor activates persistent rap1 signaling in endosomes. *J Neurosci*, **21**, 5406–16.

Xia, M., Gaufo, G. O., Wang, Q., Sreedharan, S. P. and Goetzl, E. J. (1996a) Transduction of specific inhibition of hut 78 human T cell chemotaxis by type i vasoactive intestinal peptide receptors. *J Immunol*, **157**, 1132–8.

Xia, M., Leppert, D., Hauser, S. L., Sreedharan, S. P., Nelson, P. J., Krensky, A. M. *et al.* (1996b) Stimulus specificity of matrix metalloproteinase dependence of human t cell migration through a model basement membrane. *J Immunol*, **156**, 160–7.

Zigmond, R. E. (2000) Neuropeptide action in sympathetic ganglia. Evidence for distinct functions in intact and axotomized ganglia. *Ann N Y Acad Sci*, **921**, 103–8.

Zigmond, R. E. and Sun, Y. (1997) Regulation of neuropeptide expression in sympathetic neurons. Paracrine and retrograde influences. *Ann N Y Acad Sci*, **814**, 181–97.

6 An Emerging Role for Calcitonin Gene-Related Peptide in Regulating Immune and Inflammatory Functions

Stefan Fernandez and Joseph P. McGillis

Department of Microbiology, Immunology and Molecular Genetics, University of Kentucky College of Medicine, MS 415, 800 Rose Street, Lexington, KY 40536-0084, USA

Calcitonin gene-related peptide (CGRP) is a 37 amino acid neuropeptide present primarily in nociceptive nerve fibres in the peripheral nervous system. Similar to substance P, one of its major functions is as a mediator of neurogenic inflammation. It does this by regulating both vascular and cellular functions at local sites of inflammation and in lymphoid tissue. There are two isoforms of CGRP, α and β, derived from two distinct genes, one of which also encodes the hormone calcitonin. A CGRP receptor has been recently described that requires the expression of the calcitonin receptor like receptor (CRLR) gene product as well as a member of the recently described receptor activity modulating protein (RAMP) family, RAMP 1. CGRP receptors have been identified in a number of tissues including the immune system. In the immune system, CGRP receptors have been identified on mature T and B cells and macrophages, and on developing B cells in the bone marrow. CGRP has a number of effects on lymphoid tissue, many of which are inhibitory in nature. It inhibits T cell proliferation, apparently by inhibition of IL-2 production. In macrophages, it inhibits Ag presentation and other macrophage functions including phagocytosis and the oxidative burst. It also stimulates production of a number of lymphokines including IL-6, IL-10 and TNF-α. During B cell development in the bone marrow it acts as a negative feedback inhibitor of B cell development. It does this directly by inhibiting the proliferative effect of IL-7 on early pro-B cells and indirectly by induction of inhibitory cytokines including IL-6 and TNF-α in bone marrow macrophages and other stromal cells. Overall, there is substantial experimental evidence demonstrating that CGRP can influence the function and development of inflammatory and immune cells in local microenvironments by specific receptor-mediated mechanisms.

KEY WORDS: calcitonin gene-related peptide; receptor-mediated mechanism; inflammatory mediator.

CALCITONIN GENE-RELATED PEPTIDE

Calcitonin gene-related peptide (CGRP) is a 37 amino acid neuropeptide found primarily in sensory neurons. CGRP is a member of the amylin family, which also includes amylin and adrenomedulin (ADM) (reviewed in Tache *et al.*, 1992). There is close to 20% homology between CGRP and ADM and 40% homology between CGRP and amylin. Amylin-like peptides, including CGRP, have six or seven amino acids in a common N-terminal disulphide ring that is required for biological activity, an amidated C-terminus, and some conserved amino acids in the middle of the peptide. CGRP has a disulphide linkage between Cys2 and Cys7, an amphiphathic α-helix between residues 8 and 18, a β-turn formed by residues 19 and 20, and a C-terminal amide.

CGRP was discovered as an alternative spliced mRNA product of the calcitonin gene in rat medullary thyroid carcinoma cells (Rosenfeld *et al.*, 1981). Analysis of the mRNA expression found in these cells led to the detection of two different mRNA species derived from the calcitonin gene (Amara *et al.*, 1982; Rosenfeld *et al.*, 1983). Figure 6.1 shows a scheme depicting alternative splicing of the calcitonin/CGRP gene. Alternative splicing of calcitonin and CGRP mRNAs is regulated by cis-acting elements that bind to the 3' end on the third intron (Emeson *et al.*, 1989). CGRP mRNA is composed of exons 1–3 and 5 and 6. Exon 5 contains the CGRP sequence. CGRP mRNA lacks exon 4, which contains the calcitonin coding sequence. Translation of the CGRP mRNA yields a 16 kd precursor that is proteolytically processed to yield a mature protein. Morris *et al.* (1984) reported the isolation and characterization of human CGRP, which has a high homology (89%) with rat CGRP. Immunohistochemical studies by Rosenfeld *et al.* (1983) determined that the synthesis of CGRP mRNA and calcitonin mRNA is a tissue-specific event that placed the proteins in distinct tissues. Calcitonin is found mostly in cells of the endocrine system like thyroid C cells, whereas CGRP is found in tissues from the CNS and PNS with motor and sensory functions, including dorsal root ganglion cells (Rosenfeld *et al.*, 1983). A second CGRP neuropeptide, β-CGRP, homologous to rat CGRP, was found using rat cDNA libraries (Amara *et al.*, 1985).

At the protein level, rat and human α- and β-CGRPs differ by only one and three amino acids, respectively, with no reported functional differences between the two intact forms.

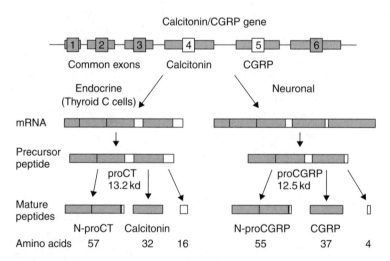

Figure 6.1   Alternative mRNA splicing of the calcitonin/CGRP gene. The calcitonin/CGRP gene is composed of six exons. CGRP mRNA is the product of the splicing of the exons 1–3 to exons 5 and 6. Exon 6 contains a poly (A) site at the 3' end. Calcitonin mRNA is composed of the exons 1–4, with the cleavage and poly (A) site at the 3' end of the fourth exon. The precursor peptide for CGRP is 12.5 kd and is proteoliticaly processed to yield the 37 amino acid long CGRP and two other peptides.

The only functional differences between the two are attributed to their cleaved fragments (Manley and Haynes, 1989; Davies *et al.*, 1992). Furthermore, when expression patterns of these two isoforms were investigated, it was found that both α- and β-CGRP mRNA are expressed in the same tissues in the brain and periphery. CGRP is synthesized in the body of sensory nerves and transported to the nerve endings, where it is released by voltage-dependent calcium uptake. Sensory nerve fibres are stimulated by noxious stimuli and certain inflammatory mediators including pro-inflammatory factors like IL-1 and prostaglandins (PGE_2) (Herbert and Holzer, 1994a,b). Release of CGRP into the extra-cellular space is often accompanied by release of Substance P (SP) (Rodrigo *et al.*, 1985). CGRP is present in nerve endings in all body surfaces, in bone and dental pulp, around all blood vessels, and in primary and secondary lymphoid tissues, including the thymus and lymph nodes (Lee *et al.*, 1985; Rodrigo *et al.*, 1985; Bjurholm *et al.*, 1988, 1990; Kimberly and Byers, 1988; Popper *et al.*, 1988; Popper and Micevych, 1989; Weihe *et al.*, 1989; Bjurholm, 1991; Kruger *et al.*, 1991). The bone marrow, site of B-cell lymphopoiesis in the adult mammal, also contains CGRP containing fibres (Bjurholm *et al.*, 1988; Hill and Elde, 1991; Buma *et al.*, 1992; Hukkanen *et al.*, 1992; Iwasaki *et al.*, 1995). CGRP is also found in secondary lymphoid tissue. Using immunohistochemical analysis of the lymph nodes, Popper *et al.* (1988) showed CGRP immunoreactivity in the medullary cord, in the blood vessels and in the deep cortex in MLN. Around 80% of neurons contain both CGRP and SP, but CGRP can be found without SP in some motor and enteric neurons (Rodrigo *et al.*, 1985). In addition, CGRP is present around all blood vessels and at all internal and external surfaces, where it can influence inflammatory and immune responses.

There are some recent reports by Wang and coworkers (Wang *et al.*, 1999; Xing *et al.*, 2000) that CGRP can also be produced by lymphocytes. These observations are based on an ability to detect an immunoreactive substance using anti-CGRP antibodies, and by RT-PCR. Over the past two decades there have been dozens of widely scattered reports suggesting that lymphocytes produce many different neuropeptides. Like the reports on CGRP, these isolated reports are based largely on RIA or IHC analysis and on RT-PCR. The caveat is that many of these studies are very preliminary in nature and have yet to be followed up. In contrast to expression of smaller neuropeptides there is more convincing evidence that lymphocytes can produce some larger proteins such as prolactin, a pituitary hormone. The ability of lymphocytes to produce larger polypeptide hormones is supported indirectly by the ability of lymphocytes to produce an array of cytokine and lymphokine mediators that share similar biochemical characteristics and processing mechanisms with larger endocrine protein hormones and cytokines. While the number of initial reports suggest that one can detect any neuropeptide mRNA desired in lymphocytes by RT-PCR, reliable and reproducible data beyond the initial reports is almost nonexistent. For this reason, the recent reports on CGRP production (and other neuropeptides) by lymphocytes must be interpreted very carefully until there is more extensive evidence and information on CGRP production by lymphocytes and on its roles and functions relative to neural CGRP.

CGRP RECEPTORS

The CGRP receptor is a seven-transmembrane domain, G-protein-associated receptor. The receptor is found throughout the body including the nervous, endocrine and cardiovascular systems (Seifert *et al.*, 1985a,b; Tschopp *et al.*, 1985; Sexton *et al.*, 1986, 1988; Sigrist *et al.*, 1986; Hirata *et al.*, 1988; Dennis *et al.*, 1989; Haegerstrand *et al.*, 1990). The CGRP receptor was originally cloned in 1993 (Chang *et al.*, 1993; Njuki *et al.*, 1993) and identified as an orphan receptor related to the calcitonin receptor and was thus named calcitonin receptor-like receptor (CRLR). Subsequently, the human homologue was identified (Aiyar *et al.*, 1996). The molecular weight of the receptor varies among various species depending on the level of glycosylation. In the human neuroblastoma cell line SK-N-MC it is about 56 kd, but is reduced to 44 kd after deglycosylation (Muff *et al.*, 1995). On rat lymphocytes, the molecular weight is close to 75 kd, while on the murine pre-B cell line 70Z/3, the molecular weight is 103 kd (McGillis *et al.*, 1991, 1993a). Differences in post-translational processing and the expression of different CGRP receptor sub-type genes account for the discrepancies in the molecular weight of the receptor. Two CGRP receptor sub-types have been proposed based on their ability to bind specific CGRP analogues. These analogues include the CGRP antagonist $CGRP_{8-37}$, which binds efficiently to receptors in guinea pig atria, but less effectively to receptors in the rat vas deferens (Dennis *et al.*, 1990) and the CGRP agonist $[Cys(acetomethoxy)]^{2,7}$-CGRP which is more selective for the rat vas deferens than guinea pig atria (Quirion *et al.*, 1992). These results have not been consistent in all studies, and will require further examination. It is possible that the issue of high and low affinity CGRP receptors is the result of accessory components to the receptor rather than distinct receptors.

Some studies suggested that the CGRP receptor could bind two different ligands: CGRP and another member of the amylin family of neuropeptides ADM. Recently, it was reported that this apparent cross-reactivity is due to the expression of a second component of the CGRP receptor system. McLatchie *et al.* (1998) reported that the type of functional receptor expressed by a cell that express CRLR depends on the expression of the receptor activity-modulating protein (RAMP), of which three different isoforms (RAMP1–3) have been identified. It was postulated that the specific RAMP protein it associates with will determine the activity of CRLR. RAMP1 is required for signalling by CGRP, while association of RAMP2 with CRLR would make the cell sensitive to ADM. The function for RAMP3 has not been established. Besides ligand-binding attributes, RAMP proteins seem to be required for transport of CRLR to the surface of the cell. The expression of RAMP proteins may help resolve the issues regarding the various reported CGRP receptors, as it may determine the affinity of the receptor to various CGRP analogues.

CGRP receptors were identified several years ago in lymphoid tissue. Nakamuta *et al.* (1986) and later Sigrist *et al.* (1986) reported the presence of the receptor in cell membrane preparations from the spleen in 1986. Since then the receptor has been found in several cell types of the lymphoid system including B and T lymphocytes, monocytes, macrophages, various lymphoid cell lines and cells from the bone marrow (Umeda and Arisawa, 1990; Abello *et al.*, 1991; Bulloch *et al.*, 1991; McGillis *et al.*, 1991, 1993a; Mullins *et al.*, 1993; Owan and Ibaraki, 1994). The lack of anti-CGRP receptor antibodies has made it difficult to identify which specific subsets of B and T cells express the receptor on their surface. Ligand-binding experiments show that the level of expression differs among different cells: bone marrow cells express CGRP receptors at a higher density (3000/cell versus 700/cell) than mature B cells (McGillis *et al.*, 1991; Mullins *et al.*, 1993). The murine cell line 70Z/3 has 20 000 receptors/cell (McGillis *et al.*, 1993a). The study of receptor mRNA expression and the receptor functionality in specific populations of developing B cells from the bone marrow shows that B cells express functional receptors from very early in their developing stage until more mature stages (McGillis *et al.*, unpublished observation).

CGRP AS AN INFLAMMATORY AND IMMUNOMODULATORY FACTOR

CGRP, INFLAMMATION AND EFFECTS ON T CELLS

CGRP was first reported to be a potent vasodilator by Brain *et al.* (1985). In these experiments, rat and human CGRP were tested for their ability to induce relaxation of rat aorta tissue *in vitro*, microvascular plasma protein leakage, persistent reddening and increased blood flow in human skin. Blood flow did not decrease in the presence of prostaglandin synthesis inhibitors like indomethacin, indicating that CGRP acts directly on endothelial cells, and not through intermediates like PGE_2 and PGI_2. In the cardiovascular system, CGRP functions to increase vasodilation and therefore decrease arteriolar blood pressure and increase muscle contractility (reviewed in Preibisz, 1993). In the renal system, CGRP stimulates renin secretion and can stimulate renal blood flow and glomerular filtration (Kurtz *et al.*, 1988; Gnaedinger *et al.*, 1989). CGRP enhances the effect of tachykinins, such as SP, on oedema via its vasodilatory effect (Newbold and Brain, 1993).

A number of functions have been attributed to CGRP, including regulation of microvasculature and local inflammatory responses (Brain *et al.*, 1985; Brain and Williams, 1985). The role of CGRP has been expanded in recent years to include important roles in regulating immune and inflammatory responses. Its effects on neutrophils includes an ability to stimulate cell adhesion (Zimmerman *et al.*, 1992), tissue infiltration (Buckley *et al.*, 1991) and granule secretion (Richter *et al.*, 1992). In eosinophils, proteolytic fragments of CGRP can act as a chemoattractant (Manley and Haynes, 1989; Davies *et al.*, 1992).

CGRP is known to have several effects on lymphocytes. CGRP inhibits proliferation of T lymphocytes through the inhibition of IL-2 production (Umeda *et al.*, 1988; Boudard and Bastide, 1991; Bulloch *et al.*, 1991; Wang *et al.*, 1992). Following antigenic activation of T cells, IL-2 production and IL-2 receptor expression are upregulated and are necessary for proliferation. CGRP-induced inhibition of IL-2 production is dependent on cAMP accumulation (a common second messenger in CGRP signalling), suggesting a connection between CGRP-dependent cAMP accumulation and downregulation of T cell proliferation (Bulloch *et al.*, 1991). CGRP has also been shown to inhibit Con-A induced proliferation of thymocytes (Bulloch *et al.*, 1991). This inhibition was suggested to be the consequence of an enhancement in the rate of apoptosis (Bulloch *et al.*, 1998). CGRP-induced apoptosis is more evident among $CD8^+/CD4^+$ cells (Bulloch *et al.*, 1998). It is interesting to note that this group of cells normally undergoes selection in the thymus, suggesting that CGRP plays a role in development and selection of T cells at this site.

The effect of CGRP on specific subsets of T-lymphocytes depends on the cell phenotype. Using T cell clones it was possible to show that while CGRP had no effect on Th2 clones, it does induce a transient, but strong, accumulation of cAMP in Th1 cells (Wang *et al.*, 1992).

However, it should be noted that these results were based on analysis of different T cell lines. The effect was specific, since the use of the CGRP antagonist $CGRP_{8-37}$ abrogated it. In contrast, Levite (1998) reported that CGRP, as well as other neuropeptides like SP and somastostatin, may have effects on both Th1 and Th2 clones, including a wide range of changes in their cytokine secretion profiles. These conclusions are limited by the lack of data from experiments done with normal cells. More studies are necessary to clarify these apparent contradictions. Recent studies do make an interesting correlation between CGRP and T cell function *in vivo*. Transgenic NOD mice expressing CGRP in their pancreas show a delay in the onset of insulin-dependent diabetes mellitus (Khachatryan *et al.*, 1997). It is possible that CGRP inhibition of Th1 cells could account for the delayed onset of the disease.

CGRP EFFECTS ON MACROPHAGES

Macrophages play several key roles in both innate and specific immune responses including phagocytosis of pathogens, generation of free radicals that kill microbes, antigen presentation to T cells and production of pro-inflammatory cytokines. The expression of functional CGRP receptors on macrophages suggested that CGRP might regulate some macrophage functions (Vignery *et al.*, 1991; Owan and Ibaraki, 1994). Nong *et al.* (1989) reported that human peripheral blood monocytes pre-incubated with nanomolar concentrations of CGRP fail to produce H_2O_2 in response to IFN-γ. A similar result was reported by Taylor *et al.* (1998) who found that the aqueous humor from the eye in rabbits contains CGRP. In this system, CGRP inhibited nitric oxide synthase (NOS2) activity in macrophages, thus inhibiting the production of nitric oxide (NO). A caveat in this report is that the concentration of CGRP found in the rabbit aqueous humor was substantially higher than the reported concentration required for CGRP to inhibit macrophage function. These results have been somewhat contradicted by reports that CGRP increases NO production in mouse peritoneal macrophages (Tang *et al.*, 1999). NO is required for the upregulation of IL-6 in the same cells. This cAMP and PKA-dependent effect is specific in that it can be blocked by $CGRP_{8-37}$. Furthermore, when NOS2 inhibitors were used, synthesis of IL-6 was stopped; suggesting that CGRP uses NO as a mediator. These differences may be explained by suggesting that CGRP's influence on macrophages depends on the surrounding microenvironment and other stimulatory signals present.

Although phagocytosis is enhanced by CGRP (Ichinose and Sawada, 1996), it has been shown that CGRP acts as an inhibitor of antigen presentation in macrophages (Nong *et al.*, 1989; Torii *et al.*, 1997). The same observation has been made regarding Langerhan cells (LC) (Asahina *et al.*, 1995a,b). More recently (Carucci *et al.*, 2000), it was shown that CGRP down regulates antigen presentation in human dendritic cells as well as CD86 expression, both of which are required for T cell activation. Reduction in antigen presentation by CGRP is probably accomplished by the down regulation of expression of B7.2 molecule on the surface of macrophages (Torii *et al.*, 1997). B7.2 binds to CD28 on T cells during activation, thus it could account for the reduction in antigen presentation. The downregulation of B7.2 appears to be the result of the increase in IL-10 production in macrophages, since neutralizing antibodies reverse this effect.

Increase in IL-10 production in CGRP-treated macrophages may have profound effects on T cell function (Torii *et al.*, 1997). As mentioned above, CGRP retains the ability to influence T cell function by skewing the balance between Th1 and Th2 in cell lines. It is possible that CGRP favours a Th1 response over a Th2 response, not only by inhibiting IL-2 production in T cells, but also by increasing IL-10 production in macrophages. Taking all together, it would appear as if the general role of CGRP in this system is to favour a cytotoxic activity (Th1 response) over the humoral Th2 response.

CGRP EFFECTS ON B LYMPHOCYTES

Studies showing that mature B cells have CGRP receptors suggests that CGRP has functional effects on mature B cells. However, little is known on how CGRP affects mature B cell function. We have examined the effects of CGRP on *in vitro* activation of small resting B lymphocytes. We found that CGRP had no effect on RNA or DNA synthesis in cells activated with anti-IgM (McGillis *et al.*, unpublished observation). These results suggest that CGRP will not inhibit antigenic activation of mature resting B lymphocytes. However, it is possible that CGRP has other regulatory effects such as influencing factors such as class switching, differentiation following antigenic activation, etc. In contrast our limited understanding of its effects on mature B lymphoyctes, our recent studies suggest that CGRP has a significant role on the development of new B cells in the bone marrow. This work was influenced by two observations, that cells in the bone marrow have higher levels of CGRP receptors (Mullins *et al.*, 1993) and the ability of CGRP to inhibit differentiation of 70Z/3 pre-B cells (McGillis *et al.*, 1993b).

Differentiation of B cells is a complex process that requires cell contact with stromal cells at early stages and depends on a number of soluble factors (Rolink and Melchers, 1992, 1993; Kincade *et al.*, 1993; Kincade, 1994; Rosenberg and Kincade, 1994). Figure 6.2 is an overview of the major stages in B cell development. An earlier impediment to studying cells at different stages was the inability to isolate them. Hardy *et al.* (1991) used antibodies to B cell proteins to subdivide B220$^+$ pre-, pro-, immature, and mature B cells into sub-fractions designated A–F (Li *et al.*, 1993). Positive regulators of B cell lymphopoiesis include IL-3, -6, -7, and -11, PBSF, Flt3, PBEF, and IGF-1 (Kincade and Medina, 1994). Some of these regulators act directly on B220$^+$ cells, and others act indirectly through effects on stromal cells. One of the most important factors in B cell development is IL-7. IL-7 was initially described based on its ability to stimulate proliferation of developing B cells (Namen *et al.*, 1988) and was subsequently found to be important for T cell development. IL-7 & IL-7 Receptor (IL-7R) knock-out mice do not have mature lymphocytes (Peschon *et al.*, 1994; von Freeden-Jeffry *et al.*, 1995). In the B cell lineage, the IL-7R is only expressed on early sIg$^-$ cells (reviewed in Candejas *et al.*, 1997). IL-7 can act as both a survival factor and as growth factor for early B cells. IL-7 is necessary at early stages for Ig rearrangement. After successful rearrangement of the heavy chain gene, there is a brief period of IL-7 dependent proliferation (Lee *et al.*, 1989). Light chain rearrangement occurs after this expansion and the expression of surface IgM marks the transition from a pre-B cell (D) to an immature B cell (E).

Several negative regulators of B cell differentiation are known, including IL-1α, IL-3, IL-13, TGF-β, IFN-γ, IFN-α, oestrogen and the neuropeptides vasoactive intestinal

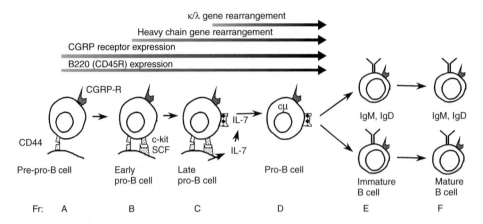

Figure 6.2   Early B cell differentiation. The diagram shows a simplified version of early B cell development. The letters on the bottom denote the nomenclature used by Hardy *et al.* (1991). The bars above the figure show the onset of some major events in B cell development. CGRP receptors are expressed at all stages in during B cell development.

peptide (VIP) and CGRP (Lee *et al.*, 1987; Dorshkind, 1988; Hayashi *et al.*, 1989; Okada *et al.*, 1989; Gimble *et al.*, 1993; Grawunder *et al.*, 1993; McGillis *et al.*, 1993b; Hirayama *et al.*, 1994; Kincade and Medina, 1994; Renard *et al.*, 1994; Shimozato and Kincade, 1997; Fernandez *et al.*, 2000). IL-4 and -10 act as both positive and negative regulators (Peschel *et al.*, 1987; Rennick *et al.*, 1987; King *et al.*, 1988; Billips *et al.*, 1990; Fine *et al.*, 1994; Elia *et al.*, 1995; Veiby *et al.*, 1997). It is possible that other regulators also have multiple influences that are context dependent – the response differs depending on other signals the cell receives. Oritani *et al.* (2000) recently described a new inhibitor of B cell development, limitin, that is of interest because it inhibits IL-7 responses.

Negative regulation of B lymphopoiesis by oestrogen and VIP provides a precedent for regulation of B lymphopoiesis by factors produced outside the immune system and bone marrow (Kincade and Medina, 1994; Shimozato and Kincade, 1997). Estrogen, which acts by inhibiting stromal cell production of soluble positive regulators, is responsible for the decrease in B lymphopoiesis observed in pregnancy (Smithson *et al.*, 1995). VIP also acts indirectly by stimulating production of IFNα in bone marrow macrophages (Shimozato and Kincade, 1997). SP which is coexpressed with CGRP in most sensory neurons, also plays a role in hematopoiesis. SP stimulates erythroid and granulocytic colonies *in vitro* (Rameshwar *et al.*, 1993), and stimulates cytokine production by bone marrow stromal cells (Rameshwar *et al.*, 1994; Rameshwar and Gascon, 1995, 1996).

The effect of CGRP on 70Z/3 cells provided the first evidence that CGRP could influence early B cell differentiation. 70Z/3 pre-B cells express a high level of CGRP receptors (~23 000 per cell) (McGillis *et al.*, 1993a). 70Z/3 cells are phenotypically similar to pre-B cells (cytoplasmic μ$^+$, κ$^-$; surface immunoglobulin$^-$ (sIg$^-$) (Paige *et al.*, 1978). Treatment with IL-1 induces κ transcription and expression of sIg. In the 70Z/3 pre-B cell line, CGRP inhibits LPS and IL-1β induced expression of surface immunoglobulin, an event that occurs on normal cells subsequent to IL-7 induced expansion. Analysis of CGRP receptor expression has shown that developing B cells, from the most immature to IgM^+/IgD^+ immunocompetent cells, express both the components of the CGRP receptor, CRLR and RAMP1 (McGillis *et al.*, unpublished observations). These receptors are functional since treatment of these cell populations with CGRP induces the expression of c-fos mRNA. Not only do developing B-cells express all the components of the CGRP receptor, they also express the 24.11 neural peptidase (CD10) on their surface (Salle *et al.*, 1993). The presence of CD10 is important in drawing a connection between B cells and neuropeptides. CD10 can cleave CGRP, suggesting that developing B cells can regulate the concentration of CGRP in their environment, and therefore their responsiveness to CGRP (Shipp *et al.*, 1991).

Recent studies show that CGRP inhibits early B cell differentiation *in vitro* by both direct and indirect mechanisms (Fernandez *et al.*, 2000, 2001 and unpublished observations). In these studies, cells from bone marrow or purified B cell progenitors were cultured in 0.3% agar with IL-7. B cells in the C fraction proliferate and form discrete colonies (Lee *et al.*, 1989). Using whole bone marrow, CGRP inhibited IL-7 induced colony formation at concentrations ranging from 10^{-11} to 10^{-7} M, with the maximal inhibition occurring at 1 nM (10^{-9} M) (Fernandez *et al.*, 2000). Using FACS purified $B220^+/IgM^-$ cells the inhibitory effect of CGRP on IL-7 induced proliferation was similar to that seen with whole bone marrow, suggesting that CGRP inhibits IL-7 responses by acting directly on responding B cells (Fernandez *et al.*, 2000). The inhibitory effect of CGRP could be blocked by the CGRP antagonist, $CGRP_{8-37}$, suggesting that it is mediated by a specific CGRP receptor. In additional studies, cultured bone marrow stromal cells capable of supporting B cell growth were treated with CGRP and their supernatants were analysed for the presence factors that could influence IL-7 responses (CGRP was neutralized by addition of anti-CGRP anti-serum). The supernatents from CGRP treated stromal cells also inhibited IL-7 induced colony formation, suggesting that CGRP also inhibits IL-7 responses indirectly by induction of inhibitory factors in the bone marrow stroma.

The most recent studies have identified two cytokines induced by CGRP capable of inhibiting IL-7 responses by B cell precursors, IL-6 and TNF-α (Fernandez *et al.*, 2001 and unpublished observation). CGRP induces IL-6 production by bone marrow macrophages. Unlike CGRP, IL-6 only inhibits IL-7 responses when using whole bone marrow for the CFU assay and not with purified B cell precursors. This suggests that the effect of IL-6 is indirect. TNF-α, in contrast, can directly inhibit IL-7 induced colony formation by purified B cell precursors. In studies on the source of CGRP induced TNF-α it was observed that TNF-α could be induced in long-term bone marrow stromal cultures where bone marrow macrophages are the predominant cell type. However, CGRP did not induce TNF-α in bone marrow derived macrophage cultures suggesting that TNF-α is induced in some other non-macrophage cell in the bone marrow matrix. In addition, there is very preliminary evidence that CGRP can induce at least one other cytokine in the bone marrow stroma that has potential to influence B cell development. While the data on how indirect effects of CGRP on B cell development is being mediated is very preliminary, it is clear that much more work will be required to determine what cytokines CGRP induces in bone marrow, their cellular sources, and their potential to influence early B cell development.

SUMMARY

CGRP has a somewhat unique history in that it was first discovered by molecular cloning and had no known functions. In the first decade after its discovery far more was known about its tissue specific expression (RNA processing of the calcitonin/CGRP gene) and localization. However, in the last 10–12 years, it has become clear that CGRP is an important inflammatory mediator and immunoregulator. It can influence a number of leukocyte processes in local microenviroments. In addition, recent evidence supports its role in the early development of B cells, and potentially in other lineages as well.

REFERENCES

Abello, J., Kaiserlian, D., Cuber, J. C., Revillard, J. P. and Chayvialle, J. A. (1991) Characterization of calcitonin gene-related peptide receptors and adenylate cyclase response in the murine macrophage cell line P388 D1. *Neuropeptides*, **19**, 43–49.

Aiyar, N., Rand, K., Elshourgagy, N. A., Zeng, A., Adamou, J. E., Bergsma, D. J. and Li, Y. (1996) A cDNA encoding the calcitonin gene-related peptide type 1 receptor. *Journal of Biological Chemistry*, **271**, 11325–11329.

Amara, S. G., Arriza, J. L., Leff, S. E., Swanson, L. W., Evans, R. M. and Rosenfeld, M. G. (1985) Expression in brain of a messenger RNA encoding a novel neuropeptide homologous to calcitonin gene-related peptide. *Science*, **229**, 1094–1097.

Amara, S. G., Jones, V., Rosenfeld, M. G., Ong, E. S. and Evans, R. M. (1982) Alternative RNA processing in calcitonin gene expression generates mRNAs encoding different polypeptide products. *Nature*, **298**, 240–244.

Asahina, A., Hosoi, J., Grabbe, S. and Granstein, R. D. (1995a) Modulation of Langerhans cell function by epidermal nerves. *Journal of Allergy and Clinical Immunology*, **96**, 1178–1182.

Asahina, A., Moro, O., Hosoi, J., Lerner, E. A., Xu, S., Takashima, A. and Granstein, R. D. (1995b) Specific induction of cAMP in Langerhans cells by calcitonin gene-related peptide: Relevance to functional effects. *Proceedings of the National Academy of Sciences USA*, **92**, 8323–8327.

Billips, L. G., Petitte, D. and Landreth, K. S. (1990) Bone marrow stromal cell regulation of B lymphopoiesis: interleukin-1 (IL-1) and IL-4 regulate stromal cell support of pre-B cell production in vitro. *Blood*, **75**, 611–619.

Bjurholm, A. (1991) Neuroendocrine peptides in bone. *International Orthopedics.*, **15**, 325–329.

Bjurholm, A., Kreicbergs, A., Brodin, E. and Schultzberb, M. (1988) Substance P- and CGRP-immunoreactive nerves in bone. *Peptides*, **9**, 166–171.

Bjurholm, A., Kreicbergs, A., Dahlberg, L. and Schultzberg, M. (1990) The occurrence of neuropeptides at different stages of DPM-induced heterotopic bone formation. *Bone and Mineral*, **10**, 95–98.

Boudard, F. and Bastide, M. (1991) Inhibition of mouse T-cell proliferation by CGRP and VIP: effects of these neuropeptides on IL-2 production and cAMP systhesis. *Journal of Neuroscience Research*, **29**, 29–41.

Brain, S. D. and Williams, T. J. (1985) Inflammatory oedema induced by synergism between calcitonin gene-releated peptide (CGRP) and mediators of increased vascular permeability. *British Journal of Pharmacology*, **86**, 855–860.

Brain, S. D., Williams, T. J., Tippins, J. R., Morris, H. R. and MacIntyre, I. (1985) Calcitonin gene-related peptide is a potent vasodilator. *Nature*, **313**, 54–56.

Buckley, T. L., Brain, S. D., Collins, P. D. and Williams, T. J. (1991) Inflammatory edema induced by interactions between IL-1 and the neuropeptide calcitonin gene-related peptide. *Journal of Immunology*, **146**, 3424–3430.

Bulloch, K., McEwen, B. S., Nordberg, J., Diwa, A. and Baird, S. (1998) Selective regulation of T-cell development and function by calcitonin gene-related peptide in thymus and spleen. *Annals of the New York Academy Sciences*, **840**, 551–562.

Bulloch, K., Radojcic, T., Yu, R., Hausman, J., Lenhard, L. and Baird, S. (1991) The distribution and function of calcitonin gene-related peptide in the mouse thymus and spleen. *Progress in Neuroendocrinimmunology*, **4**, 186–194.

Buma, P., Verxhuren, C., Versleyen, D., Van der Kraan, P. and Oestreicher, A. B. (1992) Calcitonin gene-related peptide, substance P and GAP-43/B-50 immunoreactivity in the normal and arthrotic knee joint of the mouse. *Histochemistry*, **98**, 327–339.

Candejas, S., Muegge, K. and Durum, S. K. (1997) IL-7 receptor and VDJ recombination: trophic versus mechanistic actions. *Immunity*, **6**, 501–508.

Carucci, J. A., Ignatius, R., Wei, Y., Cypess, A. M., Schaer, D. A., Pope, M., Steinman, R. M. and Mojsov, S. (2000) Calcitonin gene-related peptide decreases expression of HLA-DR and CD86 by human dendritic cells and dampens dendritic cell-driven T cell-proliferative responses via the type I calcitonin gene-related peptide receptor. *Journal of Immunology*, **164**, 3494–3499.

Chang, C. P., Pearse, R. V., O'Connell, S. and Rosenfeld, M. G. (1993) Identification of a seven transmembrane helix receptor for corticotropin-releasing factor and sauvagine in mammalian brain. *Neuron*, **11**, 1187–1195.

Davies, D., Medeiros, M. S., Keen, J., Truner, A. J. and Haynes, L. W. (1992) Endopeptidase-24.11 cleaves a chemotactic factor from alpha-calcitonin gene-related peptide. *Biochemical Pharmacology*, **43**, 1753–1756.

Dennis, T., Fournier, A., Cadieux, A., Pomerleau, F., Jolicoeur, F. B., St. Pierre, S. and Quirion, R. (1990) hCGRP8–37, a calcitonin gene-related peptide antagonist revealing calcitonin gene-related peptide receptor heterogeneity in brain and periphery. *Journal of Pharmacology and Experimental Therapeutics*, **254**, 123–128.

Dennis, T., Fournier, A., St. Pierre, S. and Quirion, R. (1989) Structure–activity profile of calcitonin gene-related peptide in peripheral and brain tissues. Evidence for receptor multiplicity. *Journal of Pharmacology and Experimental Therapeutics*, **251**, 718–725.

Dorshkind, K. (1988) IL-1 inhibits B cell differentiation in long term bone marrow cultures. *Journal of Immunology*, **141**.

Elia, J. M., Hamilton, B. L. and Riley, R. L. (1995) IL-10 inhibits IL-7 mediated murine pre-B cell growth in vitro. *Experimental Hematology*, **23**, 323–327.

Emeson, R. B., Hedjran, F., Yeakley, J. M., Guise, J. W. and Rosenfeld, M. G. (1989) Alternative production of calcitonin and CGRP mRNA is regulated at the calcitonin-specific splice acceptor. *Nature*, **341**, 76–80.

Fernandez, S., Knopf, M. A., Bjork, S. K. and McGillis, J. P. (2001) Bone marrow-derived macrophages express functional CGRP receptors and respond to CGRP by increasing transcription of c-fos and IL-6 mRNA. *Cellular Immunology*, **209**, 140–148.

Fernandez, S., Knopf, M. A. and McGillis, J. P. (2000) Calcitonin gene-related peptide (CGRP) inhibits inter-leukin-7 induced proliferation of murine B cell precursors. *Journal of Leukocyte Biology*, **67**, 669–676.

Fine, J. S., Macoscko, H. D., Grace, M. J. and Narula, S. K. (1994) Influence of IL-10 on murine CFU-pre-B formation. *Experimental Hematology*, **22**, 1188–1196.

Gimble, J. M., Medina, K., Hudson, J., Robinson, M. and Kincadek, P. W. (1993) Modulation of lymphohe-matopoiesis in long-term cultures by gamma interferon: direct and indirect action on lymphoid and stromal cells. *Experimental Hematology*, **21**, 224–230.

Gnaedinger, M. P., Uehlinger, D. E., Weidmann, P., Sha, S. G., Muff, R., Born, W., Rascher, W. and Fischer, J. A. (1989) Distinct hemodynamic and renal effects of calcitonin gene-related peptide and calcitonin in men. *American Journal of Physiology*, **257**, 870–875.

Grawunder, U., Melchers, F. and Rolink, A. (1993) Interferon-gamma arrests proliferation and causes apoptosis in stromal cell/interleukin-7-dependent normal murine pre-B cell lines and clones in vitro, but does not induce differentiation to surface immunoglobulin-positive B cells. *European Journal of Immunology*, **23**, 544–551.

Haegerstrand, A., Dalsgaard, C.-J., Jonzon, B., Larsson, O. and Nilsson, J. (1990) Calcitonin gene-related peptide stimulates proliferation of human endothelial cells. *Proceedings of the National Academy of Sciences USA*, **87**, 3299–3303.

Hardy, R. R., Carmack, C. E., Shinton, S. A., Kemp, J. D. and Hayakawa, K. (1991) Resolution and characterization of pro-B and pre-pro-B cell stages in normal mouse bone marrow. *Journal of Experimental Medicine*, **173**, 1212–1225.

Hayashi, S., Gimble, J. M., Henley, A., Ellingsworth, L. R. and Kincade, P. W. (1989) Differential effects of TGF-beta 1 on lymphohemopoiesis in long-term bone marrow cultures. *Blood*, **74**, 1711–1717.

Herbert, M. K. and Holzer, P. (1994a) Interleukin-1 enhances capsaicin-induced neurogenic vasodilatation in the rat skin. *British Journal of Pharmacology*, **111**, 681–686.

Herbert, M. K. and Holzer, P. (1994b) Nitric oxide mediates the amplification by interleukin-1a of neurogenic vasodilatation in the rat skin. *European Journal of Pharmacology*, **260**, 89–93.

Hill, E. L. and Elde, R. (1991) Distribution of CGRP-, VIP-, D βH-, SP-, and NPY-immunoreactive nerves in the periosteum of the rat. *Cell and Tissue Research*, **264**, 469–480.

Hirata, Y., Takagi, Y., Takata, S., Fukuda, Y., Yoshimi, H. and Fujita, T. (1988) Calcitonin gene-related peptide receptor in cultured vascular smooth muscle and endothelial cells. *Biochemical and Biophysical Research Communications*, **151**, 1113–1121.

Hirayama, F., Clark, S. C. and Ogawa, M. (1994) Negative regulation of early B lymphopoiesis by interleukin 3 and interleukin 1 alpha. *Proceedings of the National Academy of Sciences USA*, **91**, 469–473.

Hukkanen, M., Konttinin, Y., Rees, R. G., Gibson, S. J., Santavirta, S. and Polak, J. M. (1992) Innervation of bone from healthy and arthritic rats by substance P and calcitonin gene related peptide containing sensory fibers. *Journal of Rheumatology*, **19**, 1252–1259.

Ichinose, M. and Sawada, M. (1996) Enhancement of phagocytosis by calcitonin gene-related peptide (CGRP) in cultured mouse peritoneal macrophages. *Peptides*, **17**, 1405–1414.

Iwasaki, A., Inoue, K. and Hukuda, S. (1995) Distribution of neuropeptide-containing nerve fibers in the synovium and adjacent bone of the rat knee joint. *Clinical and Experimental Rheumatology*, **13**, 173–178.

Khachatryan, A., Guerder, S., Palluault, F., Cote, G., Solimena, M., Valentijn, K., Millet, I., Flavell, R. A. and Vignery, A. (1997) Targeted expression of the neuropeptide calcitonin gene-related peptide to beta cells prevents diabetes in NOD mice. *Journal of Immunology*, **158**, 1409–1416.

Kimberly, C. L. and Byers, M. R. (1988) Inflammation of rat molar pulp and periodontium causes increased calcitonin gene-related peptide and axonal sprouting. *Anatomical Record*, **222**, 289–300.

Kincade, P. W. (1994) B lymphopoiesis: global factors, local control. *Proceedings of the National Academy of Sciences USA*, **91**, 2888–2889.

Kincade, P. W., He, Q., Ishihara, K., Miyake, K., Lesley, J. and Hyman, R. (1993) CD44 and other interaction molecules contributing to B lymphopoiesis. *Current Topics in Microbiology and Immunology*, **184**, 215–222.

Kincade, P. W. and Medina, K. L. (1994) Pregnancy: a clue to normal regulation of B lymphopoiesis. *Immunology Today*, **15**, 539–544.

King, A. G., Wierda, D. and Landreth, K. S. (1988) Bone marrow stromal cell regulation of B-lymphopoiesis. I. The role of macrophages, IL-1, and IL-4 in pre-B cell maturation. *Journal of Immunology*, **141**, 2016–2026.

Kruger, L., Silverman, J. D., Mantyh, P. W., Sternini, C. and Brecha, N. C. (1991) Peripheral patterns of calcitonin-gene-related peptide general somatic sensory innervation: cutaneous and deep terminations. *Journal of Comparative Neurology*, **280**, 291–302.

Kurtz, A., Born, W., Lundberg, J. M., Millberg, B. I., Gnadinger, M. P., Uehlinger, D. E., Weidmann, P., Hokfelt, T., Muff, R. and Fischer, J. A. (1988) Calcitonin gene-related peptide is a stimulator of renin secretion. *Journal of Clinical Investigation*, **82**, 538–543.

Lee, G., Ellingsworth, L. R., Gillis, S., Wall, R. and Kincade, P. W. (1987) Beta transforming growth factors are potential regulators of B lymphopoiesis. *Journal of Experimental Medicine*, **166**, 1290–1299.

Lee, G., Namen, A. E., Gillis, S., Ellingsworth, L. R. and Kincade, P. W. (1989) Normal B cell precursors responsive to recombinant murine IL-7 and inhibition of IL-7 activity by transforming growth factor-beta. *Journal of Immunology*, **142**, 3875–3883.

Lee, Y., Takami, K., Kawai, Y., Girgis, S., Hillyard, C. J., MacIntyre, I., Emson, P. C. and Tohyama, M. (1985) Distribution of calcitonin gene-releated peptide in the rat peripheral nervous sytem with reference to its coexistence with substance P. *Neuroscience*, **15**, 1227–1237.

Levite, M. (1998) Neuropeptides, by direct interaction with T cells, induce secretion and break the commitment to a distinct T helper phenotype. *Proceedings of the National Academy of Sciences USA*, **95**, 12544–12549.

Li, Y. S., Hayakawa, K. and Hardy, R. R. (1993) The regulated expression of B lineage associated genes during B cell differentiation in bone marrow and fetal liver. *Journal of Experimental Medicine*, **178**, 951–960.

Manley, H. C. and Haynes, L. W. (1989) Eosinophil chemotactic response to rat CGRP-1 is increased after exposure to trypsin or guinea-pig lung particulate fraction. *Neuropeptides*, **13**, 29–34.

McGillis, J. P., Humphreys, S., Rangnekar, V. and Ciallella, J. (1993a) Modulation of B lymphocyte differentiation by calcitonin gene related peptide (CGRP) I: Characterization of high affinity CGRP receptors on murine 70Z cells. *Cellular Immunology*, **150**, 391–404.

McGillis, J. P., Humphreys, S., Rangnekar, V. and Ciallella, J. (1993b) Modulation of B lymphocyte differentiation by calcitonin gene related peptide (CGRP) II: Inhibition of LPS induced kappa light chain expression by CGRP. *Cellular Immunology*, **150**, 405–416.

McGillis, J. P., Humphreys, S. and Reid, S. (1991) Characterization of functional calcitonin gene-related peptide receptors on rat lymphocytes. *Journal of Immunology*, **147**, 3482–3489.

McLatchie, L. M., Fraser, N. J., Main, M. J., Wise, A., Brown, J., Thompson, N., Solari, R., Lee, M. G. and Foord, S. M. (1998) RAMPs regulate the transport and ligand specificity of the calcitonin-receptor like receptor. *Nature*, **393**, 333–339.

Morris, H. R., Panico, M., Etienne, T., Tippins, J., Girgis, S. I. and MacIntyre, I. (1984) Isolation and characterization of human calcitonin gene-related peptide. *Nature*, **308**, 746–748.

Muff, R., Born, W. and Fischer, J. A. (1995) Calcitonin, calcitonin gene related peptide, adrenomedullin and amylin: homologous peptides, separate receptors and overlapping biological actions. *European Journal of Endocrinology*, **133**, 17–20.

Mullins, M. S., Ciallella, J., Rangnekar, V. and McGillis, J. P. (1993) Characterization of a calcitonin gene-related peptide (CGRP) receptor on mouse bone marrow cells. *Regulatory Peptides*, **49**, 65–72.

Nakamuta, H., Fukuda, Y., Koida, M., Fufii, N., Otaka, A., Funakoshi, S., Mitsuyasu, N. and Orlowski, R. C. (1986) Binding sites of calcitonin gene-related peptide (CGRP): abundant occurrence in visceral organs. *Japananese Journal of Pharmacology*, **42**, 175–180.

Namen, A. E., Lupton, S., Hjerrild, K., Wignall, J., Mochizuki, D. Y., Schmierer, A., Mosley, B., March, C. J.,
    Urdal, D. and Gillis, S. (1988) Stimulation of B-cell progenitors by cloned murine interleukin-7. *Nature*,
    333, 571–573.

Newbold, P. and Brain, S. D. (1993) The modulation of inflammatory oedema by calcitonin gene-related peptide.
    British Journal Pharmacology, **108**, 705–710.

Njuki, F., Nicholl, C. G., Howard, A., Mak, J. C., Barnes, P. J., Girgis, S. I. and Legon, S. (1993) A new calcitonin-
    receptor-like sequence in rat pulmonary blood vessels. *Clinical Science Colchester*, **85**, 385–388.

Nong, Y. H., Titus, R. G., Ribeiro, J. M. C. and Remold, H. G. (1989) Peptides encoded by the calcitonin gene
    inhibit macrophage function. *Journal of Immunology*, **143**, 45–49.

Okada, S. T., Miura, S. J., Ito, M. Y., Sudo, T., Hayashi, S., Nishikawa, S. and Nakauchi, H. (1989) A stimulatory
    effect of recombinant murine interleukin-7 (IL-7) on B-cell colony formation and an inhibitory effect of
    IL-1 alpha. *Blood*, **74**, 1936–1941.

Oritani, K., Medina, K. L., Tomiyama, Y., Ishikawa, J., Okajima, Y., Ogawa, M., Yokota, T., Aoyama, K.,
    Takahashi, I., Kincade, P. W. and Matsuzawa, Y. (2000) Limitin: an interferon-like cytokine that preferen-
    tially influences B-lymphocyte precursors. *Nature Medicine*, **6**, 659–666.

Owan, I. and Ibaraki, K. (1994) The role of calcitonin gene-related peptide (CGRP) in macrophages: the presence
    of functional receptors and effects on proliferation and differentiation into osteoclast-like cells. *Bone and
    Mineral*, **24**, 151–164.

Paige, C. J., Kincade, P. W. and Ralph, P. (1978) Murine B cell leukemia line with inducible surface immunoglob-
    ulin expression. *Journal of Immunology*, **121**, 641–647.

Peschel, C., Green, I., Ohara, J. and Paul, W. E. (1987) Role of B cell stimulatory factor 1/interleukin 4 in clonal
    proliferation of B cells. *Journal of Immunology*, **139**, 3338–3347.

Peschon, J. J., Morrissey, P. J., Grabstein, K. H., Ramsdell, F. J., Maraskovsky, E., Gliniak, B. C., Park, L. S.,
    Ziegler, S. F., Williams, D. E. and Ware, C. B. (1994) Early lymphocyte expansion is severely impaired in
    interleukin 7 receptor-deficient mice. *Journal of Experimental Medicine*, **180**, 1955–1960.

Popper, P., Mantyh, C. R., Vigna, S. R., Maggio, J. E. and Mantyh, P. W. (1988) The localization of sensory nerve
    fibers and receptor binding sites for sensory neuropeptides in canine mesenteric lymph nodes. *Peptides*, **9**,
    257–267.

Popper, P. and Micevych, P. E. (1989) Localization of calcitonin gene-related peptide and its receptors in a striated
    muscle. *Brain Research*, **496**, 180–186.

Preibisz, J. J. (1993) Calcitonin gene-related peptide and regulation of human cardiovascular homeostasis.
    American Journal of Hypertension, **6**, 434–450.

Quirion, R., Van Rossum, D., Dumont, Y., St-Piere, S. and Fournier, A. (1992) Characterization of CGRP1 and
    CGRP2 receptor subtypes. *Annals of the New York Academy of Sciences*, **657**, 88–105.

Rameshwar, P., Ganea, D. and Gascon, P. (1993) In vitro stimulatory effect of substance P on hematopoiesis.
    Blood, **81**, 391–398.

Rameshwar, P., Ganea, D. and Gascon, P. (1994) Induction of IL-3 and granulocyte-macrophage colony-
    stimulating factor by substance P in bone marrow cells is partially mediated through the release of IL-1
    and IL-6. *Journal of Immunology*, **152**, 4044–4054.

Rameshwar, P. and Gascon, P. (1995) Substance P (SP) mediates production of stem cell factor and interleukin-1
    in bone marrow stroma: potential atuoregulatory role for these cytokines in SP receptor expression and
    induction. *Blood*, **86**, 482–490.

Rameshwar, P. and Gascon, P. (1996) Induction of negative hematopoietic regulators by neurokinin-A in bone
    marrow stroma. *Blood*, **88**, 98–106.

Renard, N., Duvert, V., Banchereau, J. and Saeland, S. (1994) Interleukin-13 inhibits the proliferation of normal
    and leukemic human B-cell precursors. *Blood*, **84**, 2253–2260.

Rennick, D., Yang, G., Muller-Sieburg, C., Smith, C., Arai, N., Takabe, Y. and Gemmell, L. (1987) Interleukin 4
    (B-cell stimulatory factor 1) can enhance or antagonize the factor-dependent growth of hemopoietic progen-
    itor cells. *Proceedings of the National Academy of Sciences USA*, **84**, 6889–6893.

Richter, J., Andersson, R., Edvinsson, L. and Gullberg, U. (1992) Calcitonin gene-related peptide (CGRP) acti-
    vates human neutrophils - inhibition by chemotactic peptide antagonist BOC-MLP. *Immunology*, **77**,
    416–421.

Rodrigo, J., Polak, J. M., Fernandez, L., Ghatei, M. A., Mulderry, P. and Bloom, S. R. (1985) Calcitonin gene-
    related peptide immunoreactive sensory and motor nerves of the rat, cat, and monkey esophagus.
    Gastroenterology, **88**, 444–451.

Rolink, A. and Melchers, F. (1992) B lymphopoiesis in the mouse. *Advances in Immunology*, **53**, 123–156.

Rolink, A. and Melchers, F. (1993) Generation and regeneration of cells of the B-lymphocyte lineage. *Current
    Opinion in Immunology*, **6**, 203–211.

Rosenberg, N. and Kincade, P. W. (1994) B-lineage differentiation in normal and transformed cells and the
    microenvironment that supports it. *Current Opinion in Immunology*, **6**, 203–211.

Rosenfeld, M. G., Amara, S. G., Roos, B. A., Ong, E. S. and Evans, R. M. (1981) Altered expression of the calcitonin gene associated with RNA polymorphism. *Nature*, **290**, 63–65.

Rosenfeld, M. G., Mermod, J. J., Amara, S. G., Swanson, L. W., Sawchenko, P. E., Rivier, J., Vale, W. W. and Evans, R. M. (1983) Production of a novel neuropeptide encoded by the calcitonin gene via tissue-specific RNA processing. *Nature*, **304**, 129–135.

Salle, G., Rodewald, H. R., Chin, B. S. and Reinherz, E. L. (1993) Inhibition of CD10/neutral endopeptidase 24.11 promotes B-cell reconstitution and maturation in vivo. *Proceedings of the National Academy of Sciences USA*, **90**, 7618–7622.

Seifert, H., Chesnut, J., De Souza, E., Rivier, J. and Vale, W. (1985a) Binding sites for calcitonin gene-related peptide in distinct areas of rat brain. *Brain Research*, **346**, 195–198.

Seifert, H., Sawchenko, P., Chesnut, J., Rivier, J., Vale, W. and Pandol, S. J. (1985b) Receptor for calcitonin gene-related peptide: Binding to exocrine pancreas mediates biological actions. *American Journal of Physiology*, **249**, G147–G151.

Sexton, P. M., McKenzie, J. S., Mason, R. T., Moseley, J. M., Martin, T. J. and Mendelsohn, F. A. O. (1986) Localization of binding sites for calcitonin gene-related peptide in rat brain by in vitro autoradiography *Neuroscience*, **19**, 1235–1245.

Sexton, P. M., McKenzie, J. S. and Mendelsohn, F. A. O. (1988) Evidence for a new subclass of calcitonin/ calcitonin gene-related peptide binding site in rat brain. *Neurochemistry International*, **12**, 323–335.

Shimozato, T. and Kincade, P. W. (1997) Indirect suppression of IL-7-responsive B cell precursors by vasoactive intestinal peptide. *Journal of Immunology*, **1997**, 5178–5184.

Shipp, M. A., Stefano, G. B., Switzer, S. N., Griffin, J. D. and Reinherz, E. L. (1991) CD10 (CALLA)/neutral endopeptidase 24.11 modulates inflammatory peptide-induced changes in neutrophil morphology, migration, and adhesion proteins and is itself regulated by neutrophil activation. *Blood*, **78**, 1834–1841.

Sigrist, S., Franco-Cereceda, A., Muff, R., Henke, H., Lundberg, J. M. and Fischer, J. A. (1986) Specific receptor and cardiovascular effects of calcitonin gene-related petide. *Endocrinology*, **119**, 381–389.

Smithson, G., Medina, K., Ponting, I. and Kincade, P. W. (1995) Estrogen suppresses stromal cell-dependent lymphopoiesis in culture. *Journal of Immunology*, **155**, 3409–3417.

Tache, Y., Holzer, P. and Rosenfeld, M. G. (1992) *Calcitonin Gene-Related Peptide, The First Decade of a Novel Pleiotropic Neuropeptide*, New York Academy of Sciences, New York.

Tang, Y., Han, C. and Wang, X. (1999) Role of nitric oxide and prostaglandins in the potentiating effects of calcitonin gene-related peptide on lipopolysaccharide-induced interleukin-6 release from mouse peritoneal macrophages. *Immunology*, **96**, 171–175.

Taylor, A. W., Yee, D. G. and Streilein, J. W. (1998) Suppression of nitric oxide generated by inflammatory macrophages by calcitonin gene-related peptide in aqueous humor. *Investigative Ophthalmology*, **39**, 1372–1378.

Torii, H., Hosoi, J., Beissert, S., Xu, S., Fox, F. E., Asahina, A., Takashima, A., Rook, A. H. and Granstein, R. D. (1997) Regulation of cytokine expression in macrophages and the Langerhans cell-like line XS52 by calcitonin gene-related peptide. *Journal of Leukocyte Biology*, **61**, 216–223.

Tschopp, F. A., Henke, H., Petermann, J. B., Tobler, P. H., Janzer, R., Hokfelt, T., Lundberg, J. M., Cuello, C. and Fisher, J. A. (1985) Calcitonin gene-related peptide and its binding sites in the human central nervous system and pituitary. *Proceedings of the National Academy Science USA*, **82**, 248–252.

Umeda, Y. and Arisawa, M. (1990) Characterization of the calcitonin gene-related peptide receptor in mouse T lymphocytes. *Neuropeptides*, **14**, 237–242.

Umeda, Y., Takamiya, M., Yoshizaki, H. and Arisawa, M. (1988) Inhibition of mitogen-stimulated T lymphocyte proliferation by calcitonin gene-related petide. *Biochemical and Biophysical Resesearch Communications*, **154**, 227–235.

Veiby, O. P., Borge, O. J., Martensson, A., Beck, E. X., Schade, A. E., Grzegorzewski, K., Lyman, S. D., Martensson, I. L. and Jacobsen, S. E. (1997) Bidirectional effect of interleukin-10 on early murine B-cell development: stimulation of flt3-lignad plus interleukin-7-dependent generation of CD19(−) ProB cells from uncommitted bone marrow progenitor cells and growth inhibition of CD19(+) ProB cells. *Blood*, **90**, 4321–4331.

Vignery, A., Wang, F. and Ganz, M. B. (1991) Macrophages express functional receptors for calcitonin-gene-related peptide. *Journal of Cellular Physiology*, **149**, 301–306.

von Freeden-Jeffry, U., Vieira, P., Lucian, L. A., McNeil, T., Burdach, S. E. and Murray, R. (1995) Lymphopenia in interleukin (IL-7) gene-deleted mice identifies IL-7 as a nonredundant cytokine. *Journal of Experimental Medicine*, **181**, 1519–1526.

Wang, F., Millet, I., Bottomly, K. and Vignery, A. (1992) Calcitonin gene-related peptide inhibits interleukin 2 production by murine T lymphocytes. *Journal of Biological Chemistry*, **267**, 21052–21057.

Wang, X., Xing, L., Xing, Y., Tang, Y. and Han, C. (1999) Identification and characterization of immunoreactive calcitonin gene-related peptide from lymphocytes of the rat. *Journal of Neuroimmunology*, **94**, 95–102.

Weihe, E., Muller, S., Fink, T. and Zentel, H. J. (1989) Tachykinins, calcitonin gene-related peptide and neuropeptide Y in nerves of the mammalian thymus: Interactions with mast cells in autonomic and sensory neuroimmunomodulation. *Neuroscience Letters*, **100**, 77–82.

Xing, L., Guo, J. and Wang, X. (2000) Induction and expression of beta-calcitonin gene-related peptide in rat T lymphocytes and its significance. *Journal of Immunology*, **165**, 4359–4366.

Zimmerman, B. J., Anderson, D. C. and Granger, D. N. (1992) Neuropeptides promote neutrophil adherence to endothelial cell monolayers. *American Journal of Physiology*, **263**, G678–G682.

7 Substance P and the Immune System

Joel V. Weinstock

Division of Gastroenterology–Hepatology, Department of Medicine, University of Iowa, Iowa City, IA 52242, USA

Substance P (SP) is an 11 amino acid molecule derived from the preprotachykinin A gene. It is a product of both nerves and leukocytes. SP-containing nerves are in thymus, bone marrow, spleen and many other organs. SP binds with high affinity to a G-protein-coupled, seven transmembrane receptor called NK1, which is its natural ligand. NK1 receptors may use one of several distinct intracellular signalling pathways depending on influences from the external environment. NK1 receptors are displayed throughout the body on vascular endothelial cells, epithelial cells, smooth muscle cells, neurons, lymphocytes, macrophages and other cell types. The NK 1 receptor and SP expression are subject to immunoregulation. SP and its receptor clearly have immune functions. They are important for mediating a process called neurogenic inflammation. They also have critical roles in several infectious, toxin and antigen-induced models of inflammation. In the thymus and bone marrow, SP may have ongoing steady-state functions. There are various reported effects of SP on cells of the immune system. Differences exist among rodents, guinea pigs and humans. Some of the biological activities attributed to SP occur only at high SP concentrations suggesting signalling pathways independent of NK1 receptors. The physiological significance of most of these observations has not yet been tested in context of inflammation or disease. There are several non-peptide NK1 receptor antagonists under clinical evaluation, which may ultimately lend insight into the critical physiological roles of SP in humans.

KEY WORDS: substance P; neurokinin 1 receptor; substance P receptor; 7 transmembrane receptor; inflammation; immunoregulation.

INTRODUCTION

Substance P (SP) was recognized as a neuropeptide for more than 30 years before it was sequenced in 1971 (Chang *et al.*, 1971). Antibodies raised against SP subsequently allowed immunohistochemistry. Developed soon after were sensitive assays (RIA and ELISA) to accurately quantify SP extracted from tissue and body fluids. In the 1970s, several reports suggested that rat neuronal tissue displayed high-affinity receptors for SP. The NK1 receptor, which is the natural high-affinity receptor for SP, was cloned and the amino acid sequence deduced in 1989 (Yokota *et al.*, 1989). The structure of the first highly specific, non-peptide NK1 antagonist was published in 1991 (Snider *et al.*, 1991). The SP receptor knockout mouse was reported in 1996 (Bozic *et al.*, 1996a). Also raised were antibodies

111

against this receptor permitting immunohistochemical and Western blot detection of NK1 receptor protein. Each of these technical advances provided important new tools to further explore the relationship of SP and its receptor to the immune system.

This chapter provides an up to date summary regarding published data pertaining to the importance of SP and its receptor in inflammation. I also recommend reading the insightful reviews by Maggi (1997).

SUBSTANCE P SYNTHESIS

Substance P, an 11 amino acid protein, is one of a series of molecules called tachykinins. The tachykinins have a common carboxyl-terminal sequence Phe-X-Gly-Leu-Met-NH2. The X is either an aromatic (Tyr or Phe) or branched aliphatic (Val or Ile) amino acid. Several mammalian tachykinins include SP, neurokinin A (substance K), neurokinin B, neuropeptide K and neuropeptide γ (Vanden Broeck *et al.*, 1999).

Two distinct genes called preprotachykinin A and B (preprotachykinin I and II) (Tac 1 and Tac 2/3) encode the tachykinins. Neurokinin B derives from preprotachykinin B, while the other four are from preprotachykinin A. The preprotachykinin A gene has seven exons that can be alternatively spliced to produce four distinct mRNA. These are called α-, β-, γ- and δ-preprotachykinin A. All four mRNA variants encode SP, which derives from exon 3. Only β-preprotachykinin A mRNA has the complete nucleotide sequence corresponding to all seven exons and, thus can generate the four preprotachykinin A gene products. The β-preprotachykinin A form is most widely expressed. The α-preprotachykinin A lacks exon 6, while the γ transcript lacks exon 4 and the δ transcript is missing exon 4 and 6 (Figure 7.1). Please see www.ncbi.nlm.nih.gov/ for the sequence data on the prepro- tachykinin genes and mRNA (Krause *et al.*, 1987; Carter and Krause, 1990; Harmar *et al.*, 1990; Helke *et al.*, 1990; Chapman *et al.*, 1993; Kako *et al.*, 1993).

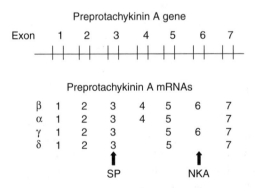

Figure 7.1   The preprotachykinin A gene has seven exons and eight introns. Substance P (SP) is encoded in exon 3, while neurokinin A (NKA) is encoded in exon 6. The gene can produce four separate splice variants of prepro- tachykinin mRNA (α, β, γ, δ). Only the β sequence has nucleotides corresponding to all seven exons. They all encode SP, but only β and γ encode NKA.

SOURCES OF SUBSTANCE P

Substance P is present near sites of inflammation, and inflammation can enhance its expression. There is a rapid increase in preprotachykinin A mRNA in mouse intestinal Peyer's patches, mesenteric lymph nodes and spleen following enteric inoculation with Salmonella (Bost, 1995). In various other inflammatory states, preprotachykinin A mRNA increases in regional ganglia (Fischer *et al.*, 1996; Castagliuolo *et al.*, 1997). Also, there is more SP in human nasal secretions (Mosimann *et al.*, 1993), induced sputum (Tomaki *et al.*, 1995) and synovial fluid during inflammation (Arnalich *et al.*, 1994). Inflamed tissue often contains more SP then that of normal control tissue (Weinstock *et al.*, 1988; Agro and Stanisz, 1993; Holzer, 1998).

Some sensory neurons, both extrinsic and intrinsic to tissue produce SP. Thus, it seems plausible that nerves are a source of the SP that can regulate inflammation. Neuronal SP is stored in vesicles and released from sensory nerves in response to various stimuli. Some factors that induce SP release include electrical nerve stimulation, leukotrienes, prostaglandins and histamine (Maggi, 1995). Capsaicin, which is the agent in hot peppers that causes the stinging sensation, induces SP depletion from sensory nerves. It also depletes several other neuropeptides and may possibly stimulate directly intestinal and bronchial epithelial cells (Veronesi *et al.*, 1999).

Substance P-containing nerves have a distribution favourable for immunoregulation. They are abundant at mucosal surfaces and are in ganglia, spinal cord, brain and skin as determined by immunohistochemistry. They also are around blood vessels and course through the tissue that neighbours surface epithelia (Maggi, 1995, 1996; van der Velden and Hulsmann, 1999). In the gut, SP is in extrinsic afferent nerve fibres and intrinsic enteric neurons (Holzer and Holzer-Petsche, 1997a,b).

Both noradrenergic and neuropeptidergic nerve fibers are adjacent to cells of the immune system in the spleen, thymus, lymph nodes, bone marrow and frequently at sites of inflammation (Madden and Felten, 1995). In the rat thymus, there are abundant nerves containing SP within the thymic capsule and interlobular septa. They also distribute among cortical thymocytes and mast cells (Lorton *et al.*, 1990; Jurjus *et al.*, 1998). The rat spleen has SP-containing nerves coursing into the trabeculae and surrounding red pulp, and into the marginal zone of the white pulp (Lorton *et al.*, 1991). In the intestine and elsewhere, there is a close association between mast cells and nerves (McKay and Bienenstock, 1994; Suzuki *et al.*, 1999), some of which produce SP (Stead *et al.*, 1987).

Using immunohistochemistry, various studies report up- or downregulation of SP expression in nerves at sites of inflammation (Swain *et al.*, 1992; Keranen *et al.*, 1995, 1996; Miura *et al.*, 1997; Chanez *et al.*, 1998; Forsgren *et al.*, 2000). It is difficult to interpret the significance of these observations, since changes in staining intensity could reflect either enhanced or impaired SP secretion, or simply technical limitations of immunohistochemistry.

Substance P is also a product of human, rat and mouse leukocytes that can be released at sites of inflammation to govern immune responses. In mice colonized with *Schistosoma mansoni*, a helminthic parasite, eggs settle in the liver and intestinal

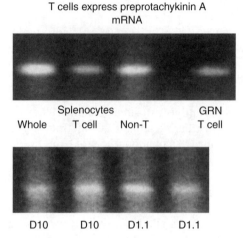

T cells express preprotachykinin A
mRNA

| | Splenocytes | | GRN |
| Whole | T cell | Non-T | T cell |

| D10 | D10 | D1.1 | D1.1 |

Figure 7.2   Animals were colonized with *Schistosoma mansoni*. At the eighth week of infection, mRNA were extracted from unfractionated dispersed splenocytes (whole), splenic T cells (T cell) or splenocytes depleted of T cells (non-T). Also extracted were mRNA from purified schistosome granuloma T cells (GRN T cell), the murine Th2 cell line D10 and the murine Th1 cell line D1.1. Preprotachykinin mRNA was detected by PCR. The T cells were isolated using magnetic beads that recognized Thy1.2.

wall. These eggs induce around themselves a focal, chronic inflammatory response called a granuloma. The granulomas comprise about 50% eosinophils, 30% macrophages, 10% T cells, 5% B cells and <1% mast cells. As the granulomas develop, they are focally destructive, removing or displacing the host tissue. The granulomas contain no nerves (Varilek *et al.*, 1991). However, they are a rich source of preprotachykinin A mRNA and SP, but not of other tachykinins. The granuloma eosinophils produce and secrete SP (Weinstock *et al.*, 1988; Weinstock and Blum, 1989). Human eosinophils may produce it also (Aliakbari *et al.*, 1987; Metwali *et al.*, 1994). Advancing technology now permits isolation of various granuloma cell subsets in large quantity and in extreme purity. It is now known that both the T cell and non-T cell elements of the granuloma and spleen in murine schistosomiasis express preprotachykinin A mRNA (Figure 7.2). The T cell lines D1.1 and D10 express it also. It recently was reported that human peripheral blood lymphocytes might make SP (De Giorgio *et al.*, 1998; Lai *et al.*, 1998).

Macrophages and dendritic cells are other potential sources. Lipopolysaccharide (LPS) induces preprotachykinin mRNA expression and SP production from rat peritoneal macrophages (Bost *et al.*, 1992). Rat and human alveolar macrophages also express preprotachykinin A mRNA and contain SP (Germonpre *et al.*, 1999). LPS stimulation enhances SP expression in these cells and in the human macrophage-like cell line U-937 (Killingsworth *et al.*, 1997; Germonpre *et al.*, 1999). Human peripheral blood monocytes and macrophages also produce SP. Capsaicin, but not LPS, induces SP secretion (Ho *et al.*, 1997). Mouse dendritic cells, grown from bone marrow, are another source (Lambrecht *et al.*, 1999).

SUBSTANCE P RECEPTOR (NK1 RECEPTOR)

SYNTHESIS AND STRUCTURE

Substance P, neurokinin A and neurokinin B have distinct biological functions because they bind and signal through different cell surface receptors. Their receptors are designated NK1, NK2 and NK3, respectively. The three receptors share homology. SP binds the NK1 receptor with high affinity ($K_i < 10^{-9}$ M). The order of affinity of the various tachykinins for NK1 receptors is SP $\gg$ neurokinin A $>$ neurokinin B. SP binds with low affinity to NK2 and NK3. Thus, the function of SP on the immune system probably is mediated mostly through interaction with the NK1 receptor.

A single gene encodes the NK1 receptor. In humans, it is located on chromosome 2 (Gerard *et al.*, 1991). It is located on chromosome 6 in mice and rats. The gene has five exons and four introns. Reports suggest that mammalian tissue also may produce a truncated isoform of the NK1 receptor that is of unknown physiological significance. Sequence data on the NK1 receptor gene (Takahashi *et al.*, 1992) and mRNA (Gerard *et al.*, 1991) in humans and in other species (Hershey and Krause, 1990; Hershey *et al.*, 1991; Sundelin *et al.*, 1992) is available at www.ncbi.nlm.nih.gov/.

The NK1 receptor protein has been deduced from the mRNA sequence. There are 407 amino acids in the human, rat and mouse NK1 receptors. The amino acid sequence of the rat is 99% identical to that of the mouse, but differs by 22 residues from that of humans (Table 7.1). These variations do not affect SP binding or signalling. They do account for the species-dependent variation in potency of the various pharmacological NK1 receptor antagonists.

The NK 1 receptor is a G-protein-coupled receptor that forms α-helices that span the cell membrane seven times (Ohkubo and Nakanishi, 1991). The receptor is anchored firmly to the plasma membrane and not released in soluble form. The binding site for SP requires regions of transmembrane domains I, II and VII, and the N-terminus (Berthold and Bartfai, 1997).

NK1 RECEPTORS AND INTRACELLULAR SIGNALLING

The intracytoplasmic C-terminal conformation determines the biological activity. In non-immune cells, the receptor can couple to several G proteins ($G_{\alpha q/11}$, $G_{\alpha s}$ and $G_{\alpha o}$) (Roush and Kwatra, 1998). Stimulation of NK1 receptors can excite various second messenger

TABLE 7.1
Comparisons of tachykinin receptor subtypes.

	NK1	*NK2*	*NK3*
Molecular weight	46 364	43 851	51 104
Amino acid residues	407	390	452
Homology in amino acid sequence	54% to NK2		55% to NK2
	66% to NK3		

pathways. It is probable that the actual signalling pathways employed vary depending on the signalling components active within a cell at the time of NK1 receptor engagement. In experimental non-immune cell systems, ligand binding can activate phospholipase C and D (Torrens *et al.*, 1998), generating I(1,4,5)P3 and increasing $[Ca^{2+}]$. Also activated are arachidonic acid release, adenylyl cyclase and phospholipase A2(1,2,3) (Grady *et al.*, 1995).

A recent study reported activation of NF-κB in a human astrocytoma cell line that was blocked by a specific NK1 antagonist (Lieb *et al.*, 1997). Also reported was that SP enhanced activation of NF-κB in murine macrophages and dendritic cells (Marriott *et al.*, 2000). NF-κB is a transcription factor that has anti-apoptotic function and promotes production of pro-inflammatory cytokines like IL-1, IL-18 and TNF-α.

Substance P can signal via other intracellular signalling pathways besides NF-κB to regulate cytokine production. SP induces transcriptional activation of IL6 in the same astrocytoma cell line. This requires activated p38 MAP kinase that functions independently of NF-κB to transduce the signal mediating IL-6 expression (Fiebich *et al.*, 2000). It also involves activation of protein kinase C and the transcriptional activator NF of IL6 (Lieb *et al.*, 1998). While there are several reports suggesting that SP can induce calcium flux in immune cells (Kavelaars *et al.*, 1993), it is unclear if these events were mediated by the NK1 receptor.

LOCATION OF THE NK1 RECEPTOR

The precise location and cellular distribution of NK1 receptors remain controversial. Most of the techniques and reagents employed over the years usually could not differentiate SP binding to NK1 receptors versus binding to other receptors. There is strong evidence for NK1 receptor expression on vascular endothelial cells (Greeno *et al.*, 1993). It also is expressed in neuronal tissue, salivary glands, renal pelvis, ureter, bladder and pulmonary microvasculature (Bowden *et al.*, 1996). The intestines have NK1 binding sites in the smooth muscle layers, submucosa, epithelium and ganglia of the enteric plexus (Goode *et al.*, 2000a).

Before the cloning of the NK1 receptor, investigators reported high affinity binding of SP to human and murine T cells (Payan *et al.*, 1984a; Stanisz *et al.*, 1987), murine B cells (Stanisz *et al.*, 1987), the human B cell line IM-9 (Payan *et al.*, 1984b, 1986) and to guinea pig macrophages (Hartung *et al.*, 1986). There were NK1-like receptors detected in germinal centres of mesenteric lymph nodes and in intestinal lymphoid tissue (Mantyh *et al.*, 1988). Since then, studies demonstrated that lymphoid organs and immunocytes can express NK1 receptors. Reports suggest that lymphocytes and macrophages can display this receptor in both human and other mammalian species (Cook *et al.*, 1994; McCormack *et al.*, 1996; Kincy-Cain and Bost, 1997; Ho *et al.*, 1997; Goode *et al.*, 1998; Germonpre *et al.*, 1999). T cells from the granulomas of murine schistosomiasis constitutively express functional NK1 receptor (Cook *et al.*, 1994) as do human intestinal lamina propria lymphocytes (Blum *et al.*, 1993a; Goode *et al.*, 2000b). It also is present on several macrophage and T cell lines (Blum *et al.*, 2001) (Li *et al.*, 1995; McCormack *et al.*, 1996) and on the rat mast cell line RBL (Cooke *et al.*, 1998).

REGULATION OF NK1 RECEPTORS

Although not yet studied in immunocytes, the physiological response of neuronal cells or transfected cell lines to SP exposure is NK1 receptor desensitization followed by gradual re-sensitization. Studies using CHO and other receptor-transfected cell lines suggest that NK1 receptor desensitization requires receptor phosphorylation by G-protein-coupled receptor kinases like GRK2 (Nishimura *et al.*, 1998). β-Arrestins uncouple the phosphorylated receptor from the heterotrimeric G proteins (Barak *et al.*, 1999; McConalogue *et al.*, 1999) terminating signal transduction. Reported is an isoform of the NK1 receptor truncated at the carboxyl-terminus that does not undergo rapid and prolonged desensitization upon exposure to SP (Li *et al.*, 1997).

Substance P also stimulates clathrin-mediated endocytosis and recycling of the NK1 receptor (Garland *et al.*, 1996), which is an important part of the process of receptor re-sensitization. Receptor transmembrane domain VII and several domains in the intracellular C-terminus are important for endocytosis (Sasakawa *et al.*, 1994; Bohm *et al.*, 1997). In acidified endosomes (Grady *et al.*, 1995), SP dissociates from its receptor and the receptor is dephosphorylated. The restored receptor is brought back to the cell surface.

The immune cell NK1 receptors are subject to regulation. In murine schistosomiasis and in various T cell lines, T cell receptor engagement, IL-12 and IL-18 all can trigger or upregulate murine T cell NK1 receptor mRNA expression and protein display (Blum *et al.*, 2001). Activation of rat macrophages with LPS also upregulates NK1 receptor mRNA expression (Bost *et al.*, 1992). IL-4 or IFN-γ can elicit increases in NK1 receptor protein and mRNA in murine peritoneal macrophages (Marriott and Bost, 2000). NK1 display is more prominent at sites of inflammation, further suggesting that this receptor is subject to upregulation during immune responses. For instance, inflammation induces heightened expression on rat pulmonary angiogenic blood vessels (Baluk *et al.*, 1997).

PHYSIOLOGICAL FUNCTION OF THE NK1 RECEPTOR IN INFLAMMATION

The wide distribution of the NK1 receptor and substantial additional evidence suggest that this receptor, and by inference SP, has many functions (Quartara and Maggi, 1997). The NK1 receptor is involved in pain transmission in peripheral nerves and the spinal cord (Mantyh *et al.*, 1997). In the CNS, it influences neuronal survival and helps regulate the emetic reflex, cardiovascular and respiratory functions, and various behavioural responses. Other functions include dilatation of blood vessels and enhancement of vascular permeability. It also has a role in intestinal secretion, motility, and neuro-neuronal communication (Holzer and Holzer-Petsche, 1997a,b).

While SP has various reported immunoregulatory functions (Maggi, 1997), the functions mediated via the NK1 receptor are less well defined. SP when used particularly at high and perhaps non-physiological concentrations can bind to receptors other than NK1.

RELEVANCE OF NK1 RECEPTORS IN INFECTIOUS
DISEASE MODELS OF INFLAMMATION

Recent experiments addressed this issue using various animal models of disease and newly developed, highly specific NK1 receptor antagonists and NK1 knock-out mice. These reports implicate the NK1 receptor as serving a critical role in immune modulation and susceptibility to infection. With the exception of studies in murine schistosomiasis, they have yet to show the relative importance of the NK1 receptors expressed on immune versus parenchymal cells pertaining to the observed pathology. NK1 receptors are on vascular endothelium, mucosal epithelial cells and other parenchymal cell types that can exert immunoregulation (Mohle *et al.*, 1998; Nakazawa *et al.*, 1999; Knolle and Gerken, 2000; Wagner and Roth, 2000).

Schistosomiasis

Murine schistosomiasis mansoni has a SP and somatostatin immunoregulatory circuit that operates within the granulomas and lymphoid organs (Figure 7.3) (Weinstock and Elliott, 1998). Schistosome granuloma cells have mRNA for both preprotachykinin A and prepro-somatostatin. Granuloma macrophages make somatostatin 1–14 (Weinstock *et al.*, 1990; Elliott *et al.*, 1998). Granuloma eosinophils (Weinstock *et al.*, 1988) and other cell types produce SP. Murine macrophage cell lines and splenic macrophages also express somato-statin. The production of somatostatin is under immunoregulatory control, since LPS, IL-10, TNF-α, IFN-γ, prostaglandin E2, vasoactive intestinal peptide and cAMP analogs each induce macrophage SOM production (Elliott *et al.*, 1998).

While various inflammatory mediators induce somatostatin expression in macrophages, SP is a potent inhibitor of somatostatin synthesis (Blum *et al.*, 1998). SP prevents

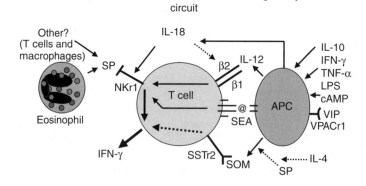

Figure 7.3   Soluble egg antigen (SEA) presented to T cells via antigen presenting cells (APC) induces NK1 receptor (NKr1) expression on T cells. IL-12 and IL-18 can do likewise independently of antigen presentation. Substance P from eosinophils and probably other immune cell types interacts with the T cell NK1 receptor to enhance IFN-γ production. Somatostatin from the APC can engage somatostatin receptor SSTr2 to downmod-ulate IFN-γ secretion. Various factors stimulate APC to make somatostatin, while substance P via the NK1 recep-tor blocks somatostatin production. However, IL-4, the Th2 cytokine, blocks this substance P effect. Dotted arrows are inhibition, while solid arrows are stimulation.

induction of somatostatin expression and downmodulates ongoing production. SP even blocks preprosomatostatin mRNA production in *Rag*-1 splenocytes devoid of T and B cells suggesting that this regulation is independent of T and B cells. SP operates through the authentic NK1 receptor, since specific NK1 receptor antagonists block this somatostatin regulation. It is likely that SP is working though a macrophage NK1 receptor.

IL-4 blocks SP from inhibiting macrophage somatostatin synthesis. Thus, somatostatin synthesis persists in the Th2 environment of the schistosome granuloma because of the production of IL-4.

Both SP and somatostatin regulate T cell IFN-γ secretion (Blum *et al.*, 1993a,b). *In vitro* and *in vivo*, SP and somatostatin are strong modulators of IFN-γ production in the granulomas and spleens of schistosome-infected mice. SP increases, while somatostatin decreases IFN-γ release. In splenic or granuloma cell cultures, they have little effect on IL-1, IL-2, IL-4, IL-5, IL-6, IL-10 or TNF-α production. IFN-γ regulation is most apparent when the cells receive suboptimal antigen stimulation. They also modulate IgG2a production, which is IFN-γ-dependent (Blum *et al.*, 1993a). This shows that the effect of SP and somatostatin on IFN-γ production is biologically significant. Intracytoplasmic analysis of IFN-γ production reveals that SP and somatostatin modulate T cell IFN-γ synthesis (unpublished observation). CD4+ T cells are the major source of IFN-γ in schistosome granulomas (Rakasz *et al.*, 1998).

IL-12 and TGF-β are two cytokines that regulate IFN-γ production. IL-12 induces Th1 cell development, and stimulates T cells and NK cells to make IFN-γ. TGF-β inhibits IFN-γ production. Schistosome granulomas produce IL-12 and TGF-β that have an important role in controlling IFN-γ synthesis (Rakasz *et al.*, 1998). In murine schistosomiasis, SP does not modulate either IL-12 or TGF-β synthesis, and it works independently of IL-12 and TGF-β to govern T cell IFN-γ secretion.

Experiments using specific receptor antagonists show that the NK1 and SSTR2 receptors mediate the effects of SP and somatostatin on IFN-γ and IgG2a secretion *in vitro* (Blum *et al.*, 1993a; Elliott *et al.*, 1999). The NK1 receptor is the only tachykinin receptor expressed in schistosome granulomas (Cook *et al.*, 1994). Also, animals treated with octreotide, a somatostatin receptor agonist or one of several SP receptor antagonists form schistosome granulomas with impaired IFN-γ dependent circuitry (Elliott and Weinstock, 1996).

Further evidence comes from experiments using NK1 mutant mice (Blum *et al.*, 1999). SP receptor knockout animals infected with *S. mansoni* form abnormal granulomas. Both dispersed granuloma cells and splenocytes from these mice show a marked impairment in IFN-γ production and IgG2a secretion compared to their wild-type controls. A newly developed T cell-selective, NK1 knockout mouse demonstrates that the NK1 receptor on the T cell actually governs this IFN-γ response (unpublished).

There are five somatostatin receptor subtypes (SSTR1-5), but only the SSTR2 receptor mediates somatostatin regulation of IFN-γ. The schistosome granuloma leukocytes only express somatostatin receptor subtype 2 (Elliott *et al.*, 1999). Granuloma CD4+ T cells express both SP (Cook *et al.*, 1994) and somatostatin receptors (Blum *et al.*, 1992; Elliott *et al.*, 1999). Other granuloma and splenic cell types also express them.

The regulation of IFN-γ at sites of inflammation is critical because IFN-γ governs macrophage activation, T helper cell subtype development and immunoglobulin class switching. Multiple antigens bombard lymphocytes at mucosal surfaces and at sites of inflammation. At these sites, SP may prompt Th1 cells to secrete IFN-γ, while somato-statin inhibits its release helping to adjust both the intensity and nature of the immune response.

Salmonellosis

Also, in a murine salmonellosis model of disease, mice treated with a SP receptor antagonist show a decreased intestinal IFNγ response to salmonella and are more susceptible to this organism (Kincy-Cain and Bost, 1996). Following exposure to intestinal *Salmonella* via oral inoculation, there is a rapid upregulation of preprotachykinin mRNA in the regional mucosal lymphoid tissues and in the spleen (Bost, 1995; Kincy-Cain and Bost, 1996). Treatment with spantide II, a tachykinin inhibitor with preference for NK1 receptors, throughout the *Salmonella* exposure worsens the clinical features of salmonellosis and decreases survival. The inhibitor does not effect the translocation rate of *Salmonella* from the intestinal lumen to regional lymph nodes.

IL-12, IFN-γ and macrophages are part of the initial response to *Salmonella* that helps limit bacterial growth and dissemination (Jouanguy *et al.*, 1999). IL-12 is a dimeric protein composed of a p40 and a P35 chain. Spantide II treatment decreases *Salmonella*-induced expression of IL-12p40 and IFN-γ mRNA in the regional lymphoid tissue (Kincy-Cain and Bost, 1996).

Macrophages produce IL-12, TGF-β and other molecules that can control IFN-γ synthesis. Murine macrophages cultured *in vitro* express NK1 receptors in response to *Salmonella* (Kincy-Cain and Bost, 1996). SP can decrease IFN-γ and LPS-induced TGFβ production from cultured peritoneal macrophages (Marriott and Bost, 1998). It may also stimulate IL12 production (Kincy-Cain and Bost, 1997). Thus, it is postulated that SP modulates production of these critical cytokines to effect *Salmonella* infection.

Trypanosomiasis

Trypanosoma brucei is a protozoan parasite that can induce inflammatory reactions in the brain. The inflammation is a meningoencephalitis with associated astrocytic proliferation. Mice treated with a SP receptor antagonist had a decrease in the inflammatory response (Kennedy *et al.*, 1997).

Trichinella spiralis

Trichinella spiralis is a helminthic parasite that induces a strong Th2-type immune response in the rat intestine. Intestinal colonization of rats with this organism induces a T cell-dependent increase in SP in the muscle-myenteric plexus (Swain *et al.*, 1992). Treatment of rats with blocking SP anti-serum (Agro and Stanisz, 1993) or a NK1 receptor antagonist (Kataeva *et al.*, 1994) affords protection from the intestinal inflammation.

RELEVANCE OF NK1 RECEPTORS IN ANIMAL MODELS OF TOXIN AND ANTIGEN-INDUCED INFLAMMATION

Clostridia difficile *toxin A*

Clostridia difficile is a bacterium that can release toxins A and B, which induce colitis in humans. The rat model involves injecting toxin A into surgically created, blind ileal loops. The toxin induces epithelial cell necrosis and neutrophil infiltration that occurs in 4 h or less. The NK1 receptor helps mediate the acute mucosal injury and inflammation induced by *C. difficile* toxin A. Mice given SP receptor antagonists (Pothoulakis *et al.*, 1994), or mice lacking the NK1 receptor (Castagliuolo *et al.*, 1998) are protected from toxin A-induced enteritis. Rats also are protected by depletion of primary afferent fibres with capsaicin injection (Mantyh *et al.*, 1996a). Adherent lamina propria mononuclear cells isolated from these acutely injured intestines appear to produce SP (Pothoulakis *et al.*, 1994), but their role in the disease process is not defined. SP could induce TNF-α release from these cultured cells. The acute nature of the injury suggests involvement of innate mechanisms of injury encompassing various cell types. The protective effect of SP blockade does not necessarily implicate direct SP and immunocyte interaction in the disease process. SP effects various intestinal parenchyma cell functions including epithelial cell secretion (Riegler *et al.*, 1999). Toxin A induces a rapid increase in intestinal epithelial cell NK1 receptor display (Pothoulakis *et al.*, 1998).

Antigenic reactions

Both NK1 and NK2 receptors have a role in reactive airway disease. In a murine model of lung inflammation, bronchoalveolar lavage lymphocytes (BAL) and alveolar macrophages have an increase in NK1 receptor mRNA content. The NK1 receptor antagonist CP-96,345 decreases leukocyte recovery by BAL after antigen challenge of sensitized mice (Kaltreider *et al.*, 1997). A guinea pig model of antigen-induced, airway disease shows similar results (Schuiling *et al.*, 1999). NK1 receptor antagonist can partly block dinitrofluorobenzene-induced, delayed-type pneumonitis in mice (Buckley and Nijkamp, 1994). In humans and guinea pigs, the NKA receptor is a potent constrictor of airways, while SP has less contractile effect (Joos *et al.*, 2000a; Tanpo *et al.*, 2000). Thus, NK2 receptors may be more important then NK1 receptors for smooth muscle contraction and bronchial constriction in asthma.

The role of SP in human allergic rhinitis was examined using nasal mucosal biopsies from allergic patients and from non-allergic controls (Okamoto *et al.*, 1993). Three hours after administration of SP to biopsies maintained in organ culture, there was increased expression of mRNA for IL-1β, IL-3, IL-5, IL-6, TNF-α and INF-γ in the tissue. This response required as little as 10^{-9} M SP, and an NK1 receptor antagonist blocked the response. The response was most evident in tissue from allergic subjects, whereas, only half the specimens from non-allergic controls responded.

Substance P and NK1 receptors may be important in interstitial cystitis. Interstitial cystitis is an inflammatory disorder of the bladder associated with urinary frequency and suprapubic pain. The bladder has SP-containing nerves in the submucosa and elsewhere

(Pang *et al.*, 1995). Mice deficient in NK1 receptors are protected from bladder inflammation in response to antigen challenge (Saban *et al.*, 2000).

NEUROGENIC INFLAMMATION, SUBSTANCE P AND NK1 RECEPTORS

Neurogenic inflammation refers to a situation in which sensory nerve irritation or stimulation increases vascular permeability and leads to plasma extravasation, tissue swelling and neutrophil infiltration in skin, lung, gut and elsewhere (Figini *et al.*, 1997). Agents like capsaicin, bradykinin or antidromic electrical stimulation can trigger the process. NK1 receptors have a central role in mediating expression of neurogenic inflammation at least in rodents and guinea pigs (McDonald *et al.*, 1996). This was demonstrated using NK1 receptor agonists and antagonists, and NK1 receptor-deficient mice. However, SP injected into normal rodent skin induces edema, but no neutrophil accumulation (Pinter *et al.*, 1999; Cao *et al.*, 1999, 2000). SP induces both edema and neutrophil infiltration in normal mouse lung (Baluk *et al.*, 1999).

The overall importance of SP was shown in several animal models. For instance, immune complexes instilled into the mouse trachea induce in less than 4 h increased pulmonary microvascular permeability and neutrophilic infiltration suggestive of neurogenic-type inflammation. Mice with disruption of the NK1 receptor gene are much less susceptible to immune complex-induced, pulmonary injury (Bozic *et al.*, 1996b).

IL-1β injected into murine skin air pouches induces rapid neutrophil migration. There is a neurogenic component to this process, since capsaicin and NK1 receptor antagonists attenuate the neutrophil accumulation. NK1 receptor knock-out mice also resist IL-1β-induced, neutrophil accumulation (Ahluwalia *et al.*, 1998). Thus, the NK1 receptor may have an important role in this cellular response to IL-1β. However, others do not find that NK1 receptors are important in IL-1β-induced, neutrophil infiltration into skin (Cao *et al.*, 2000). They attribute the discordant findings to artifical changes in the cells lining the air pouches.

Using NK1 knock-out mice and receptor antagonists, it was shown that NK1 receptors have an important role in carrageenin-induced, neutrophil accumulation in the skin and that kinins are involved in the response (Cao *et al.*, 2000). The carrageenin experiments are offered as direct evidence that endogenous tachykinins can influence neutrophil accumulation in inflamed skin. Zymosan-induced neutrophil infiltration was NK1 receptor independent.

Substance P, NK1 receptor and neurogenic inflammation may have a role in acute pancreatitis. Plasma extravasation and neutrophil infiltration characterize secretagogue- or diet-induced, pancreatitis in rodents. Animals treated with NK1 receptor antagonist or deficient in NK1 receptor are partly protected (Bhatia *et al.*, 1998; Grady *et al.*, 2000; Maa *et al.*, 2000).

Several mechanisms govern neurogenic inflammation. Stimulation of NK1 receptors on the endothelial cells of the postcapillary venules expands intercellular gaps and increase vascular permeability. Through a NF-κB and NF-AT-dependent mechanism, SP directly stimulates VCAM-1 expression on endothelial cells. This could play an important role in promoting neutrophil adherence and egress (Quinlan *et al.*, 1999). This obviously is not the whole story. Serotonin, histamine, arachidonic derivatives and mast cells also may be involved through stimulation of tachykinin release from sensory nerves (Kopp and Cicha, 1999; Joos *et al.*, 2000b) and through a multitude of other effects (Joos *et al.*, 2000a).

Engagement of SP with the endothelial cell NK1 receptor results in receptor internalization and loss from the cell surface. This may be one of the mechanisms that limit plasma leakage at sites of inflammation (Bowden *et al.*, 1994).

NEUTRAL ENDOPEPTIDASE AND SUBSTANCE P DEGRADATION

Neutral endopeptidase (NEP) (enkephalinase) is a neuropeptide-degrading, cell surface enzyme expressed on many cell types. These include neurons, leukocytes, epithelial cells and smooth muscle cells. SP is a kinetically favourable substrate. It hydrolysis SP at the Gln^6–Phe^7, Phe^7–Phe^8 and Gly^9–Leu^{10} bonds. The resulting metabolites have no biological activity (Okamoto *et al.*, 1994).

Neutral endopeptidese blockade delays SP degradation. *Trichinella spiralis* colonization of the intestine downregulates NEP activity in the ileum and decreases the rate of SP degradation in the gut (Hwang *et al.*, 1993). There are similar findings in influenza virus-induced, pulmonary inflammation (Jacoby *et al.*, 1988). Compared to wild-type controls, healthy NEP knock-out mice have more SP in the colon and more readily extravasate fluid into the intestines (Sturiale *et al.*, 1999).

Dinitrobenzene sulfonic acid given orally to rats induces colitis. NEP knock-out mice develop markedly worse intestinal inflammation and injury in response to dinitrobenzene sulfonic acid. Administration of recombinant NEP or SP receptor antagonist prevents the exacerbated inflammation (Sturiale *et al.*, 1999). This suggests that a defect in NEP expression with resulting over-expression of SP worsens inflammation.

ORGAN-SPECIFIC MAINTENANCE OF STUDY-STATE FUNCTIONS

THYMUS

The thymus is the primary site for T lymphopoiesis. It receives T cell progenitors that undergo maturation and education before mature, competent T cells are released into the peripheral circulation. Thymocyte maturation involves a complex series of events involving interactions among the T cells, dendritic cells and other non-T cell elements of this organ. The final result of this process is the release of mature T cells that have antigen receptors carefully selected to recognize particular foreign antigens without inducing autoimmune disease.

The thymus may have an endogenous SP immunoregulatory circuit. Several studies suggest that the thymus has SP peptidergic-nerve fibres (Lorton *et al.*, 1990; Jurjus *et al.*, 1998), and perhaps other cells containing SP and expressing preprotachykinin A mRNA (Piantelli *et al.*, 1990; Jurjus *et al.*, 1998). The thymus also may have binding sites for SP (van Hagen *et al.*, 1996; Reubi *et al.*, 1998). It is possible to image human thymomas with radio-labelled SP (Lastoria *et al.*, 1999) showing that the human thymus has tachykinin receptors.

Many thymic T cells are killed within the thymus through the process of apoptosis. Corticosteroids, released during stress or produced locally within the thymus (Vacchio *et al.*, 1994; Ashwell *et al.*, 1996; Iwata *et al.*, 1996), are one of the factors that can enhance thymocyte apoptosis and thymic atrophy. Experiments conducted *in vitro* and *in vivo* suggest that SP can inhibit hydrocortisone-induced apoptosis of CD4+ CD8+ thymocytes (Dimri *et al.*, 2000). SP receptor antagonist blocks the effect suggesting physiological significance.

BONE MARROW

There are nerves containing tachykinins like SP and neurokinin B in bone marrow (Goto *et al.*, 1998). Isolated bone marrow stromal cell preparations may contain fibroblasts, macrophages and endothelial cells. The stroma provides the physical support and cytokines for the growth and regulation of hematopoietic stem cells. Bone marrow stromal cells can express NK1 receptors and also produce SP. Stem cell factor and IL1-1α induce their expression (Rameshwar and Gascon, 1995). These and other observations stimulate studies on the possible effects of SP on haematopoiesis (Rameshwar *et al.*, 1994; Rameshwar, 1997).

Substance P promotes the production of granulocytic and erythroid progenitors in dispersed murine and human bone marrow cells cultured *in vitro*. Also, it enhances production of stimulatory haematopoietic growth factors like IL-3, IL-7, GM-CSF and stem cell factor (Moore *et al.*, 1988; Rameshwar *et al.*, 1994; Manske *et al.*, 1995; Rameshwar and Gascon, 1995). It is reported that the SP effect on IL-1 and IL-6 secretion in murine stroma indirectly regulates IL-3 and GM-CSF production (Rameshwar and Gascon, 1995).

Neurokinin A via the NK2 receptor has some opposing effects (Rameshwar and Gascon, 1996). Neurokinin A, via the NK2 receptor, inhibits growth of granulocyte-macrophage progenitors through stimulation of stroma cell MIP-1 and TGF-β secretion. Thus, it is postulated that SP, neurokinin A and their respective receptors are part of an interactive regulatory system controlling aspects of erythropoiesis. There currently are no reports of bone marrow dysfunction in NK1 knock-out mice.

VARIOUS REPORTED EFFECTS OF SUBSTANCE P ON CELLS OF THE IMMUNE SYSTEM

MAST CELLS

Mast cells derive from the bone marrow and reside in tissues. They frequently are associated with nerves and blood vessels (Metcalfe *et al.*, 1997). There are morphologically and functionally different subsets (McNeil and Gotis-Graham, 2000).

Substance P, but not other tachykinins, induces mast cell degranulation with the resulting release of histamine (Johnson and Erdos, 1973; Forsythe *et al.*, 2000) and other mast cell mediators (Ansel *et al.*, 1993; Amano *et al.*, 1997). However, not all mast cells respond (Arock *et al.*, 1989). It has a predilection for connective tissue-type, rather than mucosal-type mast cells. SP injected into human or rodent skin induces mast cell degranulation that promotes tissue swelling and perhaps granulocyte infiltration (Devillier *et al.*, 1986; Matsuda *et al.*, 1989; Yano *et al.*, 1989). However, the mast cell is not essential for the initial increase in vascular permeability (Kowalski *et al.*, 1990).

Most studies show that SP exerts this effect only at high concentration (μM) (Repke and Bienert, 1987; Jozaki *et al.*, 1990; Amano *et al.*, 1997) and independently of tachykinin receptors. SP is a cationic amphiphilic compound that can directly activate G proteins in mast cells like several other amphiphilic molecules (Gies *et al.*, 1993; Chahdi *et al.*, 1998). The process requires the N-terminus of SP (Repke and Bienert, 1988; Devillier *et al.*, 1989).

However, some recent reports, like those using rat peritoneal mast cells, challenge these assumptions (Janiszewski *et al.*, 1994; Ogawa *et al.*, 1999; Okada *et al.*, 1999). Also, low concentrations of SP can induce transient burst of the mast cell chloride circuit (Janiszewski *et al.*, 1994). Thus, it is possible that SP acts as an agent that sensitizes mast cells for degranulation by other factors. Moreover, SP induces murine mast cell secretion at lower concentration (10^{-8} M) if the cells are exposed to IL-4 and stem cell factor *in vitro* (Karimi *et al.*, 2000). This and another report (Ogasawara *et al.*, 1997) show that changes in the cytokine environment influence mast cell sensitivity to SP.

At least in mice, enteric nerves and mast cells participate in the regulation of SP-induced, intestinal ion secretion (Wang *et al.*, 1995). The SP receptor antagonist CP-96345 blocks the secretory response suggesting involvement of NK1 receptors.

There is a preponderance of data showing that SP can stimulate mast cells. Yet, the true physiological significance of the observed phenomena remains unsettled.

T AND B LYMPHOCYTES

Murine and human T cells can express NK1 receptors as reviewed under section 'Location of NK1 Receptors', but their expression on B cells remains controversial. The human B lymphoblast cell line does express NK1 receptors (Payan *et al.*, 1984b, 1986). Some of the biological activity attributed to SP occur only at high concentration suggesting signalling pathways independent of NK1 receptors, whereas some blockade with NK1 antagonists.

The following are biological activities of SP ascribed to lymphocytes. SP may enhance the mitogen-induced proliferation of human and murine lymphocytes (Stanisz *et al.*, 1986; Covas *et al.*, 1997) and stimulate IL2 production (Calvo *et al.*, 1992; Rameshwar *et al.*, 1993). SP may downmodulate human T cell adherence to fibronectin, and NK1 receptor blockade inhibits the process (Levite *et al.*, 1998). This suggests that SP could impair T cell extravasation from blood vessels into inflamed sites. It also is reported that SP is chemotactic for human peripheral blood T and B cells (Schratzberger *et al.*, 1997). However, experiments in healthy rats show no effect of SP or its antagonist on T or B cell migration *in vivo* (Heerwagen *et al.*, 1995).

Bost and Pascual (1992) thoroughly reviewed the evidence for SP as a growth and differentiation factor for B cells and modulator of immunoglobulin production. Exogenous administration of SP stimulates development of immunoglobulin-producing lymphocytes. SP can act directly on polyclonally activated B cells to increase immunoglobulin secretion.

As reviewed above, in murine schistosomiasis, SP stimulates T cells via a true NK1 receptor to make IFN-γ and indirectly induce IgG2a production (Weinstock and Elliott, 1998).

MONOCYTES AND MACROPHAGES

Macrophages derive from monocytes, which come from the bone marrow. They have various important functions involving host defense, immunomodulation, and tissue injury and repair. They display cell surface markers that aid in antigen presentation and influence lymphocyte activation. They produce many types of mediators like interleukins, chemokines, arachidonic acid products, nitric oxide and reactive oxygen radicals to name a few. Presented earlier in this chapter was the growing evidence that macrophages can produce SP and express an NK1 receptor inducible by inflammatory mediators. SP also may bind to low affinity, non-neurokinin sites on human monocytes. This binding can activate MAP kinase (Jeurissen et al., 1994).

Early reports suggested that SP stimulated human monocytes and macrophages to release TNF-α, IL-1 and IL-6 (Lotz et al., 1988; Chancellor-Freeland et al., 1995). SP may induce IL-3 and GM-CSF production via NK1 receptor interaction in murine bone marrow partly through induction of IL-1 and IL-6 (Rameshwar et al., 1994). Others suggest that SP does not stimulate macrophage secretion directly. Rather it sensitizes macrophages and neuroglial cells, making them more responsive to LPS (Luber-Narod et al., 1994; Berman et al., 1996; Lieb et al., 1996). It is also possible that SP interacts via a non-neurokine protein with human monocytes to induce IL-6 (Kavelaars et al., 1994).

Other studies showed that SP inhibits IFN-γ and LPS-induced, TGF-β production (Marriott and Bost, 1998) and stimulates IL-12 secretion by cultured murine macrophages (Kincy-Cain and Bost, 1997). Murine macrophages produce somatostatin in response to LPS, IL-10, IFN-γ, TNF-α and several other factors. SP acts through the NK1 receptor to inhibit somatostatin induction and to downmodulate ongoing somatostatin expression (Blum et al., 1999).

NEUTROPHILS AND EOSINOPHILS

Indirect effects

Neutrophils migrate into tissue resulting from the process called neurogenic inflammation. This results in part from the interaction of SP with the NK1 receptor on vascular endothelial cell. (See section entitled 'Neurogenic Inflammation') Also, SP at pM concentrations can induce production of neutrophil chemotactic factor from bovine bronchial epithelial cells (Von Essen et al., 1992). Moreover, it stimulates adhesion between these cells via induction of adhesion molecules (DeRose et al., 1994). The process requires an NK1 receptor.

Direct effects

There are reports of direct effects of SP on neutrophils and eosinophils, but usually not at physiological concentrations. SP can potentiate respiratory burst activity and

IL-8 production from human peripheral blood neutrophils that was induced by various stimuli (Serra *et al.*, 1994). Other effects described include induction of chemotaxis, potentiation of phagocytosis, direct stimulation of respiratory burst, augmentation of antibody-dependent cell-mediated cytotoxicity and more (Serra *et al.*, 1988; Wozniak *et al.*, 1989, 1993; Brunelleschi *et al.*, 1991, 1993; Sterner-Kock *et al.*, 1999).

Some studies also suggest that SP effects human and guinea pig eosinophils. SP can potentiate human eosinophil chemotaxis to platelet-activating factor (Numao and Agrawal, 1992) and stimulate guinea pig eosinophil peroxidase secretion (Kroegel *et al.*, 1990).

IMPORTANCE IN HUMAN DISEASE

The importance of NK1 receptors and SP in human disease is unknown. There are no diseases yet attributed to loss or over-expression of the NK1 receptor. The human intestine is an abundant source for SP. In human inflammatory bowel disease, the inflamed colon contains increased numbers of NK1 receptor positive lymphocytes (Goode *et al.*, 2000a). An autoradiographic technique demonstrated a 1000-fold upregulation of SP binding sites in the lymphoid follicles and vasculature of intestinal tissue from patients with inflammatory bowel disease (Mantyh *et al.*, 1988). There is increased expression of SP receptor in human *C. difficile*-induced colitis (Mantyh *et al.*, 1996b). Taken together, these data suggest that NK1 receptors have a role in the human intestinal inflammation.

There are several non-peptide NK1 receptor antagonists undergoing clinical evaluation in humans. Reports suggest that they may prove useful for controlling depression (Kramer *et al.*, 1998), emesis (Navari *et al.*, 1999) and exercise-induced bronchoconstriction in asthma (Ichinose *et al.*, 1996). However, recent trails in asthma have proven disappointing (Joos *et al.*, 2000a). Their role in immune modulation is mostly unexplored. There are species differences in NK1, NK2 and NK3 receptor distribution and function. We must wait and see if pre-clinical animal data translates to efficacy and physiological relevance in humans.

ACKNOWLEDGEMENTS

Grants from the National Institutes of Health (DK38327, DK07663, DK25295), the Crohn's and Colitis Foundation of America, Inc. and the Veterans Administration supported this research.

REFERENCES

Agro, A. and Stanisz, A.M. (1993) Inhibition of murine intestinal inflammation by anti-substance P antibody. *Regional Immunology* **5**, 120–126.

Ahluwalia, A., De Felipe, C., O'Brien, J., Hunt, S.P. and Perretti, M. (1998) Impaired IL-1beta-induced neutrophil accumulation in tachykinin NK1 receptor knockout mice. *British Journal of Pharmacology* **124**, 1013–1015.

Aliakbari, J., Sreedharan, S.P., Turck, C.W. and Goetzl, E.J. (1987) Selective localization of vasoactive intestinal peptide and substance P in human eosinophils. *Biochemical and Biophysical Research Communications* **148**, 1440–1445.

Amano, H., Kurosawa, M. and Miyachi, Y. (1997) Possible mechanisms of the concentration-dependent action of substance P to induce histamine release from rat peritoneal mast cells and the effect of extracellular calcium on mast-cell activation. *Allergy* **52**, 215–219.

Ansel, J.C., Brown, J.R., Payan, D.G. and Brown, M.A. (1993) Substance P selectively activates TNF-alpha gene expression in murine mast cells. *Journal of Immunology* **150**, 4478–4485.

Arnalich, F., De Miguel, E., Perez-Ayala, C., Martinez, M., Vazquez, J.J., Gijon-Banos, J. and Hernanz, A. (1994) Neuropeptides and interleukin-6 in human joint inflammation relationship between intraarticular substance P and interleukin-6 concentrations. *Neuroscience Letters* **170**, 251–254.

Arock, M., Devillier, P., Luffau, G., Guillosson, J.J. and Renoux, M. (1989) Histamine-releasing activity of endogenous peptides on mast cells derived from different sites and species. *International Archives of Allergy and Applied Immunology* **89**, 229–235.

Ashwell, J.D., King, L.B. and Vacchio, M.S. (1996) Cross-talk between the T cell antigen receptor and the glucocorticoid receptor regulates thymocyte development. *Stem Cells* **14**, 490–500 (review with 58 references).

Baluk, P., Bowden, J.J., Lefevre, P.M. and McDonald, D.M. (1997) Upregulation of substance P receptors in angiogenesis associated with chronic airway inflammation in rats. *American Journal of Physiology* **273**, L565–L571.

Baluk, P., Thurston, G., Murphy, T.J., Bunnett, N.W. and McDonald, D.M. (1999) Neurogenic plasma leakage in mouse airways. *British Journal of Pharmacology* **126**, 522–528.

Barak, L.S., Warabi, K., Feng, X., Caron, M.G. and Kwatra, M.M. (1999) Real-time visualization of the cellular redistribution of G protein-coupled receptor kinase 2 and beta-arrestin 2 during homologous desensitization of the substance P receptor. *Journal of Biological Chemistry* **274**, 7565–7569.

Berman, A.S., Chancellor-Freeland, C., Zhu, G. and Black, P.H. (1996) Substance P primes murine peritoneal macrophages for an augmented proinflammatory cytokine response to lipopolysaccharide. *Neuroimmunomodulation* **3**, 141–149.

Berthold, M. and Bartfai, T. (1997) Modes of peptide binding in G protein-coupled receptors. *Neurochemical Research* **22**, 1023–1031 (review with 57 references).

Bhatia, M., Saluja, A.K., Hofbauer, B., Frossard, J.L., Lee, H.S., Castagliuolo, I., Wang, C.C., Gerard, N., Pothoulakis, C. and Steer, M.L. (1998) Role of substance P and the neurokinin 1 receptor in acute pancreatitis and pancreatitis-associated lung injury. *Proceedings of the National Academy of Sciences of the United States of America* **95**, 4760–4765.

Blum, A.M., Elliott, D.E., Metwali, A., Li, J., Qadir, K. and Weinstock, J.V. (1998) Substance P regulates somatostatin expression in inflammation. *Journal of Immunology* **161**, 6316–6322.

Blum, A.M., Metwali, A., Mathew, R.C., Elliott, D. and Weinstock, J.V. (1993a) Substance P and somatostatin can modulate the amount of IgG2a secreted in response to schistosome egg antigens in murine schistosomiasis mansoni. *Journal of Immunology* **151**, 6994–7004.

Blum, A.M., Metwali, A., Cook, G., Mathew, R.C., Elliott, D. and Weinstock, J.V. (1993b) Substance P modulates antigen-induced, IFN-gamma production in murine Schistosomiasis mansoni. *Journal of Immunology* **151**, 225–233.

Blum, A.M., Metwali, A., Kim-Miller, M., Li, J., Qadir, K., Elliott, D.E., Lu, B., Fabry, Z., Gerard, N. and Weinstock, J.V. (1999) The substance P receptor is necessary for a normal granulomatous response in murine schistosomiasis mansoni. *Journal of Immunology* **162**, 6080–6085.

Blum, A.M., Metwali, A., Mathew, R.C., Cook, G., Elliott, D. and Weinstock, J.V. (1992) Granuloma T lymphocytes in murine schistosomiasis mansoni have somatostatin receptors and respond to somatostatin with decreased IFN-gamma secretion. *Journal of Immunology* **149**, 3621–3626.

Blum, A.M., Metwali, A., Crawford, C., Qadir, K., Elliott, D.E. and Weinstock, J.V. (2001) IL12 and antigen induce substance P receptor expression in T cells in murine schistosomiasis mansoni. *FASEB Journal*, **15**, 950–957.

Bohm, S.K., Khitin, L.M., Smeekens, S.P., Grady, E.F., Payan, D.G. and Bunnett, N.W. (1997) Identification of potential tyrosine-containing endocytic motifs in the carboxyl-tail and seventh transmembrane domain of the neurokinin 1 receptor. *Journal of Biological Chemistry* **272**, 2363–2372.

Bost, K.L. (1995) Inducible preprotachykinin mRNA expression in mucosal lymphoid organs following oral immunization with Salmonella. *Journal of Neuroimmunology* **62**, 59–67.

Bost, K.L., Breeding, S.A. and Pascual, D.W. (1992) Modulation of the mRNAs encoding substance P and its receptor in rat macrophages by LPS. *Regional Immunology* **4**, 105–112.

Bost, K.L. and Pascual, D.W. (1992) Substance P: a late-acting B lymphocyte differentiation cofactor. *American Journal of Physiology* **262**, C537–C545 (review with 52 references).

Bowden, J.J., Baluk, P., Lefevre, P.M., Vigna, S.R. and McDonald, D.M. (1996) Substance P (NK1) receptor immunoreactivity on endothelial cells of the rat tracheal mucosa. *American Journal of Physiology* **270**, L404–L414.

Bowden, J.J., Garland, A.M., Baluk, P., Lefevre, P., Grady, E.F., Vigna, S.R., Bunnett, N.W. and McDonald, D.M. (1994) Direct observation of substance P-induced internalization of neurokinin 1 (NK1) receptors at sites of inflammation. *Proceedings of the National Academy of Sciences of the United States of America* **91**, 8964–8968.

Bozic, C.R., Lu, B., Hopken, U.E., Gerard, C. and Gerard, N.P. (1996a) Neurogenic amplification of immune complex inflammation. *Science* **273**, 1722–1725.

Bozic, C.R., Gerard, N.P. and Gerard, C. (1996b) Receptor binding specificity and pulmonary gene expression of the neutrophil-activating peptide ENA-78. *American Journal of Respiratory Cell and Molecular Biology* **14**, 302–308.

Brunelleschi, S., Tarli, S., Giotti, A. and Fantozzi, R. (1991) Priming effects of mammalian tachykinins on human neutrophils. *Life Sciences* **48**, L1–L5.

Buckley, T.L. and Nijkamp, F.P. (1994) Mucosal exudation associated with a pulmonary delayed-type hypersensitivity reaction in the mouse. Role for the tachykinins. *Journal of Immunology* **153**, 4169–4178.

Calvo, C.F., Chavanel, G. and Senik, A. (1992) Substance P enhances IL-2 expression in activated human T cells. *Journal of Immunology* **148**, 3498–3504.

Cao, T., Gerard, N.P. and Brain, S.D. (1999) Use of NK(1) knockout mice to analyze substance P-induced edema formation. *American Journal of Physiology* **277**, R476–R481.

Cao, T., Pinter, E., Al Rashed, S., Gerard, N., Hoult, J.R. and Brain, S.D. (2000) Neurokinin-1 receptor agonists are involved in mediating neutrophil accumulation in the inflamed, but not normal, cutaneous microvasculature: an in vivo study using neurokinin-1 receptor knockout mice. *Journal of Immunology* **164**, 5424–5429.

Carter, M.S. and Krause, J.E. (1990) Structure, expression, and some regulatory mechanisms of the rat preprotachykinin gene encoding substance P, neurokinin A, neuropeptide K, and neuropeptide gamma. *Journal of Neuroscience* **10**, 2203–2214.

Castagliuolo, I., Keates, A.C., Qiu, B., Kelly, C.P., Nikulasson, S., Leeman, S.E. and Pothoulakis, C. (1997) Increased substance P responses in dorsal root ganglia and intestinal macrophages during Clostridium difficile toxin A enteritis in rats. *Proceedings of the National Academy of Sciences of the United States of America* **94**, 4788–4793.

Castagliuolo, I., Riegler, M., Pasha, A., Nikulasson, S., Lu, B., Gerard, C., Gerard, N.P. and Pothoulakis, C. (1998) Neurokinin-1 (NK-1) receptor is required in Clostridium difficile-induced enteritis. *Journal of Clinical Investigation* **101**, 1547–1550.

Chahdi, A., Mousli, M. and Landry, Y. (1998) Substance P-related inhibitors of mast cell exocytosis act on G-proteins or on the cell surface. *European Journal of Pharmacology* **341**, 329–335.

Chancellor-Freeland, C., Zhu, G.F., Kage, R., Beller, D.I., Leeman, S.E. and Black, P.H. (1995) Substance P and stress-induced changes in macrophages. *Annals of the New York Academy of Sciences* **771**, 472–484 (review with 39 references).

Chanez, P., Springall, D., Vignola, A.M., Moradoghi-Hattvani, A., Polak, J.M., Godard, P. and Bousquet, J. (1998) Bronchial mucosal immunoreactivity of sensory neuropeptides in severe airway diseases. *American Journal of Respiratory and Critical Care Medicine* **158**, 985–990.

Chang, M.M., Leeman, S.E. and Niall, H.D. (1971) Amino-acid sequence of substance P. *Nature – New Biology* **232**, 86–87.

Chapman, K., Lyons, V. and Harmar, A.J. (1993) The sequence of 5′ flanking DNA from the rat preprotachykinin gene; analysis of putative transcription factor binding sites. *Biochimica et Biophysica Acta* **1172**, 361–363.

Cook, G.A., Elliott, D., Metwali, A., Blum, A.M., Sandor, M., Lynch, R. and Weinstock, J.V. (1994) Molecular evidence that granuloma T lymphocytes in murine schistosomiasis mansoni express an authentic substance P (NK-1) receptor. *Journal of Immunology* **152**, 1830–1835.

Cooke, H.J., Fox, P., Alferes, L., Fox, C.C. and Wolfe, S.A., Jr. (1998) Presence of NK1 receptors on a mucosal-like mast cell line, RBL-2H3 cells. *Canadian Journal of Physiology and Pharmacology* **76**, 188–193.

Covas, M.J., Pinto, L.A. and Victorino, R.M. (1997) Effects of substance P on human T cell function and the modulatory role of peptidase inhibitors. *International Journal of Clinical and Laboratory Research* **27**, 129–134.

De Giorgio, R., Tazzari, P.L., Barbara, G., Stanghellini, V. and Corinaldesi, R. (1998) Detection of substance P immunoreactivity in human peripheral leukocytes. *Journal of Neuroimmunology* **82**, 175–181.

DeRose, V., Robbins, R.A., Snider, R.M., Spurzem, J.R., Thiele, G.M., Rennard, S.I. and Rubinstein, I. (1994) Substance P increases neutrophil adhesion to bronchial epithelial cells. *Journal of Immunology* **152**, 1339–1346.

Devillier, P., Drapeau, G., Renoux, M. and Regoli, D. (1989) Role of the N-terminal arginine in the histamine-releasing activity of substance P, bradykinin and related peptides. *European Journal of Pharmacology* **168**, 53–60.

Devillier, P., Regoli, D., Asseraf, A., Descours, B., Marsac, J. and Renoux, M. (1986) Histamine release and local responses of rat and human skin to substance P and other mammalian tachykinins. *Pharmacology* **32**, 340–347.

Dimri, R., Sharabi, Y. and Shoham, J. (2000) Specific inhibition of glucocorticoid-induced thymocyte apoptosis by substance P. *Journal of Immunology* **164**, 2479–2486.

Elliott, D.E., Blum, A.M., Li, J., Metwali, A. and Weinstock, J.V. (1998) Preprosomatostatin messenger RNA is expressed by inflammatory cells and induced by inflammatory mediators and cytokines. *Journal of Immunology* **160**, 3997–4003.

Elliott, D.E., Li, J., Blum, A.M., Metwali, A., Patel, Y.C. and Weinstock, J.V. (1999) SSTR2A is the dominant somatostatin receptor subtype expressed by inflammatory cells, is widely expressed and directly regulates T cell IFN-gamma release. *European Journal of Immunology* **29**, 2454–2463.

Elliott, D.E. and Weinstock, J.V. (1996) Granulomas in murine schistosomiasis mansoni have a somatostatin immunoregulatory circuit. *Metabolism: Clinical and Experimental* **45**, 88–90 (review with 37 references).

Fiebich, B.L., Schleicher, S., Butcher, R.D., Craig, A. and Lieb, K. (2000) The neuropeptide substance P activates p38 mitogen-activated protein kinase resulting in IL-6 expression independently from NF-κB. *Journal of Immunology* **165**, 5606–5611.

Figini, M., Emanueli, C., Grady, E.F., Kirkwood, K., Payan, D.G., Ansel, J., Gerard, C., Geppetti, P. and Bunnett, N. (1997) Substance P and bradykinin stimulate plasma extravasation in the mouse gastrointestinal tract and pancreas. *American Journal of Physiology* **272**, G785–G793.

Fischer, A., McGregor, G.P., Saria, A., Philippin, B. and Kummer, W. (1996) Induction of tachykinin gene and peptide expression in guinea pig nodose primary afferent neurons by allergic airway inflammation. *Journal of Clinical Investigation* **98**, 2284–2291.

Forsgren, S., Hockerfelt, U., Norrgard, O., Henriksson, R. and Franzen, L. (2000) Pronounced substance P inner-vation in irradiation-induced enteropathy – a study on human colon. *Regulatory Peptides* **88**, 1–13.

Forsythe, P., McGarvey, L.P., Heaney, L.G., MacMahon, J. and Ennis, M. (2000) Sensory neuropeptides induce histamine release from bronchoalveolar lavage cells in both nonasthmatic coughers and cough variant asth-matics. *Clinical and Experimental Allergy* **30**, 225–232.

Garland, A.M., Grady, E.F., Lovett, M., Vigna, S.R., Frucht, M.M., Krause, J.E. and and Bunnett, N.W. (1996) Mechanisms of desensitization and resensitization of G protein-coupled neurokinin1 and neurokinin2 recep-tors. *Molecular Pharmacology* **49**, 438–446.

Gerard, N.P., Garraway, L.A., Eddy, R.L.J., Shows, T.B., Iijima, H., Paquet, J.L. and Gerard, C. (1991) Human substance P receptor (NK-1): organization of the gene, chromosome localization, and functional expression of cDNA clones. *Biochemistry* **30**, 10640–10646.

Germonpre, P.R., Bullock, G.R., Lambrecht, B.N., Van, D.V., Luyten, W.H., Joos, G.F. and Pauwels, R.A. (1999) Presence of substance P and neurokinin 1 receptors in human sputum macrophages and U-937 cells. *European Respiratory Journal* **14**, 776–782.

Gies, J.P., Landry, Y. and Mousli, M. (1993) Receptor-independent activation of mast cells by bradykinin and related peptides. *Trends in Neurosciences* **16**, 498–499 (letter; comment).

Goode, T., O'Connell, J., Anton, P., Wong, H., Reeve, J., O'Sullivan, G.C., Collins, J.K. and Shanahan, F. (2000a) Neurokinin-1 receptor expression in inflammatory bowel disease: molecular quantitation and localization. *Gut* **47**, 387–396.

Goode, T., O'Connell, J., Ho, W.Z., O'Sullivan, G.C., Collins, J.K., Douglas, S.D. and Shanahan, F. (2000b) Differential expression of neurokinin-1 receptor by human mucosal and peripheral lymphoid cells. *Clinical and Diagnostic Laboratory Immunology* **7**, 371–376.

Goode, T., O'Connell, J., Sternini, C., Anton, P., Wong, H., O'Sullivan, G.C., Collins, J.K. and Shanahan, F. (1998) Substance P (neurokinin-1) receptor is a marker of human mucosal but not peripheral mononuclear cells: molecular quantitation and localization. *Journal of Immunology* **161**, 2232–2240.

Goto, T., Yamaza, T., Kido, M.A. and Tanaka, T. (1998) Light- and electron-microscopic study of the distribution of axons containing substance P and the localization of neurokinin-1 receptor in bone. *Cell and Tissue Research* **293**, 87–93.

Grady, E.F., Garland, A.M., Gamp, P.D., Lovett, M., Payan, D.G. and Bunnett, N.W. (1995) Delineation of the endocytic pathway of substance P and its seven-transmembrane domain NK1 receptor. *Molecular Biology of the Cell* **6**, 509–524.

Grady, E.F., Yoshimi, S.K., Maa, J., Valeroso, D., Vartanian, R.K., Rahim, S., Kim, E.H., Gerard, C., Gerard, N., Bunnett, N.W. and Kirkwood, K.S. (2000) Substance P mediates inflammatory oedema in acute pancreatitis via activation of the neurokinin-1 receptor in rats and mice. *British Journal of Pharmacology* **130**, 505–512.

Greeno, E.W., Mantyh, P., Vercellotti, G.M. and Moldow, C.F. (1993) Functional neurokinin 1 receptors for sub-stance P are expressed by human vascular endothelium. *Journal of Experimental Medicine* **177**, 1269–1276.

Harmar, A.J., Hyde, V. and Chapman, K. (1990) Identification and cDNA sequence of delta-preprotachykinin, a fourth splicing variant of the rat substance P precursor. *FEBS Letters* **275**, 22–24.

Hartung, H.P., Wolters, K. and Toyka, K.V. (1986) Substance P: binding properties and studies on cellular responses in guinea pig macrophages. *Journal of Immunology* **136**, 3856–3863.

Heerwagen, C., Pabst, R. and Westermann, J. (1995) The neuropeptide substance P does not influence the migration of B, T, CD8+ and CD4+ ('naive' and 'memory') lymphocytes from blood to lymph in the normal rat. *Scandinavian Journal of Immunology* **42**, 480–486.

Helke, C.J., Krause, J.E., Mantyh, P.W., Couture, R. and Bannon, M.J. (1990) Diversity in mammalian tachykinin peptidergic neurons: multiple peptides, receptors, and regulatory mechanisms. *FASEB Journal* **4**, 1606–1615 (review with 69 references).

Hershey, A.D., Dykema, P.E. and Krause, J.E. (1991) Organization, structure, and expression of the gene encoding the rat substance P receptor. *Journal of Biological Chemistry* **266**, 4366–4374.

Hershey, A.D. and Krause, J.E. (1990) Molecular characterization of a functional cDNA encoding the rat substance P receptor. *Science* **247**, 958–962.

Ho, W.Z., Lai, J.P., Zhu, X.H., Uvaydova, M. and Douglas, S.D. (1997) Human monocytes and macrophages express substance P and neurokinin-1 receptor. *Journal of Immunology* **159**, 5654–5660.

Holzer, P. (1998) Implications of tachykinins and calcitonin gene-related peptide in inflammatory bowel disease. *Digestion* **59**, 269–283 (review with 137 references).

Holzer, P. and Holzer-Petsche, U. (1997a) Tachykinins in the gut. Part I. Expression, release and motor function. *Pharmacology and Therapeutics* **73**, 173–217 (review with 504 references).

Holzer, P. and Holzer-Petsche, U. (1997b) Tachykinins in the gut. Part II. Roles in neural excitation, secretion and inflammation. *Pharmacology and Therapeutics* **73**, 219–263 (review with 539 references).

Hwang, L., Leichter, R., Okamoto, A., Payan, D., Collins, S.M. and Bunnett, N.W. (1993) Downregulation of neutral endopeptidase (EC 3.4.24.11) in the inflamed rat intestine. *American Journal of Physiology* **264**, G735–G743.

Ichinose, M., Miura, M., Yamauchi, H., Kageyama, N., Tomaki, M., Oyake, T., Ohuchi, Y., Hida, W., Miki, H., Tamura, G. and Shirato, K. (1996) A neurokinin 1-receptor antagonist improves exercise-induced airway narrowing in asthmatic patients. *American Journal of Respiratory and Critical Care Medicine* **153**, 936–941.

Iwata, M., Ohoka, Y., Kuwata, T. and Asada, A. (1996) Regulation of T cell apoptosis via T cell receptors and steroid receptors. *Stem Cells* **14**, 632–641 (review with 66 references).

Jacoby, D.B., Tamaoki, J., Borson, D.B. and Nadel, J.A. (1988) Influenza infection causes airway hyperresponsiveness by decreasing enkephalinase. *Journal of Applied Physiology* **64**, 2653–2658.

Janiszewski, J., Bienenstock, J. and Blennerhassett, M.G. (1994) Picomolar doses of substance P trigger electrical responses in mast cells without degranulation. *American Journal of Physiology* **267**, C138–C145.

Jeurissen, F., Kavelaars, A., Korstjens, M., Broeke, D., Franklin, R.A., Gelfand, E.W. and Heijnen, C.J. (1994) Monocytes express a non-neurokinin substance P receptor that is functionally coupled to MAP kinase. *Journal of Immunology* **152**, 2987–2994.

Johnson, A.R. and Erdos, E.G. (1973) Release of histamine from mast cells by vasoactive peptides. *Proceedings of the Society for Experimental Biology and Medicine* **142**, 1252–1256.

Joos, G.F., Germonpre, P.R. and Pauwels, R.A. (2000a) Role of tachykinins in asthma. *Allergy* **55**, 321–337 (review with 209 references).

Joos, G.F., Germonpre, P.R. and Pauwels, R.A. (2000b) Neural mechanisms in asthma. *Clinical and Experimental Allergy* **30** (Suppl 1), 60–65 (review with 22 references).

Jouanguy, E., Doffinger, R., Dupuis, S., Pallier, A., Altare, F. and Casanova, J.L. (1999) IL-12 and IFN-gamma in host defense against mycobacteria and salmonella in mice and men. *Current Opinion in Immunology* **11**, 346–351 (review with 45 references).

Jozaki, K., Kuriu, A., Waki, N., Adachi, S., Yamatodani, A., Tarui, S. and Kitamura, Y. (1990) Proliferative potential of murine peritoneal mast cells after degranulation induced by compound 48/80, substance P, tetradecanoylphorbol acetate, or calcium ionophore A23187. *Journal of Immunology* **145**, 4252–4256.

Jurjus, A.R., More, N. and Walsh, R.J. (1998) Distribution of substance P positive cells and nerve fibers in the rat thymus. *Journal of Neuroimmunology* **90**, 143–148.

Kako, K., Munekata, E., Hosaka, M., Murakami, K. and Nakayama, K. (1993) Cloning and sequence analysis of mouse cDNAs encoding preprotachykinin A and B. *Biomedical Research* **14**, 253–259.

Kaltreider, H.B., Ichikawa, S., Byrd, P.K., Ingram, D.A., Kishiyama, J.L., Sreedharan, S.P., Warnock, M.L., Beck, J.M. and Goetzl, E.J. (1997) Upregulation of neuropeptides and neuropeptide receptors in a murine model of immune inflammation in lung parenchyma. *American Journal of Respiratory Cell and Molecular Biology* **16**, 133–144.

Karimi, K., Redegeld, F.A., Blom, R. and Nijkamp, F.P. (2000) Stem cell factor and interleukin-4 increase responsiveness of mast cells to substance P. *Experimental Hematology* **28**, 626–634.

Kataeva, G., Agro, A. and Stanisz, A.M. (1994) Substance-P-mediated intestinal inflammation: inhibitory effects of CP 96,345 and SMS 201–995. *Neuroimmunomodulation* **1**, 350–356.

Kavelaars, A., Broeke, D., Jeurissen, F., Kardux, J., Meijer, A., Franklin, R., Gelfand, E.W. and Heijnen, C.J. (1994) Activation of human monocytes via a non-neurokinin substance P receptor that is coupled to Gi protein, calcium, phospholipase D, MAP kinase, and IL-6 production. *Journal of Immunology* **153**, 3691–3699.

Kavelaars, A., Jeurissen, F., Frijtag Drabbe, K.J., Herman, V., Roijen, J., Rijkers, G.T. and Heijnen, C.J. (1993) Substance P induces a rise in intracellular calcium concentration in human T lymphocytes in vitro: evidence of a receptor-independent mechanism. *Journal of Neuroimmunology* **42**, 61–70.

Kennedy, P.G., Rodgers, J., Jennings, F.W., Murray, M., Leeman, S.E. and Burke, J.M. (1997) A substance P antagonist, RP-67,580, ameliorates a mouse meningoencephalitic response to Trypanosoma brucei brucei. *Proceedings of the National Academy of Sciences of the United States of America* **94**, 4167–4170.

Keranen, U., Jarvinen, H., Kiviluoto, T., Kivilaakso, E. and Soinila, S. (1996) Substance P- and vasoactive intestinal polypeptide-immunoreactive innervation in normal and inflamed pouches after restorative proctocolectomy for ulcerative colitis. *Digestive Diseases and Sciences* **41**, 1658–1664.

Keranen, U., Kiviluoto, T., Jarvinen, H., Back, N., Kivilaakso, E. and Soinila, S. (1995) Changes in substance P-immunoreactive innervation of human colon associated with ulcerative colitis. *Digestive Diseases and Sciences* **40**, 2250–2258.

Killingsworth, C.R., Shore, S.A., Alessandrini, F., Dey, R.D. and Paulauskis, J.D. (1997) Rat alveolar macrophages express preprotachykinin gene-I mRNA-encoding tachykinins. *American Journal of Physiology* **273**, L1073–L1081.

Kincy-Cain, T. and Bost, K.L. (1996) Increased susceptibility of mice to Salmonella infection following in vivo treatment with the substance P antagonist, spantide II. *Journal of Immunology* **157**, 255–264.

Kincy-Cain, T. and Bost, K.L. (1997) Substance P-induced IL-12 production by murine macrophages. *Journal of Immunology* **158**, 2334–2339.

Knolle, P.A. and Gerken, G. (2000) Local control of the immune response in the liver. *Immunological Reviews* **174**, 21–34 (review with 96 references).

Kopp, U.C. and Cicha, M.Z. (1999) PGE2 increases substance P release from renal pelvic sensory nerves via activation of N-type calcium channels. *American Journal of Physiology* **276**, R1241–R1248.

Kowalski, M.L., Sliwinska-Kowalska, M. and Kaliner, M.A. (1990) Neurogenic inflammation, vascular permeability, and mast cells. II. Additional evidence indicating that mast cells are not involved in neurogenic inflammation. *Journal of Immunology* **145**, 1214–1221.

Kramer, M.S., Cutler, N., Feighner, J., Shrivastava, R., Carman, J., Sramek, J.J., Reines, S.A., Liu, G., Snavely, D., Wyatt-Knowles, E., Hale, J.J., Mills, S.G., MacCoss, M., Swain, C.J., Harrison, T., Hill, R.G., Hefti, F., Scolnick, E.M., Cascieri, M.A., Chicchi, G.G., Sadowski, S., Williams, A.R., Hewson, L., Smith, D. and Rupniak, N.M. (1998) Distinct mechanism for antidepressant activity by blockade of central substance P receptors. *Science* **281**, 1640–1645 (see comments).

Krause, J.E., Chirgwin, J.M., Carter, M.S., Xu, Z.S. and Hershey, A.D. (1987) Three rat preprotachykinin mRNAs encode the neuropeptides substance P and neurokinin A. *Proceedings of the National Academy of Sciences of the United States of America* **84**, 881–885.

Kroegel, C., Giembycz, M.A. and Barnes, P.J. (1990) Characterization of eosinophil cell activation by peptides. Differential effects of substance P, melittin, and FMET-Leu-Phe. *Journal of Immunology* **145**, 2581–2587.

Lai, J.P., Douglas, S.D. and Ho, W.Z. (1998) Human lymphocytes express substance P and its receptor. *Journal of Neuroimmunology* **86**, 80–86.

Lambrecht, B.N., Germonpre, P.R., Everaert, E.G., Carro-Muino, I., De Veerman, M., de Felipe, C., Hunt, S.P., Thielemans, K., Joos, G.F. and Pauwels, R.A. (1999) Endogenously produced substance P contributes to lymphocyte proliferation induced by dendritic cells and direct TCR ligation. *European Journal of Immunology* **29**, 3815–3825.

Lastoria, S., Palmieri, G., Muto, P. and Lombardi, G. (1999) Functional imaging of thymic disorders. *Annals of Medicine* **31** (Suppl 2), 63–69 (review with 32 references).

Levite, M., Cahalon, L., Hershkoviz, R., Steinman, L. and Lider, O. (1998) Neuropeptides, via specific receptors, regulate T cell adhesion to fibronectin. *Journal of Immunology* **160**, 993–1000.

Li, H., Leeman, S.E., Slack, B.E., Hauser, G., Saltsman, W.S., Krause, J.E., Blusztajn, J.K. and Boyd, N.D. (1997) A substance P (neurokinin-1) receptor mutant carboxyl-terminally truncated to resemble a naturally occurring receptor isoform displays enhanced responsiveness and resistance to desensitization. *Proceedings of the National Academy of Sciences of the United States of America* **94**, 9475–9480.

Li, Y.M., Marnerakis, M., Stimson, E.R. and Maggio, J.E. (1995) Mapping peptide-binding domains of the substance P (NK-1) receptor from P388D1 cells with photolabile agonists. *Journal of Biological Chemistry* **270**, 1213–1220.

Lieb, K., Fiebich, B.L., Berger, M., Bauer, J. and Schulze-Osthoff, K. (1997) The neuropeptide substance P acti-vates transcription factor NF-kappa B and kappa B-dependent gene expression in human astrocytoma cells. *Journal of Immunology* **159**, 4952–4958.

Lieb, K., Fiebich, B.L., Busse-Grawitz, M., Hull, M., Berger, M. and Bauer, J. (1996) Effects of substance P and selected other neuropeptides on the synthesis of interleukin-1 beta and interleukin-6 in human monocytes: a re-examination. *Journal of Neuroimmunology* **67**, 77–81.

Lieb, K., Schaller, H., Bauer, J., Berger, M., Schulze-Osthoff, K. and Fiebich, B.L. (1998) Substance P and hista-mine induce interleukin-6 expression in human astrocytoma cells by a mechanism involving protein kinase C and nuclear factor-IL-6. *Journal of Neurochemistry* **70**, 1577–1583.

Lorton, D., Bellinger, D.L., Felten, S.Y. and Felten, D.L. (1990) Substance P innervation of the rat thymus. *Peptides* **11**, 1269–1275.

Lorton, D., Bellinger, D.L., Felten, S.Y. and Felten, D.L. (1991) Substance P innervation of spleen in rats: nerve fibers associate with lymphocytes and macrophages in specific compartments of the spleen. *Brain, Behavior and Immunity* **5**, 29–40 (review with 77 references).

Lotz, M., Vaughan, J.H. and Carson, D.A. (1988) Effect of neuropeptides on production of inflammatory cytokines by human monocytes. *Science* **241**, 1218–1221.

Luber-Narod, J., Kage, R. and Leeman, S.E. (1994) Substance P enhances the secretion of tumor necrosis factor-alpha from neuroglial cells stimulated with lipopolysaccharide. *Journal of Immunology* **152**, 819–824.

Maa, J., Grady, E.F., Yoshimi, S.K., Drasin, T.E., Kim, E.H., Hutter, M.M., Bunnett, N.W. and Kirkwood, K.S. (2000) Substance P is a determinant of lethality in diet-induced hemorrhagic pancreatitis in mice. *Surgery* **128**, 232–239.

Madden, K.S. and Felten, D.L. (1995) Experimental basis for neural-immune interactions. *Physiological Reviews* **75**, 77–106 (review with 373 references).

Maggi, C.A. (1995) Tachykinins and calcitonin gene-related peptide (CGRP) as co-transmitters released from peripheral endings of sensory nerves. *Progress in Neurobiology* **45**, 1–98 (review with 1250 references).

Maggi, C.A. (1996) Tachykinins in the autonomic nervous system. *Pharmacological Research* **33**, 161–170 (review with 152 references).

Maggi, C.A. (1997) The effects of tachykinins on inflammatory and immune cells. *Regulatory Peptides* **70**, 75–90 (review with 198 references).

Manske, J.M., Sullivan, E.L. and Andersen, S.M. (1995) Substance P mediated stimulation of cytokine levels in cultured murine bone marrow stromal cells. *Advances in Experimental Medicine and Biology* **383**, 53–64.

Mantyh, C.R., Gates, T.S., Zimmerman, R.P., Welton, M.L., Passaro, E.P.J., Vigna, S.R., Maggio, J.E., Kruger, L. and Mantyh, P.W. (1988) Receptor binding sites for substance P, but not substance K or neuromedin K, are expressed in high concentrations by arterioles, venules, and lymph nodules in surgical specimens obtained from patients with ulcerative colitis and Crohn disease. *Proceedings of the National Academy of Sciences of the United States of America* **85**, 3235–3239.

Mantyh, C.R., Pappas, T.N., Lapp, J.A., Washington, M.K., Neville, L.M., Ghilardi, J.R., Rogers, S.D., Mantyh, P.W. and Vigna, S.R. (1996a) Substance P activation of enteric neurons in response to intraluminal Clostridium difficile toxin A in the rat ileum. *Gastroenterology* **111**, 1272–1280.

Mantyh, C.R., Maggio, J.E., Mantyh, P.W., Vigna, S.R. and Pappas, T.N. (1996b) Increased substance P receptor expression by blood vessels and lymphoid aggregates in Clostridium difficile-induced pseudomembranous colitis. *Digestive Diseases and Sciences* **41**, 614–620 (review with 22 references).

Mantyh, P.W., Rogers, S.D., Honore, P., Allen, B.J., Ghilardi, J.R., Li, J., Daughters, R.S., Lappi, D.A., Wiley, R.G. and Simone, D.A. (1997) Inhibition of hyperalgesia by ablation of lamina I spinal neurons expressing the substance P receptor. *Science* **278**, 275–279 (see comments).

Marriott, I. and Bost, K.L. (1998) Substance P diminishes lipopolysaccharide and interferon-gamma-induced TGF-beta 1 production by cultured murine macrophages. *Cellular Immunology* **183**, 113–120.

Marriott, I. and Bost, K.L. (2000) IL-4 and IFN-γ up-regulate substance P receptor expression in murine peri-toneal macrophages. *Journal of Immunology* **165**, 182–191.

Marriott, I., Mason, M.J., Elhofy, A. and Bost, K.L. (2000) Substance P activates NF-kappaB independent of ele-vations in intracellular calcium in murine macrophages and dendritic cells. *Journal of Neuroimmunology* **102**, 163–171.

Matsuda, H., Kawakita, K., Kiso, Y., Nakano, T. and Kitamura, Y. (1989) Substance P induces granulocyte infil-tration through degranulation of mast cells. *Journal of Immunology* **142**, 927–931.

McConalogue, K., Dery, O., Lovett, M., Wong, H., Walsh, J.H., Grady, E.F. and Bunnett, N.W. (1999) Substance P-induced trafficking of beta-arrestins. The role of beta-arrestins in endocytosis of the neurokinin-1 recep-tor. *Journal of Biological Chemistry* **274**, 16257–16268.

McCormack, R.J., Hart, R.P. and Ganea, D. (1996) Expression of NK-1 receptor mRNA in murine T lympho-cytes. *Neuroimmunomodulation* **3**, 35–46.

McDonald, D.M., Bowden, J.J., Baluk, P. and Bunnett, N.W. (1996) Neurogenic inflammation. A model for study-ing efferent actions of sensory nerves. *Advances in Experimental Medicine and Biology* **410**, 453–462 (review with 68 references).

McKay, D.M. and Bienenstock, J. (1994) The interaction between mast cells and nerves in the gastrointestinal tract. *Immunology Today* **15**, 533–538 (review with 45 references).

McNeil, H.P. and Gotis-Graham, I. (2000) Human mast cell subsets – distinct functions in inflammation? *Inflammation Research* **49**, 3–7 (review with 51 references).

Metcalfe, D.D., Baram, D. and Mekori, Y.A. (1997) Mast cells. *Physiological Reviews* **77**, 1033–1079 (review with 593 references).

Metwali, A., Blum, A.M., Ferraris, L., Klein, J.S., Fiocchi, C. and Weinstock, J.V. (1994) Eosinophils within the healthy or inflamed human intestine produce substance P and vasoactive intestinal peptide. *Journal of Neuroimmunology* **52**, 69–78.

Miura, S., Serizawa, H., Tsuzuki, Y., Kurose, I., Suematsu, M., Higuchi, H., Shigematsu, T., Hokari, R., Hirokawa, M., Kimura, H. and Ishii, H. (1997) Vasoactive intestinal peptide modulates T lymphocyte migra-tion in Peyer's patches of rat small intestine. *American Journal of Physiology* **272**, G92–G99.

Mohle, R., Rafii, S. and Moore, M.A. (1998) The role of endothelium in the regulation of hematopoietic stem cell migration. *Stem Cells* **16** (Suppl 1), 159–165 (review with 36 references).

Moore, R.N., Osmand, A.P., Dunn, J.A., Joshi, J.G. and Rouse, B.T. (1988) Substance P augmentation of CSF-1-stimulated in vitro myelopoiesis. A two-signal progenitor restricted, tuftsin-like effect. *Journal of Immunology* **141**, 2699–2703.

Mosimann, B.L., White, M.V., Hohman, R.J., Goldrich, M.S., Kaulbach, H.C. and Kaliner, M.A. (1993) Substance P, calcitonin gene-related peptide, and vasoactive intestinal peptide increase in nasal secretions after allergen challenge in atopic patients. *Journal of Allergy and Clinical Immunology* **92**, 95–104.

Nakazawa, A., Watanabe, M., Kanai, T., Yajima, T., Yamazaki, M., Ogata, H., Ishii, H., Azuma, M. and Hibi, T. (1999) Functional expression of costimulatory molecule CD86 on epithelial cells in the inflamed colonic mucosa. *Gastroenterology* **117**, 536–545 (see comments).

Navari, R.M., Reinhardt, R.R., Gralla, R.J., Kris, M.G., Hesketh, P.J., Khojasteh, A., Kindler, H., Grote, T.H., Pendergrass, K., Grunberg, S.M., Carides, A.D. and Gertz, B.J. (1999) Reduction of cisplatin-induced eme-sis by a selective neurokinin-1-receptor antagonist. L-754,030 Antiemetic Trials Group. *New England Journal of Medicine* **340**, 190–195 (see comments).

Nishimura, K., Warabi, K., Roush, E.D., Frederick, J., Schwinn, D.A. and Kwatra, M.M. (1998) Characterization of GRK2-catalyzed phosphorylation of the human substance P receptor in Sf9 membranes. *Biochemistry* **37**, 1192–1198.

Numao, T. and Agrawal, D.K. (1992) Neuropeptides modulate human eosinophil chemotaxis. *Journal of Immunology* **149**, 3309–3315.

Ogasawara, T., Murakami, M., Suzuki-Nishimura, T., Uchida, M.K. and Kudo, I. (1997) Mouse bone marrow-derived mast cells undergo exocytosis, prostanoid generation, and cytokine expression in response to G protein-activating polybasic compounds after coculture with fibroblasts in the presence of c-kit ligand. *Journal of Immunology* **158**, 393–404.

Ogawa, K., Nabe, T., Yamamura, H. and Kohno, S. (1999) Nanomolar concentrations of neuropeptides induce his-tamine release from peritoneal mast cells of a substrain of Wistar rats. *European Journal of Pharmacology* **374**, 285–291.

Ohkubo, H. and Nakanishi, S. (1991) Molecular characterization of the three tachykinin receptors. *Annals of the New York Academy of Sciences* **632**, 53–62 (review with 33 references).

Okada, T., Hirayama, Y., Kishi, S., Miyayasu, K., Hiroi, J. and Fujii, T. (1999) Functional neurokinin NK-1 recep-tor expression in rat peritoneal mast cells. *Inflammation Research* **48**, 274–279.

Okamoto, A., Lovett, M., Payan, D.G. and Bunnett, N.W. (1994) Interactions between neutral endopeptidase (EC 3.4.24.11) and the substance P (NK1) receptor expressed in mammalian cells. *Biochemical Journal* **299**, 683–693.

Okamoto, Y., Shirotori, K., Kudo, K., Ishikawa, K., Ito, E., Togawa, K., and Saito, I. (1993) Cytokine expression after the topical administration of substance P to human nasal mucosa. The role of substance P in nasal allergy. *Journal of Immunology* **151**, 4391–4398.

Pang, X., Marchand, J., Sant, G.R., Kream, R.M. and Theoharides, T.C. (1995) Increased number of substance P positive nerve fibres in interstitial cystitis. *British Journal of Urology* **75**, 744–750.

Payan, D.G., Brewster, D.R., Missirian-Bastian, A. and Goetzl, E.J. (1984a) Substance P recognition by a subset of human T lymphocytes. *Journal of Clinical Investigation* **74**, 1532–1539.

Payan, D.G., Brewster, D.R. and Goetzl, E.J. (1984b) Stereospecific receptors for substance P on cultured human IM-9 lymphoblasts. *Journal of Immunology* **133**, 3260–3265.

Payan, D.G., McGillis, J.P. and Organist, M.L. (1986) Binding characteristics and affinity labeling of protein con-
    stituents of the human IM-9 lymphoblast receptor for substance P. *Journal of Biological Chemistry* **261**,
    14321–14329.
Piantelli, M., Maggiano, N., Larocca, L.M., Ricci, R., Ranelletti, F.O., Lauriola, L. and Capelli, A. (1990)
    Neuropeptide-immunoreactive cells in human thymus. *Brain, Behavior and Immunity* **4**, 189–197.
Pinter, E., Brown, B., Hoult, J.R. and Brain, S.D. (1999) Lack of evidence for tachykinin NK1 receptor-mediated
    neutrophil accumulation in the rat cutaneous microvasculature by thermal injury. *European Journal of
    Pharmacology* **369**, 91–98.
Pothoulakis, C., Castagliuolo, I., LaMont, J.T., Jaffer, A., O'Keane, J.C., Snider, R.M. and Leeman, S.E. (1994)
    CP-96,345, a substance P antagonist, inhibits rat intestinal responses to Clostridium difficile toxin A but not
    cholera toxin. *Proceedings of the National Academy of Sciences of the United States of America* **91**,
    947–951.
Pothoulakis, C., Castagliuolo, I., Leeman, S.E., Wang, C.C., Li, H., Hoffman, B.J. and Mezey, E. (1998)
    Substance P receptor expression in intestinal epithelium in clostridium difficile toxin A enteritis in rats.
    American Journal of Physiology **275**, G68–G75.
Quartara, L. and Maggi, C.A. (1997) The tachykinin NK1 receptor. Part I: ligands and mechanisms of cellular
    activation. *Neuropeptides* **31**, 537–563 (review with 154 references).
Quinlan, K.L., Song, I.S., Naik, S.M., Letran, E.L., Olerud, J.E., Bunnett, N.W., Armstrong, C.A., Caughman,
    S.W. and Ansel, J.C. (1999) VCAM-1 expression on human dermal microvascular endothelial cells is
    directly and specifically up-regulated by substance P. *Journal of Immunology* **162**, 1656–1661.
Rakasz, E., Blum, A.M., Metwali, A., Elliott, D.E., Li, J., Ballas, Z.K., Qadir, K., Lynch, R. and Weinstock, J.V.
    (1998) Localization and regulation of IFN-gamma production within the granulomas of murine schistoso-
    miasis in IL-4-deficient and control mice. *Journal of Immunology* **160**, 4994–4999.
Rameshwar, P. (1997) Substance P: a regulatory neuropeptide for hematopoiesis and immune functions. *Clinical
    Immunology and Immunopathology* **85**, 129–133 (review with 57 references).
Rameshwar, P., Ganea, D. and Gascon, P. (1994) Induction of IL-3 and granulocyte-macrophage colony-
    stimulating factor by substance P in bone marrow cells is partially mediated through the release of IL-1 and
    IL-6. *Journal of Immunology* **152**, 4044–4054.
Rameshwar, P. and Gascon, P. (1995) Substance P (SP) mediates production of stem cell factor and interleukin-1
    in bone marrow stroma: potential autoregulatory role for these cytokines in SP receptor expression and
    induction. *Blood* **86**, 482–490.
Rameshwar, P. and Gascon, P. (1996) Induction of negative hematopoietic regulators by neurokinin-A in bone
    marrow stroma. *Blood* **88**, 98–106.
Rameshwar, P., Gascon, P. and Ganea, D. (1993) Stimulation of IL-2 production in murine lymphocytes by sub-
    stance P and related tachykinins. *Journal of Immunology* **151**, 2484–2496.
Repke, H. and Bienert, M. (1987) Mast cell activation – a receptor-independent mode of substance P action?
    FEBS Letters **221**, 236–240.
Repke, H. and Bienert, M. (1988) Structural requirements for mast cell triggering by substance P-like peptides.
    Agents and Actions **23**, 207–210.
Reubi, J.C., Horisberger, U., Kappeler, A. and Laissue, J.A. (1998) Localization of receptors for vasoactive
    intestinal peptide, somatostatin, and substance P in distinct compartments of human lymphoid organs. *Blood*
    92, 191–197.
Riegler, M., Castagliuolo, I., So, P.T., Lotz, M., Wang, C., Wlk, M., Sogukoglu, T., Cosentini, E., Bischof, G.,
    Hamilton, G., Teleky, B., Wenzl, E., Matthews, J.B. and Pothoulakis, C. (1999) Effects of substance P on
    human colonic mucosa in vitro. *American Journal of Physiology* **276**, G1473–G1483.
Roush, E.D. and Kwatra, M.M. (1998) Human substance P receptor expressed in Chinese hamster ovary cells
    directly activates G(alpha q/11), G(alpha s), G(alpha o). *FEBS Letters* **428**, 291–294.
Saban, R., Saban, M.R., Nguyen, N.B., Lu, B., Gerard, C., Gerard, N.P. and Hammond, T.G. (2000)
    Neurokinin-1 (NK-1) receptor is required in antigen-induced cystitis. *American Journal of Pathology* **156**,
    775–780.
Sasakawa, N., Ferguson, J.E., Sharif, M. and Hanley, M.R. (1994) Attenuation of agonist-induced desensitization
    of the rat substance P receptor by microinjection of inositol pentakis- and hexakisphosphates in Xenopus
    laevis oocytes. *Molecular Pharmacology* **46**, 380–385.
Schratzberger, P., Reinisch, N., Prodinger, W.M., Kahler, C.M., Sitte, B.A., Bellmann, R., Fischer-Colbrie, R.,
    Winkler, H. and Wiedermann, C.J. (1997) Differential chemotactic activities of sensory neuropeptides for
    human peripheral blood mononuclear cells. *Journal of Immunology* **158**, 3895–3901.
Schuiling, M., Zuidhof, A.B., Zaagsma, J. and Meurs, H. (1999) Involvement of tachykinin NK1 receptor in the
    development of allergen-induced airway hyperreactivity and airway inflammation in conscious, unrestrained
    guinea pigs. *American Journal of Respiratory and Critical Care Medicine* **159**, 423–430.

Serra, M.C., Bazzoni, F., Della, B.V., Greskowiak, M. and Rossi, F. (1988) Activation of human neutrophils by substance P. Effect on oxidative metabolism, exocytosis, cytosolic Ca^{2+} concentration and inositol phosphate formation. *Journal of Immunology* **141**, 2118–2124.

Serra, M.C., Calzetti, F., Ceska, M. and Cassatella, M.A. (1994) Effect of substance P on superoxide anion and IL-8 production by human PMNL. *Immunology* **82**, 63–69.

Snider, R.M., Constantine, J.W., Lowe, J.A., Longo, K.P., Lebel, W.S., Woody, H.A., Drozda, S.E., Desai, M.C., Vinick, F.J. and Spencer, R.W. (1991) A potent nonpeptide antagonist of the substance P (NK1) receptor. *Science* **251**, 435–437.

Stanisz, A.M., Befus, D. and Bienenstock, J. (1986) Differential effects of vasoactive intestinal peptide, substance P, and somatostatin on immunoglobulin synthesis and proliferations by lymphocytes from Peyer's patches, mesenteric lymph nodes, and spleen. *Journal of Immunology* **136**, 152–156.

Stanisz, A.M., Scicchitano, R., Dazin, P., Bienenstock, J. and Payan, D.G. (1987) Distribution of substance P receptors on murine spleen and Peyer's patch T and B cells. *Journal of Immunology* **139**, 749–754.

Stead, R.H., Tomioka, M., Quinonez, G., Simon, G.T., Felten, S.Y. and Bienenstock, J. (1987) Intestinal mucosal mast cells in normal and nematode-infected rat intestines are in intimate contact with peptidergic nerves. *Proceedings of the National Academy of Sciences of the United States of America* **84**, 2975–2979.

Sterner-Kock, A., Braun, R.K., van der Vliet, A., Schrenzel, M.D., McDonald, R.J., Kabbur, M.B., Vulliet, P.R. and Hyde, D.M. (1999) Substance P primes the formation of hydrogen peroxide and nitric oxide in human neutrophils. *Journal of Leukocyte Biology* **65**, 834–840.

Sturiale, S., Barbara, G., Qiu, B., Figini, M., Geppetti, P., Gerard, N., Gerard, C., Grady, E.F., Bunnett, N.W. and Collins, S.M. (1999) Neutral endopeptidase (EC 3.4.24.11) terminates colitis by degrading substance P. *Proceedings of the National Academy of Sciences of the United States of America* **96**, 11653–11658.

Sundelin, J.B., Provvedini, D.M., Wahlestedt, C.R., Laurell, H., Pohl, J.S. and Peterson, P.A. (1992) Molecular cloning of the murine substance K and substance P receptor genes. *European Journal of Biochemistry* **203**, 625–631.

Suzuki, R., Furuno, T., McKay, D.M., Wolvers, D., Teshima, R., Nakanishi, M. and Bienenstock, J. (1999) Direct neurite-mast cell communication in vitro occurs via the neuropeptide substance P. *Journal of Immunology* **163**, 2410–2415.

Swain, M.G., Agro, A., Blennerhassett, P., Stanisz, A. and Collins, S.M. (1992) Increased levels of substance P in the myenteric plexus of Trichinella-infected rats. *Gastroenterology* **102**, 1913–1919.

Takahashi, K., Tanaka, A., Hara, M. and Nakanishi, S. (1992) The primary structure and gene organization of human substance P and neuromedin K receptors. *European Journal of Biochemistry* **204**, 1025–1033.

Tanpo, T., Nabe, T., Yasui, K., Kamiki, T. and Kohno, S. (2000) Participation of neuropeptides in antigen-induced contraction of guinea pig bronchi via NK(2) but not NK(1) receptor stimulation. *Pharmacology* **60**, 169–174.

Tomaki, M., Ichinose, M., Miura, M., Hirayama, Y., Yamauchi, H., Nakajima, N. and Shirato, K. (1995) Elevated substance P content in induced sputum from patients with asthma and patients with chronic bronchitis. *American Journal of Respiratory and Critical Care Medicine* **151**, 613–617.

Torrens, Y., Beaujouan, J.C., Saffroy, M., Glowinski, J. and Tence, M. (1998) Functional coupling of the NK1 tachykinin receptor to phospholipase D in chinese hamster ovary cells and astrocytoma cells. *Journal of Neurochemistry* **70**, 2091–2098.

Vacchio, M.S., Papadopoulos, V. and Ashwell, J.D. (1994) Steroid production in the thymus: implications for thymocyte selection. *Journal of Experimental Medicine* **179**, 1835–1846.

van Hagen, P.M., Breeman, W.A., Reubi, J.C., Postema, P.T., van den Anker-Lugtenburg, P.J., Kwekkeboom, D.J., Laissue, J., Waser, B., Lamberts, S.W., Visser, T.J. and Krenning, E.P. (1996) Visualization of the thymus by substance P receptor scintigraphy in man. *European Journal of Nuclear Medicine* **23**, 1508–1513.

van der Velden, V.H. and Hulsmann, A.R. (1999) Autonomic innervation of human airways: structure, function, and pathophysiology in asthma. *Neuroimmunomodulation* **6**, 145–159 (review with 266 references).

Vanden Broeck, J., Torfs, H., Poels, J., Van Poyer, W., Swinnen, E., Ferket, K. and De Loof, A. (1999) Tachykinin-like peptides and their receptors. A review. *Annals of the New York Academy of Sciences* **897**, 374–387 (review with 94 references).

Varilek, G.W., Weinstock, J.V., Williams, T.H. and Jew, J. (1991) Alterations of the intestinal innervation in mice infected with Schistosoma mansoni. *Journal of Parasitology* **77**, 472–478.

Veronesi, B., Carter, J.D., Devlin, R.B., Simon, S.A. and Oortgiesen, M. (1999) Neuropeptides and capsaicin stimulate the release of inflammatory cytokines in a human bronchial epithelial cell line. *Neuropeptides* **33**, 447–456.

Von Essen, S.G., Rennard, S.I., O'Neill, D., Ertl, R.F., Robbins, R.A., Koyama, S. and Rubinstein, I. (1992) Bronchial epithelial cells release neutrophil chemotactic activity in response to tachykinins. *American Journal of Physiology* **263**, L226–L231.

Wagner, J.G. and Roth, R.A. (2000) Neutrophil migration mechanisms, with an emphasis on the pulmonary vasculature. *Pharmacological Reviews* **52**, 349–374 (review with 373 references).

Wang, L., Stanisz, A.M., Wershil, B.K., Galli, S.J. and Perdue, M.H. (1995) Substance P induces ion secretion in
mouse small intestine through effects on enteric nerves and mast cells. *American Journal of Physiology* **269**,
G85–G92.

Weinstock, J.V. and Blum, A.M. (1989) Tachykinin production in granulomas of murine schistosomiasis mansoni.
Journal of Immunology **142**, 3256–3261.

Weinstock, J.V., Blum, A., Walder, J. and Walder, R. (1988) Eosinophils from granulomas in murine schistosomiasis mansoni produce substance P. *Journal of Immunology* **141**, 961–966.

Weinstock, J.V., Blum, A.M. and Malloy, T. (1990) Macrophages within the granulomas of murine Schistosoma
mansoni are a source of a somatostatin 1-14-like molecule. *Cellular Immunology* **131**, 381–390.

Weinstock, J.V. and Elliott, D. (1998) The substance P and somatostatin interferon-gamma immunoregulatory circuit. *Annals of the New York Academy of Sciences* **840**, 532–539 (review with 19 references).

Wozniak, A., Betts, W.H., McLennan, G. and Scicchitano, R. (1993) Activation of human neutrophils by
tachykinins: effect on formyl-methionyl-leucyl-phenylalanine- and platelet-activating factor-stimulated
superoxide anion production and antibody-dependent cell-mediated cytotoxicity. *Immunology* **78**, 629–634.

Wozniak, A., McLennan, G., Betts, W.H., Murphy, G.A. and Scicchitano, R. (1989) Activation of human neutrophils by substance P: effect on FMLP-stimulated oxidative and arachidonic acid metabolism and on antibody-dependent cell-mediated cytotoxicity. *Immunology* **68**, 359–364.

Yano, H., Wershil, B.K., Arizono, N. and Galli, S.J. (1989) Substance P-induced augmentation of cutaneous vascular permeability and granulocyte infiltration in mice is mast cell dependent. *Journal of Clinical
Investigation* **84**, 1276–1286.

Yokota, Y., Sasai, Y., Tanaka, K., Fujiwara, T., Tsuchida, K., Shigemoto, R., Kakizuka, A., Ohkubo, H. and
Nakanishi, S. (1989) Molecular characterization of a functional cDNA for rat substance P receptor. *Journal
of Biological Chemistry* **264**, 17649–17652.

8 Nerve–Mast Cell Interactions – Partnership in Health and Disease

Hanneke P.M. van der Kleij[1], Michael G. Blennerhassett[2] and John Bienenstock[3]

[1]*Department of Pharmacology and Pathophysiology, Utrecht University, Utrecht, The Netherlands*
[2]*Gastrointestinal Diseases Research Unit, Department of Medicine, Queen's University, Kingston, Ontario, Canada*
[3]*Department of Pathology and Molecular Medicine, McMaster University, Hamilton, Ontario, Canada*

Mast cells often lie in close apposition to nerves in most tissues of the body. Bi-directional communication pathways involving many mast cell mediators including histamine, serotonin, cytokines and products of arachidonic acid metabolism, variously stimulate and regulate neuronal function. Similarly, neuropeptides and nerve growth factors stimulate secretory and other activity in mast cells. In this way, mast cells can act both as sentinel sensory afferent receptor cells for antigens, toxins, etc. and bring about local and central homeostatic responses. Emotional and behavioural activities in the central nervous system can bring about mast cell responses both centrally and peripherally. In this manner, mast cell–nerve interactions can be thought of as a clear and potent example of neuroimmune communication. They may be involved in physiologic, or pathologic processes in inflammation and disease, as well as in responses as varied as Pavlovian conditioning and reactions to stress.

It is becoming clearer that these types of interactions between the nervous and immune systems are extensively involved in the regulation of physiologic processes as well as those involved in disease mechanisms.

KEY WORDS: nerve–mast cell interactions; disease mechanism; neuroimmune communication.

INTRODUCTION

Histological studies reveal an intimate association between mast cells and neurons in both the peripheral and central nervous system (Stead *et al.*, 1989; Purcell and Atterwill, 1995; Undem *et al.*, 1995; Botchkarev *et al.*, 1997). In virtually all tissues of the body, mast cells are found in close proximity to nerve fibres. While the literature mainly focuses on the anatomical and morphological link between the two, this proximity also represents a functional link between the immune and nervous systems, whereby mast cells appear to act as bi-directional carriers of information (Bienenstock *et al.*, 1989; Bauer and Razin, 2000).

Mast cells influence their local environment and in turn are influenced by it. Mast cells are sensitized to antigens by the binding of antibodies through specific receptors on their surface, but can also be caused to secrete by other types of molecules because of their electrical charge or their chemical nature. Therefore, association with the nervous system allows mast cells to act as sensory receptors for a variety of newly encountered or potentially noxious substances. They are therefore an ideal cellular transducer, acting to pass information on through afferent nerves to local tissues by axon reflexes, as well as to the spinal cord and thence, the brain.

In various studies, tissue mast cells invariably show ultrastructural evidence of some degree of activation even in normal healthy conditions, suggesting that these cells are constantly providing information to the nervous system. The fact that they are located at sites under constant exposure to the external environment, such as the skin, respiratory and gastrointestinal tract, emphasizes the significance of these associations.

Mast cells are motile cells, and so may be viewed as sensory receptors with the unique capacity to migrate to and from sites where nervous tissue exists, or may be undergoing developmental or regenerative changes. Once in place, they may give information about the local environment, injury and potentially injurious substances to the nervous system, and thus promote appropriate efferent action. While some factors are known to be chemotactic for mast cells, this area of our understanding is limited and little is known about their traffic. It is not known whether mast cells will remain sessile once associated with nerves, and whether (or if, and under what circumstances), they will dissociate from nerves.

As opposed to sensory receptors, mast cells can also act as efferent targets and can equally be associated with local effector function, through release of their potent preformed and newly synthesized mediator molecules. It is in this way that they are involved in the regulation of vascular tone through the effects on capillaries and small blood vessels. Presumably, in the same way they are also involved in migraine, a complex example of the integration of responses to stress, environmental factors, the nervous system and the tone of the superficial and cerebral blood supply (Theoharides *et al.*, 1995; Theoharides, 1996).

Sensory neurons play a role in neurogenic inflammation involving changes in functioning due to inflammatory mediators which results in an enhanced release of neuropeptides from the sensory nerve endings (Barnes, 1991; Campbell *et al.*, 1998). The classical role of the mast cell in hypersensitivity reactions is well known and extensively studied, involving the interaction of allergens with IgE (Galli, 1993). However, it is becoming apparent that the mast cell and its mediators play an important role in neurogenic inflammation by affecting neuronal functioning. Neurogenic inflammation has been shown to occur in different tissues, including the skin, airways, urinary tract and the digestive system. Furthermore, the role of mast cells and the nervous system is becoming apparent in delayed type hypersensitivity reactions as well as in non-IgE mediated reaction for instance in the airways and intestine (Buckley and Nijkamp, 1994; Kraneveld *et al.*, 1998).

In this chapter, an overview of the various studies that have been performed in this field will be presented and discussed. Although we will only discuss nerve–mast cell communication, it should be mentioned that this is just a small part of a set of neuro-immune interactions with extensive documentation in both rodents and humans. Besides mast cells, neural contact can also occur between nerves and eosinophils or plasma cells (Arizono *et al.*, 1990; Purcell and Atterwill, 1995). In addition, neuropeptides released from sensory nerves can directly modulate the function of Langerhans cells. Among these neuropeptides, the tachykinins have been shown to modulate immune cell functions such as cytokine production, antigen presentation and cell proliferation (Stanisz *et al.*, 1987; Scholzen *et al.*, 1998). Therefore, mast cell–nerve exchanges are a most interesting example of a complex network of neuroimmune communication in the body.

How essential are these communication pathways we describe? We have to assume that these only add to other basic sensory and efferent methods of nervous communication and are inessential in the sense of survival of the organism. However, the nerve–mast cell association is preserved in phylogeny, since even frogs appear to retain these structural and even functional adaptions (Monteforte *et al.*, 2001). In mutant animals in which mast cells are not found (e.g. W/W^v or Sl/Sl^d mice), there are various minor physiologic aberrations, including delayed responses to parasite infections, but in general, transgenic knockout animals and mutants without mast cells have similar life spans to their background littermates under laboratory conditions. It seems therefore reasonable to conclude that mast cell–nerve communication has evolved as an important non-essential regulatory mechanism. This system, then additionally informs the brain of events in the periphery, and allows for an adaptable set of reactive effector responses.

MAST CELLS

Mast cells are widely distributed throughout the body in connective tissues (Botchkarev *et al.*, 1997), particularly around blood vessels and nerves. They are abundant in the submucosa of the digestive tract (Wershil and Galli, 1991), in oral and nasal mucosa (Otsuka *et al.*, 1985), respiratory mucosal surfaces (Kaliner, 1987), and skin (Scholzen *et al.*, 1998). Mast cells are detected even in the brain (Silver *et al.*, 1996; Matsumoto *et al.*, 2001) – no tissue, in fact, has been shown to be devoid of the presence of mast cells.

Mast cells are involved in the regulation of their own overall tissue cell mass, since mast cell degranulation leads to an overall increase in mast cells (Marshall *et al.*, 1990). The generation and secretion of diverse mast-cell derived factors are involved, such as GM-CSF, SCF and NGF, all of which are known to promote mast cell growth (Rottem *et al.*, 1994; Valent, 1995).

Mast cells are increased in tissues undergoing inflammation, where they may have an intimate involvement in repair processes. Mast cells also have a phagocytic function, which might contribute to host defense (Galli and Wershil, 1996) especially where tissue repair and fibrosis is occurring (Bienenstock *et al.*, 1987a; Hebda *et al.*, 1993). Mast cells have various cytokines stored in their granules that are stimulatory to fibroblasts (Hultner *et al.*, 2000). These also support immunological defense strategies against parasites (Metcalfe *et al.*, 1997). In addition, they contain serine proteases that may be involved in remodelling of the extracellular matrix during healing (Hebda *et al.*, 1993). Mast cells are most often found in association with blood vessels everywhere and are a major source of glycosoaminoglycans such as heparin which have major effects on coagulation and other physiological systems (Wedemeyer *et al.*, 2000). In addition, many of its mediators have profound direct effects on vascular tone and permeability, for example, histamine, serotonin and many products of arachidonic acid metabolism.

Mast cells can be divided into various subpopulations with distinct phenotypes. Two main subsets, connective tissue type mast cells (CTMC) and mucosal mast cells (MMC) are recognized as distinct mast cell populations with different phenotypical and functional characteristics (Befus *et al.*, 1987; Galli, 1990). Many more, phenotypically different subsets have been described in rodents (Katz *et al.*, 1985). In spite of their differences, both CTMC and MMC are considered to be derived from a common precursor in the bone marrow. Mast cell progenitor cells translocate from bone marrow to mucosal and connective tissues to locally undergo differentiation into mature forms. They possess a remarkable degree of plasticity, so that even apparently fully differentiated CTMC will transform their phenotype to MMC if transplanted into an intestinal mucosal environment (Kitamura *et al.*, 1987). In contrast to many other cell types, mast cells are absent from the blood and their final maturation takes place in the tissue (Galli, 1993). Their development and survival essentially depends on stem cell factor (SCF) and its receptor c-kit (Galli *et al.*, 1995). Besides SCF, cytokines such as IL-3, IL-4 and IL-10 influence mast cell growth and differentiation (Rennick *et al.*, 1995), as does nerve growth factor (Denburg, 1990; Matsuda *et al.*, 1991).

Mast cells are versatile cells capable of synthesis of a large number of pro- and anti-inflammatory mediators including cytokines, products of arachidonic acid metabolism, growth factors including NGF, SCF, serotonin, histamine, etc. These rich sources of mediators can be pre-stored or newly synthesized upon stimulation. Pre-stored mediators, such as histamine, serine proteases, proteoglycans, sulphatases and TNF-α, are released within minutes after degranulation of the cell (Church and Levi-Schaffer, 1997). After this primary response, a second wave of newly synthesized mediators are released and include PGD_2, LTC_4, LTD_4 and LTE_4. In the late phase allergic response, cytokines (IL-4, IL-5, IL-6, IL-8, IL-13 and TNF-α) are induced and secreted (Church and Levi-Schaffer, 1997). We will focus upon two important mast cell mediators, TNF-α and tryptase, below.

Expression of this host of cytokines supports the logical proposal for a role for mast cells in host defence. As examples, this includes IgE-dependent immune responses to certain parasites, in natural immunity to bacterial infections as well as in inflammatory diseases. Stimulation of the enteric nervous system by mast cell activation is also likely to play an important role in mast-cell mediated host defense (Echtenacher *et al.*, 1996; Malaviya *et al.*, 1996), and in general, mast cell–nerve interactions have been interpreted as important neuronal tissue repair mechanisms following injury (Gottwald *et al.*, 1998; Murphy *et al.*, 1999).

MORPHOLOGIC EVIDENCE FOR MAST CELL–NERVE ASSOCIATION

Nerve–mast cell associations have been reported within peripheral, myelinated nerves, unmyelinated nerves, neurofibromata and neuromata. A morphometric study in infected and in healthy rat intestine, showed that mast cells and nerves were closely and invariably approximated in rat intestinal villi (Bienenstock *et al.*, 1987b). Electron microscopy showed evident membrane/membrane association between mucosal mast cells and nerves with dense core vesicles at the points of contact. The nerves in contact with mast cells contained either substance P, CGRP, or both. The association appeared not to be random (Yonei *et al.*, 1985), and was also described in the human gastrointestinal tract (Stead *et al.*, 1989).

Similar observations have been made in a variety of different tissues in many species. Other than the intestine, nerve–mast cell associations are also found in rat trachea and peripheral lung tissue (Undem *et al.*, 1995), skin (Egan *et al.*, 1998), urinary bladder (Letourneau *et al.*, 1996), brain (Keller and Marfurt, 1991) and several other tissues (Olsson, 1968; Newson *et al.*, 1983). Rozniecki *et al.* (1999) provided evidence for morphological, anatomical and functional interactions of dura mast cells with cholinergic and peptidergic neurons containing substance P and CGRP. Mast cells have been shown to be abundant in the dura and they contain a substantial proportion of total brain histamine.

MAST CELL ACTIVATION

Mast cells can be activated by IgE-dependent and independent mechanisms. Classically, they are associated with hypersensitivity reactions, involving the interaction with IgE (Galli, 1993). However, mast cells also play a prominent role in non-IgE mediated hypersensitivity reactions (Kraneveld *et al.*, 2000; Ramirez-Romero *et al.*, 2000). The sensitivity of mast cells to activation by non-immunological stimuli such as polycationic compounds, complement proteins, superoxide anions or neuropeptides is dependent on the population of the mast cells examined (Johnson and Krenger, 1992).

Tachykinins can induce mast cell activation via a receptor-dependent mechanism. Activation of the neurokinin receptors is dependent on the C-terminal domain of the tachykinins (Mousli *et al.*, 1990a). C-terminal fragments of substance P cause histamine release from the mouse mast cell line MC/9 via an NK-2 receptor mediated pathway (Krumins and Broomfield, 1993). Cooke *et al.* (1998) demonstrated that RBL-2H3 cells, a mucosal mast cell line, express the high affinity NK-1 binding sites for substance P on their surface. While it is widely accepted that NK-1 receptors are not generally expressed on mast cells, little is known about their expression in inflammation. Mantyh *et al.* (1995, 1996) have shown that NK-1 receptors were significantly upregulated in inflamed tissues, on epithelium, blood vessels and in lymphoid accumulations. Karimi *et al.* (1999, 2000) showed that SP (in the micromolar range) causes dose-dependent degranulation in bone marrow-derived murine mast cells (BMMC) primed with IL-4 and SCF. We ourselves have recently found that murine BMMC cultured with IL-4 and SCF are induced to express NK-1 receptors (Van der Kleij, Ma, Bienenstock, unpublished).

It has also been demonstrated that the structurally conserved C-terminal sequences of substance P and neurokinin A can both interact with a common region of the NK-1 receptor (Bremer *et al.*, 2000). The effects of tachykinins result from signal transduction initiated by the interaction of tachykinins with specific receptors on effector cells. Three distinct subtypes of mammalian receptors have been identified and denoted as NK-1,

NK-2 and NK-3, which have the highest affinity for substance P, Neurokinin A and B, respectively (Solway and Leff, 1991; Advenier et al., 1997; Joos et al., 2000a). The tachykinin NK-3 receptors are predominantly present in the central and peripheral nervous system (Grady et al., 1996). Tachykinin NK-1 receptors are localized on smooth muscle cells, submucosal glands, blood vessels and inflammatory cells. The NK-2 receptor is also found on smooth muscle. Animal studies have shown that the NK-2 receptors are involved in bronchoconstriction, whereas activation of tachykinin NK-1 receptors leads to neurogenic inflammation (McDonald et al., 1996).

The tachykinins substance P and neurokinin A can, on the other hand, induce mast cell activation via a receptor-independent mechanism. Non-immunological stimuli such as the basic secretagogues trigger mast cell exocytosis though a mechanism called the peptidergic pathway. This family of polycationic compounds, include positively charged peptides such as substance P, various amines such as compound 48/80 and naturally occuring amines (Metcalfe et al., 1997). Instead of interacting with a membrane bound receptor, the basic secretagogues appear to directly activate and bind to pertussis toxin-sensitive GTP-binding proteins (G-proteins) through the N-terminal domain located in the inner surface of the plasma membrane (Mousli et al., 1990b; Emadi-Khiav et al., 1995). Stimulation of G-proteins will activate a signal transduction pathway (the peptidergic pathway) eventually leading to mast cell mediator production and release.

There is significant potential for this pathway of mast cell activation to influence the net contribution of mast cells to both tissue physiology and pathology. For example, Janiszewski et al. (1994) used patch clamp electrophysiology to show that mast cells did not respond to an initial application of very low concentrations of substance P (in the picomolar range), but that both activation and delayed degranulation occurred after a second exposure. Therefore, mast cells can be primed when exposed to physiologically relevant low concentrations of substance P, and lower their thresholds to subsequent activation.

TNF-α

TNF-α is one of the main preformed mediators immediately released upon mast cell degranulation. In addition, newly synthetized TNF-α can be secreted by mast cells within 30 min following certain stimuli (Gordon and Galli, 1990). Furthermore, TNF-α is also able itself to induce mast cell degranulation. This, and the fact that TNF-α has been shown to affect sensory neurons, makes it an interesting potential mediator in nerve–mast cell communication.

TNF-α may play a major beneficial role in host defense by mast cells against bacterial infections. Malaviya et al. (1996) showed that mast cells may be essential for defense against bacteria in a cecal puncture model, and that they mediate bacterial clearance by initiating neutrophil influx. They propose that the recruitment of circulating leukocytes is dependent on the mast cell mediator TNF-α. Echtenachter et al. (1996) showed that reconstitution of mast cell deficient W/Wv with mast cells prevented death from bacterial peritonitis, as did the injection of TNF-α. Maier et al. (1998) showed that subdiaphragmatic vagotomy prevented the pyrexia induced by intraperitoneal injection of endotoxin and that this was associated with the generation of TNF-α (see Figure 8.1). It is tempting to put this

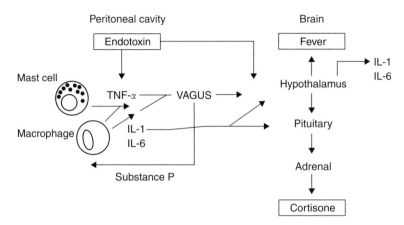

Figure 8.1   Schematic outline of the work of Maier *et al.* (1998). This work has shown that if endotoxin is introduced into the peritoneal cavity in vagotomized animals, the fever response of the brain is markedly diminished
with consequent downstream effects on cortisol production. This response appears to involve both mast cells and
nerves in a complex set of interactions.

information together with the role of mast cells in innate defense against bacteria, but the
crucial experiments have yet to be performed.

   TNF-α is involved in changing neuronal cell function as it can modulate the susceptibility of neurons to an electrical stimulus (Hattori *et al.*, 1994). The sensitizing effect of
TNF-α seems to primarily target C-fibres (Junger and Sorkin, 2000). *In vitro* incubation of
rat sensory nerves with TNF-α, enhanced the response of C-fibres to capsaicin (Nicol
et al., 1997). According to Junger and Sorkin (2000), TNF-α causes a subpopulation of
C fibres to develop spontaneous activity, which also results in local release of neuropeptides, like substance P and CGRP, from afferent fibres. This is the first time that TNF-α has
been shown to be directly capable of releasing neuropeptides from C fibres, in addition to
their priming ability, proposed by others. These authors suggest that this acute sensitization
is due to a fast mechanism such as binding to TNF receptors or activation of a constitutive
COX enzyme leading to eicosanoid synthesis.

   TNF-α can enhance the release of arachidonic acid and the synthesis of eicosanoids, in
particular prostaglandin E2 (PGE2) which acts directly on the sensory neuron (Nicol and
Cui, 1994). Nicol *et al.* (1997) also observed that TNF promoted the induction of COX-2
in sensory neurons. Their results demonstrate that selective inhibition of COX-2 prevents
the TNF-induced lowering of threshold to activation by capsaicin. However, these authors
found that TNF-α had a delayed time course of action. The discrepancy in findings
suggests that there might be more than one process involved.

   While little is known so far about the presence of the TNF receptors TNFR1 and TNFR2
on sensory nerve endings, a study by Aranguez *et al.* (1995) indicates that mouse astrocytes express TNFR1. Furthermore, rat microglia transcribe mRNA for both TNFR1 and
TNFR2 (Dopp *et al.*, 1997). These results indicate that neuronal tissue probably expresses
both TNF receptors, and implies that mast cell and nerve communication may be mediated
at least in part, by TNF-α.

TRYPTASE

The major mast cell protease is tryptase, a serine protease that is abundantly present in all mast cells, and is stored in a fully active form in the granules. Among the subsets of human mast cells, tryptase can comprise up to 25% of total cellular protein. The functions of tryptase include mitogenic actions on fibroblasts, smooth muscle cells and epithelial cells, and stimulation of ICAM-1 expression by epithelial cells (Walls *et al.*, 1995).

A recent finding of great importance is that proteases such as mast cell tryptase act on protease activated receptors (PAR), in which the peptide ligand is physically part of the receptor molecule (Dery *et al.*, 1998). Protease activity activates the PAR receptors in an irreversible manner by cleaving within the extracellular N-terminus a tethered ligand domain that then binds to the receptor (e.g. D'Andrea *et al.*, 2000; Buresi *et al.*, 2001; Fiorucci *et al.*, 2001).

Mast cell tryptase has inflammatory effects on many cells, mediated by the cleavage and activation of PAR2. These include neurons and glia in the central nervous system and in the enteric nervous system, where myenteric neuron PAR2 expression has been detected by RT-PCR. Tryptase has recently been shown to cleave PAR2 on primary spinal afferent neurons, causing the release of substance P, activation of the NK-1 receptor and amplification of inflammation and thermal and mechanical hyperalgesia (Defea *et al.*, 2000).

Corvera *et al.* (1999) demonstrated that purified tryptase stimulated calcium mobilization in myenteric neurons. They hypothesize that tryptase excites neurons through PAR2, because activation of PAR2 with trypsin or peptide agonists strongly desensitized the response to tryptase. In addition, a tryptase inhibitor suppressed calcium mobilization in response to degranulation of mast cells.

Recent investigations with the use of tryptase inhibitors have implicated tryptase as a mediator in the pathology of various allergic and inflammatory conditions including arthritis, rhinitis and most notably asthma. Very recently, Krishna *et al.* (2001) showed that the selective tryptase inhibitor APC366 inhibited antigen-induced early and late asthmatic responses and bronchial hyper-responsiveness in a sheep model of allergic asthma. Again in the respiratory system, the intratracheal administration of AMG12637, a potent inhibitor of human mast cell tryptase, inhibited the development of airway hyper-responsiveness in allergen challenged guinea pigs. In both proximal and distal bronchi of non-sensitized humans, the reactivity to histamine was significantly increased by previous incubation with tryptase (1 μg/ml) (Berger *et al.*, 1999). This effect was completely abrogated in the presence of the protease inhibitor benzamidine. The inhibitor APC366 has furthermore been shown to inhibit IgE-dependent histamine release in a dose dependent manner, with about 70% inhibition being achieved at a dose of 300 μM. This study was performed with human synovial mast cells, showing that inhibitors of tryptase could be of therapeutic value in arthritis (He *et al.*, 2001).

PRIMING

Priming is a process that increases cellular responsiveness to subsequent stimulation. In other words, an initial event lowers the threshold for response to a subsequent event, such as (re)-exposure to a given agonist or mode of stimulation. Priming appears to be a broad based biological process and has been reported in many cell types. Basophils and mast cells have been reported to be primed in this sense by many different cytokine growth factors for subsequent different agonists (Bischoff *et al.*, 1991). SCF, for instance, can act as a priming agent in some circumstances (Bischoff and Dahinden, 1992). Coleman *et al.* (1993) observed that SCF and also IL-3 upregulated responses to IgE-dependent stimuli. Since the expression of the IgE receptors was not altered, the mechanism behind this remains unclear.

Priming may be a prominent aspect of nerve–mast cell interactions. Karimi *et al.* (2000) showed that SCF and IL-4 primed bone marrow-derived murine mast cells to show increased responsiveness to subsequent challenge by substance P. Although relatively high levels of substance P are necessary on a single challenge to induce mast cell degranulation ($>10^{-5}$ M), it is possible for repeated doses of very low (picomolar) concentrations of substance P to act via the priming process to cause mast cell degranulation. Janiszewski *et al.* (1994) reported that mast cells responded electrophysiologically to very low concentrations of substance P but *without* degranulation. However, degranulation occurred with subsequent exposure to the same low dose.

This raises the question whether substance P is more a priming substance, than a substance causing direct degranulation. It is interesting that cellular activation followed by mediator release is not the same as degranulation of the cell. Exocytosis is the most obvious event associated with secretion of the mediator molecules contained in granules. However, secretion can occur without evidence of degranulation, and even molecules stored within the same granules can be released and secreted in a discriminatory pattern. For example, serotonin can be released separately from histamine (Theoharides *et al.*, 1985). Moreover, low doses of substance P can cause synthesis and secretion of TNF-α from mast cells, in the absence of degranulation (Ansel *et al.*, 1993; Marshall *et al.*, 1999).

Both stimulatory and inhibitory effects are seen with priming. Thus, NGF caused the synthesis and secretion of IL-6 and PGE2 by rat peritoneal mast cells but inhibited the release of TNF-α (Marshall *et al.*, 1999). Further, differential synthesis and release of arachidonic acid metabolites, prostaglandins and leukotrienes as a result have been reported (Payan *et al.*, 1984).

Structural evidence from intact tissue supports a physiological role for the differential release of mast cell mediators. An ultrastructural study by Ratliff *et al.* (1995) showed that mast cells in close proximity to unmyelinated nerve fibres had granules showing ultrastructural features of activation or piecemeal degranulation, a process associated with differential secretion (Theoharides *et al.*, 1985). A subtle role for this process in regulation of inflammation is supported by evidence *in vitro* that peritoneal mast cells can synthesize and release IL-6 without mast cell degranulation, as monitored by histamine release (Leal-Berumen *et al.*, 1994). In fact, the presence of nerve-associated activated mast cells that do not display anaphylactic degranulation, while found routinely in all tissues, has been suggested as a characteristic feature of interstitial cystitis (Theoharides *et al.*, 1985; Letourneau *et al.*, 1996).

The concept of priming also applies to neurons. Nicol *et al.* (1997) demonstrated that prior exposure to TNF-α can enhance the sensitivity of sensory neurons to the effects of capsaicin. However, Sorkin *et al.* (1997) showed that TNF-α is not able to evoke peptide release from peripheral afferent terminals, but enhances capsaicin-evoked release of neuropeptides. Thus, TNF-α may exert a priming, rather than a direct stimulatory effect on sensory activity. Considering that mast cells closely approximated to nerves will be exposed to locally released neuropeptides, we propose that priming enhances the sensory effectiveness of the mast cell–nerve physiological unit.

NEURALLY MEDIATED ACTIVATION OF MAST CELLS

Electrical stimulation of nerves has been shown to either cause ultrastructural changes in associated mast cell granules or actual degranulation, supporting the idea that mast cells are indeed in direct communication with nerves. McDonald (1988) showed that vagal stimulation in the rat caused neurogenic inflammation in the trachea. Leff (1982) have shown that vagal stimulation caused an enhancement of the secretion of histamine from mast cells after challenge of allergic dog lungs. Dimitriadou et al. (1991) showed that electrical stimulation of the ipsilateral trigeminal nerve caused activation of the dura mater mast cells, as evidenced by piece meal degranulation. Recently, electrical stimulation of the frog hypoglossal nerve was shown to cause mast cells to undergo progressive time-dependent activation (Monteforte et al., 2001). Gottwald et al. (1995) reported that electrical stimulation of the vagus caused moderate to marked oedema in the jejunum of stimulated rats in comparison to control rats. Here, the four-fold increase in tissue histamine levels in the jejunum was suggested to reflect mast cell activation. Bani-Sacchi et al. (1986) observed that field stimulation of rat ileum resulted in histamine release and an attenuation in mast cell granularity. This effect was decreased by atropine or tetrodotoxin, a nerve cell toxin. In the brain, granule changes consistent with activation of dura mast cells were observed after sensory afferent stimulation, and increased levels of tissue serotonin were recorded after sympathetic stimulation (Dimitriadou et al., 1991).

On the other hand, Miura et al. (1990) showed that antigen challenge to sensitized cats caused increased bronchial resistance and an increase in plasma histamine levels. However, in animals pre-treated with cholinergic and adrenergic blockers, bilateral electrical stimulation of the vagus nerve caused complete inhibition both of the generation of bronchial resistance and elevation of plasma histamine levels. These data show that the non-adrenergic non-cholinergic (NANC) system in cats is able to inhibit mast cell degranulation induced by antigen.

Many of the experiments that have shown mast cell degranulation upon electrical stimulation have also shown that these effects were inhibited by atropine or prior treatment with capsaicin, a substance which permanently depletes sensory nerves of substance P and destroys unmyelinated sensory axons. For instance, vagal stimulation does not cause neurogenic inflammation in the airways of rats treated at birth with capsaicin. Furthermore, antidromic nerve stimulation causes mast cell degranulation in the skin, but is absent following neonatal capsaicin treatment (Kowalski et al., 1997).

In contrast, Baraniuk et al. (1990) proposed that electrically induced neurogenic inflammation in the superficial dermis of the rat skin is a direct response to neuronal release of neuropeptides and that mast cell degranulation is not involved. Their study showed that mast cell activation did not occur except on prolonged stimulation. Another study performed in the airways, indicates that both neurogenic increased vascular permeability and plasma exudation into the airway lumen resulted from activation of capsaicin-sensitive sensory nerves without the association of mast cell activation (Kowalski *et al.*, 1997). However, Yano *et al.* (1989) demonstrated that mast cell deficient mice did not develop ear oedema or inflammation after substance P injection; in contrast, mice reconstituted with mast cells did show oedema. This suggests that substance P needs a mast cell target to cause vascular permeability.

In conclusion, there is some disagreement among available studies on the participation of mast cells in neurogenic inflammation. This may reflect differences in the duration of stimulation, the time schedule that was used, the various tissues and species-specific factors. Overall, there is good evidence supporting the involvement of blood vessels, nerves and mast cells in neurogenic inflammation, and the direct regulatory effect of nerves upon mast cells.

NERVE GROWTH FACTOR (NGF)

NGF was the first discovered member of the family of neurotrophins in the 1950s, now including brain-derived neurotrophic factor (BDNF) and neurotrophins 3–5. NGF is the best characterized neurotrophic protein and is required for survival and differentiation of neuronal cell types in both the peripheral and central nervous system (Aloe *et al.*, 1994). For example, removal of circulating NGF has been shown to result in death of sympathetic neurons (Sofroniew *et al.*, 2001). The biological activities of NGF are mediated by binding to two receptors: trkA, a tyrosine kinase receptor, and p75, a low affinity receptor.

In addition to neurons, non-neuronal cells such as mast cells (Leon *et al.*, 1994), T-cells (Mizuma *et al.*, 1999), B-cells (Solomon *et al.*, 1998), eosinophils, lymphocytes (Barouch *et al.*, 2000), fibroblasts (Hattori *et al.*, 1996) and epithelial cells (Fox *et al.*, 2001) can synthesize NGF. Many of these inflammatory cells express the high affinity NGF receptor which allows NGF to promote inflammatory mediator release. Several of these inflammatory mediators such as IL-1, IL-4, IL-5, TNF-α and IFN can, in turn, induce the release of NGF (Yoshida *et al.*, 1992; Hattori *et al.*, 1996). Therefore, NGF seems to be a mediator with functions on both immune and nerve cells and is likely an important factor integrating communication between the nervous and immune systems.

NGF acts as a chemoattractant and thereby causes an increase in the number of mast cells as well their degranulation (Marshall *et al.*, 1990, 1999; Horigome *et al.*, 1994). NGF receptors on mast cells act as autoreceptors, regulating mast cell NGF synthesis and release, while at the same time being sensitive to NGF from the environment. Inflammation can lead to an enhanced production and release of NGF. In turn, NGF induced the expression of neuropeptides and lowered the threshold of neurons for firing (Lindsay and Harmar, 1989). In mice, Braun *et al.* (2001) have recently shown that nasal treatment of mice with NGF induced airway hyper-responsiveness to subsequent electrical field stimulation. In an earlier study, Braun *et al.* (1998) showed that nasal treatment of mice with anti-NGF prevented the development of airway hyper-responsiveness. As well, NGF-transgenic mice that overexpress NGF in Clara cells showed bronchial hyper-reactivity in comparison to wild-type mice (Hoyle *et al.*, 1998). These data suggest that NGF by itself can induce airway hyper-responsiveness in the absence of airway inflammation in mice.

Neurogenic inflammation involves a change in function of sensory neurons due to inflammatory mediators, thereby inducing an enhanced release of peptides from sensory nerves. NGF is able to augment neurogenic inflammation and can upregulate the synthesis of products of the PTT gene (Vedder *et al.*, 1993) which codes for several tachykinins such as substance P and NKA. Furthermore, NGF changes the properties of peripheral sensory nerve endings by inducing an accumulation of second messengers or by sensitizing nerve terminals (Woolf *et al.*, 1996).

In vivo administration of NGF into neonatal rats caused a great increase in the number and size of mast cells in the peripheral tissues (Aloe and Levi-Montalcini, 1979). Studies have provided evidence that mast cells, similar to nerve cells, express the trkA NGF receptor, suggesting that mast cells are receptive to NGF (Nilsson *et al.*, 1997; Tam *et al.*, 1997). Furthermore, NGF has been shown to induce degranulation and histamine release from mast cells (Pearce and Thompson, 1986; Aloe, 1988). To complete the circle, mast cells are capable of producing NGF (Leon *et al.*, 1994) and mRNA for NGF is expressed in adult rat peritoneal mast cells. Medium conditioned by peritoneal mast cells has been shown to contain biologically active NGF. Therefore, it is not surprising that injection of NGF causes mast cell proliferation in part by mast cell degranulation (Marshall *et al.*, 1990).

A study by Bonini *et al.* (1996) showed an increase in serum NGF in humans with allergic diseases. The more severe the disease, the higher the NGF values found in the tissues of allergic patients. In asthma patients, a significant increase in the neurotrophins NGF, BDNF and NT-3 has been found in the BAL fluid 18 h after allergen challenge (Virchow *et al.*, 1998). NGF levels are increased in the nasal secretions within 10 min after allergen challenge in allergic rhinitis patients (Sanico *et al.*, 2000). It could be hypothesized that mast cells are a major source for NGF in allergic diseases, although a variety of other cells including T and B cells, eosinophils, lymphocytes and epithelial cells are also capable of synthesizing NGF. The case in support of mast cell involvement is supported by evidence from several allergic animal models where substance P synthesis is increased (Hunter *et al.*, 2000): the increase seems to be mimicked by the administration of NGF 24 h after application, implying that mast cell activation induces NGF release, thereby inducing the increase in substance P.

THE BRAIN, STRESS AND THE IMMUNE SYSTEM

The nervous and immune systems are the major adaptive systems of the body (Elenkov *et al.*, 2000). Teleologic arguments suggest that they are in contact with each other to maintain homeostasis. Several pathways have been shown to link the brain and the immune system, such as the autonomic nervous system acting via direct neural influences and second, the neuroendocrine humoral outflow via the pituitary. Furthermore, the sympathetic nervous system provides another important regulatory pathway between brain and immune systems (Elenkov *et al.*, 2000).

The sympathetic nervous system and the hypothalamic–pituitary–adrenal (HPA) axis are the peripheral and central limbs of the stress-response system, whose main function is to maintain basal and stress-related homeostasis (Zelazowski *et al.*, 1992). The key components of this system are located in the brain stem. The stress system is even active when the body is at rest, responding to many signals including those from cytokines produced by immune-mediated inflammatory reactions, such as TNF-α, IL-1 and IL-6 (Sternberg *et al.*, 1992). Activation of the system changes cardiovascular function, accelerates motor reflexes, increases the tolerance to pain and affects immune function (e.g. Dhabhar and McEwen, 1996).

The adaptive changes to stressors are both behavioural and physical. Initially, the body responds with an adaptive response to the stressor. For example, acute stress actually stimulates immune responsiveness, whereas chronic stress inhibits it (Dhabhar and McEwen, 1997). Once a certain threshold has been exceeded, a reaction takes place that involves the brain, the HPA axis and the sympathetic nervous system (Calogero *et al.*, 1992). Corticotropin-releasing hormone (CRH), secreted by the pituitary gland, is a major regulator of the HPA axis and cortisone synthesis, and acts as a coordinator of the stress response.

Mast cells in the central nervous system may participate in the regulation of inflammatory responses through interactions with the HPA axis. Matsumoto *et al.* (2001) showed that in the dog, degranulation of mast cells evoked HPA activation in response to histamine release. In this study, dogs were passively sensitized with IgE and challenged with specific antigen centrally or peripherally. Both routes resulted in cortisone release from the adrenal glands. The effect could be mimicked by intracranial injection of the mast cell secretagogue compound 48/80, and blocked by CRH antibodies or histamine H1 blockers but not H2 blockers. These results suggest that intracranial mast cells may act as allergen sensors, and that the activated adrenocortical response may represent a host defense reaction to prevent anaphylaxis.

CRH is also thought to be involved peripherally in tissue responses to stress in the skin, respiratory tract and intestine. Many, if not all the recorded changes have involved mast cells and neuronal activation, the latter being often mediated by neurotensin and/or substance P.

Theoharides and coworkers were the first to show that CRH could act as a mast cell degranulating agent in the dura mater (Theoharides *et al.*, 1995), the skin (Theoharides *et al.*, 1998a) and the bladder (Theoharides *et al.*, 1998a,b). Others have extended these observations. Pothoulakis *et al.* (1998) have studied the intestinal responses to stress. They concluded that increased intestinal motility, mucus hypersecretion and intestinal chloride ion secretion, as judged by increased short circuit current generation by intestinal tissue in using chambers, was mediated by CRH. They could inhibit these mostly by intracerebral or systemic injection of a specific CRH inhibitor. Furthermore, both mast cells and nerves were involved since the effects were inhibited by NK-1 antagonists and mast cell stabilizers. Perdue and coworkers have shown that ganglionic blockers, and inhibition of cholinergic as well as sympathetic activity also blocked these effects (Santos *et al.*, 1999). They did not occur in mast cell deficient rats, and equally they occurred in a chronic stress model (Santos and Perdue, 2000). Moreover, increased intestinal permeability to macromolecules occurred after stress and was also abrogated with these pharmacologic inhibitors. The sequence of events that occurs, and how these various systemic effects are mediated, are just beginning to be explored. These studies, taken together, show that the physiologic effects of psychological stress are often largely mediated by CRH, released either centrally or peripherally, and that mast cell–nerve interactions are important components of this response.

DELAYED TYPE HYPERSENSITIVITY

Delayed type hypersensitivity (DTH) is an expression of cell-mediated immunity and plays a major role in the pathology and chronic aspects of many inflammatory disorders. DTH reactions are primarily studied in the skin. Since pre-treatment of human or guinea pig skin with capsaicin enhanced the DTH reaction at the site of treatment, it was suggested that capsaicin-sensitive neurons modulate DTH via release of neuropeptides (Girolomoni and Tigelaar, 1990). These neuropeptides can directly affect Langerhans cells, mast cells, endothelial cells and infiltrating immune cells thereby effectively modulating skin and immune cell functions such as cell proliferation, cytokine production or antigen presentation (Scholzen *et al.*, 1998).

Ultraviolet (UV) B radiation has been shown to suppress DTH responses in both humans and experimental animals (Ullrich, 1995). Hart *et al.* (1997) have reported the involvement of histamine in UVB-induced suppression in mice of DTH responses. Mast cells were then shown to be the source of UVB-induced histamine (Hart *et al.*, 1998), emphasizing the importance of the mast cell in the mechanism by which UVB inhibit DTH responses in mice. Furthermore, TNF-α, reported to be derived from mast cells, has been shown to be a major cytokine implicated in signalling the immunosuppressive effects of UVB (Alard *et al.*, 1999). In addition, the sensory neuropeptide CGRP appears to be required for this mechanism since it has been shown to trigger cutaneous mast cells to release TNF-α (Yoshikawa and Streilein, 1990) (see Figure 8.2).

Recently, it has become apparent that DTH reactions are also present in the lung (Buckley and Nijkamp, 1994; van Houwelingen *et al.*, 1999) and the intestine (Kraneveld *et al.*, 1995) of sensitized animals after local antigen challenge. DTH reactions in the lung and the intestine are preferably called non-IgE mediated reactions because responses take place almost immediately after hapten challenge whereas DTH in general requires more than 12 h to develop.

Low molecular weight substances (< 5000 Da) are the most common agents causing non-IgE mediated asthma (Friedman-Jimenez *et al.*, 2000). Sensitization and local challenge with the low molecular weight compound dinitrofluorobenzene (DNFB), picrylchloride or toluene diisocyanate (TDI) has been shown to induce bronchoconstriction, airway hyperreactivity, cellular accumulation, mast cell activation and increased vascular permeability in the mouse airways. This is not associated with an increase in hapten specific IgE. Although a small percentage of the total inflammatory cell infiltrate are T-lymphocytes, they are required for the reaction. Ek *et al.* demonstrated that substance P and neurokinin A are released in response to epicutaneous application of the allergen oxazolone, a hapten known to cause hypersensitivity reactions in skin and airways (Ek and Theodorsson, 1990).

Furthermore, mast cells play a prominent role in non-IgE mediated hypersensitivity reactions (Johnson and Krenger, 1992; Kraneveld *et al.*, 1998; Ramirez-Romero *et al.*, 2000). First, mast cell degranulation was observed in DNFB-sensitized mice directly after challenge, while non-IgE hypersensitivity responses such as tracheal hyperreactivity and neutrophil infiltration were absent from WBB6F1-W/W^v and Sl/Sl^d mast cell deficient

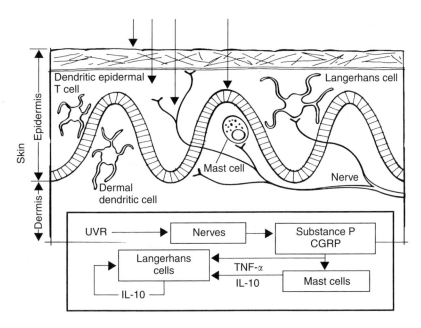

Figure 8.2   This schematic figure of skin responses to UVB radiation and the consequent immune suppression is dependent on mast cells and nerves and also Langerhans cells (Streilein *et al.*, 1999).

mice (Kraneveld *et al.*, 1998). In addition, preliminary results indicate that mast cell reconstitution restores the DNFB/DNS-induced pulmonary hypersensitivity reaction in mice. These data demonstrate that mast cells are required for the development of DNFB-induced tracheal hyper-reactivity and cellular accumulation into the airways.

Buckley and Nijkamp (1994) showed that DNFB-induced tracheal hyper-reactivity and cellular infiltration were inhibited by capsaicin-induced depletion of excitatory non-adrenergic non-cholinergic neuropeptides, showing that sensory nerves play a role in the pathogenesis of non-IgE mediated asthma. Preliminary results in the non-IgE mediated murine asthma model, using NK-1 receptor transgenic knockout animals, showed that the NK-1 receptor is crucial for the development of non-IgE mediated pulmonary reactions (Van der Kleij *et al.*, submitted). Thus, studies performed in murine non-IgE mediated models point to important roles for both excitatory NANC nerves and mast cells.

THE MAST CELL IN THE NERVOUS SYSTEM

The mast cell–nerve communication pathway has been implicated as an important element in the pathophysiology of various diseases. Stimuli can affect the primary afferent sensory nerve response, the integrative response into the CNS, the response within the autonomic ganglion, and the response at the postganglionic autonomic site within the tissue. For instance, allergen challenge and mast cell degranulation lead to substantial changes in neuronal function (Weinreich *et al.*, 1995). Changes in neuronal activity lead to symptoms as sneezing, coughing, irritation, itching and cutaneous flare reactions, key symptoms of allergy providing direct evidence that neurons are involved in allergic diseases such as asthma and gastroenteritis (Bienenstock *et al.*, 1991; Bienenstock, 1992). Thus, allergen challenge can lead to a simple reflex that can affect neuronal activity on a variety of levels in this pathway.

The most evident neuronal response to mast cell activation is the stimulation of sensory nerves, releasing tachykinins such as substance P, CGRP and Neurokinin A to cause neurogenic inflammation (Holzer, 1988; Barnes, 1992). The stimulation of C-fibres by a range of chemical and physical factors results in afferent neuronal conduction eliciting parasympathetic reflexes and antidromic impulses travelling along the peripheral nerve terminal. Such communication from one nerve to another, without passing through a cell body, is called the axon reflex and results in local release of tachykinin and CGRP from C-fibre terminals (Solway and Leff, 1991).

Axon reflexes account for many of the local physiological responses to antigen in sensitized lung (Lundberg *et al.*, 1983; Sestini *et al.*, 1989) and gut tissues (Delbro *et al.*, 1982; Baird and Cuthbert, 1987), and have long been recognized to be involved in local vasodilatation in the skin. Antidromic stimulation of guinea pig vagal sensory fibres results in contractions of the airway smooth muscle, and is mediated by tachykinins (Undem *et al.*, 1990). Further studies indicate that neuropeptide release can also be induced via direct depolarization of the terminal and, possibly, via a chemically mediated mechanism independent of electrical excitation (White, 1997).

Mast cells have been shown to be capable of affecting nerve function (Purcell and Atterwill, 1995; Levi-Montalcini *et al.*, 1996). Arachidonic acid metabolites, serotonin, nitric oxide, nerve growth factor and histamine have been suggested to affect neuronal function (Purcell and Atterwill, 1995; Silver *et al.*, 1996). Furthermore, NANC nerve endings express receptors for histamine (H1 and H3) and serotonin (5-HT2a) (Imamura *et al.*, 1996; Sekizawa *et al.*, 1998; Nemmar *et al.*, 1999), and histamine H1 receptor expression at least is upregulated in primary NANC nerves in inflammation (Kashiba *et al.*, 1999). Mediator release from mast cells can be induced by neurotrophic factors such as NGF (Horigome *et al.*, 1993) and substance P (Devillier *et al.*, 1986; Rozniecki *et al.*, 1999). Stimulation of nerve fibres of humans and mammals induces mast cells to degranulate and release histamine and other mediators such as serotonin (Reynier-Rebuffel *et al.*, 1994) and TNF-α (Ansel *et al.*, 1993). Thus, mast cell mediators can, on the one hand, sensitize afferent C fibres by lowering their threshold (Hua and Yaksh, 1993) and, on the other hand, cause release of substance P and CGRP from unmyelinated fibres (Michaelis *et al.*, 1998).

Relatively few studies have focused on the role of mast cell activation on parasympathetic nerve activity compared to the sympathetic nervous system. Studies conducted *in vitro* and in individual organs will be evaluated below.

IN VITRO STUDIES

Because mast cells are often found in the proximity of nerve endings, and may be activated by a number of neuropeptides, including tachykinins, mast cells are strongly implicated in neurogenic inflammation. The tissue site as well as the local environment affects the state of relative activation. Close proximity to nerves of all kinds allow mast cells to communicate with and be communicated with by the nervous system. Burnstock (1986) has shown that cells within 200 nm of a nerve can be affected by an action potential moving down a nerve fibre. This lends more meaning to the likelihood of two-way communication. Morphometric analysis has shown mast cells associated with nerves along their axial path, for example, in the intestine (Blennerhassett *et al.*, 1992; Blennerhassett and Bienenstock, 1998).

A tissue culture model of murine sympathetic neurons cultured with rat basophilic leukemia cells (RBL-2H3; a model of mucosal mast cells) was developed to study functional interactions between mast cells and peripheral nerves. Time-lapse microscopy showed that 60–100% of the mast cells acquired neurite contact within 17 h of co-culture (Blennerhassett *et al.*, 1991), a process facilitated by development of lasting contact between nerves and mast cells when the intervening distance was less than $36 \pm 4 \mu m$. NGF synthesized by mast cells is likely to contribute to this trophic effect (Leon *et al.*, 1994). Further studies showed that activation of mast cells *in vitro* had the capacity of altering neuronal physiology by inducing depolarization and decreasing membrane resistance (Janiszewski *et al.*, 1990).

Contact between nerves and mast cells also had developmental consequences *in vitro*, since attached RBL-2H3 cells ceased to divide and showed an increase in granules compared to control cells without contact, an indication of maturation of the mast cell (Blennerhassett and Bienenstock, 1998). Thus, sympathetic neurons have the ability to change the state of RBL-2H3 cells in co-culture. Contact between mast cells and neurites, once formed, was maintained up to 120 h, while the associated neurites often branched after contact. Electron microscopic examination of the sites of contact showed on average <20 nm distance between apposed membranes. While no specialized structures were seen either in the neurite or the closely approximated mast cell, dense cored neurosecretory vesicles accumulated in the neurite endings apposed to the mast cell membrane (Blennerhassett *et al.*, 1991).

Suzuki and coworkers, have shown direct communication between mast cells and superior cervical neurons in co-culture (Suzuki *et al.*, 1999; Ohshiro *et al.*, 2000). Calcium-binding dyes were used to show activation and signalling with confocal microscopy. Scorpion venom and bradykinin were used to activate the neuritis, which in turn led to calcium increases in the attached RBL-2H3 cells. The ability to block with an NK-1 receptor antagonist showed that the signalling was dependent on substance P. Furthermore, application of an anti-IgE antibody showed initial RBL-2H3 activation followed by neurite activation of the associated nerve fibre, providing evidence for bi-directional communication.

BRAIN

Mast cells are resident in the brain of many species (Silver *et al.*, 1996), can also move though the brain in the absence of inflammation, apparently entering via penetrating blood vessels. Brain mast cells are mainly perivascular but are abundant in the thalamus and hypothalamus, with only a small number of cells in the cerebral cortex and basal ganglia (Purcell and Atterwill, 1995). Large numbers of tryptase containing mast cells have been described surrounding the pituitary gland (Cromlish *et al.*, 1987). These mast cells can respond to antigen, and regulate CRH secretion via histamine effects (Matsumoto *et al.*, 2001). There is, however, considerable variability in the number and distribution of mast cells among individuals, species, strains between the same species, and sexes (Silver *et al.*, 1996). Moreover, the localization of mast cells in the developing rat brain differs from that of the adult animal (Lambracht-Hall *et al.*, 1990a).

The physiological significance of mast cells in brain function and/or metabolism is unclear. However, they can modulate neuroendocrine control systems (Purcell and Atterwill, 1995) and they could play a role in the regulation of meningeal blood flow and vessel permeability (Mares *et al.*, 1979). A functional interaction between brain mast cells and neurons may be an important neurobiological process, since mast cells were found in close association with neurons containing substance P and CGRP, neuropeptides that can in turn activate dura mast cells (Lambracht-Hall *et al.*, 1990b; Keller and Marfurt, 1991; Ottosson and Edvinsson, 1997).

Histamine secretion from brain mast cells, in response to local neurotrophic factors and neuropeptides, may contribute to the aetiology of neuroinflammatory conditions, such as multiple sclerosis and Alzheimer's disease (Theoharides, 1996). Mast cells and CRH have thought to play a significant role in migraine (Theoharides *et al.*, 1995). Substance P and compound 48/80 each have been shown to induce histamine secretion from rat brain mast cells in a concentration dependent manner (Purcell and Atterwill, 1995). Also, the growth factors NGF and BDNF trigger histamine release from rat brain mast cells suggesting that these mast cells contain the TrkA and TrkB receptors (Horigome *et al.*, 1993).

The ability of brain mast cells to secrete histamine and possibly other mediators following exposure to these neurotrophic factors, supports their role in the neuro-immune cross-talk. One of the most striking examples of the intimate relationship and functional importance of mast cells and the brain occurs as a result of studies of the courting behaviour of doves. Silverman *et al.* (1996) have demonstrated that a population of mast cells is present in large numbers in the medial habenula (MH) of the ring dove after a brief period of courtship. There were fewer mast cells in birds housed in isolation, and mast cell numbers were further reduced in long-term castrates (Zhuang *et al.*, 1993). This suggests that the appearance of mast cells is related to the behavioural state of the animal and proposes a novel mechanism for interactions between the nervous and the immune system.

AIRWAYS

Efferent and afferent autonomic nerves regulate many aspects of human and animal airway function. The parasympathetic nervous system is the dominant neuronal pathway for airway smooth muscle tone. Stimulation of cholinergic nerves causes bronchoconstriction, mucus secretion, and bronchial vasodilatation. At this point, there is no convincing evidence for cholinergic dysfunction leading to pulmonary diseases.

The NANC nervous system is an important neural network in the lung. Inhibitory NANC nerves contain vasoactive intestinal peptide (VIP) and nitric oxide (NO), potent relaxants of the airways that counteract bronchoconstriction. VIP also has been suggested to inhibit mediator release from mast cells by elevating the level of intracellular cyclic

AMP (Friedman and Kaliner, 1987). Undem *et al.* (1983) reported that VIP inhibited antigen-induced histamine release. In cats, as mentioned before, Miura *et al.* (1990) demonstrated that NANC inhibitory electrical nerve stimulation inhibited antigen-induced bronchoconstriction and histamine secretion from histamine containing cells. Although dysfunction of inhibitory NANC nerves has also been proposed in asthma, no differences in inhibitory NANC responses have been found between asthmatics and healthy subjects (van der Velden and Hulsmann, 1999; Joos *et al.*, 2000b).

In addition to inhibitory NANC efferent systems in the airways, there is also a NANC afferent nervous system that protects the airways against inhaled irritants and chemical particles (Barnes, 1991). These excitatory NANC nerves play a regulatory role in airways. Their activation results in maintenance of pulmonary homeostasis via reflex pathways such as bronchoconstriction, mucus secretion and cough. Excitatory NANC nerves innervate the airways of humans and other mammalian species (Joos *et al.*, 2000a). Excitatory NANC nerves (so-called sensory nerves) are mainly localized between and beneath the airway epithelium. These sensory nerve fibres are also present in the vicinity of blood vessels and submucosal glands, and make direct contact with the smooth muscle layer and local tracheobronchial ganglion cells. The excitatory NANC nerves can be activated by different stimuli, that affect the chemosensitive C-fibre afferents in the airways and can lead to the local release of neuropeptides via axon reflex mechanisms (Advenier *et al.*, 1999). The tachykinins and calcitonin gene-related peptide (CGRP) are the predominant excitatory NANC-neuropeptides in the airways (Solway and Leff, 1991). These neuropeptides contract airway smooth muscle, dilate bronchial arteries, increase vascular permeability, increase mucus production and modulate ganglionic transmission. In addition, they are certain to influence the recruitment, proliferation and activation of inflammatory cells such as mast cells. As recent specific evidence for this, Forsythe *et al.* (2000) have demonstrated that substance P and NKA induce histamine release from human airway mast cells.

We have discussed elsewhere in this chapter that electrical stimulation of the vagus nerve causes an increase in neurogenic inflammation in the lung (McDonald *et al.*, 1988). Moreover, antigen causes a secretory response in rat trachea via a mast cell and nerve dependent interaction (Sestini *et al.*, 1989).

Many animal studies show that capsaicin-induced depletion of excitatory NANC neuropeptides prevents airway hyper-responsiveness and pulmonary inflammation. These data support the concept that E-NANC nerves are an important factor in the pathway leading to characteristics as hyper-responsiveness and inflammation as seen in asthma.

URINARY TRACT

In the urinary bladder, an intimate relationship has been demonstrated between mast cells and peptidergic sensory nerves. As shown by Keith and Saban (1995), mast cells of mucosal and connective tissue type were found within nerves and ganglia and were in close contact with individual nerve fibres displaying substance P- and CGRP-like immunoreactivity in healthy tissue. Elbadawi (1997) demonstrated the presence of mast cells, undergoing piecemeal degranulation, in close proximity to intrinsic nerves in the bladder wall. In the bladder, axon reflexes can cause non-infectious neurogenic inflammation (Lundberg *et al.*, 1984) via the influx of inflammatory cells from the vascular system.

Interstitial cystitis (IC) has continued to be an unresolved problem in clinical urology. Mast cell migration has been described, and increased mast cell numbers have been reported in a subset of IC patients (Spanos *et al.*, 1997). Substance P-reactive nerves have been observed close to mast cells in bladder suburothelium as well as in mucosa and sub-mucosa (Elbadawi, 1997; Spanos *et al.*, 1997). Increased concentrations of histamine seem to be a consistent finding in biopsies of bladder with IC (el Mansoury *et al.*, 1994), and histamine release may stimulate afferent C-fibres to release neuropeptides through an antidromic local reflex loop (axon reflex) causing changes in associated venules and capillaries (Elbadawi, 1997).

Using an animal model, Saban *et al.* (2000) showed that NK-1 receptor $-/-$ mice failed to present bladder inflammatory cell infiltrate or oedema in response to antigen challenge, showing protection from inflammation. This work supports evidence for the role of sub-stance P and the NK-1 receptor in the development of interstitial cystitis. Overall, there is good data supporting the involvement of neurogenic inflammation in the pathogenesis of interstitial cystitis in humans.

SKIN

In man and other mammals, the dermis is richly innervated by primary afferent sensory nerves, postganglionic cholinergic parasympathetic nerves and postganglionic adrenergic and cholinergic sympathetic nerves (Rossi and Johansson, 1998). Botchkarev *et al.* used *in situ* histochemistry to show that these nerve fibres form close contacts with mast cells in the mouse skin. Neuropeptides, released by cutaneous nerves, have been shown to activate a number of target cells including Langerhans cells, endothelial cells and mast cells (Ansel *et al.*, 1997).

Neuropeptides, such as substance P and CGRP, exhibit a variety of pro-inflammatory effects in the skin, where their release in response to nociceptive stimulation by pain, mechanical and chemical irritants mediate skin responses to infection, injury and wound healing.

Skin mast cells can respond to trauma, releasing a variety of inflammatory mediators through an immediate sensory nerve-stimulated response or via an axon reflex, inducing the release of substance P from peripheral nerve endings, in turn leading to more mast cell degranulation (Church *et al.*, 1991).

Substance P is recognized to be one of the main neuropeptides responsible for the skin reaction characterized by erythema, pain and swelling (Scholzen *et al.*, 1998). Substance P can cause the release of histamine (Church *et al.*, 1991) and TNF-α (Ansel *et al.*, 1993) from skin mast cells which in turn leads to vasodilatation. Mast cells and nerves are involved in the expression of delayed hypersensitivity and also in the immune suppression induced by UV radiation (see section on delayed hypersensitivity and also Figure 8.2).

Topical application of capsaicin results in a significant reduction in the number of mast cells and the appearance of degranulated mast cells in the skin (Bunker *et al.*, 1991). Since capsaicin itself does not release histamine from mast cells, these data suggest that peptides released from neurons could cause mast cell degranulation in the skin, pointing to a nerve–mast cell communication occurring in the skin.

THE GASTROINTESTINAL TRACT

The gastrointestinal lamina propria is densely innervated, and mast cells are commonly observed near these nerves. Careful analysis showed that 50–75% of mast cells are closely apposed to nerves in both rodents and humans (Stead, 1992). Electron microscopy showed nerve terminals containing predominantly small clear neurosecretory vesicles were observed in direct contact with the plasma membrane of mucosal mast cells in the rat ileum (Newson et al., 1983). Increased numbers of mast cells are a hallmark of inflammatory or allergic conditions, and increased numbers of mast cell–nerve contacts appear in the infected/inflamed condition. For instance, nematode infection of the rat intestine causes the frequency of mast cell–nerve associations in the intestinal mucosa to increase above that of the normal counterparts (Hebda et al., 1993). In the course of inflammation induced by a nematode (Nippostrongylus brasiliensis) in rat intestine, imaging and histochemistry we used to quantitate changes in nerves. New, small diameter neurites staining for B50, found in new growth axons, increased in number and persisted well beyond the time that inflammation could be detected. There was a very strong correlation between mast cell numbers and the number of small diameter neurites (Stead et al., 1991).

Nerve stimulation can directly cause mast cell degranulation in the intestine and release of peptidergic neurotransmitters is the most likely mechanism. Electrical stimulation of the vagus has been shown to either cause intestinal mast cell degranulation (Bani-Sacchi et al., 1986) or an increase in mast cell histamine content which did not occur after sub-diaphragmatic vagotomy (Gottwald et al., 1995). Shananan et al. (1985) showed that substance P caused mediator release from intestinal mucosal mast cells. Mast cell mediators also appear to have an effect on the nerves in the intestine. For example, Jiang et al. (2000) showed that serotonin and histamine, released from the mast cells after intestinal anaphylaxis, stimulated mesenteric afferents via 5-HT(3) and histamine H(1) receptors. Mesenteric afferent nerve discharge increased approximately 1 min after luminal antigen challenge and was attenuated by serotonin and histamine receptor antagonists. As already mentioned, tryptase, the major serine protease in mast cells, activates nerves via the PAR2 receptor (Corvera et al., 1999).

Overall, it can reasonably be concluded that nerves and mast cells form a physiological unit which maintains and regulates homeostasis of diverse aspects of intestinal function, in health, in response to stress and in response to injuries and environmental pathogens. We present some examples below of functional nerve–mast cell interactions in the gastrotestinal tract, and further consideration of the extensive evidence for this is presented elsewhere within this volume.

MAST CELLS AND THE INTESTINAL EPITHELIUM

This section is purposefully shortened and is dealt with in detail in other contributions in this book (see Perdue, and also Cooke).

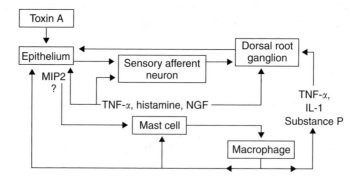

Figure 8.3   This figure outlines a number of complex interactions which occur as a consequence to clostridial toxin A in the intestine. This is largely based on the work of Castagliuolo *et al.* (1994). It outlines the interactions between mast cells and nerves and their involvement in response to the toxin.

Perdue *et al.* (1991) determined the existence of an integral nerve-to-mast cell and mast cell-to-nerve connection during intestinal anaphylaxis. A role for the mast cell-to-nerve connection was established by an increase in the epithelial short circuit current (I_{sc}) after antigen challenge, which was inhibited by antagonists to histamine, serotonin, cyclo-oxygenase and nerve conduction. The intestinal secretory response to antigenic stimulation was reduced in mast cell-deficient (W/Wv) mice compared to their littermates and could be inhibited by different mast cell antagonists in +/+ but not in W/Wv mice, pointing to functional mast-cell-to-nerve connections. Importantly, they also showed that electrical field stimulation causing neural activation induced changes in I_{sc} dependent on the presence of mast cells. In further study of this communication using sensitized guinea pig intestine, Cooke *et al.* (1998) showed that the exposure to specific antigen caused acetyl-choline release at the same time as the secretory response, and that this was blocked by atropine. These data, then, provided evidence for a bi-directional communication between nerves and mast cells in the regulation of ion transport in the gastrointestinal tract, similar to results obtained in the rat lung by Sestini *et al.* (1989).

The diarrhoea caused by overgrowth of *Clostridium difficile* is an intriguing example of the importance of nerve–mast cell communication in the intestine (see Figure 8.3). This enteritis is a significant cause of morbidity after antibiotic therapy, and involves a secretory diarrhoea due to the production of bacterial toxin A. Castagliuolo *et al.* (1994) showed that toxin A activated capsaicin-sensitive intestinal afferent nerves, which in turn led to mast cell activation and that this could be blocked by NK1 receptor antagonists. This effect was quite distinct from the actions of the non-inflammatory cholera toxin, this illustrates the pathological consequences of dysregulation of the nerve–mast cell interaction.

MAST CELLS AND THE PATHOLOGY OF OESOPHAGITIS

Nerve–mast cell interactions in oesophagitis may provide acute protection acutely, but also participate in development of chronic disease. Protection against the damaging effects of refluxed gastric contents may involve the interactions of mast cell-derived products with both substance P-containing nerves and blood vessels. In the opossum oesophagus, experimental addition of acid that mimics the exposure to refluxed gastric contents caused increased local blood flow due to mast cell activation and histamine release (Paterson, 1998). This protective reflex may, however, have harmful consequences in conditions of chronic acid reflux disease: in this model and in humans, even a single exposure to acid causes oesophageal shortening due to the sustained contraction of the longitudinal smooth muscle (Muinuddin and Paterson, 2001; White *et al.*, 2001). Since this occurs within minutes of acid exposure and can be blocked by mast cell stabilizing agents (Paterson, 1998), Paterson *et al.* (2002) proposed that acid-induced activation of sub-epithelial mast cells acted on local neural circuits containing substance P to contract longitudinal smooth muscle via NK2 receptors. With repeated stimulation, this is suggested to participate in the development of hiatal hernia and chronic inflammatory disease (White *et al.*, 2001).

CONCLUSIONS

Topographical associations between mast cells and nerves have now been recorded in most tissues in the body. These now include the stomach, the small and large intestine, blood vessels, mesentery, gall bladder, diaphragm, pleura, skin, urinary bladder and myocardium, and have even been recorded in the frog tongue. Many of these morphologic associations have been shown to occur between substance P and CGRP containing neurons and mast cells of all subtypes.

The potential for cross-talk between mast cells and nerves is therefore extensive and it is not surprising that electrical stimulation of nerves can result in both activation/degranulation as well as inhibitory signals, apparently depending on the type of nerve found in association with particular mast cells.

The role of this bi-directional communication between the nervous and the immune systems appears to be multi-factorial. As a start, the maintenance of local homeostasis in skin, blood vessels and mucosal tissue seems well established. Mast cells are situated in a particularly appropriate position to act as sensors of environmental change and, in this respect, their positioning in skin and mucosal tissues make them sentinels for detection of antigens to which the host has been exposed or noxious chemicals and other substances.

The communication with the nervous system allows the peripheral and central nervous system to become involved in the regulation of defense mechanisms and inflammation and in response to infection. The further involvement of mast cell–nerve communication in the regulation of blood flow, tissue remodelling and repair is a natural extrapolation from this information. Clearly, the involvement of mast cells with nerves can affect nerve growth, conduction, transmission and the central nervous system, as well as the regulation of neuro-endocrine systems such as those involved the regulation of adrenal hormone secretion (e.g. cortisol).

The surprising involvement of nerve–mast cell communication in responses to stress, in the neuro-endocrine involvement referred to above, and even the role of mast cells in the courting behaviour of doves, all point to extensive development of communication between nervous and immune systems. It remains to be determined how many of these interactions are essential for life, since it is not apparent that mast cell-deficient mice or rats with mutations in the c-kit pathway have shorter than normal life spans. While many of these pathways could therfore be redundant, the absence of mast cells does render these animals more sensitive to infections in experimental models. Despite this observation, it is reasonable to assume that mast cell–nerve bi-directional communication has arisen and developed in evolution as a complementary set of pathways to others which also exist. While not essential for life, they may be important in adding to the diversity of responses by individuals to environmental change. Lastly, the reader should recognize that while mast cell–nerve interactions have been focused on in this chapter as an example of neuro-immune interaction, similar pathways may, and likely do, exist in which other cells of the immune system may have analogous modes of interaction.

ACKNOWLEDGEMENTS

The authors gratefully acknowledge the assistance of the Canadian Institute for Health Research (CIHR).

REFERENCES

Advenier C, Joos G, Molimard M, Lagente V, Pauwels R (1999) Role of tachykinins as contractile agonists of human airways in asthma. *Clin Exp Allergy* 29: 579–584.

Advenier C, Lagente V, Boichot E (1997) The role of tachykinin receptor antagonists in the prevention of bronchial hyperresponsiveness, airway inflammation and cough. *Eur Respir J* 10: 1892–1906.

Alard P, Niizeki H, Hanninen L, Streilein JW (1999) Local ultraviolet B irradiation impairs contact hypersensitivity induction by triggering release of tumor necrosis factor-alpha from mast cells. Involvement of mast cells and Langerhans cells in susceptibility to ultraviolet B. *J Invest Dermatol* 113: 983–990.

Aloe L (1988) The effect of nerve growth factor and its antibody on mast cells in vivo. *J Neuroimmunol* 18: 1–12.

Aloe L, Levi-Montalcini R (1979) Nerve growth factor induced overgrowth of axotomized superior cervical ganglia in neonatal rats. Similarities and differences with NGF effects in chemically axotomized sympathetic ganglia. *Arch Ital Biol* 117: 287–307.

Aloe L, Skaper SD, Leon A, Levi-Montalcini R (1994) Nerve growth factor and autoimmune diseases. *Autoimmunity* 19: 141–150.

Ansel JC, Armstrong CA, Song I, Quinlan KL, Olerud JE, Caughman SW, Bunnett NW (1997) Interactions of the skin and nervous system. *J Invest Dermatol Symp Proc* 2: 23–26.

Ansel JC, Brown JR, Payan DG, Brown MA (1993) Substance P selectively activates TNF-alpha gene expression in murine mast cells. *J Immunol* 150: 4478–4485.

Aranguez I, Torres C, Rubio N (1995) The receptor for tumor necrosis factor on murine astrocytes: characterization, intracellular degradation, and regulation by cytokines and Theiler's murine encephalomyelitis virus. *Glia* 13: 185–194.

Arizono N, Matsuda S, Hattori T, Kojima Y, Maeda T, Galli SJ (1990) Anatomical variation in mast cell nerve associations in the rat small intestine, heart, lung, and skin. Similarities of distances between neural processes and mast cells, eosinophils, or plasma cells in the jejunal lamina propria. *Lab Invest* 62: 626–634.

Baird AW, Cuthbert AW (1987) Neuronal involvement in type 1 hypersensitivity reactions in gut epithelia. *Br J Pharmacol* 92: 647–655.

Bani-Sacchi T, Barattini M, Bianchi S, Blandina P, Brunelleschi S, Fantozzi R, Mannaioni PF, Masini E (1986) The release of histamine by parasympathetic stimulation in guinea-pig auricle and rat ileum. *J Physiol* 371: 29–43.

Baraniuk JN, Kowalski ML, Kaliner MA (1990) Relationships between permeable vessels, nerves, and mast cells in rat cutaneous neurogenic inflammation. *J Appl Physiol* 68: 2305–2311.

Barnes PJ (1991) Neurogenic inflammation in airways. *Int Arch Allergy Appl Immunol* 94: 303–309.

Barnes PJ (1992) Neurogenic inflammation and asthma. *J Asthma* 29: 165–180.

Barouch R, Appel E, Kazimirsky G, Braun A, Renz H, Brodie C (2000) Differential regulation of neurotrophin expression by mitogens and neurotransmitters in mouse lymphocytes. *J Neuroimmunol* 103: 112–121.

Bauer O, Razin E (2000) Mast cell–nerve interactions. *News Physiol Sci* 15: 213–218.

Befus AD, Dyck N, Goodacre R, Bienenstock J (1987) Mast cells from the human intestinal lamina propria. Isolation, histochemical subtypes, and functional characterization. *J Immunol* 138: 2604–2610.

Berger P, Compton SJ, Molimard M, Walls AF, N'Guyen C, Marthan R, Tunon-de-Lara JM (1999) Mast cell tryptase as a mediator of hyperresponsiveness in human isolated bronchi. *Clin Exp Allergy* 29: 804–812.

Bienenstock J (1992) Cellular communication networks. Implications for our understanding of gastrointestinal physiology. *Ann N Y Acad Sci* 664: 1–9.

Bienenstock J, Croitoru K, Ernst PB, Stanisz AM (1989) Nerves and neuropeptides in the regulation of mucosal immunity. *Adv Exp Med Biol* 257: 19–26.

Bienenstock J, MacQueen G, Sestini P, Marshall JS, Stead RH, Perdue MH (1991) Mast cell/nerve interactions in vitro and in vivo. *Am Rev Respir Dis* 143: S55–S58.

Bienenstock J, Tomioka M, Stead R, Ernst P, Jordana M, Gauldie J, Dolovich J, Denburg J (1987a) Mast cell involvement in various inflammatory processes. *Am Rev Respir Dis* 135: S5–S8.

Bienenstock J, Perdue M, Stanisz A, Stead R (1987b) Neurohormonal regulation of gastrointestinal immunity. *Gastroenterology* 93: 1431–1434.

Bischoff SC, Baggiolini M, de Weck AL, Dahinden CA (1991) Interleukin 8-inhibitor and inducer of histamine and leukotriene release in human basophils. *Biochem Biophys Res Commun* 179: 628–633.

Bischoff SC, Dahinden CA (1992) c-kit ligand: a unique potentiator of mediator release by human lung mast cells. *J Exp Med* 175: 237–244.

Blennerhassett MG, Bienenstock J (1998) Sympathetic nerve contact causes maturation of mast cells in vitro. *J Neurobiol* 35: 173–182.

Blennerhassett MG, Janiszewski J, Bienenstock J (1992) Sympathetic nerve contact alters membrane resistance of cells of the RBL-2H3 mucosal mast cell line. *Am J Respir Cell Mol Biol* 6: 504–509.

Blennerhassett MG, Tomioka M, Bienenstock J (1991) Formation of contacts between mast cells and sympathetic neurons in vitro. *Cell Tissue Res* 265: 121–128.

Bonini S, Lambiase A, Bonini S, Angelucci F, Magrini L, Manni L, Aloe L (1996) Circulating nerve growth factor levels are increased in humans with allergic diseases and asthma. *Proc Natl Acad Sci USA* 93: 10955–10960.

Botchkarev VA, Eichmuller S, Peters EM, Pietsch P, Johansson O, Maurer M, Paus R (1997) A simple immuno-fluorescence technique for simultaneous visualization of mast cells and nerve fibers reveals selectivity and hair cycle–dependent changes in mast cell–nerve fiber contacts in murine skin. *Arch Dermatol Res* 289: 292–302.

Braun A, Appel E, Baruch R, Herz U, Botchkarev V, Paus R, Brodie C, Renz H (1998) Role of nerve growth factor in a mouse model of allergic airway inflammation and asthma. *Eur J Immunol* 28: 3240–3251.

Braun A, Quarcoo D, Schulte-Herbruggen O, Lommatzsch M, Hoyle G, Renz H (2001) Nerve growth factor induces airway hyperresponsiveness in mice. *Int Arch Allergy Immunol* 124: 205–207.

Bremer AA, Leeman SE, Boyd ND (2000) The common C-terminal sequences of substance P and neurokinin A contact the same region of the NK-1 receptor. *FEBS Lett* 486: 43–48.

Buckley TL, Nijkamp FP (1994) Airways hyperreactivity and cellular accumulation in a delayed-type hypersensitivity reaction in the mouse. Modulation by capsaicin-sensitive nerves. *Am J Respir Crit Care Med* 149: 400–407.

Bunker CB, Cerio R, Bull HA, Evans J, Dowd PM, Foreman JC (1991) The effect of capsaicin application on mast cells in normal human skin. *Agents Actions* 33: 195–196.

Buresi MC, Schleihauf E, Vergnolle N, Buret A, Wallace JL, Hollenberg MD, MacNaughton WK (2001) Protease-activated receptor-1 stimulates $Ca(2+)$-dependent $Cl(-)$ secretion in human intestinal epithelial cells. *Am J Physiol Gastrointest Liver Physiol* 281: G323–G332.

Burnstock G (1986) The changing face of autonomic neurotransmission. *Acta Physiol Scand* 126: 67–91.

Calogero AE, Sternberg EM, Bagdy G, Smith C, Bernardini R, Aksentijevich S, Wilder RL, Gold PW, Chrousos GP (1992) Neurotransmitter-induced hypothalamic-pituitary-adrenal axis responsiveness is defective in inflammatory disease-susceptible Lewis rats: in vivo and in vitro studies suggesting globally defective hypothalamic secretion of corticotropin-releasing hormone. *Neuroendocrinology* 55: 600–608.

Campbell EA, Gentry CT, Patel S, Panesar MS, Walpole CS, Urban L (1998) Selective neurokinin-1 receptor antagonists are anti-hyperalgesic in a model of neuropathic pain in the guinea-pig. *Neuroscience* 87: 527–532.

Castagliuolo I, LaMont JT, Letourneau R, Kelly C, O'Keane JC, Jaffer A, Theoharides TC, Pothoulakis C (1994) Neuronal involvement in the intestinal effects of Clostridium difficile toxin A and Vibrio cholerae enterotoxin in rat ileum. *Gastroenterology* 107: 657–665.

Church MK, el Lati S, Caulfield JP (1991) Neuropeptide-induced secretion from human skin mast cells. *Int Arch Allergy Appl Immunol* 94: 310–318.

Church MK, Levi-Schaffer F (1997) The human mast cell. *J Allergy Clin Immunol* 99: 155–160.

Coleman JW, Holliday MR, Kimber I, Zsebo KM, Galli SJ (1993) Regulation of mouse peritoneal mast cell secretory function by stem cell factor, IL-3 or IL-4. *J Immunol* 150: 556–562.

Cooke HJ, Fox P, Alferes L, Fox CC, Wolfe SA, Jr. (1998) Presence of NK1 receptors on a mucosal-like mast cell line, RBL-2H3 cells. *Can J Physiol Pharmacol* 76: 188–193.

Corvera CU, Dery O, McConalogue K, Gamp P, Thoma M, Al Ani B, Caughey GH, Hollenberg MD, Bunnett NW (1999) Thrombin and mast cell tryptase regulate guinea-pig myenteric neurons through proteinase-activated receptors-1 and -2. *J Physiol* 517(Pt 3): 741–756.

Cromlish JA, Seidah NG, Marcinkiewicz M, Hamelin J, Johnson DA, Chretien M (1987) Human pituitary tryptase: molecular forms, NH2-terminal sequence, immunocytochemical localization, and specificity with prohormone and fluorogenic substrates. *J Biol Chem* 262: 1363–1373.

D'Andrea MR, Rogahn CJ, Andrade-Gordon P (2000) Localization of protease-activated receptors-1 and -2 in human mast cells: indications for an amplified mast cell degranulation cascade. *Biotech Histochem* 75: 85–90.

Defea K, Schmidlin F, Dery O, Grady EF, Bunnett NW (2000) Mechanisms of initiation and termination of signalling by neuropeptide receptors: a comparison with the proteinase-activated receptors. *Biochem Soc Trans* 28: 419–426.

Delbro D, Fandriks L, Lisander B, Andersson SA (1982) Gastric atropine-sensitive excitation by peripheral vagal stimulation after hexamethonium. Antidromic activation of afferents? *Acta Physiol Scand* 114: 433–440.

Denburg JA (1990) Cytokine-induced human basophil/mast cell growth and differentiation in vitro. *Springer Semin Immunopathol* 12: 401–414.

Dery O, Corvera CU, Steinhoff M, Bunnett NW (1998) Proteinase-activated receptors: novel mechanisms of signaling by serine proteases. *Am J Physiol* 274: C1429–C1452.

Devillier P, Regoli D, Asseraf A, Descours B, Marsac J, Renoux M (1986) Histamine release and local responses of rat and human skin to substance P and other mammalian tachykinins. *Pharmacology* 32: 340–347.

Dhabhar FS, McEwen BS (1996) Stress-induced enhancement of antigen-specific cell-mediated immunity. *J Immunol* 156: 2608–2615.

Dhabhar FS, McEwen BS (1997) Acute stress enhances while chronic stress suppresses cell-mediated immunity in vivo: a potential role for leukocyte trafficking. *Brain Behav Immun* 11: 286–306.

Dimitriadou V, Buzzi MG, Moskowitz MA, Theoharides TC (1991) Trigeminal sensory fiber stimulation induces morphological changes reflecting secretion in rat dura mater mast cells. *Neuroscience* 44: 97–112.

Dopp JM, Mackenzie-Graham A, Otero GC, Merrill JE (1997) Differential expression, cytokine modulation, and specific functions of type-1 and type-2 tumor necrosis factor receptors in rat glia. *J Neuroimmunol* 75: 104–112.

Echtenacher B, Mannel DN, Hultner L (1996) Critical protective role of mast cells in a model of acute septic peritonitis. *Nature* 381: 75–77.

Egan CL, Viglione-Schneck MJ, Walsh LJ, Green B, Trojanowski JQ, Whitaker-Menezes D, Murphy GF (1998) Characterization of unmyelinated axons uniting epidermal and dermal immune cells in primate and murine skin. *J Cutan Pathol* 25: 20–29.

Ek L, Theodorsson E (1990) Tachykinins and calcitonin gene-related peptide in oxazolone-induced allergic contact dermatitis in mice. *J Invest Dermatol* 94: 761–763.

el Mansoury M, Boucher W, Sant GR, Theoharides TC (1994) Increased urine histamine and methylhistamine in interstitial cystitis. *J Urol* 152: 350–353.

Elbadawi A (1997) Interstitial cystitis: a critique of current concepts with a new proposal for pathologic diagnosis and pathogenesis. *Urology* 49: 14–40.

Elenkov IJ, Wilder RL, Chrousos GP, Vizi ES (2000) The sympathetic nerve–an integrative interface between two supersystems: the brain and the immune system. *Pharmacol Rev* 52: 595–638.

Emadi-Khiav B, Mousli M, Bronner C, Landry Y (1995) Human and rat cutaneous mast cells: involvement of a G protein in the response to peptidergic stimuli. *Eur J Pharmacol* 272: 97–102.

Fiorucci S, Mencarelli A, Palazzetti B, Distrutti E, Vergnolle N, Hollenberg MD, Wallace JL, Morelli A, Cirino G (2001) Proteinase-activated receptor 2 is an anti-inflammatory signal for colonic lamina propria lymphocytes in a mouse model of colitis. *Proc Natl Acad Sci USA* 98: 13936–13941.

Forsythe P, McGarvey LP, Heaney LG, MacMahon J, Ennis M (2000) Sensory neuropeptides induce histamine release from bronchoalveolar lavage cells in both nonasthmatic coughers and cough variant asthmatics. *Clin Exp Allergy* 30: 225–232.

Fox AJ, Patel HJ, Barnes PJ, Belvisi MG (2001) Release of nerve growth factor by human pulmonary epithelial cells: role in airway inflammatory diseases. *Eur J Pharmacol* 424: 159–162.

Friedman-Jimenez G, Beckett WS, Szeinuk J, Petsonk EL (2000) Clinical evaluation, management, and prevention of work-related asthma. *Am J Ind Med* 37: 121–141.

Friedman MM, Kaliner MA (1987) Human mast cells and asthma. *Am Rev Respir Dis* 135: 1157–1164.

Galli SJ (1990) New insights into "the riddle of the mast cells": microenvironmental regulation of mast cell development and phenotypic heterogeneity. *Lab Invest* 62: 5–33.

Galli SJ (1993) New concepts about the mast cell. *N Engl J Med* 328: 257–265.

Galli SJ, Tsai M, Wershil BK, Tam SY, Costa JJ (1995) Regulation of mouse and human mast cell development, survival and function by stem cell factor, the ligand for the c-kit receptor. *Int Arch Allergy Immunol* 107: 51–53.

Galli SJ, Wershil BK (1996) The two faces of the mast cell. *Nature* 381: 21–22.

Girolomoni G, Tigelaar RE (1990) Capsaicin-sensitive primary sensory neurons are potent modulators of murine delayed-type hypersensitivity reactions. *J Immunol* 145: 1105–1112.

Gordon JR, Galli SJ (1990) Mast cells as a source of both preformed and immunologically inducible TNF-alpha/cachectin. *Nature* 346: 274–276.

Gottwald T, Coerper S, Schaffer M, Koveker G, Stead RH (1998) The mast cell–nerve axis in wound healing: a hypothesis. *Wound Repair Regen* 6: 8–20.

Gottwald TP, Hewlett BR, Lhotak S, Stead RH (1995) Electrical stimulation of the vagus nerve modulates the histamine content of mast cells in the rat jejunal mucosa. *Neuroreport* 7: 313–317.

Grady EF, Baluk P, Bohm S, Gamp PD, Wong H, Payan DG, Ansel J, Portbury AL, Furness JB, McDonald DM, Bunnett NW (1996) Characterization of antisera specific to NK1, NK2, and NK3 neurokinin receptors and their utilization to localize receptors in the rat gastrointestinal tract. *J Neurosci* 16: 6975–6986.

Hart PH, Grimbaldeston MA, Swift GJ, Jaksic A, Noonan FP, Finlay-Jones JJ (1998) Dermal mast cells determine susceptibility to ultraviolet B-induced systemic suppression of contact hypersensitivity responses in mice. *J Exp Med* 187: 2045–2053.

Hart PH, Jaksic A, Swift G, Norval M, el Ghorr AA, Finlay-Jones JJ (1997) Histamine involvement in UVB- and cis-urocanic acid-induced systemic suppression of contact hypersensitivity responses. *Immunology* 91: 601–608.

Hattori A, Hayashi K, Kohno M (1996) Tumor necrosis factor (TNF) stimulates the production of nerve growth factor in fibroblasts via the 55-kDa type 1 TNF receptor. *FEBS Lett* 379: 157–160.

Hattori A, Iwasaki S, Murase K, Tsujimoto M, Sato M, Hayashi K, Kohno M (1994) Tumor necrosis factor is markedly synergistic with interleukin 1 and interferon-gamma in stimulating the production of nerve growth factor in fibroblasts. *FEBS Lett* 340: 177–180.

He S, Gaca MD, Walls AF (2001) The activation of synovial mast cells: modulation of histamine release by tryptase and chymase and their inhibitors. *Eur J Pharmacol* 412: 223–229.

Hebda PA, Collins MA, Tharp MD (1993) Mast cell and myofibroblast in wound healing. *Dermatol Clin* 11: 685–696.

Holzer P (1988) Local effector functions of capsaicin-sensitive sensory nerve endings: involvement of tachykinins, calcitonin gene-related peptide and other neuropeptides. *Neuroscience* 24: 739–768.

Horigome K, Bullock ED, Johnson EM, Jr. (1994) Effects of nerve growth factor on rat peritoneal mast cells. Survival promotion and immediate-early gene induction. *J Biol Chem* 269: 2695–2702.

Horigome K, Pryor JC, Bullock ED, Johnson EM, Jr. (1993) Mediator release from mast cells by nerve growth factor. Neurotrophin specificity and receptor mediation. *J Biol Chem* 268: 14881–14887.

Hoyle GW, Graham RM, Finkelstein JB, Nguyen KP, Gozal D, Friedman M (1998) Hyperinnervation of the airways in transgenic mice overexpressing nerve growth factor. *Am J Respir Cell Mol Biol* 18: 149–157.

Hua XY, Yaksh TL (1993) Pharmacology of the effects of bradykinin, serotonin, and histamine on the release of calcitonin gene-related peptide from C-fiber terminals in the rat trachea. *J Neurosci* 13: 1947–1953.

Hultner L, Kolsch S, Stassen M, Kaspers U, Kremer JP, Mailhammer R, Moeller J, Broszeit H, Schmitt E (2000) In activated mast cells, IL-1 up-regulates the production of several Th2-related cytokines including IL-9. *J Immunol* 164: 5556–5563.

Hunter DD, Myers AC, Undem BJ (2000) Nerve growth factor-induced phenotypic switch in guinea pig airway sensory neurons. *Am J Respir Crit Care Med* 161: 1985–1990.

Imamura M, Smith NC, Garbarg M, Levi R (1996) Histamine H3-receptor-mediated inhibition of calcitonin gene-related peptide release from cardiac C fibers. A regulatory negative-feedback loop. *Circ Res* 78: 863–869.

Janiszewski J, Bienenstock J, Blennerhassett MG (1990) Activation of rat peritoneal mast cells in coculture with sympathetic neurons alters neuronal physiology. *Brain Behav Immun* 4: 139–150.

Janiszewski J, Bienenstock J, Blennerhassett MG (1994) Picomolar doses of substance P trigger electrical responses in mast cells without degranulation. *Am J Physiol* 267: C138–C145.

Jiang W, Kreis ME, Eastwood C, Kirkup AJ, Humphrey PP, Grundy D (2000) 5-HT(3) and histamine H(1) receptors mediate afferent nerve sensitivity to intestinal anaphylaxis in rats. *Gastroenterology* 119: 1267–1275.

Johnson D, Krenger W (1992) Interactions of mast cells with the nervous system – recent advances. *Neurochem Res* 17: 939–951.

Joos GF, Germonpre PR, Pauwels RA (2000a) Role of tachykinins in asthma. *Allergy* 55: 321–337.

Joos GF, Germonpre PR, Pauwels RA (2000b) Neural mechanisms in asthma. *Clin Exp Allergy* 30 (Suppl 1): 60–65.

Junger H, Sorkin LS (2000) Nociceptive and inflammatory effects of subcutaneous TNFalpha. *Pain* 85: 145–151.

Kaliner M (1987) Mast cell mediators and asthma. *Chest* 91: 171S–176S.

Karimi K, Redegeld FA, Blom R, Nijkamp FP (2000) Stem cell factor and interleukin-4 increase responsiveness of mast cells to substance P. *Exp Hematol* 28: 626–634.

Karimi K, Redegeld FA, Heijdra B, Nijkamp FP (1999) Stem cell factor and interleukin-4 induce murine bone marrow cells to develop into mast cells with connective tissue type characteristics in vitro. *Exp Hematol* 27: 654–662.

Kashiba H, Fukui H, Morikawa Y, Senba E (1999) Gene expression of histamine H1 receptor in guinea pig primary sensory neurons: a relationship between H1 receptor mRNA-expressing neurons and peptidergic neurons. *Brain Res Mol Brain Res* 66: 24–34.

Katz HR, Stevens RL, Austen KF (1985) Heterogeneity of mammalian mast cells differentiated in vivo and in vitro. *J Allergy Clin Immunol* 76: 250–259.

Keith IM, Jin J, Saban R (1995) Nerve–mast cell interaction in normal guinea pig urinary bladder. *J Comp Neurol* 363: 28–36.

Keller JT, Marfurt CF (1991) Peptidergic and serotoninergic innervation of the rat dura mater. *J Comp Neurol* 309: 515–534.

Kitamura Y, Kanakura Y, Sonoda S, Asai H, Nakano T (1987) Mutual phenotypic changes between connective tissue type and mucosal mast cells. *Int Arch Allergy Appl Immunol* 82: 244–248.

Kowalski ML, Didier A, Lundgren JD, Igarashi Y, Kaliner MA (1997) Role of sensory innervation and mast cells in neurogenic plasma protein exudation into the airway lumen. *Respirology* 2: 267–274.

Kraneveld AD, Buckley TL, Heuven-Nolsen D, van Schaik Y, Koster AS, Nijkamp FP (1995) Delayed-type hypersensitivity-induced increase in vascular permeability in the mouse small intestine: inhibition by depletion of sensory neuropeptides and NK1 receptor blockade. *Br J Pharmacol* 114: 1483–1489.

Kraneveld AD, James DE, de Vries A, Nijkamp FP (2000) Excitatory non-adrenergic–non-cholinergic neuropeptides: key players in asthma. *Eur J Pharmacol* 405: 113–129.

Kraneveld AD, Muis T, Koster AS, Nijkamp FP (1998) Role of mucosal mast cells in early vascular permeability changes of intestinal DTH reaction in the rat. *Am J Physiol* 274: G832–G839.

Krishna MT, Chauhan A, Little L, Sampson K, Hawksworth R, Mant T, Djukanovic R, Lee T, Holgate S (2001) Inhibition of mast cell tryptase by inhaled APC 366 attenuates allergen-induced late-phase airway obstruction in asthma. *J Allergy Clin Immunol* 107: 1039–1045.

Krumins SA, Broomfield CA (1993) C-terminal substance P fragments elicit histamine release from a murine mast cell line. *Neuropeptides* 24: 5–10.

Lambracht-Hall M, Dimitriadou V, Theoharides TC (1990a) Migration of mast cells in the developing rat brain. *Brain Res Dev Brain Res* 56: 151–159.

Lambracht-Hall M, Konstantinidou AD, Theoharides TC (1990b) Serotonin release from rat brain mast cells in vitro. *Neuroscience* 39: 199–207.

Leal-Berumen I, Conlon P, Marshall JS (1994) IL-6 production by rat peritoneal mast cells is not necessarily preceded by histamine release and can be induced by bacterial lipopolysaccharide. *J Immunol* 152: 5468–5476.

Leff A (1982) Pathogenesis of asthma. Neurophysiology and pharmacology of bronchospasm. *Chest* 81: 224–229.

Leon A, Buriani A, Dal Toso R, Fabris M, Romanello S, Aloe L, Levi-Montalcini R (1994) Mast cells synthesize, store, and release nerve growth factor. *Proc Natl Acad Sci USA* 91: 3739–3743.

Letourneau R, Pang X, Sant GR, Theoharides TC (1996) Intragranular activation of bladder mast cells and their association with nerve processes in interstitial cystitis. *Br J Urol* 77: 41–54.

Levi-Montalcini R, Skaper SD, Dal Toso R, Petrelli L, Leon A (1996) Nerve growth factor: from neurotrophin to neurokine. *Trends Neurosci* 19: 514–520.

Lindsay RM, Harmar AJ (1989) Nerve growth factor regulates expression of neuropeptide genes in adult sensory neurons. *Nature* 337: 362–364.

Lundberg JM, Brodin E, Hua X, Saria A (1984) Vascular permeability changes and smooth muscle contraction in relation to capsaicin-sensitive substance P afferents in the guinea-pig. *Acta Physiol Scand* 120: 217–227.

Lundberg JM, Martling CR, Saria A (1983) Substance P and capsaicin-induced contraction of human bronchi. *Acta Physiol Scand* 119: 49–53.

Maier SF, Goehler LE, Fleshner M, Watkins LR (1998) The role of the vagus nerve in cytokine-to-brain communication. *Ann N Y Acad Sci* 840: 289–300.

Malaviya R, Ikeda T, Ross E, Abraham SN (1996) Mast cell modulation of neutrophil influx and bacterial clearance at sites of infection through TNF-alpha. *Nature* 381: 77–80.

Mantyh CR, Maggio JE, Mantyh PW, Vigna SR, Pappas TN (1996) Increased substance P receptor expression by blood vessels and lymphoid aggregates in Clostridium difficile-induced pseudomembranous colitis. *Dig Dis Sci* 41: 614–620.

Mantyh CR, Vigna SR, Bollinger RR, Mantyh PW, Maggio JE, Pappas TN (1995) Differential expression of substance P receptors in patients with Crohn's disease and ulcerative colitis. *Gastroenterology* 109: 850–860.

Mares V, Bruckner G, Biesold D (1979) Mast cells in the rat brain and changes in their number under different light regimens. *Exp Neurol* 65: 278–283.

Marshall JS, Gomi K, Blennerhassett MG, Bienenstock J (1999) Nerve growth factor modifies the expression of inflammatory cytokines by mast cells via a prostanoid-dependent mechanism. *J Immunol* 162: 4271–4276.

Marshall JS, Stead RH, McSharry C, Nielsen L, Bienenstock J (1990) The role of mast cell degranulation products in mast cell hyperplasia. I. Mechanism of action of nerve growth factor. *J Immunol* 144: 1886–1892.

Matsuda H, Kannan Y, Ushio H, Kiso Y, Kanemoto T, Suzuki H, Kitamura Y (1991) Nerve growth factor induces development of connective tissue-type mast cells in vitro from murine bone marrow cells. *J Exp Med* 174: 7–14.

Matsumoto I, Inoue Y, Shimada T, Aikawa T (2001) Brain mast cells act as an immune gate to the hypothalamic–pituitary–adrenal axis in dogs. *J Exp Med* 194: 71–78.

McDonald DM, Bowden JJ, Baluk P, Bunnett NW (1996) Neurogenic inflammation. A model for studying efferent actions of sensory nerves. *Adv Exp Med Biol* 410: 453–462.

McDonald DM, Mitchell RA, Gabella G, Haskell A (1988) Neurogenic inflammation in the rat trachea. II. Identity and distribution of nerves mediating the increase in vascular permeability. *J Neurocytol* 17: 605–628.

Metcalfe DD, Baram D, Mekori YA (1997) Mast cells. *Physiol Rev* 77: 1033–1079.

Michaelis M, Vogel C, Blenk KH, Arnarson A, Janig W (1998) Inflammatory mediators sensitize acutely axotomized nerve fibers to mechanical stimulation in the rat. *J Neurosci* 18: 7581–7587.

Miura M, Inoue H, Ichinose M, Kimura K, Katsumata U, Takishima T (1990) Effect of nonadrenergic noncholinergic inhibitory nerve stimulation on the allergic reaction in cat airways. *Am Rev Respir Dis* 141: 29–32.

Mizuma H, Takagi K, Miyake K, Takagi N, Ishida K, Takeo S, Nitta A, Nomoto H, Furukawa Y, Furukawa S (1999) Microsphere embolism-induced elevation of nerve growth factor level and appearance of nerve growth factor immunoreactivity in activated T- lymphocytes in the rat brain. *J Neurosci Res* 55: 749–761.

Monteforte R, De Santis A, Chieffi BG (2001) Morphological changes in frog mast cells induced by nerve stimulation in vivo. *Neurosci Lett* 315: 77–80.

Mousli M, Bronner C, Landry Y, Bockaert J, Rouot B (1990a) Direct activation of GTP-binding regulatory proteins (G-proteins) by substance P and compound 48/80. *FEBS Lett* 259: 260–262.

Mousli M, Bronner C, Bockaert J, Rouot B, Landry Y (1990b) Interaction of substance P, compound 48/80 and mastoparan with the alpha-subunit C-terminus of G protein. *Immunol Lett* 25: 355–357.

Muinuddin A, Paterson WG (2001) Initiation of distension-induced descending peristaltic reflex in opossum esophagus: role of muscle contractility. *Am J Physiol Gastrointest Liver Physiol* 280: G431–G438.

Murphy PG, Borthwick LS, Johnston RS, Kuchel G, Richardson PM (1999) Nature of the retrograde signal from injured nerves that induces interleukin-6 mRNA in neurons. *J Neurosci* 19: 3791–3800.

Nemmar A, Delaunois A, Beckers JF, Sulon J, Bloden S, Gustin P (1999) Modulatory effect of imetit, a histamine H3 receptor agonist, on C-fibers, cholinergic fibers and mast cells in rabbit lungs in vitro. *Eur J Pharmacol* 371: 23–30.

Newson B, Dahlstrom A, Enerback L, Ahlman H (1983) Suggestive evidence for a direct innervation of mucosal mast cells. *Neuroscience* 10: 565–570.

Nicol GD, Cui M (1994) Enhancement by prostaglandin E2 of bradykinin activation of embryonic rat sensory neurones. *J Physiol* 480(Pt 3): 485–492.

Nicol GD, Lopshire JC, Pafford CM (1997) Tumor necrosis factor enhances the capsaicin sensitivity of rat sensory neurons. *J Neurosci* 17: 975–982.

Nilsson G, Forsberg-Nilsson K, Xiang Z, Hallbook F, Nilsson K, Metcalfe DD (1997) Human mast cells express functional TrkA and are a source of nerve growth factor. *Eur J Immunol* 27: 2295–2301.

Ohshiro H, Suzuki R, Furuno T, Nakanishi M (2000) Atomic force microscopy to study direct neurite-mast cell (RBL) communication in vitro. *Immunol Lett* 74: 211–214.

Olsson Y (1968) Mast cells in the nervous system. *Int Rev Cytol* 24: 27–70.

Otsuka H, Denburg J, Dolovich J, Hitch D, Lapp P, Rajan RS, Bienenstock J, Befus D (1985) Heterogeneity of metachromatic cells in human nose: significance of mucosal mast cells. *J Allergy Clin Immunol* 76: 695–702.

Ottosson A, Edvinsson L (1997) Release of histamine from dural mast cells by substance P and calcitonin gene-related peptide. *Cephalalgia* 17: 166–174.

168   AUTONOMIC NEUROIMMUNOLOGY

Paterson WG (1998) Role of mast cell-derived mediators in acid-induced shortening of the esophagus. *Am J Physiol* 274: G385–G388.

Paterson WG, Miller DV, Zhang Y (2002) Substance P nerves mediate acid-induced esophageal shortening. *Can J Gastroenterol* 16: 52A.

Payan DG, Levine JD, Goetzl EJ (1984) Modulation of immunity and hypersensitivity by sensory neuropeptides. *J Immunol* 132: 1601–1604.

Pearce FL, Thompson HL (1986) Some characteristics of histamine secretion from rat peritoneal mast cells stimulated with nerve growth factor. *J Physiol* 372: 379–393.

Perdue MH, Masson S, Wershil BK, Galli SJ (1991) Role of mast cells in ion transport abnormalities associated with intestinal anaphylaxis. Correction of the diminished secretory response in genetically mast cell-deficient W/Wv mice by bone marrow transplantation. *J Clin Invest* 87: 687–693.

Pothoulakis C, Castagliuolo I, Leeman SE (1998) Neuroimmune mechanisms of intestinal responses to stress. Role of corticotropin-releasing factor and neurotensin. *Ann N Y Acad Sci* 840: 635–648.

Purcell WM, Atterwill CK (1995) Mast cells in neuroimmune function: neurotoxicological and neuropharmacological perspectives. *Neurochem Res* 20: 521–532.

Ramirez-Romero R, Gallup JM, Sonea IM, Ackermann MR (2000) Dihydrocapsaicin treatment depletes peptidergic nerve fibers of substance P and alters mast cell density in the respiratory tract of neonatal sheep. *Regul Pept* 91: 97–106.

Ratliff TL, Klutke CG, Hofmeister M, He F, Russell JH, Becich MJ (1995) Role of the immune response in interstitial cystitis. *Clin Immunol Immunopathol* 74: 209–216.

Rennick D, Hunte B, Holland G, Thompson-Snipes L (1995) Cofactors are essential for stem cell factor-dependent growth and maturation of mast cell progenitors: comparative effects of interleukin- 3 (IL-3), IL-4, IL-10, and fibroblasts. *Blood* 85: 57–65.

Reynier-Rebuffel AM, Mathiau P, Callebert J, Dimitriadou V, Farjaudon N, Kacem K, Launay JM, Seylaz J, Abineau P (1994) Substance P, calcitonin gene-related peptide, and capsaicin release serotonin from cerebrovascular mast cells. *Am J Physiol* 267: R1421–R1429.

Rossi R, Johansson O (1998) Cutaneous innervation and the role of neuronal peptides in cutaneous inflammation: a minireview. *Eur J Dermatol* 8: 299–306.

Rottem M, Hull G, Metcalfe DD (1994) Demonstration of differential effects of cytokines on mast cells derived from murine bone marrow and peripheral blood mononuclear cells. *Exp Hematol* 22: 1147–1155.

Rozniecki JJ, Dimitriadou V, Lambracht-Hall M, Pang X, Theoharides TC (1999) Morphological and functional demonstration of rat dura mater mast cell–neuron interactions in vitro and in vivo. *Brain Res* 849: 1–15.

Saban R, Saban MR, Nguyen NB, Lu B, Gerard C, Gerard NP, Hammond TG (2000) Neurokinin-1 (NK-1) receptor is required in antigen-induced cystitis. *Am J Pathol* 156: 775–780.

Sanico AM, Stanisz AM, Gleeson TD, Bora S, Proud D, Bienenstock J, Koliatsos VE, Togias A (2000) Nerve growth factor expression and release in allergic inflammatory disease of the upper airways. *Am J Respir Crit Care Med* 161: 1631–1635.

Santos J, Perdue MH (2000) Stress and neuroimmune regulation of gut mucosal function. *Gut* 47 (Suppl 4): iv49–iv51.

Santos J, Saunders PR, Hanssen NP, Yang PC, Yates D, Groot JA, Perdue MH (1999) Corticotropin-releasing hormone mimics stress-induced colonic epithelial pathophysiology in the rat. *Am J Physiol* 277: G391–G399.

Scholzen T, Armstrong CA, Bunnett NW, Luger TA, Olerud JE, Ansel JC (1998) Neuropeptides in the skin: interactions between the neuroendocrine and the skin immune systems. *Exp Dermatol* 7: 81–96.

Sekizawa S, Tsubone H, Kuwahara M, Sugano S (1998) Does histamine stimulate trigeminal nasal afferents? *Respir Physiol* 112: 13–22.

Sestini P, Dolovich M, Vancheri C, Stead RH, Marshall JS, Perdue M, Gauldie J, Bienenstock J (1989) Antigen-induced lung solute clearance in rats is dependent on capsaicin-sensitive nerves. *Am Rev Respir Dis* 139: 401–406.

Shanahan F, Denburg JA, Fox J, Bienenstock J, Befus D (1985) Mast cell heterogeneity: effects of neuroenteric peptides on histamine release. *J Immunol* 135: 1331–1337.

Silver R, Silverman AJ, Vitkovic L, Lederhendler II (1996) Mast cells in the brain: evidence and functional significance. *Trends Neurosci* 19: 25–31.

Sofroniew MV, Howe CL, Mobley WC (2001) Nerve growth factor signaling, neuroprotection, and neural repair. *Annu Rev Neurosci* 24: 1217–1281.

Solomon A, Aloe L, Pe'er J, Frucht-Pery J, Bonini S, Bonini S, Levi-Schaffer F (1998) Nerve growth factor is preformed in and activates human peripheral blood eosinophils. *J Allergy Clin Immunol* 102: 454–460.

Solway J, Leff AR (1991) Sensory neuropeptides and airway function. *J Appl Physiol* 71: 2077–2087.

Sorkin LS, Xiao WH, Wagner R, Myers RR (1997) Tumour necrosis factor-alpha induces ectopic activity in nociceptive primary afferent fibres. *Neuroscience* 81: 255–262.

Spanos C, Pang X, Ligris K, Letourneau R, Alferes L, Alexacos N, Sant GR, Theoharides TC (1997) Stress-induced bladder mast cell activation: implications for interstitial cystitis. *J Urol* 157: 669–672.

Stanisz A, Scicchitano R, Stead R, Matsuda H, Tomioka M, Denburg J, Bienenstock J (1987) Neuropeptides and immunity. *Am Rev Respir Dis* 136: S48–S51.

Stead RH (1992) Innervation of mucosal immune cells in the gastrointestinal tract. *Reg Immunol* 4: 91–99.

Stead RH, Dixon MF, Bramwell NH, Riddell RH, Bienenstock J (1989) Mast cells are closely apposed to nerves in the human gastrointestinal mucosa. *Gastroenterology* 97: 575–585.

Stead RH, Kosecka-Janiszewska U, Oestreicher AB, Dixon MF, Bienenstock J (1991) Remodeling of B-50 (GAP-43)- and NSE-immunoreactive mucosal nerves in the intestines of rats infected with Nippostrongylus brasiliensis. *J Neurosci* 11: 3809–3821.

Sternberg EM, Chrousos GP, Wilder RL, Gold PW (1992) The stress response and the regulation of inflammatory disease. *Ann Intern Med* 117: 854–866.

Streilein JW, Alard P, Niizeki H (1999) Neural influences on induction of contact hypersensitivity. *Ann N Y Acad Sci* 885: 196–208.

Suzuki R, Furuno T, McKay DM, Wolvers D, Teshima R, Nakanishi M, Bienenstock J (1999) Direct neurite-mast cell communication in vitro occurs via the neuropeptide substance P. *J Immunol* 163: 2410–2415.

Tam SY, Tsai M, Yamaguchi M, Yano K, Butterfield JH, Galli SJ (1997) Expression of functional TrkA receptor tyrosine kinase in the HMC-1 human mast cell line and in human mast cells. *Blood* 90: 1807–1820.

Theoharides TC (1996) The mast cell: a neuroimmunoendocrine master player. *Int J Tissue React* 18: 1–21.

Theoharides TC, Kops SK, Bondy PK, Askenase PW (1985) Differential release of serotonin without comparable histamine under diverse conditions in the rat mast cell. *Biochem Pharmacol* 34: 1389–1398.

Theoharides TC, Pang X, Letourneau R, Sant GR (1998b) Interstitial cystitis: a neuroimmunoendocrine disorder. *Ann N Y Acad Sci* 840: 619–634.

Theoharides TC, Singh LK, Boucher W, Pang X, Letourneau R, Webster E, Chrousos G (1998a) Corticotropin-releasing hormone induces skin mast cell degranulation and increased vascular permeability, a possible explanation for its proinflammatory effects. *Endocrinology* 139: 403–413.

Theoharides TC, Spanos C, Pang X, Alferes L, Ligris K, Letourneau R, Rozniecki JJ, Webster E, Chrousos GP (1995) Stress-induced intracranial mast cell degranulation: a corticotropin-releasing hormone-mediated effect. *Endocrinology* 136: 5745–5750.

Ullrich SE (1995) The role of epidermal cytokines in the generation of cutaneous immune reactions and ultraviolet radiation-induced immune suppression. *Photochem Photobiol* 62: 389–401.

Undem BJ, Dick EC, Buckner CK (1983) Inhibition by vasoactive intestinal peptide of antigen-induced histamine release from guinea-pig minced lung. *Eur J Pharmacol* 88: 247–250.

Undem BJ, Myers AC, Barthlow H, Weinreich D (1990) Vagal innervation of guinea pig bronchial smooth muscle. *J Appl Physiol* 69: 1336–1346.

Undem BJ, Riccio MM, Weinreich D, Ellis JL, Myers AC (1995) Neurophysiology of mast cell-nerve interactions in the airways. *Int Arch Allergy Immunol* 107: 199–201.

Valent P (1995) Cytokines involved in growth and differentiation of human basophils and mast cells. *Exp Dermatol* 4: 255–259.

van der Velden V, Hulsmann AR (1999) Autonomic innervation of human airways: structure, function, and pathophysiology in asthma. *Neuroimmunomodulation* 6: 145–159.

van Houwelingen A, van der Avoort LA, Heuven-Nolsen D, Kraneveld AD, Nijkamp FP (1999) Repeated challenge with dinitrobenzene sulphonic acid in dinitrofluorobenzene-sensitized mice results in vascular hyperpermeability in the trachea: a role for tachykinins. *Br J Pharmacol* 127: 1583–1588.

Vedder H, Affolter HU, Otten U (1993) Nerve growth factor (NGF) regulates tachykinin gene expression and biosynthesis in rat sensory neurons during early postnatal development. *Neuropeptides* 24: 351–357.

Virchow JC, Julius P, Lommatzsch M, Luttmann W, Renz H, Braun A (1998) Neurotrophins are increased in bronchoalveolar lavage fluid after segmental allergen provocation. *Am J Respir Crit Care Med* 158: 2002–2005.

Walls AF, He S, Teran LM, Buckley MG, Jung KS, Holgate ST, Shute JK, Cairns JA (1995) Granulocyte recruitment by human mast cell tryptase. *Int Arch Allergy Immunol* 107: 372–373.

Wedemeyer J, Tsai M, Galli SJ (2000) Roles of mast cells and basophils in innate and acquired immunity. *Curr Opin Immunol* 12: 624–631.

Weinreich D, Undem BJ, Taylor G, Barry MF (1995) Antigen-induced long-term potentiation of nicotinic synaptic transmission in the superior cervical ganglion of the guinea pig. *J Neurophysiol* 73: 2004–2016.

Wershil BK, Galli SJ (1991) Gastrointestinal mast cells. New approaches for analyzing their function in vivo. *Gastroenterol Clin North Am* 20: 613–627.

White DM (1997) Release of substance P from peripheral sensory nerve terminals. *J Peripher Nerv Syst* 2: 191–201.

White RJ, Zhang Y, Morris GP, Paterson WG (2001) Esophagitis-related esophageal shortening in opossum is associated with longitudinal muscle hyperresponsiveness. *Am J Physiol Gastrointest Liver Physiol* 280: G463–G469.

Woolf CJ, Ma QP, Allchorne A, Poole S (1996) Peripheral cell types contributing to the hyperalgesic action of nerve growth factor in inflammation. *J Neurosci* 16: 2716–2723.

Yano H, Wershil BK, Arizono N, Galli SJ (1989) Substance P-induced augmentation of cutaneous vascular permeability and granulocyte infiltration in mice is mast cell dependent. *J Clin Invest* 84: 1276–1286.

Yonei Y, Oda M, Nakamura M (1985) Evidence for direct interaction between the cholinergic nerve and mast cells in rat colonic mucosa. An electron microscope cytochemical and autoradiographic study. *J Clin Electron Microsc* 18: 560–561.

Yoshida K, Kakihana M, Chen LS, Ong M, Baird A, Gage FH (1992) Cytokine regulation of nerve growth factor-mediated cholinergic neurotrophic activity synthesized by astrocytes and fibroblasts. *J Neurochem* 59: 919–931.

Yoshikawa T, Streilein JW (1990) Tumor necrosis factor-alpha and ultraviolet B light have similar effects on contact hypersensitivity in mice. *Reg Immunol* 3: 139–144.

Zelazowski P, Smith MA, Gold PW, Chrousos GP, Wilder RL, Sternberg EM (1992) In vitro regulation of pituitary ACTH secretion in inflammatory disease susceptible Lewis (LEW/N) and inflammatory disease resistant Fischer (F344/N) rats. *Neuroendocrinology* 56: 474–482.

Zhuang X, Silverman AJ, Silver R (1993) Reproductive behavior, endocrine state, and the distribution of GnRH-like immunoreactive mast cells in dove brain. *Horm Behav* 27: 283–295.

9 Submandibular Gland Factors and Neuroendocrine Regulation of Inflammation and Immunity

Paul Forsythe[1], Rene E. Déry[1], Ronald Mathison[2],
Joseph S. Davison[2] and A. Dean Befus[1]

[1]*Pulmonary Research Group, Department of Medicine, Faculty of
Medicine, University of Alberta, Edmonton, Alberta, Canada, T6G 2S2*
[2]*Department of Physiology and Biophysics, Faculty of Medicine,
University of Calgary, Calgary, Alberta, Canada, T2N 4N1*

Endocrine secretion from salivary glands was first reported almost 50 years ago. Since then it has emerged that endocrine factors from the submandibular gland (SMG) have regulatory effects on the immune system. It has generally been believed that the immunomodulatory effects of the SMG are mediated predominately by growth factors released into saliva and blood. However, more recent studies, described here, indicate that small peptides of salivary gland origin are also capable of modulating inflammatory reactions.

We have established that decentralization of the superior cervical ganglia reduces the magnitude of allergic inflammation in the airways and anaphylactic and endotoxic hypotension in the rat. This anti-inflammatory activity is dependent on an intact SMG and reconstitution of sialadenectomized rats with soluble extracts from the SMG has identified two polypeptides with anti-inflammatory activities. The sequences of both these polypeptides were found within the prohormone, submandibular gland rat 1. It is clear that androgens, thyroid and adrenocortical hormones as well as the autonomic nervous system influence the production and release of biologically active polypeptides from the SMG. Evidence is presented that suggests the SMG is a fully integrated component of the neuroimmunoendocrine system and plays a significant role in maintaining systemic homeostasis in response to stress.

KEY WORDS: submandibular gland; neuroendocrine regulation; inflammation; bioactive peptides.

INTRODUCTION

Secretory glands have classically been divided into two distinct types, exocrine and endocrine. Exocrine glands secrete their products via the gland's duct system into the external environment, such as the lumen of the gastrointestinal tract. In contrast endocrine glands release their products internally, initially into the interstitial fluid and then into the blood stream. However, there have been reports of endocrine secretion from glands traditionally thought to be purely exocrine organs (reviewed in Isenman *et al.*, 1999).

Endocrine secretion from salivary glands was first reported almost 50 years ago. Since then it has emerged that endocrine factors from salivary glands aid in the integrity of oesophageal and gastrointestinal mucosa, promote hepatic regeneration and mammary gland tumorigenesis, are essential for the maintenance of the reproductive system and have regulatory effects on the immune system. It is clear that the submandibular gland (SMG) is under multifactorial neuroendocrine control in a similar manner to other glands of the digestive tract such as the pancreas, which is well known for both endocrine and exocrine function (Isenman *et al.*, 1999). Despite the accumulation of this knowledge, salivary glands are still generally regarded as exocrine organs concerned purely with aiding diges-tion and maintaining oral health. This perception is clearly wrong, what follows is a description of endocrine factors released from the SMG that indicate the gland is involved in controlling biological functions not directly associated with the digestive tract. The mechanisms that control the production and release of such factors will be discussed and evidence presented which suggests the SMG is a fully integrated component of the neuroimmunoendocrine system that plays a significant role in maintaining systemic homeostasis under stressful conditions.

STRUCTURE OF THE SUBMANDIBULAR GLAND

The SMG is comprised of four major epithelial compartments: acinar cells, intercalated ducts, granular convoluted tubule (GCT) cells and striated excretory ducts. The acini are connected to each other by intercalated ducts that lead to the GCT, which in turn join into the striated secretory ducts (Mathison *et al.*, 1994b). The acinar cells secrete the important digestive enzyme amylase, and produce saliva. The intercalated duct cells include stem cells that produce acinar and GCT cells during development. Cells of the striated excretory ducts regulate water content and ionic composition of saliva. GCT cells represent the major source of biologically active polypeptides produced in the SMG (Barka, 1980) and have served as a convenient source for the isolation and purification of polypeptides such as nerve growth factor (NGF). The diversity and quantity of hormonal systems that exert control over GCT cell development and content of biologically active polypeptides are an indication of the potential importance of these cells in homeostatic mechanisms.

BIOLOGICALLY ACTIVE POLYPEPTIDES

A large number of biologically active polypeptides have been identified in the SMG, many of which have been localized to the GCT (Barka, 1980; Gresik, 1994). These factors, released as both exocrine and endocrine agents, can be classified into three major groups: (i) growth factors such as nerve growth factor (NGF), epidermal growth factor (EGF) and transforming growth factor-β (TGF-β); (ii) processing enzymes such as kallikrein-like proteinases and renin; and (iii) regulatory peptides including glugagon, insulin, erythro-poietin, somatostatin, angiotensin II, vasoactive intestinal peptide and neuropeptide Y.

TABLE 9.1
Effects of some immunomodulatory factors of the SMG
(adapted from Sabbadini and Berczi, 1998, see text for details).

SMG factor	Immunological effect
NGF	Increased mast cell numbers *in vivo* Increased histamine release from mast cells Chemotaxis and stimulation of phagocytic activity of neutrophils Stimulation of lymphocytes proliferation Increased IgM and IgG production Anti-inflammatory activity *in vivo*
EGF/TGF-α	Stimulation of lymphocyte proliferation Regulation of IL-1 and IL-4 production Increased IFN-γ production Reduced activity of suppressor T cells Macrophage chemotaxis and increased phagocytosis Anti-inflammatory activity *in vivo*
TGF-β	Activation of monocytes Inhibition of lymphocyte proliferation Inhibition of Ig production Enhancement of IgA production
Kallikrein	Stimulation of B and T cell proliferation Switch of Ig production by LPS-stimulated lymphocytes Enhancement of IgE production

In the following sections we will describe a number of well-characterized SMG derived peptides many of which have biological functions relevant to immunoregulation (Table 9.1).

NERVE GROWTH FACTOR

NGF was first described in the mouse SMG and identified as a neurotrophic agent. NGF has properties that suggest it plays a role in immunoregulation (Levi-Montalcini *et al.*, 1996; Levi-Montalcini, 1998). NGF increases the number and size of mast cells in tissues of neonatal mice (Aloe and Levi-Montalcini, 1977), and stimulates histamine release from mast cells, both *in vivo* and *in vitro* (Marshall *et al.*, 1990; Pearce and Thompson, 1986; Bischoff and Dahinden, 1992). NGF also increases phagocytosis and chemotaxis of neutrophils, promotes the development of haemopoietic colonies and stimulates lymphocyte growth *in vitro* (Gee *et al.*, 1983; Boyle *et al.*, 1985; Amico-Roxas *et al.*, 1989; Kannan *et al.*, 1991; Brodie and Gelfand, 1992). Other effects include the upregulation of IgM and IgG$_4$ production by human B cells (Otten *et al.*, 1989; Kimata *et al.*, 1991). NGF stimulates the growth of sympathetic ganglia that innervate immune organs and may aid in neuronal control of the immune system (Levi-Montalcini, 1998). While the effects described suggest a pro-inflammatory role for NGF, it can suppress *in vivo* inflammatory reactions in several models (Amico-Roxas *et al.*, 1989). The reason for this disparity is unclear.

TRANSFORMING GROWTH FACTOR-β

TGF-β is a homodimeric polypeptide originally purified from human platelets. It acts as a multifunctional regulator of cell growth and differentiation in a wide variety of normal and neoplastic systems (Massague et al., 1992). TGF-β shares structural and functional homology with EGF. These factors also share a receptor (Gill et al., 1987). Binding of TGF-β or EGF to the 170 kDa plasma membrane receptor on the target cell stimulates intrinsic tyrosine kinase activity. The subsequent phosphorylation cascade leads to increased proliferation and differentiation of skin tissues, corneal epithelium, lung and tracheal epithelium (Partridge et al., 1988; Nordlund et al., 1991; Tsutsumi et al., 1993). TGF-β exerts a range of immunomodulatory effects, acting as a chemoattractant for monocytes, neutrophils and lymphocytes and activating monocytes to secrete cytokines and growth factors (Wahl et al., 1987; Adams et al., 1991; Brandes et al., 1991). TGF-β is a stimulatory factor in the early stages of inflammation, but later supports resolution of inflammation and contributes to healing. CD8[+] T cells are stimulated by TGF-β, while activated CD4[+] T cells are suppressed (Salgame et al., 1991). B cell proliferation and the production of IgG and IgM are also suppressed by TGF-β (Kehrl et al., 1991).

EPIDERMAL GROWTH FACTOR

Exocrine release of EGF is one of the most widely studied aspects of SMG function. EGF has regenerative and reparative properties on oral, oesophageal and gastric mucosa. The importance of EGF in tissue repair is emphasized by diverse effects that include gastric epithelial cell division, DNA synthesis, collagen synthesis, matrix deposition, neovascularization and stimulation of protein and hyaluronic acid synthesis in epithelial cells (Konturek et al., 1995). Through these multiple activities SMG-derived EGF promotes the healing of gastric ulcers and tongue lesions and maintains the integrity of oesophageal mucosa (Wingren et al., 1989; Noguchi et al., 1991b).

EGF can be released into the blood stream and has actions that extend beyond the oral and gastrointestinal mucosa to peripheral organs. Through the circulation EGF interacts with receptors to play a role in organ to organ communication. Several studies support the possibility of an endocrine role for SMG-derived EGF. Endocrine effects attributed to EGF include enhancing cellular proliferation in the regenerating liver following partial hepatectomy in the rat (Jones et al., 1995), enhancing bone remodelling (Dolce et al., 1994), stimulating the meiotic phase of spermatogenesis in the male mouse (Tsutsumi et al., 1986) as well as maintaining uterine growth, fertility and mammary gland development in the female (Noguchi et al., 1991a; Tsutsumi et al., 1993). EGF has been detected in nearly all mammalian biological fluids, including saliva, blood and milk of human, rat and mouse.

As murine SMG are the most abundant sources of EGF most studies have been carried out in the mouse. The ductular granules and their content of EGF are discharged into the saliva after stimulation with α-adrenergic agonists (Byyny et al., 1974; Garrett et al., 1991). Under this stimulation, high concentrations of EGF are found in the mouse saliva. Stimulation of salivary secretion of EGF by α-adrenergic agonists, cyclocytidine or by sympathetic nerve stimulation results in a marked increase in the plasma levels of the peptide. However, the adrenergic-induced increase in plasma EGF does not occur in sialadenectomized mice. Adrenergic stimulation of mice causes plasma levels of EGF to rise much higher in males than in females. However the increase in EGF plasma levels takes 60 min to reach its maximal level and suggests there is no direct endocrine secretion from the salivary compartment (Byyny et al., 1974; Barka et al., 1978; Grau et al., 1994). In adult rats, growth factor released from the SMG into the saliva can undergo systemic distribution via absorption in the oral cavity and the GI tract (Purushotham et al., 1995).

Immunoregulatory functions of EGF include stimulation of T cell proliferation, upregulation of interferon-γ production and reduction of T suppressor cell activity (Acres et al., 1985; Aune, 1985; Johnson and Torres, 1985). EGF also stimulates macrophage chemotaxis and phagocytosis (Laskin et al., 1980, 1981). These actions suggest that EGF stimulates the immune system and is pro-inflammatory. However, as with NGF, immunosuppressive activity of EGF has been reported (Roberts et al., 1976).

PROTEASES AND CONVERTASES

The polypeptides produced by the SMG are synthesized as inactive precursors that have the potential to become active following proteolytic processing. The SMG contain an array of enzymes capable of hydrolysing peptide bonds including members of the prohormone convertase and kallikrein families (Barka, 1980; Farhadi et al., 1997). These proteases modulate the biological activities of growth factors and regulatory peptides. In addition to being vital for the production of biologically active polypeptides, it is also apparent that at least one family of proteases found in the SMG may be directly involved in the modulation of immune responses. These enzymes, the kallikreins, will be considered here in more detail.

KALLIKREINS

The SMG represents the major site of kallikrein gene expression in the rat. In rat and mouse SMG, tissue kallikrein seems to be present predominately in granules of GCT cells and in a short section of intralobe striated duct cells. Kallikrein is secreted into the saliva and blood. In the mouse, 14 genes have have the potential to encode functional kallikrein proteins. At least three of these are EGF binding proteins and can cleave EGF to its active form (Farhadi et al., 1997).

The rat SMG expresses at least six kallikrein genes (Wines et al., 1989). These include rK2 (tonin, restricted to the SMG) rK7, rK8, rK9 (prostatic protease), rK10 and rK1 (true tissue kallikrein). There is a differential pattern of SMG gene expression for the six rat kallikrein genes that reflects androgen dependence for five of them and androgen independence for true kallikrein (rK1). However, in the anterior pituitary gland rK1 is regulated by oestrogens. Similarly, with the exception of rK1, rK expression in the SMG is responsive to thyroid hormones (van Leeuwen et al., 1984, 1987; Clements et al., 1990). Glandular or tissue kallikreins differ in molecular weight and enzymatic activity from plasma kallikreins. Mouse glandular kallikrein strongly enhances spontaneous and mitogen-induced proliferation of lymphocytes (Hu et al., 1992). This function is independent of EGF, as non-EGF binding forms of kallikrein also exhibit this activity. Serine proteinase inhibitors can block the response suggesting it is related to enzymatic activity of kallikreins. This is in accordance with reports that trypsin, chymotrypsin, thrombin and other proteinases mitogenically stimulate many kinds of cells including lymphocytes. *In vitro*, the addition of kallikrein and other serine proteases to B cells stimulated with LPS and IL-4 enhances the production of IgE, IgG$_1$ and IgG$_3$ (Matsushita and Katz, 1993). Rat glandular kallikrein has also been reported to suppress the DTH response to picryl chloride in mice (Mathison, 1998).

Mounting evidence suggests that in addition to the production of vasoactive peptides in the stroma of secreting glands, tissue kallikreins are involved in a wide range of physiological processes, including the cleavage of prohormones or other precursors, such as atrial natriuretic peptide (ANP), EGF and NGF. rK1 is involved in the processing of kininogens into kinins such as bradykinin, whereas rK2, tonin, can release angiotensin II, a powerful vasoconstrictor directly from angiotensin (Boucher *et al.*, 1977). The kallikreins rK7, rK8 and rK10 exhibit different substrate specificity and different susceptibilities to inhibitors from those of rK1 and may be involved in the processing of peptides other than kinins and angiotensin (Gauthier *et al.*, 1992). Exocrine and endocrine secretions of kallikrein are greatly increased after adrenergic stimulation. Studies indicate that direct endocrine release and/or reabsorption from the duct lumen represents a major route for entry of kallikrein into the circulation and suggests that at least part of rat blood glandular kallikrein is of SMG origin (Penschow and Coghlan, 1993).

VASOACTIVE FACTORS

The salivary ducts in most species are sites of fluid and electrolyte transport. Several studies provide evidence that salivary electrolyte homeostasis is under multifactorial autonomic and endocrine control, including acetylcholine, norepinephrine, angiotensin, gastrointestinal hormones, VIP, substance P, ACTH and aldosterone (reviewed in Cohen and Carpenter, 1975; Cook *et al.*, 1994). The SMG contains well-known vasoactive factors. Rat SMG contains relatively large amounts of preproendothelin-3 mRNA suggesting that the vasoconstrictor, endothelin 3, may play a role in controlling SMG fluid circulation (Matsumoto *et al.*, 1989; Shiba *et al.*, 1992). ANP is involved in natriuretic function regulation of blood pressure and body fluid circulation. It is present in secretory granules of rat GCT cells and is released upon administration of sympathomimetic stimulants and water deprivation (Jankowski *et al.*, 1996).

UNIQUE SMG PEPTIDES

In rodents, there is increasing evidence that SMG derived peptides acting as classic endocrine factors play a role in behavioural and physiological integration especially reproduction, biochemical homeostasis and development. The role of such mediators in immunomodulation is also becoming apparent. However, as most of the mediators released by the SMG are found in other peripheral organs and have well-known biological roles, questions remain regarding the role of the SMG as a major endocrine contributor to homeostatic mechanisms and whether there are any biological factors unique to the organ. The discovery of the submandibular rat 1 (SMR1) protein and its derived peptides, which are predominately synthesised, processed and secreted by the rat SMG, has opened the possibility that endocrine signalling from this organ does have a unique role.

SMR1

The sequence of the SMR1 protein was deduced from the cDNA sequence of the SMR1-VA1 gene, which encodes for the prohormone-like protein in rat SMG (Rosinski-Chupin *et al.*, 1988). The SMR1 polypeptide is found almost exclusively in the salivary glands and prostate and is one of several peptides generated by the variable coding sequence (VCS) multigene family that has been localized to chromosome 14 bands p21–p22 (Rosinski-Chupin *et al.*, 1995; Courty *et al.*, 1996). The gene family has at least 10 members (Table 9.2). Three of these, VCSA1 (SMR1 gene), VCSA2 and VCSA3 belong to the VCSA subclass and are found exclusively in the rat. These genes encode SMR1 and SMR1-related polypeptides that contain potential recognition sites for proteolytic enzymes and can all be considered as potential preprohormones (Rosinski-Chupin *et al.*, 1993). The structure of the VCSA1 gene is similar to the structure of several genes encoding prohormones such as genes for preprothyrotrophin-releasing hormone, preproenkephalins or preproopiomelanocortin (Rosinski-Chupin and Rougeon, 1990). It is not known whether this structure reflects the existence of a common ancestor, or convergent evolution. Seven genes belong to the VCSB subclass. The B subclass is found in several species and encodes a family of proline rich proteins found in rats, mice and humans (Rougeot *et al.*, 1998).

A major characteristic of the VCS gene family is the presence of a hypervariable region inside the coding sequence (Courty *et al.*, 1996). In intraspecies pairwise comparisons, a higher level of sequence divergence is observed in the hypervariable region than in the adjacent exonic or intronic sequences. Furthermore, most of the mutations at the nucleotide level lead to amino acid substitutions. As a consequence, the VCS family encodes proteins that are diverse in amino acid content, structure and probably function.

TABLE 9.2
VCS genes in the rat mouse and human (adapted from Rougeot *et al.*, 1998).

Species	Gene	Protein	Tissue expression	Androgen regulation
Rat	VCSA1	SMR1	SMG prostate	✓
	VCSA2	SMR1-VA2	SMG	✗
	VCSA3	SMR1-VA3	?	?
	VCSB1	PR-VB1	Parotid SMG?	✗
Mouse	VCS1	MSG1	Parotid SMG? Skin?	✗
	VCS2	MSG2a MSG2d MSG2g	SMG	✓
Human	PB	P-B	SMG	✗
	PBI	P-B1	?	?
	BPLP	BPLP	Lacrimal gland	?

Like many SMG-derived polypeptides, the accumulation of mature SMR1 peptides appears to be dependent on the integrity of the hypothalamic–pituitary–gonad axis. In 4-week-old hypophysectomized or gonadectomized male rats the levels of mature peptide are greatly reduced, being close to two orders of magnitude less than in the SMG of sham operated animals (Rosinski-Chupin et al., 1988).

Processing of SMR1

The SMR1 prohormone contains an amino-terminal putative secretory signal sequence and a tetrapeptide (QHNP), located between dibasic amino acids, that constitutes the most common signal for prohormone processing. The proteolytic processing of SMR1 has been partially characterized by Rougeot et al. (1998). Cleavage at pairs of arginine residues close to the amino-terminus of the SMR1 polypeptide generates three structurally related peptides. The undecapeptide ($_{23}$VRGPRRQHNPR$_{33}$) is generated by selective endoproteolysis at the Arg33–Arg34 bond and at the signal sequence. The hexapeptide ($_{28}$RQHNPR$_{33}$) and the pentapeptide ($_{29}$QHNPR$_{33}$) are generated by selective cleavages at both the Arg27–Arg28 and Arg33–Arg34 bonds.

The biosynthesis of these peptides is subject to distinct regulatory pathways depending on the organ, sex and age of the rat. The peptides are differentially distributed within the SMG and in resting or epinephrine-elicited salivary secretions, suggesting distinct proteolytic pathways are involved in their maturation. In the male rat SMG, the hexapeptide and the undecapeptide are found at increasing levels at 6–10 weeks of postnatal life (Rougeot et al., 1998). This corresponds to the differentiation of acinar cells where the SMR-1 protein has been shown to localize. The 6-week-old rat contains mostly the undecapeptide form, while the hexapeptide predominates in 10-week-old animals. In 14-week-old female SMG, the undecapeptide is the major form. Protease activities are very low in the glands of newborn rats and mice and a rapid increase in enzyme activity coincides with the onset of puberty. Therefore, the changes in ratio of processed peptides may be explained by the sex hormone-dependent accumulation of SMR-1, processing enzymes or both, with fully active processing enzymes only being present in the male rat from 10 weeks onward.

Although generated in the gland of both the male and female rats under basal conditions, the undecapeptide is only released into the saliva of the male. The hexapeptide is produced in large amounts in the gland of adult male rats and released into the saliva under both resting and epinephrine stimulated conditions. The pentapeptide appears only in the male saliva and is present mostly under stimulated conditions. Administration of epinephrine also induces the release of the hexapeptide into the blood stream (Figure 9.1). Therefore, it appears that the SMG can act as both an exocrine and endocrine organ for the SMR1-derived peptides (Rougeot et al., 1998).

Although no biological function has been ascribed to these peptides, Rougeot et al. (1997) were able to identify specific binding sites of the pentapeptide at physiological concentrations using radiolabelled peptide coupled to quantitative image analysis of whole rat body sections. The peptide was detected in the renal outer medulla, bone and dental tissue, glandular gastric mucosa and pancreatic lobules. Pentapeptide binding was localized to selective portions of the male rat nephron and in bone exclusively accumulated within the trabecular bone, remodeling unit. Based on these studies it has been suggested that the

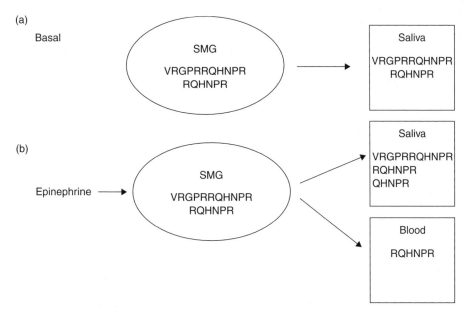

Figure 9.1   Differential processing and release of SMR1-derived peptides in resting (a) and epinephrine-stimulated (b) SMG of male rats.

SMR1-derived pentapeptide might participate as an endocrine factor in maintaining directly or indirectly systemic mineral homeostasis, between at least four systems: kidney, bone, tooth and circulation (Rougeot *et al.*, 1997).

THE SMG AS A NEUROENDOCRINE ORGAN

Receptors for several neurohormonal agonists have been demonstrated in the SMG. Pharmacological and immunohistochemical studies have indicated the presence of SMG receptors for acetylcholine (M1 and M3 muscarinic receptors), norepinephrine (α1, α2, β1 and β2 receptors), purines (P2 receptors), VIP, substance P (NK1 receptors), nucleotides and ANP (reviewed in Emmelin *et al.*, 1976; Lundberg *et al.*, 1988). SMG receptor populations differ according to the species and the cell population. Substance P receptors are expressed in the SMG of the rat, but not in the mouse SMG. End piece cells but not duct cells respond to substance P in the rat SMG. Substance P and other tachykinins are extremely potent sialogogues in the rat but not in the cat.

The neurotransmitters and neuromodulators, localized mostly in parasympathetic and sensory nerve terminals (substance P, calcitonin gene-related peptide (CGRP), neuropeptide Y, galanin-like peptide, cholecystokinin (CCK), pituitary adenylate cyclase activating peptide (PACAP-38) and vasoactive intestinal peptide (VIP)) and in close contact with secretory acini, ducts and blood vessels are potential transmitter/autocoids that act to regulate fluid and secretory responses of rat SMG.

HORMONAL CONTROL OF THE SMG

It has been clearly demonstrated that androgens, thyroid and adrenocortical hormones are necessary for normal development of SMG and for production of biologically active polypeptides (Kumegawa *et al.*, 1977; Gresik, 1980; Sagulin and Roomans, 1989). These hormonal compounds acting alone or synergistically exert crucial effects on developmental cellular growth, differentiation and on synthesis and secretion processes as well as mineral transport in the SMG.

Sexual dimorphism is the most striking aspect of hormonal control of the SMG. This is reflected in the androgen-induced increase in both the number of GCT cells and their content of biologically active peptides (Gresik, 1994). Assays have revealed much greater concentrations of many polypeptides (e.g. EGF, NGF, kallikrein and renin) in the glands of male compared to female rodents (Aloe and Levi-Montalcini, 1980a; Wagner *et al.*, 1990; Bhoola *et al.*, 1992). The levels of these factors are androgen dependent. The NGF content of the SMG is reduced by castration and increased by administration of testosterone to female or castrated male animals. NGF levels also increase during pregnancy and lactation (Ishii and Shooter, 1975). Similar observations have been made regarding EGF levels in the SMG. The gland of the female contains about 1/10 the EGF detected in the male. Administration of testosterone to both male and female animals leads to complete reversal of the drastic reduction in EGF levels observed following hypophysectomy (Hiramatsu *et al.*, 1994). While the sex difference in glandular amount of EGF is attributed to the level of endogenous androgens, there is also evidence that EGF synthesis and secretion are controlled by other classical hormones such as thyroxine, adrenocorticosteroids, progesterone and pituitary growth hormone acting singly and/or synergistically.

The receptors for glucocorticoids are localized to the GCT cells but not to acinar cells. Glucocorticoids greatly influence the growth, differentiation and secretory activity of GCT cells and are important in morphogenesis of the embryonic mouse salivary gland. EGF and NGF levels increase in SMG after treatment with glucocorticoids, suggesting that the production of bioactive peptides by the SMG come under the control of the HPA axis.

Androgen receptors have been detected in mouse and rat SMG. Specific androgen binding is higher in homogenates of female SMG compared to males in both rats and mice (Kyakumoto *et al.*, 1985, 1986) despite the fact that glands of male mice contain three times as much total androgen receptor capacity as those of females. The reason for this apparent conflict was determined by Kyakamoto *et al.*, who showed that in males 74% of receptor capacity was in the nuclei (occupied) while in females 94% was in the cytosol (unoccupied). Castration results in a female distribution of the receptor that can be reversed by testosterone administration (Kyakumoto *et al.*, 1985). There is a significant decrease in plasma lutenizing hormone (LH) levels following salindectomy but this procedure does not significantly modify hypophyseal LH content (Boyer *et al.*, 1990). The decrease in plasma LH causes changes in Leydig cells and reduces testosterone production. This observation indicates that while testosterone plays a crucial role in maintenance of the SMG, there is also a feedback relationship between the SMG and the testis via hypophyseal LH secretion.

Several studies have established that in addition to androgens, thyroid hormones are important in the development and maintenance of the GCT cells (Gresik, 1994). Aloe and Levi-Montalcini (1980b) reported that thyroxine caused a precocious differentiation of

GCT cells. Thyroid hormones have also been demonstrated to increase the levels of NGF and EGF in the SMG of the adult female mouse (Verhoeven, 1979; Walker *et al.*, 1981; Walker, 1986; Cabello and Wrutniak, 1989; Fujieda *et al.*, 1993). The SMG of T*fm*/Y mice lacks specific androgen binding capacity (Verhoeven, 1979) and subsequently these mice are deficient in NGF and EGF (Gresik *et al.*, 1980). Administration of thyroxine but not testosterone greatly increases the level of both growth factors in the SMG of T*fm*/Y mice. Thyroxine also restores EGF levels to normal in hypothyroid animals. Removal of the pituitary gland (hypophysectomy) induces marked atrophy of the SMG and reduces both *in vitro* lymphocyte-stimulating activity and *in vivo* immunosuppressive activity. The important role of thyroid hormones and androgens in maintaining SMG function is further emphasized by the fact that thyroid stimulating hormone (TSH) and LH restore the ability of the SMG to modulate lymphocytes in hypophysectomized rats (Nagy *et al.*, 1992). Feedback between the SMG and the HPA axis has been suggested, based on observations that NGF and EGF are able to stimulate release of ACTH and glucocorticoids (Otten *et al.*, 1979; Luger *et al.*, 1988) while, in contrast, TGF-β depresses acetylcholine induced CRH release from the hypothalamus (Raber *et al.*, 1997). The status of the SMG as a fully integrated component of the body's endocrine system is confirmed by these demonstrations of SMG dependence on pituitary function and the feedback relationships with other glands (Figure 9.2).

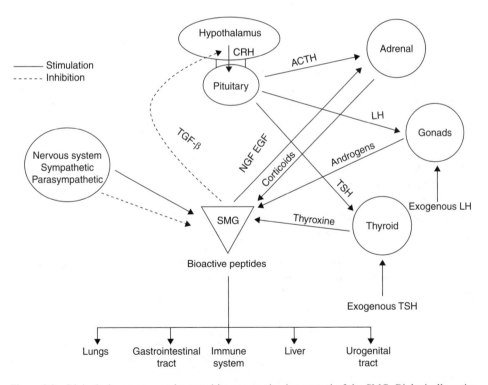

Figure 9.2   Biological responses orchestrated by neuroendocrine control of the SMG. Biologically active factors are released from the SMG into the saliva or blood following appropriate stimulation by hormones or nerves. These factors initiate tissue repair and control inflammation in the lungs, gastrointestinal tract liver and urogenital tract (see text for details).

AUTONOMIC CONTROL OF THE SMG

In addition to endocrine control, the autonomic nervous system (ANS) also participates in homeostatic responses to inflammation through the regulation of cardiovascular function and stimulation of release of certain stress hormones such as glucagon and renin. ANS also regulates aspects of SMG biology. Prolonged electrical stimulation of the sympathetic branch of the ANS via the superior cervical ganglion (SCG) causes enlargement of SMG by increasing both cell size and number (Isenman et al., 1999). Similarly the β-adrenoreceptor agonist isoprenaline (IPR) induces hypertrophic and hyperplastic enlargements of rodent SMG and induces the expression of a number of genes.

SMG are innervated by the sympathetic and parasympathetic systems and both branches regulate the volume and composition of saliva. The parasympathetic system stimulates the glandular acinus leading to the secretion of large volumes of saliva with low concentrations of biologically active polypeptides. β-Adrenergic stimulation increases the synthesis and release of these polypeptides. α-Adrenergic agents stimulate the secretion of growth factors and homeostatic proteases by the GCT cells and lead to the appearance of large amounts of kallikrein, NGF, EGF and renin in the saliva, whereas β-adrenergic agents exert a significantly lesser effect (Martinez et al., 1975; Tsukitani and Mori, 1986). The release of EGF is stimulated by both α- and β-adrenergic mechanisms although the α-adrenergic response is more intense and prolonged. The sympathetic nerves also regulate gene expression for NGF and EGF (Mathison et al., 1992b, 1994b).

The parasympathetic and sympathetic nerves differentially regulate the compartments into which kallikrein is secreted. Local parasympathetic nerve stimulation selectively releases kallikrein into the saliva (exocrine secretion) while activation of the sympathetic nerves increases kallikrein levels in saliva and blood through a process mediated by adrenergic receptors (Orstavik et al., 1982). Kallikreins released from SMG exert both local and distant effects in that they dilate veins of the glands as well as decrease systemic blood pressure in response to heat stress. The increase in the release of kallikrein and EGF from SMG upon activation of the sympathetic nerves, suggests that autonomic regulation of inflammatory responses includes stimulation of the release of glandular factors involved in cardiovascular control and tissue repair.

Cystatin S is a cysteine proteinase inhibitor that regulates proteolysis by endogenous and/or exogenous cysteine proteases such as cathepsins. Rat cystatin S gene expression is tissue specific and occurs temporally during normal development (Shaw et al., 1990). The steady state level reaches a maximum at 28 days of age and is not observed in the adult animal reaching barely detectable levels at 32 days of age. However, cystatin S can be induced in adult SMG by IPR (Shaw et al., 1990; Chaparro et al., 1998). Data suggests that expression of the rat cystatin S gene is also controlled by tissue-specific factors. This is reflected in a much greater IPR induced increase of cystatin S mRNA in the SMG compared to the parotid gland. Induction of cystatin mRNA is also more pronounced in the SMG of female compared to male rats. This difference is evident in 15-day-old animals and is therefore thought to be linked to gender genotype rather than to circulating levels of steroid hormones (Shaw et al., 1990).

Autonomic regulation of protease inhibitor levels represents another control mechanism for the production of biologically active polypeptides in the SMG. Investigations of the changes in the ratios of polypeptides, proteases and protease inhibitors following hormonal or neuronal stimulation of the SMG may provide important insights into this complex regulatory system.

THE CERVICAL SYMPATHETIC TRUNK–SUBMANDIBULAR GLAND AXIS

In the early 1900s Pavlov's classical experiments demonstrated autonomic regulation of salivary gland function, yet the significance of the cervical sympathetic trunk–submandibular gland (CST–SMG) axis as an effector of neuroendocrine immune regulation has only recently come to light (Mathison et al., 1992b).

Axons project down the cervical sympathetic trunk to the inferior and superior cervical ganglia (SCG). The postganglionic neurons leaving the inferior cervical ganglia predominately innervate the lungs and heart, while axons leaving the SCG provide sympathetic innervation to the upper thorax, neck and skull as well as to facial structures (Cardinali and Romeo, 1990). The relevance of the SCG to the neuroendocrine system is indicated by the number of endocrine organs found in these areas, including the cervical lymph nodes, pineal, thyroid, parathyroid and salivary glands.

Surgical sympathectomy has been one approach used to study neural regulation of the immune system. These surgical denervations involve either SCG ganglionectomy (SCGx) or severing the connection between the inferior and superior cervical ganglia, so called decentralization. Unilateral removal of the SCG enhances contact hypersensitivity and delayed type reactions in the denervated submandibular lymph nodes (Alito et al., 1987). These altered responses of the immune system indicate a direct modulation of inflammatory events by the sympathetic nervous system.

AN IMMUNOMODULATORY ROLE FOR THE CST–SMG AXIS

We have used rats sensitized by infection with the nematode Nippostrongylus brasiliensis to investigate the modulatory role of the sympathetic nervous system in pulmonary inflammation. Within 8 h of intravenous challenge with sensitizing allergen, these rats develop a pronounced influx of macrophages and neutrophils into the lumen of the airway (Ramaswamy et al., 1990). The cellular responses can be used as a read out to determine the effects of surgical SCGx and decentralization. We have shown these surgical interventions have dramatic anti-inflammatory effects on life-threatening anaphylaxis and subsequent pulmonary inflammation when compared to sham operated animals (Ramaswamy et al., 1990). Macrophage and neutrophil influx into the lumen of the airways is markedly reduced. Peripheral blood neutrophils from treated animals exhibit a decreased phagocytotic ability and respiratory burst (Carter et al., 1992). Chemotaxis to N-formyl-methionyl-leucyl-phenyalanine is also depressed, while TNF production by alveolar macrophages is similarly compromised (Mathison et al., 1994a).

In marked contrast to the anti-inflammatory effects of SCGx and decentralization, the hypotensive effects of endotoxin are increased following these interventions in an animal model of endotoxic shock (Mathison et al., 1993). While the cells responsible for the modified responses to endotoxin in decentralized or SCGx animals have not been identified, neutrophils, platelets and monocytes/macrophages could all be involved. Neutrophils are known to play a significant role in determining the severity of hypersensitivity reactions, endotoxemia and postoperative hypoxia in the heart (Grisham et al., 1988; Kraemer and Mullane, 1989; Wilkinson et al., 1989). In addition, a correlation exists between the severity of the response to endotoxin and neutrophil activity to nitroblue tetrazolium (Waddell et al., 1992). Platelets may attenuate the cytolytic activity of neutrophils through H_2O_2 scavenging by a glutathione cycle dependent process (Dallegri et al., 1989) and the monocyte/macrophage has a tremendous capacity to release mediators implicated in the immunophysiological effects of endotoxemia, for example, H_2O_2 and TNF (Tsunawaki et al., 1988; Zuckerman et al., 1989).

At the time we first made these observations they highlighted the involvement of the SCG in modulation of responses to endotoxic and anaphylactic shock. However, given that the field of innervation of the SCG is limited to the upper thoracic and head regions it was deemed unlikely that the nerves affected by SCGx were innervating the organs primarily responsible for the immunophysiological reactions. These results suggested the involvement of an intermediary gland or organ.

Through subsequent experiments we identified this intermediary as the SMG. Prior removal of the SMG prevented the depression of pulmonary inflammation and downregulation of macrophage and neutrophil function seen following surgical denervation (Mathison et al., 1992b). These observations suggest that the SMG is a direct source of factors, which downregulate inflammation or control the release of anti-inflammatory factors from elsewhere in the body. Under normal circumstances the SCG exerts an inhibitory influence preventing the release of these factors from the gland while removal of the inhibitory tone through sympathetic decentralization or SCGx allows for an increased anti-inflammatory action as observed in the animal model of pulmonary inflammation. The enhanced hypotensive effects of endotoxin observed following sympathetic decentralization and SCGx were also observed following sialadenectomy suggesting that the CST–SMG axis normally protects against potential hypotensive effects of endotoxin (Mathison et al., 1993).

Taken together, these results suggested a model in which postganglionic fibres arising from the SCG innervate cells within the SMG that synthesize both anti-hypotensive and anti-inflammatory factors. Under normal circumstances sympathetic signalling differentially regulates the release of these factors, stimulating the hypotensive factor while inhibiting the release of the anti-inflammatory agents (Figure 9.3).

CST–SMG control of mast cell function

Mast cell function is also modulated by the CST–SMG axis. However, the regulatory mechanism appears distinct from that involved in controlling neutrophil and macrophage responses (Mathison et al., 1992a). TNF-dependent cytotoxic activity of peritoneal mast cells (PMC) is reduced in rats following decentralization but not SCGx. This suggests that the neural regulation of mast cell function probably occurs at the level of the SCG, unlike neutrophils and macrophages where neural structures within the thoracic spinal cord are responsible for depressing function. Removal of the SMG also inhibits TNF production by

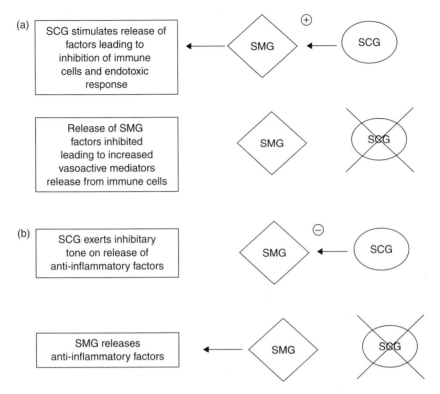

Figure 9.3   Immunomodulatory actions of the CST–SMG axis. (a) Superior cervical ganglion decentralization increases the hypotensive responses to endotoxin. This suggests that under normal circumstances the cervical sympathetic nerves stimulates the SMG to release factors that downregulate the response of immune cells to endotoxin. (b) Superior cervical ganglion decentralization downregulates inflammatory responses in sensitized rats. This suggests that intact sympathetic nerves suppress the release of anti-inflammatory agents from the SMG.

PMC indicating that salivary glands constitutively release a factor that upregulates mast cell function. However, since a combination of sialadenectomy and decentralization has no effect on PMC cytotoxicity, glands or organs other than the SMG probably participate in the regulation of mast cell activity. Taken as a whole these observations indicate that there are multiple mechanisms by which the CST–SMG can regulate immunological functions.

NOVEL IMMUNOREGULATORY PEPTIDES

In an attempt to determine the SMG-derived factors responsible for the modulation of endotoxic and anaphylaxtic reactions, extracts of SMG subjected to molecular weight cut-off filtration and high-performance liquid chromatography (HPLC) purification were tested for their ability to reduce the severity of endotoxin induced hypotension. Our initial expectation was that the agents involved would be well-characterized regulatory factors such as NGF and EGF. However, utilizing these methods two novel peptides, which could attenuate the severity of endotoxin-induced hypotension, were isolated from purified extracts, sequenced and synthesized. These peptides were: a pentapeptide, submandibular gland peptide S (SGP-S), with the sequence SGEGV and a hexapeptide, with the sequence TDIFEGG, named SGP-T (Mathison *et al.*, 1997; Mathison, 1998). When given intravenously, SGP-T can attenuate hypotension during cardiovascular anaphylaxis, inhibit the disruption of intestinal myoelectric activity of the intestine and development of diarrhoea during intestinal anaphylaxis and downregulate neutrophil chemotaxis (Mathison *et al.*, 1997, 1998). These activities were also evident in the C-terminal fragment of the peptide, FEG. The D-isomeric form of FEG, denoted feG, was an effective inhibitor in both models of anaphylaxis when administered orally. SGP-S has yet been assessed in a biological assay other than the endotoxic shock model.

SGP-T was identified as a carboxy-terminal fragment (residues 138–144) of the SMR1 protein, while SGP-S is found closer to the amino terminal of the same polypeptide (Mathison, 1998) (Figure 9.4). The enzymatic cleavage processes that generate SGP-T have not been identified. However, non-arginine-dependent serine proteases are abundant in salivary glands and could generate this peptide (Berg *et al.*, 1992).

It has generally been believed that the immunomodulatory effects of the SMG are mediated predominately by growth factors released into saliva and blood. However, more recent studies, described here, indicate that small peptides of salivary gland origin are also capable of modulating inflammatory reactions. The possibility that other pro-hormones in the salivary gland may yield small peptides capable of immunomodulation must be considered. Salivary glands contain chromogranin A and B, members of a family of highly acidic proteins, the chromogranins (Letic-Gavrilovic *et al.*, 1989). Chromogranins are prohormones and biological functions have been ascribed to many chromogranin-derived peptides. There are reports of antibacterial and antifungal activities and immunoregulatory functions including induction of monocyte chemotaxis and modulation of mast cell activity (Reinisch *et al.*, 1993; Forsythe *et al.*, 1997; Kong *et al.*, 1998; Lugardon *et al.*, 2000). In rats, chromogranin A has been localized to the GCT and chromogranin-like immunoreactivity is detected in the saliva following stimulation with acetylcholine and noradrenaline (Kanno *et al.*, 1999). An immunomodulatory role for chromogranins in the SMG should be considered.

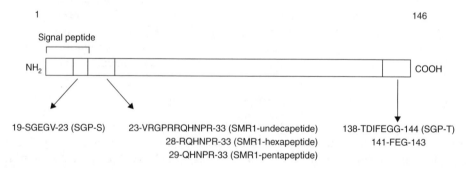

Figure 9.4   Schematic representation of the SMR1 preprotein and the peptides derived from it through proteolytic cleavage.

NEUROIMMUNOLOGY AND STRESS

From the earliest times in history, it has been a commonly held belief that stress and emotions can exert significant effects on the health of an individual. It is therefore surprising that research in this area was long neglected by the scientific and medical community who generally dismissed the concept as unfounded. However, recent advances in neurobiology, endocrinology, immunology and psychiatry have helped determine the biochemical pathways through which the immune, nervous and endocrine systems communicate. The term used to describe this communication, neuroimmunomodulation, was coined by N.H. Spector in 1987 and has been described as 'a modern reflection in neurosciences and immunosciences of the ideas and experience of philosophers and ingenious observers of ancient Egypt, Greece, China, India, and other civilisations, that the mind is involved in the defence against disease'. In addition to promoting a holistic concept of mind and body, these advances have increased awareness of the neuroimmunoendocrine system and stimulated new areas of research such as psychoneuroimmunology. A major focus of psychoneuroimmunology has been a study of the effects of stress on the immune system.

It is generally considered that stressful experiences suppress the ability of the immune system to respond to challenge and thus increase susceptibility to infectious and neoplastic disease (Borysenko, 1982; Kort, 1994). Accordingly, studies have confirmed the immunosuppressive effects of stress. Immune responses are blunted in humans experiencing excessive physical stress and in caregivers exposed to psychological stress. However, there is evidence that stress may also enhance immune function, suggesting a complex relationship between stress and the immune system (Dhabhar, 1998; Dhabhar *et al.*, 2000). The contrasting responses of the immune system to stress may be related to the type of stress inducing agent or the duration of stress (acute versus chronic).

When the body is challenged physically or psychologically, short term activation of the neuroendocrine and ANS promote adaptation and survival during the period of challenge. This has been termed allostasis, literally 're-establishing stability through change' (Sterling and Eyer, 1988). During allostasis physiological systems operate at higher or lower levels than during 'normal' homeostasis; for example, emotional distress may lead to elevated heart rate and blood pressure, as well as elevated glucocorticoid levels that in turn suppress the production of inflammatory cytokines (McEwen, 1998). Providing allostatic responses are shut off when they are no longer needed, the body is able to adapt to and survive the immediate challenge and does not suffer many long term consequences. However, if the same response systems are activated over a longer period of time, or remain active when no longer needed, these adaptive changes may cause damage or exacerbate disease processes.

A short-term increase in immune surveillance, preparing an organism for potential immune challenges arising from the action of stress inducing agents, makes sense from an evolutionary point of view. In aggressive or stressful situations, for example, during fighting or when under attack from a predator, it would be advantageous for an organism to have an enhanced capacity to respond to immunological challenges such as a wound. Indeed, antigen-specific cell mediated immunity can be significantly enhanced through activation of the physiological stress response (Dhabhar *et al.*, 1995).

Psychological stressors have been associated with activation of the sympathetic nervous system and the HPA axis. The hormonal responses of the HPA axis are thought to represent a physiological reaction to excessive stimulation, particularly stimulation associated with stressful situations. During this reaction, the hypothalamus produces corticotrophin-releasing hormone (CRH) which triggers the anterior pituitary gland to secrete ACTH which in turn triggers the adrenal cortex to secrete corticosteroids. Many of the effects of the central stress response on the immune system occur through the anti-inflammatory actions of these glucocorticoids (Sternberg, 2000).

The overall effect of the HPA axis on immune/inflammatory responses is to prevent them from proceeding unchecked by suppressing the responses as soon as they begin. A variety of human autoimmune/inflammatory diseases have recently been associated with blunted HPA axis responses. These include rheumatoid arthritis, atopic dermatitis, allergic asthma, chronic fatigue syndrome and fibromyalgia (Demitrack et al., 1991; Cash et al., 1992; Crofford et al., 1994; Buske-Kirschbaum et al., 1997).

While under-activity of the HPA axis responses predispose to over-activity of the immune system and enhanced susceptibility to inflammation, upregulation of the HPA axis, as occurs during stress, predisposes to enhanced susceptibility to infectious disease. Thus, excess secretion of glucocorticoids during chronic stress activation of the HPA axis tends to shift the cytokine response patterns away from a pro-inflammatory pattern and towards an anti-inflammatory pattern, by relatively greater suppression of pro-inflammatory cytokines and less suppression or even stimulation of anti-inflammatory cytokines. The resultant effect on the immune response is to make immune cells less effective in responding to infectious agents. The effects of stress on the course of infection can depend on the pathogenesis of the illness. In pathology caused by excess of the infectious agent, activation of the stress response and increased glucocorticoid production enhances pathology. However, if the pathology is related to an excessive inflammatory response of the host to the infectious agent, the disease course may be ameliorated by the increased HPA axis and glucocorticoid response (Brown et al., 1993; Hermann et al., 1993; Sternberg, 2000). These paradoxes of stress hormone action were noted as early as 1936, when Hans Selye first introduced the concept of stress.

Stress is a ready example of how a psychological response to the environment manifests itself in physiological changes. However, all emotional feelings and the corresponding emotional expressions generated by the somato-motor system are integrated components of behaviour in humans and animals and are important for the regulation of social behaviour and survival. While it may be difficult to devise hypothesis testing experimental approaches to determine the role of these factors in disease, it should be considered that such emotions may also influence an individual's health and immune system.

STRESS AND THE SUBMANDIBULAR GLAND

The SMG may be one effector of the immunomodulatory actions of stress. Fighting is known to increase plasma NGF in mice, while other types of stress such as cold-water swim, escapable and inescapable foot shock or physical restraint, do not effect plasma levels of this growth factor. Plasma NGF levels are related to the number of fighting episodes

(Alleva *et al.*, 1986, 1993; Maestripieri *et al.*, 1990). The NGF released from the SMG of adult male mice following intraspecific fighting is biologically active and causes degranulation of peritoneal mast cells. Administration of NGF antibodies or sialadenectomy prior to fighting blocks mast cell degranulation (De Simone *et al.*, 1990). Bilateral removal of the SMG leads initially to a significant increase in aggressive behaviour during social encounters. These behavioural changes significantly decreased during subsequent social encounters while defensive behaviours and elements of arrested flight increased progressively. These results suggested that sialadenectomy, perhaps by removing salivary NGF interferes with the ability of mice to cope in stressful situations (De Simone *et al.*, 1990).

The exposure of rats to saturated ether vapour for 2 min has been used as a model of acute stress. In adult male rats secretion of mature SMR1 peptides is rapidly stimulated in response to exposure to ether fumes (2 min), while no detectable increase in release is observed following long term exposure to stressful ambient temperatures (120 min at 4 or 37°C) (Khansari *et al.*, 1990). The response to acute stress stimuli in conscious male rats results in an SMR1 increase similar to that observed under pharmacologically induced adrenergic stimulation, suggesting that the surge in circulating SMR1 is mediated by an endogenous adrenergic secretory response to acute stress. It would be of interest to examine changes in SMR1 levels induced by behaviourally relevant stress such as fighting among males. It has been shown that adrenalectomy fails to block NGF release in fighting mice and injections of ACTH or extracts of adrenal gland or hypophysis do not result in NGF release into the bloodstream (Aloe *et al.*, 1986). This suggests the HPA axis is not involved in NGF release by the SMG. However, there are likely to be mechanisms that control differential expression of biologically active peptides in response to various stimuli and the role of the HPA axis in modulating SMR1 release should be examined.

SMG endocrine secretion is mainly observed under stress conditions, that is, strong pharmacological or neural stimuli inflammatory stimuli or aggressive behaviour. This supports the possibility that SMG-derived factors might participate in reestablishment of dynamic homeostatic responses to severe physiological situations after injury trauma or infection. The principal role of the SMG may be as a component of an allostatic system rather than contributing to the regulation of resting homeostasis.

CONCLUSIONS

As described above, the immunoregulatory effects of the SMG have been observed in many species. However, the role of the human SMG in this regard is unclear. The qualitative and quantitative differences in protein content, coupled with the fact that many of the biologically active peptides can be synthesized elsewhere in the body, prompts the question; how necessary is the SMG for immune homeostasis? The possible redundancy of the glands' immunomodulatory actions is unclear, even in rodents. However, there are a number of observations that encourage further investigation of immuoregulation by the SMG in humans.

Sjögren's syndrome (SS) is an inflammatory disease of the salivary and lacrimal glands. The disease results in failure of the glands leading to xerostomia and keratoconjunctivitis sicca (St. Clair, 1993; Rhodus, 1999). While SS can occur as an isolated disease, it is often associated with systemic immune disorders such as systemic lupus erythematous, rheumatoid arthritis and sarcoidosis (Lois et al., 1998; McDonagh and Isenberg, 2000). These associations suggest that immunoregulation by salivary glands plays an important role in modulating systemic immunity. It is also interesting to note that airway-like inflammation, characterized by lymphocyte infiltration, an increase in mast cell numbers and basement membrane thickening, has been detected in the minor salivary glands of humans with bronchial asthma (Wallaert et al., 1994).

Strenuous exercise is followed by lymphopenia, neutrophilia, depressed NK cell function and lymphocyte proliferative responses to mitogens and impaired natural immunity (Nieman, 1999; Pedersen et al., 1999a,b). These exercise-induced immune changes may provide the physiological basis of altered resistance to infections. The mechanisms underlying exercise-induced immune changes are thought to be multifactorial and include neuroendocrinological and metabolic mechanisms. The contribution of the SMG to the immunological changes observed following exercise is unknown. However, salivary composition does change following strenuous exercise, salivary IgA levels are decreased, while amylase, EGF and total protein have all been demonstrated to increase (Nexo et al., 1988; McDowell et al., 1992; Mackinnon et al., 1993). Further investigation of salivary and plasma levels of other SMG derived regulatory peptides may provide evidence for involvement of the gland in the immunological changes which follow exercise induced stress.

As stated previously, the gene encoding SMR1 (VCSA1) is found only in rats, while human salivary glands express genes from the VCSB subfamily. However it is possibly that these, or as yet undiscovered, genes encode a human homologue of the SMR1. The observation that human neutrophils respond to, and thus may express receptors for, the tri-peptide FEG lends support to this hypothesis. The use of degenerative primers for SMR1 mRNA to screen for human homologues and the development of sensitive assay systems for FEG or SGP-T are two approaches which may lead to the discovery of novel anti-inflammatory peptides in man.

From the information presented above it is clear the SMG constitutes a fully integrated component of the body's neuroimmunoendocrine network. The gland is under control of the CNS and regulated by the hypothalamus and sympathetic nervous system. The extent of the role played by the SMG in controlling systemic immune function remains to be seen. However, investigation of the SMG as a stress response organ rather than a homeostatic endocrine gland may provide valuable insights into mechanisms underlying psychological and neuroendocrine control of the immune system.

REFERENCES

Acres, R.B., Lamb, J.R. and Feldman, M. (1985) Effects of platelet-derived growth factor and epidermal growth factor on antigen-induced proliferation of human T-cell lines. Immunology 54, 9–16.

Adams, D.H., Hathaway, M., Shaw, J., Burnett, D., Elias, E. and Strain, A.J. (1991) Transforming growth factor-beta induces human T lymphocyte migration in vitro. J. Immunol. 147, 609–612.

Alito, A.E., Romeo, H.E., Bajer, R., Chuluyan, H.E., Braun, M. and Cardinali, D.P. (1987) Autonomic nervous system regulation of murine immune responses as assessed by local surgical sympathetic and parasympathetic denervation. Acta Physiol. Pharmacol. Latinoam. 37, 305–319.

Alleva, E., Aloe, L. and Bigi, S. (1993) An updated role for nerve growth factor in neurobehavioural regulation of adult vertebrates. Rev. Neurosci. 4, 41–62.

Aloe, L., Alleva, E., Bohm, A. and Levi-Montalcini, R. (1986) Aggressive behavior induces release of nerve growth factor from mouse salivary gland into the bloodstream. *Proc. Natl. Acad. Sci. USA* **83**, 6184–6187.

Aloe, L. and Levi-Montalcini, R. (1977) Mast cells increase in tissues of neonatal rats injected with the nerve growth factor. *Brain Res.* **133**, 358–366.

Aloe, L. and Levi-Montalcini, R. (1980a) Comparative studies on testosterone and L-thyroxine effects on the synthesis of nerve growth factor in mouse submaxillary salivary glands. *Exp. Cell Res.* **125**, 15–22.

Aloe, L. and Levi-Montalcini, R. (1980b) Enhanced differentiation of sexually dimorphic organs in L-thyroxine treated Tfm mice. *Cell Tissue Res.* **205**, 19–29.

Amico-Roxas, M., Caruso, A., Leone, M.G., Scifo, R., Vanella, A. and Scapagnini, U. (1989) Nerve growth factor inhibits some acute experimental inflammations. *Arch. Int. Pharmacodyn. Ther.* **299**, 269–285.

Aune, T.M. (1985) Inhibition of soluble immune response suppressor activity by growth factors. *Proc. Natl. Acad. Sci. USA* **82**, 6260–6264.

Barka, T., Gresik, E.W. and Van der Noen, H. (1978) Stimulation of secretion of epidermal growth factor and amylase by cyclocytidine. *Cell Tissue Res.* **186**, 269–278.

Barka, T. (1980) Biologically active polypeptides in submandibular glands. *J. Histochem. Cytochem.* **28**, 836–859.

Berg, T., Wassdal, I. and Sletten, K. (1992) Immunohistochemical localization of rat submandibular gland esterase B (homologous to the RSKG-7 kallikrein gene) in relation to other serine proteases of the kallikrein family. *J. Histochem. Cytochem.* **40**, 83–92.

Bhoola, K.D., Figueroa, C.D. and Worthy, K. (1992) Bioregulation of kinins: kallikreins, kininogens, and kininases. *Pharmacol. Rev.* **44**, 1–80.

Bischoff, S.C. and Dahinden, C.A. (1992) Effect of nerve growth factor on the release of inflammatory mediators by mature human basophils. *Blood* **79**, 2662–2669.

Borysenko, J.Z. (1982) Behavioral–physiological factors in the development and management of cancer. *Gen. Hosp. Psychiatry* **4**, 69–74.

Boucher, R., Demassieux, S., Garcia, R. and Genest, J. (1977) Tonin, angiotensin II system. A review. *Circ. Res.* **41**, 26–29.

Boyer, R., Escola, R., Bluet-Pajot, M.T. and Arancibia, S. (1990) Ablation of submandibular salivary glands in rats provokes a decrease in plasma luteinizing hormone levels correlated with morphological changes in Leydig cells. *Arch. Oral Biol.* **35**, 661–666.

Boyle, M.D., Lawman, M.J., Gee, A.P. and Young, M. (1985) Nerve growth factor: a chemotactic factor for polymorphonuclear leukocytes in vivo. *J. Immunol.* **134**, 564–568.

Brandes, M.E., Mai, U.E., Ohura, K. and Wahl, S.M. (1991) Type I transforming growth factor-beta receptors on neutrophils mediate chemotaxis to transforming growth factor-beta. *J. Immunol.* **147**, 1600–1606.

Brodie, C. and Gelfand, E.W. (1992) Functional nerve growth factor receptors on human B lymphocytes. Interaction with IL-2. *J. Immunol.* **148**, 3492–3497.

Brown, D.H., Sheridan, J., Pearl, D. and Zwilling, B.S. (1993) Regulation of mycobacterial growth by the hypothalamus–pituitary–adrenal axis: differential responses of *Mycobacterium bovis* BCG-resistant and -susceptible mice. *Infect. Immun.* **61**, 4793–4800.

Buske-Kirschbaum, A., Jobst, S., Psych, D., Wustmans, A., Kirschbaum, C., Rauh, W. and Hellhammer, D. (1997) Attenuated free cortisol response to psychosocial stress in children with atopic dermatitis. *Psychosom. Med.* **59**, 419–426.

Byyny, R.I., Orth, D.N., Cohen, S. and Doyne, E.S. (1974) Epidermal growth factor: effects of androgens and adrenergic agents. *Endocrinology* **95**, 776–782.

Cabello, G. and Wrutniak, C. (1989) Thyroid hormone and growth: relationships with growth hormone effects and regulation. *Reprod. Nutr. Dev.* **29**, 387–402.

Cardinali, D.P. and Romeo, H.E. (1990) Peripheral neuroendocrine interrelationships in the cervical region. *NIPS* **5**, 100–104.

Carter, L., Ferrari, J.K., Davison, J.S. and Befus, D. (1992) Inhibition of neutrophil chemotaxis and activation following decentralization of the superior cervical ganglia. *J. Leukoc. Biol.* **51**, 597–602.

Cash, J.M., Crofford, L.J., Gallucci, W.T., Sternberg, E.M., Gold, P.W., Chrousos, G.P. and Wilder, R.L. (1992) Pituitary–adrenal axis responsiveness to ovine corticotropin releasing hormone in patients with rheumatoid arthritis treated with low dose prednisone. *J. Rheumatol.* **19**, 1692–1696.

Chaparro, O., Yu, W.H. and Shaw, P.A. (1998) Effect of sympathetic innervation on isoproterenol-induced cystatin S gene expression in rat submandibular glands during early development. *Dev. Neurosci.* **20**, 65–73.

Clements, J.A., Matheson, B.A. and Funder, J.W. (1990) Tissue-specific developmental expression of the kallikrein gene family in the rat. *J. Biol. Chem.* **265**, 1077–1081.

Cohen, S. and Carpenter, G. (1975) Human epidermal growth factor: isolation and chemical and biological properties. *Proc. Natl. Acad. Sci. USA* **72**, 1317–1321.

Cook, D.I., Van Lennep, E.W., Roberts, M.L. and Young, J.A. (1994) Secretion by the major salivary glands. In: Johnson, L.R. (Ed.) *Physiology of the Gastrointestinal Tract.*, pp. 1061–1117. New York: Raven Press.

Courty, Y., Singer, M., Rosinski-Chupin, I. and Rougeon, F. (1996) Episodic evolution and rapid divergence of members of the rat multigene family encoding the salivary prohormone-like protein SMR1. *Mol. Biol. Evol.* **13**, 758–766.

Crofford, L.J., Pillemer, S.R., Kalogeras, K.T., Cash, J.M., Michelson, D., Kling, M.A., Sternberg, E.M., Gold, P.W., Chrousos, G.P. and Wilder, R.L. (1994) Hypothalamic–pituitary–adrenal axis perturbations in patients with fibromyalgia. *Arthritis Rheum.* **37**, 1583–1592.

Dallegri, F., Ballestrero, A., Ottonello, L. and Patrone, F. (1989) Platelets as inhibitory cells in neutrophil-mediated cytolysis. *J. Lab. Clin. Med.* **114**, 502–509.

De Simone, R., Alleva, E., Tirassa, P. and Aloe, L. (1990) Nerve growth factor released into the bloodstream following intraspecific fighting induces mast cell degranulation in adult male mice. *Brain Behav. Immun.* **4**, 74–81.

Demitrack, M.A., Dale, J.K., Straus, S.E., Laue, L., Listwak, S.J., Kruesi, M.J., Chrousos, G.P. and Gold, P.W. (1991) Evidence for impaired activation of the hypothalamic–pituitary–adrenal axis in patients with chronic fatigue syndrome. *J. Clin. Endocrinol. Metab.* **73**, 1224–1234.

Dhabhar, F.S., Miller, A.H., McEwen, B.S. and Spencer, R.L. (1995) Effects of stress on immune cell distribution. Dynamics and hormonal mechanisms. *J. Immunol.* **154**, 5511–5527.

Dhabhar, F.S. (1998) Stress-induced enhancement of cell-mediated immunity. *Ann. N. Y. Acad. Sci.* **840**, 359–372.

Dhabhar, F.S., Satoskar, A.R., Bluethmann, H., David, J.R. and McEwen, B.S. (2000) Stress-induced enhancement of skin immune function: a role for gamma interferon. *Proc. Natl. Acad. Sci. USA* **97**, 2846–2851.

Dolce, C., Anguita, J., Brinkley, L., Karnam, P., Humphreys-Beher, M., Nakagawa, Y., Keeling, S. and King, G. (1994) Effects of sialoadenectomy and exogenous EGF on molar drift and orthodontic tooth movement in rats. *Am. J. Physiol.* **266**, E731–E738.

Emmelin, N., Schneyer, C.A. and Schneyer, L.H. (1976) The pharmacology of salivary secretion. In: Holton, P. (Ed.) *Pharmacology of Gastrointestinal Motility and Secretion*, pp. 1–39. Oxford: Pergamon Press.

Farhadi, H., Pareek, S., Day, R., Dong, W., Chretien, M., Bergeron, J.J., Seidah, N.G. and Murphy, R.A. (1997) Prohormone convertases in mouse submandibular gland: co-localization of furin and nerve growth factor. *J. Histochem. Cytochem.* **45**, 795–804.

Forsythe, P., Curry, W.J., Johnston, C.F., Harriott, P., MacMahon, J. and Ennis, M. (1997) The modulatory effects of WE-14 on histamine release from rat peritoneal mast cells. *Inflamm. Res.* **46** (Suppl 1), S13–S14.

Fujieda, M., Murata, Y., Hayashi, H., Kambe, F., Matsui, N. and Seo, H. (1993) Effect of thyroid hormone on epidermal growth factor gene expression in mouse submandibular gland. *Endocrinology* **132**, 121–125.

Garrett, J.R., Suleiman, A.M., Anderson, L.C. and Proctor, G.B. (1991) Secretory responses in granular ducts and acini of submandibular glands in vivo to parasympathetic or sympathetic nerve stimulation in rats. *Cell Tissue Res.* **264**, 117–126.

Gauthier, F., Moreau, T., Gutman, N., el Moujahed, A. and Brillard-Bourdet, M. (1992) Functional diversity of proteinases encoded by genes of the rat tissue kallikrein family. *Agents Actions Suppl.* **38**, 42–50.

Gee, A.P., Boyle, M.D., Munger, K.L., Lawman, M.J. and Young, M. (1983) Nerve growth factor: stimulation of polymorphonuclear leukocyte chemotaxis in vitro. *Proc. Natl. Acad. Sci. USA* **80**, 7215–7218.

Gill, G.N., Bertics, P.J. and Santon, J.B. (1987) Epidermal growth factor and its receptor. *Mol. Cell. Endocrinol.* **51**, 169–186.

Grau, M., Rodriguez, C., Soley, M. and Ramirez, I. (1994) Relationship between epidermal growth factor in mouse submandibular glands, plasma, and bile: effects of catecholamines and fasting. *Endocrinology* **135**, 1854–1862.

Gresik, E.W. (1980) Postnatal developmental changes in submandibular glands of rats and mice. *J. Histochem. Cytochem.* **28**, 860–870.

Gresik, E.W. (1994) The granular convoluted tubule (GCT) cell of rodent submandibular glands. *Microsc. Res. Tech.* **27**, 1–24.

Gresik, E.W., Chung, K.W., Barka, T. and Schenkein, I. (1980) Immunocytochemical localization of nerve growth factor, submandibular glands of Tfm/Y mice. *Am. J. Anat.* **158**, 247–250.

Grisham, M.B., Everse, J. and Janssen, H.F. (1988) Endotoxemia and neutrophil activation in vivo. *Am. J. Physiol.* **254**, H1017–H1022.

Hermann, G., Tovar, C.A., Beck, F.M., Allen, C. and Sheridan, J.F. (1993) Restraint stress differentially affects the pathogenesis of an experimental influenza viral infection in three inbred strains of mice. *J. Neuroimmunol.* **47**, 83–94.

Hiramatsu, M., Kashimata, M., Takayama, F. and Minami, N. (1994) Developmental changes in and hormonal modulation of epidermal growth factor concentration in the rat submandibular gland. *J. Endocrinol.* **140**, 357–363.

Hu, Z.Q., Murakami, K., Ikigai, H. and Shimamura, T. (1992) Enhancement of lymphocyte proliferation by mouse glandular kallikrein. *Immunol. Lett.* **32**, 85–89.

Isenman, L., Liebow, C. and Rothman, S. (1999) The endocrine secretion of mammalian digestive enzymes by exocrine glands. *Am. J. Physiol.* **276**, E223–E232.

Ishii, D.N. and Shooter, E.M. (1975) Regulation of nerve growth factor synthesis in mouse submaxillary glands by testosterone. *J. Neurochem.* **25**, 843–851.

Jankowski, M., Petrone, C., Tremblay, J. and Gutkowska, J. (1996) Natriuretic peptide system in the rat submaxillary gland. *Regul. Pept.* **62**, 53–61.

Johnson, H.M. and Torres, B.A. (1985) Peptide growth factors PDGF, EGF, and FGF regulate interferon-gamma production. *J. Immunol.* **134**, 2824–2826.

Jones, D.E., Jr, Tran-Patterson, R., Cui, D.M., Davin, D., Estell, K.P. and Miller, D.M. (1995) Epidermal growth factor secreted from the salivary gland is necessary for liver regeneration. *Am. J. Physiol.* **268**, G872–G878.

Kannan, Y., Ushio, H., Koyama, H., Okada, M., Oikawa, M., Yoshihara, T., Kaneko, M. and Matsuda, H. (1991) 2.5S nerve growth factor enhances survival, phagocytosis, and superoxide production of murine neutrophils. *Blood* **77**, 1320–1325.

Kanno, T., Asada, N., Yanase, H., Iwanaga, T., Ozaki, T., Nishikawa, Y., Iguchi, K., Mochizuki, T., Hoshino, M. and Yanaihara, N. (1999) Salivary secretion of highly concentrated chromogranin a in response to noradrenaline and acetylcholine in isolated and perfused rat submandibular glands. *Exp. Physiol.* **84**, 1073–1083.

Kehrl, J.H., Thevenin, C., Rieckmann, P. and Fauci, A.S. (1991) Transforming growth factor-beta suppresses human B lymphocyte Ig production by inhibiting synthesis and the switch from the membrane form to the secreted form of Ig mRNA. *J. Immunol.* **146**, 4016–4023.

Khansari, D.N., Murgo, A.J. and Faith, R.E. (1990) Effects of stress on the immune system [see comments]. *Immunol. Today* **11**, 170–175.

Kimata, H., Yoshida, A., Ishioka, C., Kusunoki, T., Hosoi, S. and Mikawa, H. (1991) Nerve growth factor specifically induces human IgG4 production. *Eur. J. Immunol.* **21**, 137–141.

Kong, C., Gill, B.M., Rahimpour, R., Xu, L., Feldman, R.D., Xiao, Q., McDonald, T.J., Taupenot, L., Mahata, S.K., Singh, B., O'Connor, D.T. and Kelvin, D.J. (1998) Secretoneurin and chemoattractant receptor interactions. *J. Neuroimmunol.* **88**, 91–98.

Konturek, P.C., Konturek, S.J., Brzozowski, T. and Ernst, H. (1995) Epidermal growth factor and transforming growth factor-alpha: role in protection and healing of gastric mucosal lesions. *Eur. J. Gastroenterol. Hepatol.* **7**, 933–937.

Kort, W.J. (1994) The effect of chronic stress on the immune response. *Adv. Neuroimmunol.* **4**, 1–11.

Kraemer, R. and Mullane, K.M. (1989) Neutrophils delay functional recovery of the post-hypoxic heart of the rabbit. *J. Pharmacol. Exp. Ther.* **251**, 620–626.

Kumegawa, M., Takuma, T. and Takagi, Y. (1977) Precocious induction of secretory granules by hormones in convoluted tubules of mouse submandibular glands. *Am. J. Anat.* **149**, 111–114.

Kyakumoto, S., Kurokawa, R. and Ota, M. (1985) Effect of castration and administration of testosterone on cytosol and nuclear androgen receptor in mouse submandibular gland. *Biochem. Int.* **11**, 701–707.

Kyakumoto, S., Kurokawa, R., Ohara-Nemoto, Y. and Ota, M. (1986) Sex difference in the cytosolic and nuclear distribution of androgen receptor in mouse submandibular gland. *J. Endocrinol.* **108**, 267–273.

Laskin, D.L., Laskin, J.D., Weinstein, I.B. and Carchman, R.A. (1980) Modulation of phagocytosis by tumor promoters and epidermal growth factor in normal and transformed macrophages. *Cancer Res.* **40**, 1028–1035.

Laskin, D.L., Laskin, J.D., Weinstein, I.B. and Carchman, R.A. (1981) Induction of chemotaxis in mouse peritoneal macrophages by phorbol ester tumor promoters. *Cancer Res.* **41**, 1923–1928.

Letic-Gavrilovic, A., Shibaike, S., Niina, M., Naruse, S. and Abe, K. (1989) Localization of chromogranin A and B, beta-endorphin and enkephalins in the submandibular glands of mice. *Shika. Kiso. Igakkai. Zasshi.* **31**, 453–462.

Levi-Montalcini, R., Skaper, S.D., Dal Toso, R., Petrelli, L. and Leon, A. (1996) Nerve growth factor: from neurotrophin to neurokine. *Trends Neurosci.* **19**, 514–520.

Levi-Montalcini, R. (1998) The saga of the nerve growth factor. *Neuroreport* **9**, R71–R83.

Lois, M., Roman, J., Holland, W. and Agudelo, C. (1998) Coexisting Sjogren's syndrome and sarcoidosis in the lung. *Semin. Arthritis Rheum.* **28**, 31–40.

Lugardon, K., Raffner, R., Goumon, Y., Corti, A., Delmas, A., Bulet, P., Aunis, D. and Metz-Boutigue, M.H. (2000) Antibacterial and antifungal activities of vasostatin-1, the N-terminal fragment of chromogranin A. *J. Biol. Chem.* **275**, 10745–10753.

Luger, A., Calogero, A.E., Kalogeras, K., Gallucci, W.T., Gold, P.W., Loriaux, D.L. and Chrousos, G.P. (1988) Interaction of epidermal growth factor with the hypothalamic–pituitary–adrenal axis: potential physiologic relevance. *J. Clin. Endocrinol. Metab.* **66**, 334–337.

Lundberg, J.M., Martling, C.R. and Hokfelt, T. (1988) Airways, oral cavity and salivary glands: classical trans-
    mitters and peptides in sensory and autonomic motor neurons. In: Bjorklund, A., Hokfelt, T. and Owman, C.
    (Eds) *The Peripheral Nervous System*, pp. 391–444. Amsterdam: Elsevier Science Publishers.
McEwen, B.S. (1998) Protective and damaging effects of stress mediators. *N. Engl. J. Med.*, **338**, 171–179.
Mackinnon, L.T., Ginn, E. and Seymour, G.J. (1993) Decreased salivary immunoglobulin A secretion rate after
    intense interval exercise in elite kayakers. *Eur. J. Appl. Physiol.* **67**, 180–184.
Maestripieri, D., De Simone, R., Aloe, L. and Alleva, E. (1990) Social status and nerve growth factor serum
    levels after agonistic encounters in mice. *Physiol. Behav.* **47**, 161–164.
Marshall, J.S., Stead, R.H., McSharry, C., Nielsen, L. and Bienenstock, J. (1990) The role of mast cell degranula-
    tion products in mast cell hyperplasia. I. Mechanism of action of nerve growth factor. *J. Immunol.* **144**,
    1886–1892.
Martinez, J.R., Quissell, D.O., Wood, D.L. and Giles, M. (1975) Abnormal secretory response to parasympath-
    omimetic and sympathomimetic stimulations from the submaxillary gland of rats treated with reserpine.
    J. Pharmacol. Exp. Ther. **194**, 384–395.
Massague, J., Cheifetz, S., Laiho, M., Ralph, D.A., Weis, F.M. and Zentella, A. (1992) Transforming growth
    factor-beta. *Cancer Surv.* **12**, 81–103.
Mathison, R. (1998) Submandibular gland peptides and the modulation of anaphylactic and endotoxic reactions.
    Biomed. Rev. **9**, 101–109.
Mathison, R., Bissonnette, E., Carter, L., Davison, J.S. and Befus, D. (1992a) The cervical sympathetic
    trunk–submandibular gland axis modulates neutrophil and mast cell functions. *Int. Arch. Allergy Immunol.*
    99, 419–421.
Mathison, R., Hogan, A., Helmer, D., Bauce, L., Woolner, J., Davison, J.S., Schultz, G. and Befus, D. (1992b)
    Role for the submandibular gland in modulating pulmonary inflammation following induction of systemic
    anaphylaxis. *Brain Behav. Immun.* **6**, 117–129.
Mathison, R., Befus, D. and Davison, J.S. (1993) Removal of the submandibular glands increases the acute
    hypotensive response to endotoxin. *Circ. Shock* **39**, 52–58.
Mathison, R., Carter, L., Mowat, C., Bissonnette, E., Davison, J.S. and Befus, A.D. (1994a) Temporal analysis of
    the anti-inflammatory effects of decentralization of the rat superior cervical ganglia. *Am. J. Physiol.* **266**,
    R1537–R1543.
Mathison, R., Davison, J.S. and Befus, A.D. (1994b) Neuroendocrine regulation of inflammation and tissue repair
    by submandibular gland factors. *Immunol. Today* **15**, 527–532.
Mathison, R., Tan, D., Oliver, M., Befus, D., Scott, B. and Davison, J.S. (1997) Submandibular gland peptide-T
    (SGP-T) inhibits intestinal anaphylaxis. *Dig. Dis. Sci.* **42**, 2378–2383.
Mathison, R., Lo, P., Moore, G., Scott, B. and Davison, J.S. (1998) Attenuation of intestinal and
    cardiovascular anaphylaxis by the salivary gland tripeptide FEG and its D-isomeric analog feG. *Peptides* **19**,
    1037–1042.
Mathison, R.D., Befus, A.D. and Davison, J.S. (1997) A novel submandibular gland peptide protects against endo-
    toxic and anaphylactic shock. *Am. J. Physiol.* **273**, R1017–R1023.
Matsumoto, H., Suzuki, N., Onda, H. and Fujino, M. (1989) Abundance of endothelin-3 in rat intestine, pituitary
    gland and brain. *Biochem. Biophys. Res. Commun.* **164**, 74–80.
Matsushita, S. and Katz, D.H. (1993) Biphasic effect of kallikrein on IgE and IgG1 syntheses by LPS/IL-4-
    stimulated B cells. *Cell. Immunol.* **146**, 210–214.
McDonagh, J.E. and Isenberg, D.A. (2000) Development of additional autoimmune diseases in a population of
    patients with systemic lupus erythematosus. *Ann. Rheum. Dis.* **59**, 230–232.
McDowell, S.L., Hughes, R.A., Hughes, R.J., Housh, T.J. and Johnson, G.O. (1992) The effect of exercise train-
    ing on salivary immunoglobulin A and cortisol responses to maximal exercise. *Int. J. Sports Med.* **13**,
    577–580.
Nagy, E., Berczi, I. and Sabbadini, E. (1992) Endocrine control of the immunosuppressive activity of the sub-
    mandibular gland. *Brain Behav. Immun.* **6**, 418–428.
Nexo, E., Hansen, M.R. and Konradsen, L. (1988) Human salivary epidermal growth factor, haptocorrin and
    amylase before and after prolonged exercise. *Scand. J. Clin. Lab. Invest.* **48**, 269–273.
Nieman, D.C. (1999) Nutrition, exercise, and immune system function. *Clin. Sports Med.* **18**, 537–548.
Noguchi, S., Ohba, Y. and Oka, T. (1991a) Influence of epidermal growth factor on liver regeneration after partial
    hepatectomy in mice. *J. Endocrinol.* **128**, 425–431.
Noguchi, S., Ohba, Y. and Oka, T. (1991b) Effect of salivary epidermal growth factor on wound healing of tongue
    in mice. *Am. J. Physiol.* **260**, E620–E6255.
Nordlund, L., Hormia, M., Saxen, L. and Thesleff, I. (1991) Immunohistochemical localization of epidermal
    growth factor receptors in human gingival epithelia. *J. Periodontal. Res.* **26**, 333–338.
Orstavik, T.B., Carretero, O.A. and Scicli, A.G. (1982) Kallikrein-kinin system in regulation of submandibular
    gland blood flow. *Am. J. Physiol.* **242**, H1010–H1014.

Otten, U., Baumann, J.B. and Girard, J. (1979) Stimulation of the pituitary–adrenocortical axis by nerve growth factor. *Nature* **291**, 358–366.

Otten, U., Ehrhard, P. and Peck, R. (1989) Nerve growth factor induces growth and differentiation of human B lymphocytes. *Proc. Natl. Acad. Sci. USA* **86**, 10059–10063.

Partridge, M., Gullick, W.J., Langdon, J.D. and Sherriff, M. (1988) Expression of epidermal growth factor receptor on oral squamous cell carcinoma. *Br. J. Oral Maxillofac. Surg.* **26**, 381–389.

Pearce, F.L. and Thompson, H.L. (1986) Some characteristics of histamine secretion from rat peritoneal mast cells stimulated with nerve growth factor. *J. Physiol. (Lond)* **372**, 379–393.

Pedersen, B.K., Bruunsgaard, H., Jensen, M., Krzywkowski, K. and Ostrowski, K. (1999a) Exercise and immune function: effect of ageing and nutrition. *Proc. Nutr. Soc.* **58**, 733–742.

Pedersen, B.K., Bruunsgaard, H., Jensen, M., Toft, A.D., Hansen, H. and Ostrowski, K. (1999b) Exercise and the immune system–influence of nutrition and ageing. *J. Sci. Med. Sport* **2**, 234–252.

Penschow, J.D. and Coghlan, J.P. (1993) Secretion of glandular kallikrein and renin from the basolateral pole of mouse submandibular duct cells: an immunocytochemical study. *J. Histochem. Cytochem.* **41**, 95–103.

Purushotham, K.R., Offenmuller, K., Bui, A.T., Zelles, T., Blazsek, J., Schultz, G.S. and Humphreys-Beher, M.G. (1995) Absorption of epidermal growth factor occurs through the gastrointestinal tract and oral cavity in adult rats. *Am. J. Physiol.* **269**, G867–G873.

Raber, J., Koob, G.F. and Bloom, F.E. (1997) Interferon-alpha and transforming growth factor-beta 1 regulate corticotropin-releasing factor release from the amygdala: comparison with the hypothalamic response. *Neurochem. Int.* **30**, 455–463.

Ramaswamy, K., Mathison, R., Carter, L., Kirk, D., Green, F., Davison, J.S. and Befus, D. (1990) Marked anti-inflammatory effects of decentralization of the superior cervical ganglia. *J. Exp. Med.* **172**, 1819–1830.

Reinisch, N., Kirchmair, R., Kahler, C.M., Hogue-Angeletti, R., Fischer-Colbrie, R., Winkler, H. and Wiedermann, C.J. (1993) Attraction of human monocytes by the neuropeptide secretoneurin. *FEBS Lett.* **334**, 41–44.

Rhodus, N.L. (1999) Sjogren's syndrome. *Quintessence Int.* **30**, 689–699.

Roberts, M.L., Freston, J.A. and Reade, P.C. (1976) Suppression of immune responsiveness by a submandibular salivary gland factor. *Immunology* **30**, 811–814.

Rosinski-Chupin, I., Tronik, D. and Rougeon, F. (1988) High level of accumulation of a mRNA coding for a precursor-like protein in the submaxillary gland of male rats. *Proc. Natl. Acad. Sci. USA* **85**, 8553–8557.

Rosinski-Chupin, I., Rougeot, C., Courty, Y. and Rougeon, F. (1993) Localization of mRNAs of two androgen-dependent proteins, SMR1 and SMR2, by in situ hybridization reveals sexual differences in acinar cells of rat submandibular gland. *J. Histochem. Cytochem.* **41**, 1645–1649.

Rosinski-Chupin, I., Kuramoto, T., Courty, Y., Rougeon, F. and Serikawa, T. (1995) Assignment of the rat variable coding sequence (VCS) gene family to chromosome 14. *Mamm. Genome* **6**, 153–154.

Rosinski-Chupin, I. and Rougeon, F. (1990) The gene encoding SMR1, a precursor-like polypeptide of the male rat submaxillary gland, has the same organization as the preprothyrotropin-releasing hormone gene. *DNA Cell Biol.* **9**, 553–559.

Rougeot, C., Vienet, R., Cardona, A., Le Doledec, L., Grognet, J.M. and Rougeon, F. (1997) Targets for SMR1-pentapeptide suggest a link between the circulating peptide and mineral transport. *Am. J. Physiol.* **273**, R1309–R1320.

Rougeot, C., Rosinski-Chupin, I. and Rougeon, F. (1998) Novel genes and hormones in salivary glands: from the gene for the submandibular rat 1 protein (SMR1) precursor to receptor sites for SMR1 mature peptides. *Biomedi. Rev.* **9**, 17–32.

Sabbadini, E. and Berczi, I. (1998) Immunoregulation by the salivary glands. *Biomed. Rev.* **9**, 79–91.

Sagulin, G.B. and Roomans, G.M. (1989) Effects of thyroxine and dexamethasone on rat submandibular glands. *J. Dent. Res.* **68**, 1247–1251.

Salgame, P., Abrams, J.S., Clayberger, C., Goldstein, H., Convit, J., Modlin, R.L. and Bloom, B.R. (1991) Differing lymphokine profiles of functional subsets of human CD4 and CD8 T cell clones. *Science* **254**, 279–282.

Shaw, P.A., Barka, T., Woodin, A., Schacter, B.S. and Cox, J.L. (1990) Expression and induction by beta-adrenergic agonists of the cystatin S gene in submandibular glands of developing rats. *Biochem. J.* **265**, 115–120.

Shiba, R., Sakurai, T., Yamada, G., Morimoto, H., Saito, A., Masaki, T. and Goto, K. (1992) Cloning and expression of rat preproendothelin-3 cDNA. *Biochem. Biophys. Res. Commun.* **186**, 588–594.

St. Clair, E.W. (1993) New developments in Sjogren's syndrome. *Curr. Opin. Rheumatol.* **5**, 604–612.

Sterling, P. and Eyer, J. (1988) Allostasis: a new paradigm to explain arousal pathology. In: Fisher, S. and Reason, J. (Eds) *Handbook of Life Stress, Cognition and Health*, pp. 629–649. New York: John Wiley & Sons.

Sternberg, E.M. (2000) Interactions between the immune and neuroendocrine systems. In: Mayer, E.A. and Saper, C.B. (Eds) *The Biological Basis for Mind Body Interactions*, pp. 35–59. Amsterdam: Elsevier Science.

Tsukitani, K. and Mori, M. (1986) Immunohistochemistry and radioimmunassay of EGF in submandibular glands of mice treated with secretagogues. *Cell. Mol. Biol.* **32**, 677–683.

Tsunawaki, S., Sporn, M., Ding, A. and Nathan, C. (1988) Deactivation of macrophages by transforming growth factor-beta. *Nature* **334**, 260–262.

Tsutsumi, O., Kurachi, H. and Oka, T. (1986) A physiological role of epidermal growth factor in male reproductive function. *Science* **233**, 975–977.

Tsutsumi, O., Taketani, Y. and Oka, T. (1993) The uterine growth-promoting action of epidermal growth factor and its function in the fertility of mice. *J. Endocrinol.* **138**, 437–444.

van Leeuwen, B.H., Grinblat, S.M. and Johnston, C.I. (1984) Tissue-specific control of glandular kallikreins. *Am. J. Physiol.* **247**, F760–F764.

van Leeuwen, B.H., Penschow, J.D., Coghlan, J.P. and Richards, R.I. (1987) Cellular basis for the differential response of mouse kallikrein genes to hormonal induction. *EMBO J.* **6**, 1705–1713.

Verhoeven, G. (1979) Androgen binding proteins in mouse submandibular gland. *J. Steroid Biochem.* **10**, 129–138.

Waddell, S.C., Davison, J.S., Befus, A.D. and Mathison, R.D. (1992) Role for the cervical sympathetic trunk in regulating anaphylactic and endotoxic shock [published erratum appears in *J. Manipulative Physiol. Ther.* 1992 Mar–Apr, 15(3) following Table of Contents]. *J. Manipulative. Physiol. Ther.* **15**, 10–15.

Wagner, D., Metzger, R., Paul, M., Ludwig, G., Suzuki, F., Takahashi, S., Murakami, K. and Ganten, D. (1990) Androgen dependence and tissue specificity of renin messenger RNA expression in mice. *J. Hypertens.* **8**, 45–52.

Wahl, S.M., Hunt, D.A., Wakefield, L.M., McCartney-Francis, N., Wahl, L.M., Roberts, A.B. and Sporn, M.B. (1987) Transforming growth factor type beta induces monocyte chemotaxis and growth factor production. *Proc. Natl. Acad. Sci. USA* **84**, 5788–5792.

Walker, P., Weichsel, M.E., Jr, Hoath, S.B., Poland, R.E. and Fisher, D.A. (1981) Effect of thyroxine, testosterone, and corticosterone on nerve growth factor (NGF) and epidermal growth factor (EGF) concentrations in adult female mouse submaxillary gland: dissociation of NGF and EGF responses. *Endocrinology* **109**, 582–587.

Walker, P. (1986) Thyroxine increases submandibular gland nerve growth factor and epidermal growth factor concentrations precociously in neonatal mice: evidence for thyroid hormone-mediated growth factor synthesis. *Pediatr. Res.* **20**, 281–284.

Wallaert, B., Janin, A., Lassalle, P., Copin, M.C., Devisme, L., Gosset, P., Gosselin, B., Tonnel, A.B. (1994) Airway-like inflammation of minor salivary gland in bronchial asthma. *Am. J. Respir. Crit. Care Med.* **150**, 802–809.

Wilkinson, J.R., Crea, A.E., Clark, T.J. and Lee, T.H. (1989) Identification and characterization of a monocyte-derived neutrophil-activating factor in corticosteroid-resistant bronchial asthma. *J. Clin. Invest.* **84**, 1930–1941.

Wines, D.R., Brady, J.M., Pritchett, D.B., Roberts, J.L. and MacDonald, R.J. (1989) Organization and expression of the rat kallikrein gene family. *J. Biol. Chem.* **264**, 7653–7662.

Wingren, U., Brown, T.H., Watkins, B.M. and Larson, G.M. (1989) Delayed gastric ulcer healing after extirpation of submandibular glands is sex-dependent. *Scand. J. Gastroenterol.* **24**, 1102–1106.

Zuckerman, S.H., Evans, G.F., Snyder, Y.M. and Roeder, W.D. (1989) Endotoxin-macrophage interaction: post-translational regulation of tumor necrosis factor expression. *J. Immunol.* **143**, 1223–1227.

10 Mechanisms by which Lipid Derivatives and Proteinases Signal to Primary Sensory Neurons: Implications for Inflammation and Pain

Nigel W. Bunnett[1] and Pierangelo Geppetti[2]

[1]*Departments of Surgery and Physiology, University of California, 521 Parnassus Avenue, San Francisco, CA 94143-0660, USA*
[2]*Department of Experimental and Clinical Medicine, Section of Pharmacology, University of Ferrara, Via Fossato di Mortara 19, 44100 Ferrara, Italy*

Certain primary sensory neurons which express the neuropeptides calcitonin gene-related peptide and substance P play important roles in the central transmission of painful stimuli and in inflammation of peripheral tissues. Thermal, chemical and high-threshold mechanical stimuli generate action potentials which induce the release of neuropeptides within the central nervous system, resulting in the transmission of nociceptive signals. These stimuli can also release neuropeptides from peripheral endings of these neurons to initiate neurogenic inflammation that is characterized by arteriolar vasodilatation, extravasation of plasma proteins from gaps between endothelial cells of post-capillary venules, and adhesion of leukocytes to the vascular endothelium. An understanding of the molecular mechanisms of nociception and inflammation is important since primary sensory neurons mediate the pain and inflammation of many diseases, including arthritis, migraine, asthma, and inflammatory diseases of the gastrointestinal tract. Recent progress has been made in the identification of novel membrane proteins on primary sensory neurons that permit them to sense diverse agents and to thereby mediate the central transmission of pain and neurogenic inflammation of peripheral tissues. These membrane proteins include ion channels, G-protein coupled receptors and tyrosine kinase receptors that allow neurons to respond to diverse ligands ranging from protons to lipids and proteinases. This chapter reviews two types of receptor on sensory neurons, vanilloid receptors (VRs) and proteinase-activated receptors (PARs) which play emerging roles in pain and inflammation. VR1 is a non-selective cation channel that responds to vanilloids, such as capsaicin, heat, protons and certain lipid derivatives. PARs are G-protein coupled receptors that respond to serine proteases derived from the circulation or from pro-inflammatory cells. Although VR1 and PARs are structurally distinct receptors that detect markedly different agonists, there is convergence of function in that agonists of both receptors can release neuropeptides from sensory nerve fibres within the spinal cord and peripheral tissues to induce hyperalgesia and neurogenic inflammation. The identification and characterization of novel membrane proteins on the peripheral and central projections of primary sensory neurons has provided new insights into the regulation of the nervous system, which will have important implications for our understanding and treatment of disease.

KEY WORDS: pain; inflammation; vanilloid receptors; proteinase-activated receptors; neuropeptides.

INTRODUCTION

Primary sensory neurons with cell bodies located in the dorsal root ganglia (DRG), trigeminal and vagal (nodose and jugular) ganglia consist of an heterogeneous neuronal population. A subgroup of these neurons is characterized by the expression of neuropeptides, including calcitonin gene-related peptide (CGRP) and the tachykinins, substance P (SP) and neurokinin A (NKA). Neurons expressing neuropeptides have small, dark cell bodies with unmyelinated (C) or thinly myelinated (A-δ) fibres. These neurons play important roles in the central transmission of painful stimuli and in inflammation of peripheral tissues (Figure 10.1). Thermal, chemical and high-threshold mechanical stimulation of peripheral

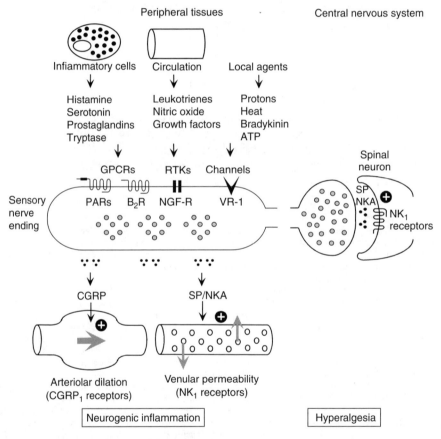

Figure 10.1   A summary of the role of primary sensory nerves in neurogenic inflammation and nociception. Agents that are released from inflammatory cells (mast cells, macrophages), derived from the circulation, or that are generated within tissues interact with a series of membrane proteins on the peripheral projections of primary sensory nerves. These proteins include GPCRs for proteinases, peptides and other mediators, receptor tyrosine kinases such as those for NGF, and ion channels such as VR1. Receptor activation triggers the release of the neuropeptides CGRP and SP from the peripheral projections of these neurons, which trigger arteriolar dilation and increase vascular permeability, the hall marks of neurogenic inflammation. Thee same stimuli may also generate action potentials, allowing central transmission and hyperalgesia.

tissues generates action potentials which are transmitted centrally where they induce the release of neuropeptides within the central nervous system, resulting in the central transmission of nociceptive signals (Otsuka and Yoshioka, 1993). These same stimuli also cause inflammation of peripheral tissues. Antidromic propagation of action potentials can release sensory neuropeptides from peripheral endings of these neurons. The local release of CGRP, SP and NKA from endings of primary sensory neurons in peripheral tissues initiates a variety of responses, collectively referred to as 'neurogenic inflammation' (Geppetti and Holzer, 1996). Neurogenic inflammation consists of a series of stereotypic responses, mainly at the vascular level, that include arteriolar vasodilatation, extravasation of plasma proteins from gaps between endothelial cells of post-capillary venules, and adhesion of leukocytes to the vascular endothelium. Additional tissues-specific components of neurogenic inflammation include various non-adrenergic non-cholinergic (NANC) motor effects, such as the atropine-resistant bronchoconstriction of the guinea pig airways, the contraction and relaxation of rodent urinary bladder, and the miotic response in rodent, rabbit and pig iris smooth muscle (Holzer, 1988; Maggi *et al.*, 1991).

An understanding of the molecular mechanisms of nociception and inflammation is important since primary sensory neurons mediate the pain and inflammation of many diseases, including arthritis, migraine, asthma and inflammatory diseases of the gastrointestinal tract. In recent years there has been considerable progress in the identification and characterization of novel membrane proteins on primary sensory neurons that permit them to sense thermal, chemical and mechanical stimuli and to thereby mediate the central transmission of pain and neurogenic inflammation of peripheral tissues. These membrane proteins include ion channels, G-protein coupled receptors (GPCRs) and tyrosine kinase receptors that allow neurons to respond to diverse ligands, ranging from protons to lipids and proteinases. In this article we will discuss two types of receptor on sensory neurons, vanilloid receptors (VRs) and proteinase-activated receptors (PARs), that allow sensory nerves to detect protons, lipids and proteinases and which play emerging roles in pain and inflammation.

VANILLOID RECEPTORS: SENSORS FOR PROTONS, HEAT AND LIPID DERIVATIVES

DISCOVERY AND MOLECULAR CLONING OF VR1

The cloning of the 'receptor' selectively activated by capsaicin and other vanilloid molecules has opened new avenues for the understanding of the sensory modalities transduced by this membrane protein (Caterina *et al.*, 1997). This receptor, termed VR1, is a non-selective cation channel that is structurally related to members of the transient receptor potential (TRP) family of ion channels (Caterina *et al.*, 1997). Two variants of the receptor have been identified, a vanilloid receptor 5'-splice variant (VR.5'sv), which differs from VR1 by elimination of the majority of the intracellular N-terminal domain and ankyrin repeat elements (Schumacher *et al.*, 2000), and an analogue of the VR1, the vanilloid receptor like 1 (VRL1) (Caterina *et al.*, 1999). Like the parent receptor, both the variants are activated by protons and vanilloid molecules.

VR1 responds to a variety of diverse stimuli, including vanilloids such as capsaicin, protons, noxious temperatures and arachidonic acid derivatives. Studies in VR1 knockout mice confirmed that VR1 is a thermoceptor that is activated by temperatures in the noxious range and specifically mediates inflammatory thermal hyperalgesia (Caterina et al., 2000; Davis et al., 2000). The expression the recombinant receptor in heterologous systems (Tominaga et al., 1998) has also allowed the definitive acceptance that VR1 is stimulated by protons (Bevan and Geppetti, 1994). Another outcome of the cloning and transfection of VR1 in heterologous expression systems has been the discovery that different arachidonic acid derivatives stimulate the channel with various degrees of potency, but with significant efficacy.

MECHANISMS OF VR1 ACTIVATION BY LIPID DERIVATIVES: ENDOGENOUS VR1 AGONISTS IN HEALTH AND DISEASE

Two almost simultaneous observations gave at least a partial answer for the lengthy search of a lipid derivative with the features of an endogenous activator of the VR1. The first report (Zygmunt et al., 1999) showed that elevated ($\sim\mu$M) concentrations of arachidoinyl-ethanolamine (anandamide), the putative endogenous ligand of cannabinoid (CB) receptors, were able to relax the rat mesenteric artery or the guinea pig basilar artery in vitro in a manner inhibited by the VR1 antagonist capsazepine (Bevan et al., 1992). The study of anandamide properties in heterologous systems expressing the recombinant VR1 indicated that this molecule, probably though an intracellular site of action, stimulates directly the VR1 (Zygmunt et al., 1999; Di Marzo et al., 2001). Further studies in DRG and trigeminal ganglion neurons showed that anandamide was able to mobilize intracellular Ca^{2+} in capsaicin-sensitive neurons via a capsazepine-dependent mechanism and that anandamide stimulated release of neuropeptides from central (dorsal spinal cord) (Tognetto et al., 2001) and peripheral (guinea pig airways) (Tognetto and Geppetti, personal observations) terminals of primary sensory neurons. In peripheral tissues, recombinant expression systems or isolated sensory neurons anandamide exerts a dual action: stimulation of an inhibitory CB1 receptor (negatively coupled to adenylyl cyclase), an effect that results in the depression of neuronal activity, and activation of VR1, which excites neurons. The factor that discriminates between these two actions is the agonist concentration: while low ($\sim$nM) concentrations of anandamide are sufficient to stimulate the CB receptors, high ($\sim\mu$M) concentrations are required to stimulate the VR1 receptor (Zygmunt et al., 1999; Di Marzo et al., 2001; Tognetto et al., 2001). Neurochemical findings that anandamide exerts a dual action on sensory nerves has been proven in a functional test where anandamide ($\sim$nM) caused a CB1-dependent inhibition of the atropine-resistant contraction induced by electrical field stimulation, as well as a direct contractile response mediated by VR1 stimulation in isolated guinea pig bronchi (Tucker et al., 2001).

Anandamide is produced and released from a series of cells, including central nervous system neurons, endothelial cells, macrophages and other cells (Devane et al., 1992; Di Marzo et al., 1994). Thus, it is possible that during inflammatory conditions macrophages or other inflammatory cells recruited in the vicinity of sensory nerve terminals release anandamide in the perineural micro-milieu in quantities high enough to stimulate the VR1.

However, it must be emphasized that whereas metabolic and pharmacokinetic factors may explain the low potency of anandamide at VR1, the elevated concentrations of anandamide required to excite sensory neurons question the physiological and pathophysiological role of this molecule as a pro-inflammatory and pain-producing agent in the diverse experimental and clinical settings (Szolcsanyi, 2000).

Another study (Hwang et al., 2000), while confirming the ability of anandamide to stimulate VR1, showed that this property is not unique to this arachidoinyl derivative. Thus, 12-(S)-hydroperoxyeicosatetraenoic acid (12-HPETE) and other related molecules are more efficacious stimulants of VR1 than anandamide in isolated membrane patches from DRG neurons (Hwang et al., 2000). Of interest for the present discussion is the finding that also the leukotriene B_4 (LTB$_4$), a major product of arachidonic acid metabolism in neutrophils, was found able to stimulate the VR1 (Hwang et al., 2000). It is well established that SP from sensory nerve endings activates the neurokinin 1 (NK1) receptor on endothelial cells of post-capillary venules to induce plasma extravasation and the adhesion of leukocytes to the vascular endothelium (Baluk et al., 1995). Leukocytes could produce LTB$_4$ in the vicinity of perivascular sensory fibres which may act on the VR1 receptor to release additional SP, thereby amplifying a sensory nerve/leukocyte pro-inflammatory loop.

INTRACELLULAR MECHANISMS REGULATING VR1 ACTIVITY AND EXPRESSION: IMPLICATIONS FOR DISEASE

Alterations in the sensitivity of VR1 to agonists or its level of expression would be expected to markedly affect the sensitivity of sensory neurons with important implications for inflammatory diseases, such as asthma and inflammatory diseases of the gastrointestinal tract. The interest in the participation of VR1 in asthma stems from epidemiological and experimental findings. Gastro-oesophageal reflux is often associated with the development of asthma. Repeated episodes of reflux of acidic digestive secretions into the airways are considered to trigger asthma attacks in these patients. During these episodes, the acidity of the breath condensate falls to pH ~ 5, and corticosteroid therapy restores a physiological pH (~7.4) (Hunt et al., 2000). Acidic media reduce ciliary beating, increase mucous viscosity and damage the tracheal mucosa (see Hunt et al., 2000). Protons may also increase bronchoconstriction via a neurogenic mechanism involving VR1 (Ricciardolo et al., 1999). This effect of protons may be due to the release of kinins, which are generated by kallikreins more easily in an acidic milieu (Ricciardolo et al., 1999). However, most of the neurogenic effects of protons in the airways, gut or urinary tract are mediated by a direct action on sensory nerve terminals via the stimulation of VR1 (Bevan and Geppetti, 1994).

Expression of VR1 is highly dependent on nerve growth factor (NGF) (Winter et al., 1988), and NGF has been found increased in the serum of patients with asthma (Bonini et al., 1996). Accumulating evidence shows that the cough threshold to capsaicin is decreased in asthmatic patients (Doherty et al., 2000). This observation supports the hypothesis that asthma is associated with 'neuronal hyper-responsiveness', a phenomenon possibly mediated by an increased sensitivity or level of expression of VR1. An upregulation of VR1 may also contribute to intestinal inflammatory diseases, since VR1 immunoreactivity is elevated in colonic nerve fibres of patients with active inflammatory bowel diseases (Yiangou et al., 2001).

It is important to understand the intracellular mechanisms that regulate the expression and sensitivity of VR1, since alterations in VR1 activity may account for the 'sensory' symptoms observed in chronic inflammatory conditions. Recent studies have focused on the interactions between tyrosine kinase receptors and GPCRs in the sensitization of VR1. Intracellular signalling pathways have been proposed to play a key role in the potentiation of the action of putative endogenous stimulants on VR1. In particular, phosphorylation of VR1 by the protein kinase Cε subtype, that results from stimulation of certain GPCRs such as the kinin B_2 receptor (Cesare et al., 1999; Premkumar and Ahern, 2000) has been recognized as a major mechanism to exaggerate the action of exogenous (capsaicin) or putative endogenous (anandamide, protons) stimulants of VR1. An additional mechanism of VR1 sensitization that involves phospholipase-γ (PLC-γ)-dependent displacement of phosphatidylinositol-4,5-bisphosphate has been proposed recently (Chuang et al., 2001). Activation of PLC-γ following GPCR (bradykinin B_2 receptor) or tyrosine kinase receptor (NGF receptor TrkA) stimulation results in the potentiation of responses to VR1 agonists. A recent study has also underlined the role of protein kinase A (PKA) in the regulation of VR1 activity (De Petrocellis et al., 2001). The ability of anandamide to activate VR1 in a heterologous expression system was greatly enhanced by pre-exposure to the PKA stimulant forskolin. Exposure to forskolin also enhanced the capsazepine-sensitive contractile response of anandamide in isolated guinea pig bronchi (De Petrocellis et al., 2001). Moreover, PKC activation by phorbol esters produced a remarkable increase in the anandamide-induced and VR-mediated contraction in isolated guinea pig bronchi (Harrison and Geppetti, personal observations).

PROTEINASE-ACTIVATED RECEPTORS: SENSORS FOR CIRCULATING AND MAST CELL PROTEINASES

PROTEINASES AS SIGNALLING MOLECULES

Although proteinases are traditionally viewed as degradatory enzymes, for instance in the gastrointestinal lumen or in lysosomes, certain proteinases can directly signal to cells by cleaving PARs, members of a new family of GPCRs (reviewed by Dery et al., 1998; Cocks and Moffatt, 2000; Coughlin, 2000; Macfarlane et al., 2001; Vergnolle et al., 2001b). Thrombin is the best example of 'signalling proteinase'. Thrombin is generated in the circulation during coagulation where it converts fibrinogen to fibrin, the basis of a blood clot. In addition, thrombin has direct receptor-mediated effects on platelets, endothelial cells, immune cells, fibroblasts, myocytes, astrocytes and neurons, which are mediated by PARs. To date, four PARs have been cloned, agonists have been identified, and mechanisms of receptor activation have been defined (Figure 10.2).

Thrombin cleaves and activates PAR1, PAR3 and PAR4 (Figure 10.2). However, the precise molecular mechanisms of activation vary between these receptors. Thrombin activates PAR1 in two stages (Vu et al., 1991). First, thrombin binds to the extracellular N-terminus of PAR1. Second, thrombin cleaves PAR1 between Arg41 and Ser42 to expose

(a) Thrombin activation of PAR1: receptor binding and cleavage

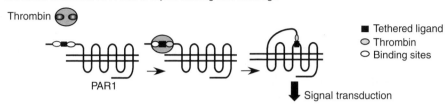

Signal transduction

(b) Thrombin activation of PAR3 and PAR4: interactions between receptors on mouse platelets

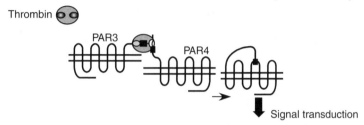

Signal transduction

(b) Trypsin, tryptase and FVIIa, FXa activation of PAR2: interaction with membrance co-factors

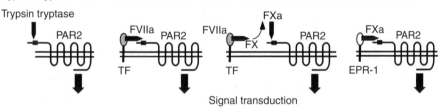

Signal transduction

Figure 10.2   Mechanisms of activation of PARs. (a) Thrombin binds to extracellular PAR1 sites and then cleaves to expose the tethered ligand, which binds and activates the cleaved receptor. (b) Mouse platelets express PAR3 and PAR4. Although murine PAR3 does not signal, it binds thrombin and localizes it close to PAR4 at the cell surface, which facilitates PAR4 cleavage and signalling. (c) Activation of PAR2. Trypsin and tryptase cleave PAR2 without binding. Factor VIIa (FVIIa) can activate PAR2 only if tethered to the cell surface by tissue factor (TF). The FVIIa and TF complex also generates factor Xa (FXa) from FX at the cell surface, which facilitates PAR2 activation. Effector cell proteinase receptor-1 (EPR-1) may also concentrate FXa at the cell surface to facilitate activation of PAR2. (Reproduced from Cottrell *et al.*, 2002.)

a new N-terminal terminus of the receptor (SFLLRN) that serves as a 'tethered ligand'. The tethered ligand interacts with domains in extracellular loop 2, which presumably alters the conformation of the receptor to permit coupling to G proteins. PAR3 also contains the thrombin binding sites (Ishihara *et al.*, 1997). The existence of the binding site permits thrombin to signal to cells expressing PAR1 or PAR3 with high potency. PAR4, which lacks thrombin binding sites, responds only to higher concentrations of thrombin (Kahn *et al.*, 1998).

Trypsin and tryptase activate PAR2 (Nystedt *et al.*, 1994; Corvera *et al.*, 1997; Molino *et al.*, 1997) (Figure 10.2). Trypsin is an enzyme that is traditionally considered to be secreted as an inactive zymogen from the pancreas and activated by enteropeptidase in the intestinal lumen, where it serves to degrade dietary proteins. Indeed, luminal trypsin can regulate enterocytes by cleaving PAR2 at the apical membrane (Kong *et al.*, 1997). Additionally, trypsinogens are prematurely activated in the inflamed pancreas, where they may also signal to acinar cells and neurons via PAR2 (Hofbauer *et al.*, 1998). However, trypsinogens are widely expressed in endothelial and epithelial cells, leukocytes and neurons, raising the possibility that extrapancreatic trypsins may signal to many cell types expressing PAR2 (Koivunen *et al.*, 1989, 1991; Koshikawa *et al.*, 1998; Alm *et al.*, 2000). The trypsin-like serine proteinase tryptase, is expressed by most human mast cells where it can comprise up to 25% of the total cellular protein (Caughey, 1995). This enzyme is released as an active, heparin-bound tetramer, which is capable of cleaving neuropeptides, procoagulant proteins and PAR2. Trypsin and tryptase do not bind to PAR2, but rather cleave directly at Arg36 and Ser37 to expose the tethered ligand SLIGKV, which then binds to extracellular loop 2 and activates the receptor. Post-translational modifications of PAR2, notably glycosylation of extracellular domains, can markedly alter the capacity of tryptase to cleave and activate this receptor (Compton *et al.*, 2000, 2001).

Synthetic peptides that correspond to the tethered ligand domains of PAR1, 2 and 4, but not PAR3, can directly activate receptors without the requirement of cleavage. These activating peptides are useful pharmacological tools that can be used to characterize receptors without the use of proteinases, which may have many other effects. Whether there are endogenous peptides capable of activating PARs remains is unknown.

ROLE OF PROTEINASES AND PARs IN HEALTH AND DISEASE

The proteolytic cleavage of PARs is an irreversible event; a proteinase cannot reactivate a cleaved receptor. Thus, PARs are disposable 'one-shot' receptors. Indeed, cleaved PARs are mostly targeted to lysosomes for degradation; any receptor that does recycle to the cell surface would no longer be responsive to a proteinase (Brass *et al.*, 1994; Böhm *et al.*, 1996). Given this irreversible and seemingly wasteful mechanism of activation it is unlikely that PARs mediate routine intercellular signaling under most circumstances. Rather, it is generally accepted that proteinases and PARs play important roles in 'emergency situations' – during coagulation or when mast cells degranulate. Proteinases can regulate many different cell types by cleaving and triggering PARs (Figure 10.3). We will focus on proteinase signalling to the nervous system.

PROTEINASE SIGNALLING TO THE NERVOUS SYSTEM

PAR1 in the nervous system

PAR1 was the first proteinase receptor identified in the nervous system and recent observations suggest an important role for thrombin and PAR1 in the nervous system under normal conditions and following injury.

PAR1 is widely expressed in the central and peripheral nervous system. PAR1 mRNA is highly expressed in the brain and DRG of the neonatal rat, although levels decline after birth (Niclou *et al.*, 1994). At postnatal day 28, PAR1 is expressed in the substantia nigra, the ventral tegmental area, the pretectal area, certain hypothalamic nuclei and certain

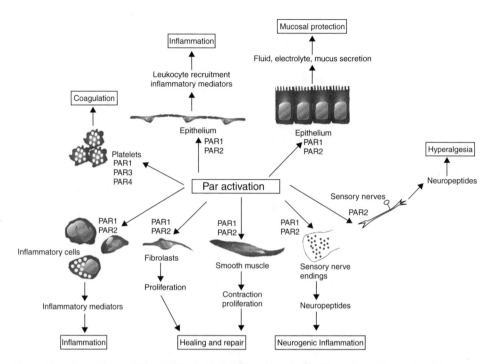

Figure 10.3 A summary of the pathophysiological affects of PAR agonists. PAR1 and PAR2 are expressed by endothelial cells, epithelial cells, neurons, myocytes, fibroblasts and inflammatory cells. Effects on these cell types trigger pathways related to inflammation, hyperalgesia, tissue repair and protection. (Reproduced from Cottrell *et al.*, 2002.)

regions of the cortex. It is localized to neurons, glial cells and ependymal cells, but the white matter and most endothelial cells do not express detectable PAR1. PAR1 mRNA and immunoreactivity have been detected in DRG neurons, some of which also express sensory neuropeptides and are thus small diameter neurons that participate in inflammation and nociception (Gill *et al.*, 1998; de Garavilla *et al.*, 2001). Another component of the nervous system that expresses PAR1 is the enteric nervous system. Thus, PAR1 is expressed in the myenteric plexus of the guinea pig small intestine and a majority of neurons in culture respond to thrombin and selective agonists of PAR1 by elevated concentrations of cytosolic Ca^{2+} (Corvera *et al.*, 1999).

An important issue is the source of the proteinase that signals to neurons through PAR1. Thrombin is generated in the circulation during coagulation. However, it is quite likely that thrombin would leak into the interstitial fluid during the edema that inevitably follows inflammation, trauma and coagulation. Alterations in the blood–brain barrier as a result of trauma could also expose neurons to circulation-derived thrombin. In these conditions, thrombin could signal to many extravascular cell types, including neurons, by cleaving PAR1. In addition, thrombin may also derive from the nervous system. Although most prothrombin is produced by the liver, prothrombin mRNA is present in the normal brain, including the olfactory bulb, the cortex and the cerebellum (Dihanich *et al.*, 1991).

A comparison of the distribution of prothrombin and PAR1 in the brain indicates that there is a distinct but in some regions overlapping localization. In many instances, prothrombin and PAR1 are expressed in similar regions, suggesting that thrombin may regulate cells expressing PAR1 in a paracrine or even autocrine manner. For example, PAR1 and prothrombin are expressed in the Pukinje cell layer of the cerebellum.

What are the consequences of activating neuronal PAR1? One of the most striking consequences of PAR1 activation is to induce morphological alterations in neurons and glial cells. PAR1 agonists cause retraction of processes by neurons and induce astrocytes to lose their stellate morphology and become flat and epithelial in shape (Gurwitz and Cunningham, 1988; Cavanaugh et al., 1990; Beecher et al., 1994). PAR1 agonists also act on astrocytes to trigger release of the vasoconstrictor endothelin-1 (Ehrenreich et al., 1993), increase release of NGF (Neveu et al., 1993), and suppress expression of the metabotropic glutamate receptor (Miller et al., 1996). PAR1 agonists are also mitogenic for astrocytes (Perraud et al., 1987). Considerable evidence suggest that PAR1 plays an important role in protecting the central nervous system from injury. Notably, thrombin and PAR1 activating peptides protect astrocytes and neurons in culture from cell death induced by hypoglycemia and oxidative stress (Vaughan et al., 1995). Experiments in vivo have confirmed these findings. Thus, the brief occlusions of the common carotid arteries of gerbils induces ischemic tolerance of hippocampal CA1 neurons which is impaired by the thrombin inhibitor hirudin (Striggow et al., 2000). PAR1 agonists also protect organotypic slice cultures of rat hippocampus from the deleterious effects of experimental ischemia (Striggow et al., 2000). Thus, thrombin may play an important role in protecting the brain during trauma, when circulating thrombin could cross the blood–brain barrier and activate PAR1. Conversely, PAR1 agonists can have deleterious effects in the nervous system. High concentrations of thrombin result in degeneration of CA1 neurons, possibly due to the sustained elevations of cytosolic Ca^{2+} (Striggow et al., 2000). PAR1 agonists can also potentiate hippocampal NMDA receptor responses in CA1 pyramidal cells, raising the possibility that potentiation of neuronal NMDA receptor function after entry of thrombin into brain may exacerbate glutamate-mediated cell death (Gingrich et al., 2000).

Less is known about the role of thrombin in DRG. Thrombin and PAR1 activating peptide suppress neurite outgrowth from cultured neurons (Gill et al., 1998). PAR1 agonists can also directly signal to these neurons by increasing cytosolic Ca^{2+} levels. The observation that PAR1 is expressed in small diameter nociceptive neurons (de Garavilla et al., 2001) that also express VR1 raises the possibility that PAR1 agonists may signal to these neurons to regulate neurogenic inflammation and nociception (Figure 10.1). In support of this hypothesis, PAR1 agonists injected into the paw of rats induces a profound and sustained edema that is markedly diminished by an antagonists of the NK1 receptor and by ablation of sensory nerves with chronic administration of VR1 agonist capsaicin (de Garavilla et al., 2001). Additionally, intravenous administration of PAR1 activating peptide induces widespread extravasation of plasma proteins in wild type but not PAR1 knockout mice, which is inhibited in certain tissues by an NK1 receptor antagonist (de Garavilla et al., 2001). Together, these findings show that thrombin activates PAR1 on sensory nerve endings to trigger the release of SP in peripheral tissues, which interacts with the NK1 receptor to cause neurogenic inflammation. PAR1 agonists also modulate pain responses. Intraplantar injection of PAR1 activating peptide into the rat increases the nociceptive threshold and withdrawal latency, leading to mechanical and thermal analgesia, whereas Intraplantar injection of thrombin causes only mechanical analgesia (Asfaha *et al.*, 2002). Thus, PAR1 agonists that can modulate nociceptive response to noxious stimuli.

PAR2 in the nervous system

Like PAR1, PAR2 is expressed in several components of the nervous system. PAR2 is expressed by neurons and astrocytes of the brain (Smith-Swintosky *et al.*, 1997). Notably, PAR2 is also expressed in DRG (Steinhoff *et al.*, 2000). A majority of these neurons that express PAR2 co-express SP and CGRP and respond to agonists of VR1, and are thus primary spinal afferent neurons that participate in neurogenic inflammation and pain. Within the enteric nervous system, PAR2 is expressed by a majority of neurons in both the myenteric and submucosal nerve plexuses (Corvera *et al.*, 1999).

The agonist of PAR2 in the nervous system remains to be unequivocally determined. One strong candidate is mast cell tryptase. It is well established that mast cells are in close proximity to the projections of primary spinal afferent neurons containing sensory neuropeptides in many peripheral tissues, including the skin, intestine and joints (Stead *et al.*, 1987). This juxtaposition of mast cells and nerve fibres is observed under normal circumstances and during inflammation and associated infiltration of mast cells. Mast cells are also in close contact with enteric nerve fibres. Thus, mast cell tryptase could signal to sensory nerves through the release of tryptase and activation of neuronal PAR2. Alternatively, other proteinases derived from neurons of astrocytes could signal to neurons through PAR2. Trypsinogens have been identified in the brain, but almost nothing is known about their mechanisms of secretion, activation and function (Koshikawa *et al.*, 1998).

There has been considerable recent interest in the capacity of proteinases to signal to primary spinal afferent neurons and regulate neurogenic inflammation and nociception (Figure 10.1). Trypsin, tryptase and PAR2 activating peptides trigger increases in the cytosolic concentration of Ca^{2+} in primary spinal afferent neurons in short-term culture, a result which suggests that these agonists can directly signal to neurons by cleaving PAR2 (Steinhoff *et al.*, 2000; Hoogerwerf *et al.*, 2001). PAR2 agonists also stimulate the release of SP and CGRP from superfused segments of the dorsal horn of the spinal cord and the urinary bladder, indicating that proteinases stimulate the release of neuropeptides from both the central and peripheral projections of these primary spinal afferent neurons (Steinhoff *et al.*, 2000). As for PAR1, the intraplantar injection of PAR2 activating peptide causes a marked and sustained edema that is substantively diminished by ablation of sensory nerves with capsaicin as well as by antagonist of the NK1 receptor and the CGRP type 1 receptor (Steinhoff *et al.*, 2000) (Figure 10.4). Together, these findings support the hypothesis that proteinases such as tryptase, that are released from mast cells in close proximity to sensory nerve fibres in the periphery, cleave PAR2 on nerve endings to induce secretion of SP and CGRP. SP can then interact with the NK1 receptor on endothelial cells of post-capillary venules to cause plasma extravasation, and CGRP then causes arteriolar vasodilatation and hyperemia.

Neurogenic mechanisms also contribute to the proinflammatory effects of proteinases in other systems. In aneasthetized guinea pigs, trypsin and PAR2 activating peptide

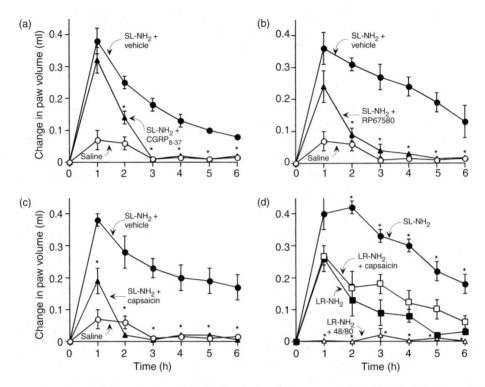

Figure 10.4   Neurogenic mechanism of PAR2-induced edema of the rat paw. (a–c) PAR2-stimulated edema of the paw. In animals treated with vehicle for the antagonists or capsaicin (controls), intraplantar injection of 500 μg SLIGRL-NH₂ (SL-NH₂) markedly increased paw volume from 1 to 6 h. Pre-treatment of animals with CGRP$_{8-37}$ (antagonist of CGRP receptor type 1), RP67580 (antagonist of NK1 receptor), or capsaicin (ablates sensory nerves) inhibited SLIGRL-NH₂-induced edema. Intraplantar injection of saline had minimal effect. *$p<0.05$ compared to rats receiving SLIGRL-NH₂ and saline vehicle. (d) Intraplantar injection of 500 μg LRGILS-NH₂ (LS-NH₂) (control peptide that does not activate PAR2) had a modest effect on paw volume, compared to SLIGRL-NH₂. The response to LRGILS-NH₂ was unaffected by capsaicin but abolished by compound 48/80 (degranulates mast cells). *$p<0.05$ compared to rats receiving LRGILS-NH₂. (Reproduced from Steinhoff et al., 2000.)

(given intratracheally or intravenously) cause bronchoconstriction that is inhibited by the combination of NK1 receptor and NK2 receptor antagonists, suggesting that contraction is mediated by release of tachykinins from sensory nerve endings (Ricciardolo et al., 2000). In several species (rat, mouse, pig), PAR2 agonists applied to the serosal surface of sheets of small and large intestine in Ussing chambers induce an increase in short circuit current across the tissues due to elevated chloride secretion (Vergnolle et al., 1998; Green et al., 2000; Cuffe et al., 2002). This elevated secretion in the pig and mouse is attenuated by neurotoxins (Green et al., 2000; Cuffe et al., 2002), suggesting that PAR2 agonists stimulate neurons of the submucosal plexus, which adheres to the serosal surface of these preparations, to induce release of agents that then regulate enterocytes. Here the elevated chloride secretion would be expected to be associated with increased fluid secretion into the intestinal lumen, which may be protective. Neuropeptides released from primary spinal afferent neurons can also serve a protective function. For example, CGRP can protect against inflammatory challenges, in some cases due to elevated blood flow. Thus, the protective effects of PAR2 agonists in experimentally induced colitis is attenuated by a CGRP antagonist and by ablation of sensory neurons with capsaicin (Fiorucci *et al.*, 2001). Thus, PAR2 may protect against inflammation through release of CGRP from sensory nerve fibres in the colon. PAR2 agonists also stimulate the secretion of gastric mucus and protect against gastric mucosa injury induced by indomethacin and acid–ethanol (Kawabata *et al.*, 2001). Antagonists of the CGRP1 and NK2, but not the NK1, receptors inhibits PAR2-mediated mucus secretion and prevents the protective actions of PAR2 agonists. Thus, PAR2 activation induces the cytoprotective secretion of gastric mucus by stimulating the release of CGRP and tachykinins from sensory neurons.

The activation of PAR2 on the peripheral endings of sensory nerves also affects pain transmission. Notably, intraplantar injections of PAR2 agonists, as doses that do not cause detectable inflammation of the paw, induce a profound and sustained thermal and mechanical hyperalgesia, and causes activation of fos in spinal neurons (Vergnolle *et al.*, 2001a) (Figure 10.5). Thus, PAR2 agonists are capable of inducing hyperalgesia that is not secondary to their inflammatory actions. The hyperalgesia observed in mice to trypsin, tryptase and activating peptides is absent from animals lacking PAR2, indicating specific involvement of this receptor. The mechanism of this hyperalgesia appears to involve release of SP and activation of the NK1 receptor, since the hyperalgesia response to PAR2 activating peptide is absent from mice lacking either preprotachykinin A, which encodes SP and NKA, or the NK1 receptor. However, the mechanism does not involved release of tachykinins in peripheral tissues, since PAR2-dependent hyperalgesia is unaffected by the systemic administration of agonists that cannot penetrate the nervous system. Rather, PAR2-induced hyperalgesia is suppressed by systemic administration of NK1 receptor antagonists that can cross the blood–brain barrier or by intrathecal injection of NK1 receptor antagonists. Thus, PAR2-mediated somatic hyperalgesia depends on the release of SP in the dorsal horn and activation of the NK1 receptor on spinal neurons (Vergnolle *et al.*, 2001a).

Proteinases can also induce visceral hyperalgesia by triggering PAR2 on sensory nerve endings. Rectal distention induces abdominal contractions that are indicative of pain. The intracolonic administration of sub-inflammatory doses of PAR2 agonists markedly enhances contractile responses to high volumes of distention, a response that is attenuated by peripherally acting NK1 R antagonists (Coelho *et al.*, 2002). PAR2 may also contribute to the pain that is associated with pancreatitis. Pancreatic inflammation is associated with premature activation of trypsin, which is released within the inflamed tissue. Thus, it is possible that trypsin could signal to sensory nerves within the pancreas to induce the pain that is commonly associated with this disease. In support of this hypothesis, injection of PAR2 activating peptide into the pancreatic duct can activate and sensitize pancreas-specific afferent neurons *in vivo*, as measured by Fos expression in the dorsal horn of the spinal cord (Hoogerwerf *et al.*, 2001). These observations suggest that PAR2 contributes to nociceptive signalling and may provide a novel link between inflammation and pain.

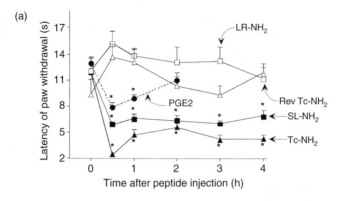

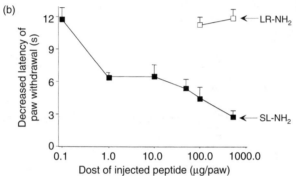

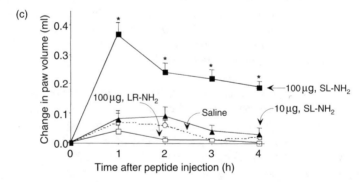

Figure 10.5 PAR2-induced thermal hyperalgesia and edema of the rat paw. (a) Withdrawal latencies were measured after intraplantar injection of 10 μg/paw PAR2 agonists (SL-NH$_2$, Tc-NH$_2$) or control peptides (LR-NH$_2$, Rev-Tc) or 0.3 μg/paw PGE2. *$p < 0.05$ compared with basal measurement (time 0) in each group. (b) Hyperalgesia (decrease in withdrawal latency) 1 h after intraplantar injection of different doses of SL-NH$_2$ and LR-NH$_2$. (c) PAR2 agonist-induced oedema, measured before and after injection of SL-NH$_2$, LR-NH$_2$ or 0.9% saline. *$p < 0.05$ compared with rats receiving the same dose of control peptide. Importantly, hyperalgesia was observed even with doses of PAR2 agonists that failed to trigger detectable inflammation. (Reproduced from Vergnolle et al., 2001a.)

CONCLUSIONS

The identification and characterization of novel membrane proteins on the peripheral and central projections of primary sensory neurons has provided new insights into the regulation of the nervous system, which will have important implications for our understanding and treatment of disease. Certain diseases are characterized by symptoms that may result directly (migraine, neuropathic pain) or indirectly (asthma, chronic obstructive pulmonary disease, irritable bowel syndrome) from activation of sensory neurons by an underlying tissue damage or enduring inflammation. The discovery that diverse agents, such as protons, lipids and proteases, can specifically signal to sensory nerves by distinct receptors presents the opportunity to block this signalling by use of selective antagonists. Thus, antagonists of VR1 and PARs may be useful for the treatment of a wide range of diseases that affect the sensory nervous system. The development of VR1 antagonist is considered a valuable step in the search for new drugs for the treatment of painful diseases. However, VR1 antagonists could also be useful agents in other diseases. For example, VR1 antagonists could be useful for the treatment of asthma that is associated with acid reflux into the airway.

Exciting recent findings indicate a convergence of signalling between GPCRs, tyrosine kinase receptors and ion channels. Thus, agonists of GPCRs (bradykinin, proteases) and tyrosine kinase receptors (NGF) can upregulate the function of VR1. These findings suggest that pro-inflammatory agents such as bradykinin, proteases or NGF, which derive from the circulation or proinflammatory cells may enhance responses to agonists of VR1. This VR1-mediated 'neuronal hyper-responsiveness' may represent the basis for the hypersensitivity observed in certain diseases. For example, there is hypersensitivity to capsaicin or citric acid in patients with asthma and chronic obstructive pulmonary disease (Doherty et al., 2000). The upregulation of VR1 in colonic nerve fibres of patients with active inflammatory bowel disease may also result in hypersensitivity (Yiangou et al., 2001). The recent discovery that tryptase, the major proteinase contained and released from mast cells, excites VR1-positive sensory neurons by cleaving PAR2 (Steinhoff et al., 2000; Vergnolle et al., 2001a), supports the hypothesis that PAR2 may participate in the upregulation or sensitization of VR1 and in the 'neuronal hyper-responsiveness' observed in chronic inflammatory diseases.

REFERENCES

Alm, A.K., Gagnemo-Persson, R., Sorsa, T. and Sundelin, J. (2000). Extrapancreatic trypsin-2 cleaves proteinase-activated receptor-2. *Biochem Biophys Res Commun*, **275**, 77–83.

Asfaha, S., Brussee, V., Chapman, K., Zochodne, D.W. and Vergnolle, N. (2002). Proteinase-activated receptor-1 agonists attenuate nociception in response to noxious stimuli. *Br J Pharmacol*, **135**, 1101–1106.

Baluk, P., Bertrand, C., Geppetti, P., McDonald, D.M. and Nadel, J.A. (1995). NK1 receptors mediate leukocyte adhesion in neurogenic inflammation in the rat trachea. *Am J Physiol*, **268**, L263–L269.

Beecher, K.L., Andersen, T.T., Fenton, J.W.N. and Festoff, B.W. (1994). Thrombin receptor peptides induce shape change in neonatal murine astrocytes in culture. *J Neurosci Res*, **37**, 108–115.

Bevan, S. and Geppetti, P. (1994). Protons: small stimulants of capsaicin-sensitive sensory nerves. *Trends Neurosci*, **17**, 509–512.

Bevan, S., Hothi, S., Hughes, G., James, I.F., Rang, H.P., Shah, K. et al. (1992). Capsazepine: a competitive antagonist of the sensory neurone excitant capsaicin. *Br J Pharmacol*, **107**, 544–552.

Böhm, S.K., Khitin, L.M., Grady, E.F., Aponte, G., Payan, D.G. and Bunnett, N.W. (1996). Mechanisms of desensitization and resensitization of proteinase-activated receptor-2. *J Biol Chem*, **271**, 22003–22016.

Bonini, S., Lambiase, A., Angelucci, F., Magrini, L., Manni, L. and Aloe, L. (1996). Circulating nerve growth factor levels are increased in humans with allergic diseases and asthma. *Proc Natl Acad Sci USA*, **93**, 10955–10960.

Brass, L.F., Pizarro, S., Ahuja, M., Belmonte, E., Blanchard, N., Stadel, J.M. *et al.* (1994). Changes in the structure and function of the human thrombin receptor during receptor activation, internalization, and recycling. *J Biol Chem*, **269**, 2943–2952.

Caterina, M.J., Leffler, A., Malmberg, A.B., Martin, W.J., Trafton, J., Petersen-Zeitz, K.R. *et al.* (2000). Impaired nociception and pain sensation in mice lacking the capsaicin receptor. *Science*, **288**, 306–313.

Caterina, M.J., Rosen, T.A., Tominaga, M., Brake, A.J. and Julius, D. (1999). A capsaicin-receptor homologue with a high threshold for noxious heat. *Nature*, **398**, 436–441.

Caterina, M.J., Schumacher, M.A., Tominaga, M., Rosen, T.A., Levine, J.D. and Julius, D. (1997). The capsaicin receptor: a heat-activated ion channel in the pain pathway. *Nature*, **389**, 816–824.

Caughey, G.H. (1995). In *Mast Cell Proteases in Immunology and Biology*. Caughey, G.H. (ed.), Marcel Dekker, Inc., New York, pp. 305–329.

Cavanaugh, K.P., Gurwitz, D., Cunningham, D.D. and Bradshaw, R.A. (1990). Reciprocal modulation of astrocyte stellation by thrombin and protease nexin-1. *J Neurochem*, **54**, 1735–1743.

Cesare, P., Moriondo, A., Vellani, V. and McNaughton, P.A. (1999). Ion channels gated by heat. *Proc Natl Acad Sci USA*, **96**, 7658–7663.

Chuang, H.H., Prescott, E.D., Kong, H., Shields, S., Jordt, S.E., Basbaum, A.I. *et al.* (2001). Bradykinin and nerve growth factor release the capsaicin receptor from PtdIns(4,5)P2-mediated inhibition. *Nature*, **411**, 957–962.

Cocks, T.M. and Moffatt, J.D. (2000). Protease-activated receptors: sentries for inflammation? *Trends Pharmacol Sci*, **21**, 103–108.

Coelho, A.M., Vergnolle, N., Guiard, B., Fioramonti, J. and Bueno, L. (2002). Proteinases and proteinase-activated receptor 2: a possible role to promote visceral hyperalgesia in rats. *Gastroenterology*, **122**, 1035–1047.

Compton, S.J., Cairns, J.A., Palmer, K.J., Al-Ani, B., Hollenberg, M.D. and Walls, A.F. (2000). A polymorphic protease-activated receptor 2 (PAR2) displaying reduced sensitivity to trypsin and differential responses to PAR agonists. *J Biol Chem*, **275**, 39207–39212.

Compton, S.J., Renaux, B., Wijesuriya, S.J. and Hollenberg, M.D. (2001). Glycosylation and the activation of proteinase-activated receptor 2 (PAR(2)) by human mast cell tryptase. *Br J Pharmacol*, **134**, 705–718.

Corvera, C.U., Dery, O., McConalogue, K., Bohm, S.K., Khitin, L.M., Caughey, G.H. *et al.* (1997). Mast cell tryptase regulates rat colonic myocytes through proteinase-activated receptor 2. *J Clin Invest*, **100**, 1383–1393.

Corvera, C.U., Dery, O., McConalogue, K., Gamp, P., Thoma, M., Al-Ani, B. *et al.* (1999). Thrombin and mast cell tryptase regulate guinea-pig myenteric neurons through proteinase-activated receptors-1 and -2. *J Physiol*, **517**, 741–756.

Cottrell, G.S., Coehlo, A.-M. and Bunnett, N.W. (2002). Protease-activated receptors: the role of cell-surface proteolysis in signaling. *Essays in Biochemistry*, in press.

Coughlin, S.R. (2000). Thrombin signalling and protease-activated receptors. *Nature*, **407**, 258–264.

Cuffe, J.E., Bertog, M., Velazquez-Rocha, S., Dery, O., Bunnett, N. and Korbmacher, C. (2002). Basolateral PAR-2 receptors mediate KCl secretion and inhibition of Na^+ absorption in the mouse distal colon. *J Physiol*, **539**, 209–222.

Davis, J.B., Gray, J., Gunthorpe, M.J., Hatcher, J.P., Davey, P.T., Overend, P. *et al.* (2000). Vanilloid receptor-1 is essential for inflammatory thermal hyperalgesia. *Nature*, **405**, 183–187.

de Garavilla, L., Vergnolle, N., Young, S.H., Ennes, H., Steinhoff, M., Ossovskaya, V.S. *et al.* (2001). Agonists of proteinase-activated receptor 1 induce plasma extravasation by a neurogenic mechanism. *Br J Pharmacol*, **133**, 975–987.

De Petrocellis, L., Harrison, S., Bisogno, T., Tognetto, M., Brandi, I., Smith, G.D. *et al.* (2001). The vanilloid receptor (VR1)-mediated effects of anandamide are potently enhanced by the cAMP-dependent protein kinase. *J Neurochem*, **77**, 1660–1663.

Dery, O., Corvera, C.U., Steinhoff, M. and Bunnett, N.W. (1998). Proteinase-activated receptors: novel mechanisms of signaling by serine proteases. *Am J Physiol*, **274**, C1429–C1452.

Devane, W.A., Hanus, L., Breuer, A., Pertwee, R.G., Stevenson, L.A., Griffin, G. *et al.* (1992). Isolation and structure of a brain constituent that binds to the cannabinoid receptor. *Science*, **258**, 1946–1949.

Dihanich, M., Kaser, M., Reinhard, E., Cunningham, D. and Monard, D. (1991). Prothrombin mRNA is expressed by cells of the nervous system. *Neuron*, **6**, 575–581.

Di Marzo, V., Fontana, A., Cadas, H., Schinelli, S., Cimino, G., Schwartz, J.C. *et al.* (1994). Formation and inactivation of endogenous cannabinoid anandamide in central neurons. *Nature*, **372**, 686–691.

Di Marzo, V., Melck, D., Orlando, P., Bisogno, T., Zagoory, O., Bifulco, M. *et al.* (2001). Palmitoylethanolamide inhibits the expression of fatty acid amide hydrolase and enhances the anti-proliferative effect of anandamide in human breast cancer cells. *Biochem J*, **358**, 249–255.

Doherty, M.J., Mister, R., Pearson, M.G. and Calverley, P.M. (2000). Capsaicin responsiveness and cough in asthma and chronic obstructive pulmonary disease. *Thorax*, **55**, 643–649.

Ehrenreich, H., Costa, T., Clouse, K.A., Pluta, R.M., Ogino, Y., Coligan, J.E. *et al.* (1993). Thrombin is a regulator of astrocytic endothelin-1. *Brain Res*, **600**, 201–207.

Fiorucci, S., Mencarelli, A., Palazzetti, B., Distrutti, E., Vergnolle, N., Hollenberg, M.D. *et al.* (2001). Proteinase-activated receptor 2 is an anti-inflammatory signal for colonic lamina propria lymphocytes in a mouse model of colitis. *Proc Natl Acad Sci USA*, **98**, 13936–13941.

Geppetti, P. and Holzer, P. (1996). *Neurogenic Inflammation*, CRC Press Inc., Boca Raton, FL.

Gill, J.S., Pitts, K., Rusnak, F.M., Owen, W.G. and Windebank, A.J. (1998). Thrombin induced inhibition of neurite outgrowth from dorsal root ganglion neurons. *Brain Res*, **797**, 321–327.

Gingrich, M.B., Junge, C.E., Lyuboslavsky, P. and Traynelis, S.F. (2000). Potentiation of NMDA receptor function by the serine protease thrombin. *J Neurosci*, **20**, 4582–4595.

Green, B.T., Bunnett, N.W., Kulkarni-Narla, A., Steinhoff, M. and Brown, D.R. (2000). Intestinal type 2 proteinase-activated receptors: expression in opioid-sensitive secretomotor neural circuits that mediate epithelial ion transport. *J Pharmacol Exp Ther*, **295**, 410–416.

Gurwitz, D. and Cunningham, D.D. (1988). Thrombin modulates and reverses neuroblastoma neurite outgrowth. *Proc Natl Acad Sci USA*, **85**, 3440–3444.

Hofbauer, B., Saluja, A.K., Lerch, M.M., Bhagat, L., Bhatia, M., Lee, H.S. *et al.* (1998). Intra-acinar cell activation of trypsinogen during caerulein-induced pancreatitis in rats [in process citation]. *Am J Physiol*, **275**, G352–G362.

Holzer, P. (1988). Local effector functions of capsaicin-sensitive sensory nerve endings: involvement of tachykinins, calcitonin gene-related peptide and other neuropeptides. *Neuroscience*, **24**, 739–768.

Hoogerwerf, W.A., Zou, L., Shenoy, M., Sun, D., Micci, M.A., Lee-Hellmich, H. *et al.* (2001). The proteinase-activated receptor 2 is involved in nociception. *J Neurosci*, **21**, 9036–9042.

Hunt, J.F., Fang, K., Malik, R., Snyder, A., Malhotra, N., Platts-Mills, T.A. *et al.* (2000). Endogenous airway acidification. Implications for asthma pathophysiology. *Am J Respir Crit Care Med*, **161**, 694–699.

Hwang, S.W., Cho, H., Kwak, J., Lee, S.Y., Kang, C.J., Jung, J. *et al.* (2000). Direct activation of capsaicin receptors by products of lipoxygenases: endogenous capsaicin-like substances. *Proc Natl Acad Sci USA*, **97**, 6155–6160.

Ishihara, H., Connolly, A.J., Zeng, D., Kahn, M.L., Zheng, Y.W., Timmons, C. *et al.* (1997). Protease-activated receptor 3 is a second thrombin receptor in humans. *Nature*, **386**, 502–506.

Kahn, M.L., Zheng, Y.-W., Huang, W., Bigornia, V., Zeng, D., Moff, S. *et al.* (1998). A dual thrombin receptor system for platelet activation. *Nature*, **394**, 690–694.

Kawabata, A., Kinoshita, M., Nishikawa, H., Kuroda, R., Nishida, M., Araki, H. *et al.* (2001). The protease-activated receptor-2 agonist induces gastric mucus secretion and mucosal cytoprotection. *J Clin Invest*, **107**, 1443–1450.

Koivunen, E., Huhtala, M.-L. and Stenman, U.-H. (1989). Human ovarian tumor-associated trypsin. Its purification and characterization from mucinous cyst fluid and identification as an activator of pro-urokinase. *J Biol Chem*, **264**, 14095–14099.

Koivunen, E., Saksela, O., Itkonen, O., Osman, S., Huhtala, M.-L. and Stenman, U.-H. (1991). Human colonic carcinoma, fibrosarcoma and leukemia cell lines produce tumor-associated trypsinogen. *Int J Cancer*, **47**, 592–596.

Kong, W., McConalogue, K., Khitin, L.M., Hollenberg, M.D., Payan, D.G., Bohm, S.K. *et al.* (1997). Luminal trypsin may regulate enterocytes through proteinase-activated receptor 2. *Proc Natl Acad Sci USA*, **94**, 8884–8889.

Koshikawa, N., Hasegawa, S., Nagashima, Y., Mitsuhashi, K., Tsubota, Y., Miyata, S. *et al.* (1998). Expression of trypsin by epithelial cells of various tissues, leukocytes, and neurons in human and mouse. *Am J Pathol*, **153**, 937–944.

Macfarlane, S.R., Seatter, M.J., Kanke, T., Hunter, G.D. and Plevin, R. (2001). Proteinase-activated receptors. *Pharmacol Rev*, **53**, 245–282.

Maggi, C.A., Patacchini, R., Santicioli, P. and Giuliani, S. (1991). Tachykinin antagonists and capsaicin-induced contraction of the rat isolated urinary bladder: evidence for tachykinin-mediated cotransmission. *Br J Pharmacol*, **103**, 1535–1541.

Miller, S., Sehati, N., Romano, C. and Cotman, C.W. (1996). Exposure of astrocytes to thrombin reduces levels of the metabotropic glutamate receptor mGluR5. *J Neurochem*, **67**, 1435–1447.

Molino, M., Barnathan, E.S., Numerof, R., Clark, J., Dreyer, M., Cumashi, A. *et al.* (1997). Interactions of mast cell tryptase with thrombin receptors and PAR-2. *J Biol Chem*, **272**, 4043–4049.

Neveu, I., Jehan, F., Jandrot-Perrus, M., Wion, D. and Brachet, P. (1993). Enhancement of the synthesis and secretion of nerve growth factor in primary cultures of glial cells by proteases: a possible involvement of thrombin. *J Neurochem*, **60**, 858–867.

Niclou, S., Suidan, H.S., Brown-Luedi, M. and Monard, D. (1994). Expression of the thrombin receptor mRNA in rat brain. *Cell Mol Biol (Noisy-le-grand)*, **40**, 421–428.

Nystedt, S., Emilsson, K., Wahlestedt, C. and Sundelin, J. (1994). Molecular cloning of a potential proteinase activated receptor [see comments]. *Proc Natl Acad Sci USA*, **91**, 9208–9212.

Otsuka, M. and Yoshioka, K. (1993). Neurotransmitter functions of mammalian tachykinins. *Physiol Rev*, **73**, 229–308.

Perraud, F., Besnard, F., Sensenbrenner, M. and Labourdette, G. (1987). Thrombin is a potent mitogen for rat astroblasts but not for oligodendroblasts and neuroblasts in primary culture. *Int J Dev Neurosci*, **5**, 181–188.

Premkumar, L.S. and Ahern, G.P. (2000). Induction of vanilloid receptor channel activity by protein kinase C. *Nature*, **408**, 985–990.

Ricciardolo, F.L., Rado, V., Fabbri, L.M., Sterk, P.J., Di Maria, G.U. and Geppetti, P. (1999). Bronchoconstriction induced by citric acid inhalation in guinea pigs: role of tachykinins, bradykinin, and nitric oxide. *Am J Respir Crit Care Med*, **159**, 557–562.

Ricciardolo, F.L., Steinhoff, M., Amadesi, S., Guerrini, R., Tognetto, M., Trevisani, M. *et al.* (2000). Presence and bronchomotor activity of protease-activated receptor-2 in guinea pig airways. *Am J Respir Crit Care Med*, **161**, 1672–1680.

Schumacher, M.A., Jong, B.E., Frey, S.L., Sudanagunta, S.P., Capra, N.F. and Levine, J.D. (2000). The stretch-inactivated channel, a vanilloid receptor variant, is expressed in small-diameter sensory neurons in the rat. *Neurosci Lett*, **287**, 215–218.

Smith-Swintosky, V.L., Cheo-Isaacs, C.T., D'Andrea, M.R., Santulli, R.J., Darrow, A.L. and Andrade-Gordon, P. (1997). Protease-activated receptor-2 (PAR-2) is present in the rat hippocampus and is associated with neurodegeneration. *J Neurochem*, **69**, 1890–1896.

Stead, R.H., Tomioka, M., Quinonez, G., Simon, G.T., Felten, S.Y. and Bienenstock, J. (1987). Intestinal mucosal mast cells in normal and nematode-infected rat intestines are in intimate contact with peptidergic nerves. *Proc Natl Acad Sci USA*, **84**, 2975–2979.

Steinhoff, M., Vergnolle, N., Young, S.H., Tognetto, M., Amadesi, S., Ennes, H.S. *et al.* (2000). Agonists of proteinase-activated receptor 2 induce inflammation by a neurogenic mechanism. *Nat Med*, **6**, 151–158.

Striggow, F., Riek, M., Breder, J., Henrich-Noack, P., Reymann, K.G. and Reiser, G. (2000). The protease thrombin is an endogenous mediator of hippocampal neuroprotection against ischemia at low concentrations but causes degeneration at high concentrations. *Proc Natl Acad Sci USA*, **97**, 2264–2269.

Szolcsanyi, J. (2000). Anandamide and the question of its functional role for activation of capsaicin receptors. *Trends Pharmacol Sci*, **21**, 203–204.

Tognetto, M., Amadesi, S., Harrison, S., Creminon, C., Trevisani, M., Carreras, M. *et al.* (2001). Anandamide excites central terminals of dorsal root ganglion neurons via vanilloid receptor-1 activation. *J Neurosci*, **21**, 1104–1109.

Tominaga, M., Caterina, M.J., Malmberg, A.B., Rosen, T.A., Gilbert, H., Skinner, K. *et al.* (1998). The cloned capsaicin receptor integrates multiple pain-producing stimuli. *Neuron*, **21**, 531–543.

Tucker, R.C., Kagaya, M., Page, C.P. and Spina, D. (2001). The endogenous cannabinoid agonist, anandamide stimulates sensory nerves in guinea-pig airways. *Br J Pharmacol*, **132**, 1127–1135.

Vaughan, P.J., Pike, C.J., Cotman, C.W. and Cunningham, D.D. (1995). Thrombin receptor activation protects neurons and astrocytes from cell death produced by environmental insults. *J Neurosci*, **15**, 5389–5401.

Vergnolle, N., Bunnett, N.W., Sharkey, K.A., Brussee, V., Compton, S.J., Grady, E.F. *et al.* (2001a). Proteinase-activated receptor-2 and hyperalgesia: a novel pain pathway. *Nat Med*, **7**, 821–826.

Vergnolle, N., Wallace, J.L., Bunnett, N.W. and Hollenberg, M.D. (2001b). Protease-activated receptors in inflammation, neuronal signaling and pain. *Trends Pharmacol Sci*, **22**, 146–152.

Vergnolle, N., Macnaughton, W.K., Al-Ani, B., Saifeddine, M., Wallace, J.L. and Hollenberg, M.D. (1998). Proteinase-activated receptor 2 (PAR2)-activating peptides: identification of a receptor distinct from PAR2 that regulates intestinal transport. *Proc Natl Acad Sci USA*, **95**, 7766–7771.

Vu, T.K., Hung, D.T., Wheaton, V.I. and Coughlin, S.R. (1991). Molecular cloning of a functional thrombin receptor reveals a novel proteolytic mechanism of receptor activation. *Cell*, **64**, 1057–1068.

Winter, J., Forbes, C.A., Sternberg, J. and Lindsay, R.M. (1988). Nerve growth factor (NGF) regulates adult rat cultured dorsal root ganglion neuron responses to the excitotoxin capsaicin. *Neuron*, **1**, 973–981.

Yiangou, Y., Facer, P., Dyer, N.H., Chan, C.L., Knowles, C., Williams, N.S. *et al.* (2001). Vanilloid receptor 1 immunoreactivity in inflamed human bowel. *Lancet*, **357**, 1338–1339.

Zygmunt, P.M., Petersson, J., Andersson, D.A., Chuang, H., Sorgard, M., Di Marzo, V. *et al.* (1999). Vanilloid receptors on sensory nerves mediate the vasodilator action of anandamide. *Nature*, **400**, 452–457.

11 Functional Consequences of Neuroimmune Interactions in the Intestinal Mucosa

Johan D. Söderholm* and Mary H. Perdue

Intestinal Disease Research Program, McMaster University, Hamilton, Ontario, Canada

A single layer of epithelial cells lines the gastrointestinal tract forming a critical first-line barrier between the external environment and the body proper. Acute changes in epithelial physiology (e.g. increased fluid secretion to wash away noxious substances), mediated by neuroendocrine or immune signals, are necessary and beneficial to host defence against enteric pathogens and other threats from the intestinal lumen. However, an inability to down-regulate this response may lead to pathophysiological events. In this chapter, we will review the structure and function of the intestinal epithelium, and present recent advances in the understanding of neuroimmunophysiology of the intestinal mucosa. Because of the wealth of literature in the area, we will emphasize the interaction between enteric nerves and mucosal mast cells in the regulation of epithelial function. It has become clear that the role of mast cells in the intestinal mucosa is not only to react to antigens via bound IgE antibodies, but also to actively control the barrier and transport properties of the epithelium. Studies in animal models of food allergy, nematode infections, and stress have provided evidence that changes in mucosal physiology are due to the direct action of mast cell mediators on epithelial receptors and/or indirect action via nerves/neurotransmitters. Moreover, chronic activation of mast cells in models of hypersensitivity and stress suggest a role for mast cells in initiating mucosal inflammation. There is also mounting evidence for the importance of nerve–mast cell interactions in gastrointestinal diseases in humans. An increased understanding of the mechanisms involved in pathogenic events may lead to new treatment modalities for various gastrointestinal disorders.

KEY WORDS: barrier function; enteric nerves; food allergy; ion secretion; mast cells; stress.

ABBREVIATIONS: CNS, central nervous system; CD 23, the low-affinity IgE receptor (FcεRII); ^{51}Cr-EDTA, ^{51}chromium-labelled ethylene-diamine tetra-acetic acid; CRH, corticotropin-releasing hormone; CRS, cold restraint stress; CTMC, connective tissue mast cells; DPDPE, [D-Pen^2,D-Pen^5]-enkephalin; FAE, follicle-associated epithelium; HRP, horseradish peroxidase; 5-HT, 5-hydroxytryptamine; IBD, inflammatory bowel disease; IBS, irritable bowel syndrome; IFN, interferon; IL, interleukin; IMMC, intestinal mucosa mast cells; I_{sc}, short-circuit current; LT, leukotriene; MALT, mucosa-associated lymphoid tissue; MHC, major histocompatibility complex; NO, nitric oxide; OVA, ovalbumin; PAF, platelet-activating factor; PAR-2, type 2 protease-activated receptor; PD, transepithelial potential difference; PG, prostaglandin; RMCP II, type II rat mast cell protease; RS, restraint stress; TNF, tumour necrosis factor; VIP, vasoactive intestinal polypeptide; WAS, water avoidance stress; Ws/Ws rats, genetically mast cell-deficient rats; W/W^V mice, genetically mast cell-deficient mice; ZO, zonula occludens.

---

* Current affiliation: Division of Surgery, Department of Biomedicine and Surgery, Linköping University, Sweden.

INTRODUCTION

The microenvironment of the intestinal mucosa is complex. Luminal contents are rich in bacteria, as well as in food antigens and other noxious material that has been ingested, and the defence systems of the intestinal mucosa fight a constant battle to avoid a state of ongoing immune activation. The single-cell layer of the epithelium lining the gastrointestinal tract forms a critical first-line barrier between the external environment and the body proper. Passage of phlogistic products across a transiently or intrinsically leaky intestinal barrier stimulates resident lamina propria macrophages, mast cells, eosinophils and lymphocytes to secrete inflammatory mediators. A normal host can down-regulate this response when the triggering event has resolved, whereas in a susceptible host ongoing enhanced permeability may result in chronic inflammation. In recent years, it has become clear that modulation of innate gut mucosal defence systems by neuroimmune signalling is important in the regulation of the mucosal inflammatory response, and that disturbances at various levels of this regulation can result in intestinal dysfunction and/or diseases. In this chapter we will focus on physiological effects and pathophysiological changes in epithelial secretion and barrier function induced by neuroimmune interactions.

STRUCTURE AND FUNCTION OF THE INTESTINAL EPITHELIUM

The intestinal epithelium has diverse and seemingly contradictory functions. At the same time as it plays a major role in the digestion and absorption of nutrients, it constitutes the organism's most important barrier between the internal and external environment, with a surface area of approximately $300 \, \text{m}^2$. In addition, it has become evident that the epithelium actively participates in immune responses and contributes to both innate and specific immunity.

Cell types in the intestinal epithelium

Enterocytes, that is, absorptive epithelial cells, constitute 75–80% of epithelial cells in the small bowel, forming a single continuous layer of columnar cells. The enterocyte surface has tightly packed microvilli, increasing the surface area 20–40 times. Anchored to the microvilli are large carbohydrates, the major component of the glycocalyx, which also contains enzymes necessary for the digestion of nutrients. At their apices, adjacent enterocytes are interconnected by three types of junctional complexes: tight junctions, adherence junctions and desmosomes. The cells are anchored to the basal lamina by hemidesmosomes. The basal lamina is composed of collagen, laminin, fibronectin and glycosaminoglycans, and has numerous round or oval pores, facilitating contact with the lamina propria.

    Goblet cells (20% of epithelial cells), are the mucus-secreting cells. Mucus plays an important role in host defence, acting as a lubricant that assists in distal propulsion of luminal contents, and as a physical barrier between the luminal contents and intestinal epithelial cells, for example, by trapping bacteria. The mucus layer also harbours lysozyme and secretory IgA (Clamp and Creeth, 1984), which enhances its defence properties. In the large bowel, mucus also represents a nutrient source for anaerobic bacteria.

Paneth cells (3.5% of epithelial cells), which are located in the crypts, contain various secretory granules containing lysozyme, phospholipases, and other antimicrobial proteins, called defensins (Ouellette, 1999). Their function is thought to be the prevention of proliferation of microorganisms in the crypts.

Enteroendocrine cells are distributed throughout the epithelium, and release gastrointestinal hormones (for example, secretin, somatostatin, neurotensin, peptide YY, etc.) in response to changes in the external environment and enteric nerves.

M cells (microfold or membranous cells) are not present in normal epithelium, but constitute ~10% of the cells in the follicle associated epithelium (FAE), that is, covering lymphoid follicles. M cells are specialized for the uptake of particles, microorganisms, and macromolecules (Owen, 1999). They are in contact with macrophages and lymphocytes, which are enfolded within the cells. The function of M cells is the transfer of antigen into lymphoid tissue (Sanderson and Walker, 1993). The origin of the M cells is not clearly understood, but they seem to be a phenotypic conversion of enterocytes in the FAE, modulated by lymphocytes and cytokines (Kraehenbuhl and Neutra, 2000). Caco-2 cells in culture can be converted to M cells by co-culture with lymphocytes expressing B cell markers (Kerneis *et al.*, 1997). The number of M cells is modified by short-term exposure to bacteria (Smith *et al.*, 1987; Sansonetti and Phalipon, 1999).

In addition, cells with immune properties are present within the epithelium. The most common of these are *intraepithelial lymphocytes* (IEL), which are located below the tight junctions between enterocytes. Compared with other lymphocytes, the IEL have unique surface markers (such as the $\gamma\delta$ T-cell receptor) and are relatively unreactive (Beagley and Husband, 1998; Kagnoff, 1998).

Dendritic cells have been identified below and within intestinal epithelia (Maric *et al.*, 1996; Iwasaki and Kelsall, 1999). These cells with their long processes may be involved in antigen sampling and presentation (Rescigno *et al.*, 2001).

Other immune/inflammatory cells (such as mast cells, neutrophils, macrophages) are also present within the epithelium depending on the state of immune activation in the mucosa. Trafficking of immune cells and localization to the epithelium during inflammation is related to expression of adhesion molecules on endothelial and epithelial cells as well as epithelial release of chemokines (for example, interleukin (IL)-8 and leukotriene (LT)B_4).

Ion and fluid transport

Transport of nutrients, fluid and ions across the epithelium can be divided into absorption (toward the bloodstream) and secretion (into the lumen). These functions are spatially distinct along the villus–crypt axis, with nutrient and fluid absorption occurring mainly in the upper part of the villi, and the crypt region being the site of electrolyte and water secretion. Epithelial transport functions are directly or indirectly dependent on the sodium pump (Na^+/K^+ ATPase) located in the basolateral membrane of enterocytes. This energy-requiring ion pump creates an electrochemical gradient across the cell membrane due to the low intracellular Na^+ concentration. This gradient facilitates the uptake into the cell of Na^+ together with various nutrients via Na^+-coupled cotransporters (e.g. for glucose, amino acids, vitamins, etc.). Water movement follows the osmotic gradient generated by the absorption of ions and nutrients. The $Na^+/K^+/2Cl^-$ entry ion transporter in the basolateral membrane of crypt cells creates an electrochemical gradient for Cl^- (high intracellular concentration) that facilitates rapid secretion of Cl^- through channels in the apical membrane when the cells are stimulated by secretagogues such as neurotransmitters.

Mucus secretion

The major constituents of intestinal mucus are high molecular-weight proteoglycans called mucins. Mucins consist of a central protein core with large numbers of oligosaccharides, accounting for up to 60–80% of the molecular mass, attached to specific regions of the core. The oligosaccharides are variable and complex, and the degree and type of glycolsylation of mucins is central to their function. The patterns of glycolsylation are tissue specific within the gastrointestinal tact. The mucins can be divided into two groups: secreted gel-forming mucins (mainly in the form encoded by the gene denoted MUC 2) and membrane bound mucins (predominantly encoded by MUC 4). Goblet cells of the surface epithelium store mucin in apically located granules. Mucin is secreted at a low baseline rate to maintain the mucus coat over the epithelium. In response to stimulation, such as by cholinergic nerves (Neutra *et al.*, 1984), goblet cells can accelerate their discharge of mucins substantially. The discharge of mucins in response to stimulation is either by compound exocytosis, resulting in deep apical membrane cavitations, or by an accelerated single-granule exocytosis (Forstner, 1995). Colonic mucin release has been shown to be stimulated by signals from the enteric nervous system (acetylcholine, vasoactive intestinal polypeptide [VIP]), enteroendocrine cells (peptide YY, 5-hydroxytryptamine [5-HT]), and resident immune cells (mast cell mediators, IL-1β, nitric oxide [NO]) (Plaisancie *et al.*, 1997).

The intestinal barrier

The intestinal mucosa is continuously exposed to a heavy load of molecules with antigenic potential from ingested food, resident bacteria, invading viruses, etc. The epithelial cell layer, interconnected by the tight junctions, constitutes the principal part of the barrier, and restricts both transcellular and paracellular permeation of molecules. In addition, the epithelium exerts important defence by secretion of fluid and mucus, together with secretory IgA, into the lumen, serving to dilute, wash away and bind noxious substances. The normal intestinal epithelium allows small quantities of intact protein antigens to cross and interact with cells of the immune system, as an important surveillance of the luminal contents. However, excessive or inappropriate exposure of immune cells to antigens leads to diseases of the gastrointestinal tract or other organ systems (Sanderson and Walker, 1993). A disturbance of intestinal barrier function has been suggested as an etiologic factor in Crohn's disease (Meddings, 1997; Söderholm *et al.*, 1999) and food allergy (Crowe and Perdue, 1992; Bjarnason *et al.*, 1995), and in several other disease states an increased mucosal permeability is implicated in pathogenesis and development of complications (e.g. viral and bacterial gastroenteritis, ulcerative colitis, and multiple organ dysfunction syndrome in patients with sepsis and trauma).

Junctional complexes, first described by Farquhar and Palade (1963), join adjacent enterocytes to each other and mainly consist of: *tight junctions*, continuous cell–cell contacts at the apical end of the lateral cell interspace, creating a highly regulated rate-limiting barrier to the diffusion of solutes, and also acting in the separation of the different lipid and protein components of the apical and basolateral parts of the cell membrane; and *adherence junctions*, another continuous circumferential contact zone (positioned below the tight junctions) formed by adhesion molecules of the cadherin family and their cytoplasmic binding proteins α-, β-, and γ-catenin. It has been suggested that the tight junctions and adherence junctions form a single functional unit (Nusrat *et al.*, 2000). In addition, the junctional complexes include *desmosomes*, which form spot-like dense adhesions between epithelial cells, and are linked to the cytokeratin intermediate filaments. They are distributed over the entire lateral cell surfaces, but are often found to be concentrated in a circumferential band below the adherence junctions. *Gap junctions*, are specialized regions for cell–cell communication mediating electrotonic coupling, allowing ionic currents to pass between cells.

By transmission electron microscopy, tight junctions appear as a series of focal contacts ('kisses') between the plasma membrane of adjacent epithelial cells. Freeze fracture electron microscopy has revealed that these contacts correspond to continuous branching fibrils on the apical-most part of the lateral membranes of the enterocytes. Fibrils on one cell interact with the fibrils on the adjacent cell to close the intercellular space and define the paracellular permeability characteristics. Over the last few years it has become known that these fibrils are formed by at least two types of transmembrane proteins: occludin and different variants of the 20 members of the claudin family (Mitic *et al.*, 2000). It is now thought that the varied tissue and cell-type distribution of the claudins explains the variable permeability observed among different tissues. These transmembrane proteins are connected to a multiprotein complex, with components referred to as cytoplasmic plaque proteins, including zonula occludens (ZO)-1, ZO-2, ZO-3, cingulin, and 7H6. Moreover, these proteins are linked to a perijunctional actomyosin ring, the contraction of which is controlled by ligand binding of membrane receptors (Nusrat *et al.*, 2000).

The ability of epithelia to transport substances towards or away from the lumen depends on the polarity of the cells and on the tight junctions to prevent back diffusion through the intercellular space (gate function) (Cereijido *et al.*, 1998). The tight junctions are also involved in maintaining the polarity of lipid substances by preventing lateral movement of cell membrane constituents from the apical to the basolateral region in the outer cell membrane leaflet (fence function). Previously regarded as static and hermetic, it is now clear that tight junctions are highly regulated structures, and should be looked upon as steady state situations that may be drastically changed by a variety of signalling events. Moreover, the gate and fence functions of the tight junctions appear to be regulated in different ways (Mandel *et al.*, 1993). The gate function is an important component of the intestinal barrier. Tight junctions are selectively permeable to ions (depending on charge) and allow the passage of small molecules, but are virtually impermeable to protein-sized molecules under normal conditions, thus constituting an effective barrier to macromolecules with antigenic potential.

A limited and controlled uptake of undegraded proteins by the intestinal mucosa is physiologic and essential for the surveillance of antigens in the gastrointestinal tract (Sanderson and Walker, 1993). This sampling is believed to occur primarily via the M-cells in the FAE overlying lymphoid follicles (Owen, 1999; Shao *et al.*, 2001). Once the antigen has passed through the M cell, it can interact with macrophages or CD4+ T cells localized within pockets in the M cell's membrane. Antigen is then carried to the mesenteric lymph nodes where it interacts with T and B cells that activate an IgA-secreting cell population, and in a incompletely defined manner also induce a systemic immune tolerance to the antigens, known as oral tolerance. The IgA-producing B cells enter the systemic circulation and 'home' back to the mucosa associated lymphoid tissue (MALT) within the gut and other tissues.

The relative importance of uptake of soluble macromolecules via the infrequent M cells is, however, a point of discussion (Swaan, 1998). Enterocytes also have the capacity of uptake and transcellular transport of antigens, and presentation of these via class I and II major histocompatibility complex (MHC) to lymphocytes (Shao *et al.*, 2001). In general, three types of endocytotic uptake are recognized in the intestinal epithelium (Walker and Isselbacher, 1974; Sanderson and Walker, 1993): highly specific, receptor-mediated uptake, for example, uptake of immunoglobulins and growth factors from breast milk; adsorptive endocytosis, after binding of molecules to the cell membrane; and non-specific fluid-phase endocytosis of substances in the luminal fluid. The latter two types of endocytosis are relevant for the non-specific uptake of dietary and bacterial luminal antigens. After the antigen is internalized in an apical membrane vesicle, the vesicle fuses with the early endocytotic compartment (Schaerer *et al.*, 1991). In this compartment, the antigen is sorted to different pathways: recycling to the apical membrane, delivery to lysosomes, or transcytosis. The fate of the macromolecules after endocytosis is largely determined by membrane binding. Membrane-bound proteins are sorted to vesicles that traverse the cell to the basolateral pole, whereas proteins in free solution are sorted in the tubulocisternae to vesicles that fuse with lysosomes. The microtubules play an important role in this vesicle trafficking. Normally, the bulk of protein is transferred to the lysosomal pathway and degraded to peptide fragments and amino acids. A small but significant amount of undegraded protein does, however, cross the epithelium (Gardner, 1988; Marcon-Genty *et al.*, 1989). Whether this is the effect of an escape from lysosomal breakdown or a limited paracellular uptake, or both, has not been fully settled.

Both physiologic and pathologic immune responses to antigenic proteins require antigen presentation to T cells. The antigen-presenting cells must internalise, digest, and link fragments of the antigen to surface MHC glycoproteins, that interact with a T cell receptor. Class II MHC are present on the basolateral membrane of normal human enterocytes (Sanderson and Walker, 1993). Antigen presentation by enterocytes normally results in stimulation of CD8+ suppressor T cells, in contrast to the CD4+ cell stimulation produced by antigen-presenting cells in the lamina propria. Situations with enhanced antigen exposure or processing could, however, allow enterocytes to present antigen to CD4+ T cells (Shao *et al.*, 2001).

THE ENTERIC NERVOUS SYSTEM

For a comprehensive description of the enteric nervous system and its components, see Chapter 3. In brief, the enteric nervous system is made up of the submucosal plexus and the myenteric plexus, containing neurons mainly regulating epithelial transport functions and intestinal motility, respectively. In addition, there are a large number of primary sensory nerves, which upon activation release neuropeptides. The submucosal plexus consists of sensory, inter-, secreto-motor, and vasodilator neurons connected in complex microcircuits that mainly regulate mucosal ion transport and local blood flow. Thus, the lamina propria is densely innervated with nerve fibres containing a considerable number of different neurochemicals and neuropeptides that have direct and indirect effects on epithelial function.

IMMUNE CELLS IN THE INTESTINAL MUCOSA

The intestinal tract is the largest immune organ in the body, encompassing the lymphoid aggregates (Peyer's patches), and diffusely scattered immune cells throughout the mucosa and epithelium. All the major effector cells of immune reactions (lymphocytes, eosinophils, mast cells, neutrophils, macrophages, and dendritic cells) are present in or near the intestinal epithelial layer. An emerging concept is that also non-immune cells of the intestinal mucosa, classically viewed as bystanders, can display effector and modulatory functions in immunologic and inflammatory responses. Besides epithelial cells (Berin et al., 1999a), this group includes mesenchymal (fibroblasts, myofibroblasts, smooth muscle cells), endothelial and nerve cells, and in addition, non-cellular components such as the extracellular matrix containing matrix metalloproteinases (Fiocchi, 1997).

NERVE–IMMUNE INTERACTIONS IN THE INTESTINAL MUCOSA

In the intestinal mucosa, close approximations between lymphoid sites and nerve fibres have been found (Felten et al., 1987; Ottaway et al., 1987; Stead et al., 1987). Distances as short as 20–200 nm have been reported between peptidergic nerves and mast cells, lymphocytes, and plasma cells in the lamina propria (Stead et al.,1987; Feher et al., 1992). Direct membrane–membrane contacts between mast cells, eosinophils or plasma cells, and nerves often occur (Bienenstock et al., 1987), and it has been suggested that all of the inflammatory cells in the intestinal mucosa of rats may be structurally innervated (Stead, 1992). Leukocytes may also be transiently innervated, as suggested by synaptic-like contact between sympathetic nerve fibres and lymphocytes in the spleen (Felten et al., 1987). Moreover, neurotransmitters, neuropeptides, and neuroendocrine hormones can affect immune function and conversely cytokines can affect neural function. Many leukocytes express receptors for neuropeptides and neuroendocrine hormones, and the expression of these receptors is energy demanding (Pascual et al., 1999). This strongly suggests that the receptor expression is important and is a means to receive neuronal signals that modulate immune function. Thus, there seem to be anatomical and functional prerequisites for neuroimmune interaction in the intestinal mucosa. Neurally mediated epithelial ion secretion is well established (Cooke, 1998) and, more recently, immunomodulation of epithelial barrier function has become evident (McKay and Baird, 1999). Recent data are now also showing that epithelial cells can produce neuroregulatory factors, indicating that the communication between nerves, immune cells and epithelium may be tri-directional (Berin et al., 1999a).

The most well-studied functional consequences of neuro-immuno-epithelial interactions are studies of nerve–mast cell-mediated changes in epithelial function in models of food allergy, nematode infections, and stress.

MUCOSAL MAST CELLS

Mast cells are a heterogeneous population of cells, and their characteristics have been best studied in rodents. Mature mast cells are, in rodents, divided into intestinal mucosal mast cells (IMMC) and connective tissue mast cells (CTMC), which are distinct in several aspects (Lin and Befus, 1999). IMMC hyperplasia is more thymus-dependent than CTMC growth. Moreover, IMMC are smaller (12 μm in diameter), contain fewer granules per cell, have a smooth plasma membrane lacking microvilli and lobulated nuclei. Whereas heparin is the predominant proteoglycan in CTMC, chondroitin sulphate dominates in IMMC. The mast cell phenotypes also differ in mast cell proteases (IMMC: rat mast cell protease [RMCP]-II in rats, mouse MCP-1 and -2 in mice), arachidonic acid metabolite pathways (IMMC: prostaglandin [PG]D_2, LTB_4, LTC_4), and histamine content (IMMC 1/10 of CTMC). The distinct properties of IMMC also include their unique responsiveness to cytokines, secretagogues, and anti-allergic drugs. There are no currently available cell surface markers to distinguish different mast cell phenotypes. The human equivalent to IMMC are called MC_T, as they express only tryptase as protease (Lin and Befus, 1999).

Mast cells are present in all layers of the gut wall throughout the entire gastrointestinal tract. It has been estimated that the lamina propria contains 20 000 mast cells/mm^3 (Befus, 1994). The mast cells are, however, predominantly distributed in close proximity to nerves and in perivascular areas, in humans (Dvorak et al., 1992) as well as in rodents (Befus, 1994). Emerging evidence indicates that the close association between mast cells and enteric nerves is bi-directional and important in the regulation of intestinal transport functions.

Mucosal mast cells respond to both IgE-dependent (antigen) and non-IgE-dependent (bacterial toxins, neurotransmitters, etc.) stimulation, and release a wide variety of bioactive mediators that differ in their potency and biological activities. Many of these mediators have the ability to alter mucosal function, as shown in Ussing chamber studies with intestinal tissue or cultured epithelial cell monolayers (Table 11.1). It is noteworthy that the spectrum of mediators and cytokines produced and released by the heterogeneous mast cells is different for each subpopulation and depends on the form of stimulation (Theoharides et al., 1982; Metcalfe et al., 1997; Lorentz et al., 2000). The mediators can be divided into preformed mediators, newly synthesized mediators, and cytokines (Lin and Befus, 1999). *Preformed mediators* are constitutively synthesized, stored in cytoplasmic granules, and released upon stimulation. The preformed mediators include biogenic amines (histamine and serotonin), proteoglycans (heparin and chondroitin sulphate), and proteases (tryptase, chymase, RMCP II, etc.). *Newly synthesized mediators* mainly consist of the metabolites from arachidonic acid (PGs, LTs, thromboxanes, and HETEs) and 2-acetylated phospholipids (platelet-activating factor [PAF]), and also include nitrogen radicals (NO) and oxygen radicals (superoxide anion, hydrogen peroxide, hydroxyl radical). A diversity of *cytokines* are produced by mast cells. Those shown to be produced

TABLE 11.1
Mast cell mediators and their effects on epithelial function.

Mediators		Ion secretion	Barrier function
Pre-formed mediators			
Biogenic amines	Histamine	×	
	5-hydroxytryptamine (5HT)	×	
Proteoglycans	Heparin		
	Chondroitin sulphate		
Protease	Tryptase	×	
	Chymase		
	Carboxypeptidase A		
	Cathepsin G protease		
	Rat mast cell protease (RMCP)		×
Newly synthesized mediators			
Lipid-derived	Prostaglandins	×	
	Leukotrienes	×	
	Platelet activating factor (PAF)	×	
	HETEs, HPETEs		
Cytokines	TNF-α	×	×
	IL-1–6, IL-8, IL-18	×	×
	IFN-γ	×	×

Mast cell mediators listed are collective data from mucosal and connective tissue mast cells, many of which have the ability to alter mucosal function as shown in Ussing chamber studies with intestinal tissues or cultured epithelial cell monolayers.

by both rodent and human mast cells include tumour necrosis factor (TNF)-α, IL-1β, IL-3, IL-4, IL-5, IL-6, IL-13, and chemokines IL-8, MIP-1α,β, RANTES, MCP-1. The regulation of mediator release from mucosal mast cells is poorly understood. It is, however, becoming increasingly evident that enteric nerves are important in this respect, and that secretion of different mediators is regulated through different mechanisms. Understanding the details in these events is important for designing new therapeutic strategies for mast cell-mediated disorders.

TECHNIQUES FOR THE STUDY OF NEUROIMMUNE MODULATION OF INTESTINAL FUNCTION

The functional properties of the intestinal epithelium can be studied *in vivo* or *in vitro*. While the *in vivo* studies are more physiological, the *in vitro* approach makes possible detailed studies of epithelial electrophysiology and mechanisms. *In vivo* techniques for assessing mucosal function involve measurements of net water flux and permeability to small marker molecules in intestinal segments.

USSING CHAMBERS

Many of the *in vitro* studies of mucosal function have employed Ussing-type chambers, the principle of which is shown in Figure 11.1. A flat sheet of mucosa is mounted between two half-chambers filled with continuously oxygenated (95% O_2, 5% CO_2) buffer. The gas ports are arranged to provide a lift system for circulating the buffer, providing efficient mixing of the fluid and reducing the thickness of the unstirred water layer to physiological values (Karlsson and Artursson, 1992). Temperature is kept at 37 °C by thermostatically controlled reservoirs. The use of two pairs of electrodes makes it possible to measure the electrophysiological properties of the epithelium during experiments.

The ability to maintain a transepithelial potential difference (PD) is a characteristic shared by all transporting epithelia and is dependent on the activity of all the electrogenic ion pumps in the epithelial cell membranes, mainly Na^+/K^+-ATPase, and on the epithelial barrier function (Armstrong, 1987). Electrophysiologically, the two can be theoretically separated: the short circuit current (I_{sc}), that is, the current needed to make the PD 0 mV, is a function of the activity of the ion pumps, and the conductance (the inverse of electrical resistance) reflects the passive movement of ions through the paracellular routes, that is, mainly the tight junctions. One pair of electrodes, preferably connected via agar bridges, is placed adjacent to the tissue for measurements of spontaneous PD, and current passing electrodes are placed in the periphery of the chambers. The current needed to nullify the PD is the I_{sc}. The basal PD or I_{sc} can be used as a measure of tissue viability, since active ion transport requires energy production, generally in the form of ATP. From the change in PD, by passing the current (I) through the epithelium, the transepithelial conductance, or its inverse, transepithelial resistance can be determined by Ohm's law. This calculation relies on simplified equivalent circuit models of epithelia, viewing the epithelium as a parallel circuit consisting of paracellular and transcellular pathways. Microelectrode impalement studies have shown that this approach in most cases is a reasonable approximation of the epithelium (Gordon *et al.*, 1989; Madara, 1998).

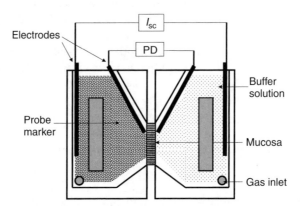

Figure 11.1   Schematic illustration showing the principle of Ussing-type chambers used for studies of intestinal mucosal function. The mucosal sheet is mounted between the two half-chambers. The gas inlet for oxygenation is situated low in the chamber to produce circulation of the fluid via 'gas lift'. Two pairs of electrodes are used, one pair for measurements of transepithelial potential difference (PD), and the other pair for supplying current to short circuit the system (I_{sc}), which also allows calculation of the transepithelial resistance. See text for further details.

Ion secretion can be studied by monitoring I_{sc}, as an indicator of active luminally-directed ion secretion. Antigens or pharmacological agents added to either side of the chamber, or neural activation by transmural electrical stimulation may alter active ion transport, which is shown by a change in I_{sc}. Ussing chambers also allow the researcher to conduct studies of bi-directional flux of radioactive isotopes, both to measure ion transport (e.g. isotopes of chloride and sodium) or permeability (isotopically labelled probe molecules). Studies with specific ion-free buffers also give information on the identity of ions involved in secretory responses.

To determine intestinal permeability, various marker molecules are added, usually to the mucosal (luminal) buffer, and their fluxes across the tissue to the serosal buffer are measured. These fluxes reflect mainly paracellular permeability if small probes are used (e.g. [51]chromium-ethylene-diamine tetraacetic acid [EDTA], small sugars/dextrans) or transcellular permeability if macromolecular probes are used (e.g. horseradish peroxidase, ovalbumin). Measurements in the presence (or absence) of inhibitors of nerve transmission (e.g. tetrodotoxin) and/or mast cell-degranulation (e.g. doxantrazole) and with various agonists and antagonists for neuropeptides (e.g. substance P), and mast cell activators (e.g. anti-IgE) have resulted in the accumulation of a large amount of data on the role of nerves and mast cells in the regulation of mucosal function.

NERVE–MAST CELL INTERACTION IN FOOD HYPERSENSITIVITY

Food allergy has mainly been studied in rodent models, using rats or mice sensitized to a protein antigen, either actively by systemic injection of the antigen together with an adjuvant to stimulate IgE production, or passively by injection of IgE-containing serum from actively sensitized animals. Mucosal function can then be studied during hypersensitivity reactions induced by antigen challenge to the intestinal lumen. Many studies, including those in our laboratory have typically used ovalbumin (OVA) as the sensitizing antigen.

ION SECRETION

Early studies in OVA-sensitized rats revealed abnormalities in jejunal ion and water absorption after luminal antigen challenge (Perdue et al., 1984a), whereas mucus release and goblet cells were unaffected (Perdue et al., 1984b). The functional changes were associated with reduced numbers of granulated mast cells and decreased histamine content in mucosal homogenates. The use of mast cell-stabilizing agents (Perdue and Gall, 1985, 1986; Crowe et al., 1990) and mast cell-deficient animals (Perdue et al., 1991) provided further evidence for mast cell involvement in the regulation of epithelial ion and water transport. Doxantrazole, which stabilizes both CTMC and IMMC types of mast cells, prevented antigen-induced changes in epithelial function (Perdue and Gall, 1985, 1986), whereas sodium cromoglycate, which stabilizes only CTMC, failed to inhibit the response (Perdue and Gall, 1985), suggesting that intestinal ion secretion in response to luminal antigen challenge is mainly mediated by activation of mucosal mast cells. Intestine removed from sensitized rats and mounted in Ussing chambers rapidly responded to luminal OVA challenge with an increase in I_{sc}. This I_{sc} response was due mainly to Cl^- secretion, and was inhibited by mast cell-stabilizing agents (Perdue and Gall, 1986). Further support for mast cells as mediators came from studies of co-cultured human colonic epithelial cells (HCA-7) and peritoneal mast cells from sensitized guinea pigs (Baird et al., 1987), where antigen challenge increased Cl^- secretion only when mast cells were added to the serosal side of the epithelium (where mast cells normally are located in vivo) and only when mast cells were from sensitized animals. Conclusive evidence for the role of mast cells in the regulation of intestinal secretion was provided by studies using genetically mast cell-deficient mice (W/W^V). These mice lack a normal receptor (c-kit tyrosine kinase) for stem cell growth factor (c-kit ligand), which is necessary for the differentiation of mast cells; thus, the mice are completely deficient in functional intestinal mast cells. In intestinal preparations from W/W^V mice, antigen-stimulated responses were minimal compared with congenic control $(+/+)$ mice (Perdue et al., 1991). Reconstitution of mast cells in W/W^V mice, by injection of bone marrow containing precursor cells from $+/+$ mice, restored the I_{sc} responses to antigen to levels observed in challenged $+/+$ mice (Table 11.2). Taken together, these studies indicate that mast cell-dependent mechanisms are primarily responsible for the ion secretion associated with intestinal anaphylaxis.

A neural component in the mucosal response to antigen-challenge in sensitized rats was shown by Crowe et al. (1990, 1993a). In vivo (Crowe et al., 1990), as well as in vitro (Crowe et al., 1993a), the antigen-induced increase in jejunal ion secretion was significantly inhibited by the neurotoxin, tetrodotoxin. Moreover, rats that had undergone neonatal treatment with capsaicin (to deplete substance P-containing nerves) prior to sensitiszation to OVA had a diminished I_{sc} response to antigen challenge. The involvement of neuropeptides in antigen-induced ion secretion was confirmed in a more recent study (McKay et al., 1996). In jejunal mucosa mounted in Ussing chambers, neuropeptide Y and tetrodotoxin inhibited the I_{sc} response to luminal, but not serosal, antigen challenge. Conversely, substance P counteracted neuropeptide Y, and restored the luminal antigen-induced secretory response to pre-treatment values. It was also shown that substance P can mediate direct nerve–mast cell communication in vitro (Suzuki et al., 1999), and that substance P-induced ion secretion in intestinal tissue is mediated via NK-1 receptor-stimulation and is partly dependent on enteric nerves and mast cells (Wang et al., 1995).

TABLE 11.2
Effect of mast cells on I_{sc} responses to antigen and nerve stimulation.

Mouse group	Response to antigen ΔI_{sc} (μA/cm^2)	Response to nerve stimulation ΔI_{sc} (μA/cm^2)
Control $+/+$	38 ± 5	43 ± 3
Mast cell-deficient W/W^V	$17 \pm 2*$	$16 \pm 1*$
Mast cell-reconstituted W/W^V	37 ± 3	42 ± 4

Jejunal tissues from normal mice $(+/+)$, mast cell-deficient mice (W/W^V), and mast cell-reconstituted W/W^V mice $(+/+$ bone marrow to $W/W^V)$ were studied in Ussing chambers; values represent mean $\pm$ SEM; $n = 12$ in each group. The I_{sc} response to antigen is the maximal change in I_{sc} within the first 5 min after addition of antigen; the I_{sc} response to nerve stimulation is the peak I_{sc} after electrical transmural stimulation of enteric nerves in the tissue. $*p < 0.05$ compared with $+/+$ controls.

This suggests that neuropeptides play a regulatory role in immunophysiology by acting at neural and epithelial sites in the intestinal mucosa. Experiments with W/W^V mice also demonstrated a functional interaction between mast cells and nerves in the regulation of intestinal ion transport (Perdue *et al.*, 1991; Wang *et al.*, 1995). In sensitized $+/+$ mice, the antigen-induced I_{sc} response was inhibited by tetrodotoxin, which was not the case in W/W^V mice. Moreover, increases in I_{sc} in response to electrical transmural stimulation of nerves were approximately 50% in W/W^V versus $+/+$ mice, and were inhibited by antagonists of mast cell mediators in $+/+$ but not W/W^V mice. Again, the diminished I_{sc} response in W/W^V mice was restored to normal levels by reconstitution of mast cells (Table 11.2). Thus, stimulation of nerves induced the release of mast cell mediators that amplified the secretory response of the epithelium. These data provide evidence for the functional importance of communication between nerves and mast cells in the regulation of jejunal ion transport in food hypersensitivity.

Studies from other laboratories have illustrated that nerve–mast cell interactions involving several types of enteric neurons are important in models of food allergy. The role of cholinergic neurons in mediating Cl^- secretion in anaphylaxis has been studied in muscle-stripped segments of distal colon from guinea pigs immunized to bovine β-lactoglobulin. Antigen challenge in Ussing chambers evoked a concentration-dependent increase in I_{sc} in tissues from sensitized, but not naive animals. The I_{sc} response to β-lactoglobulin was associated with a concentration-dependent increase in acetylcholine release and was inhibited by atropine and tetrodotoxin. This finding suggested that antigen-induced secretory responses result, in part, from activation of cholinergic neurons that utilize muscarinic synapses for transfer of signals to the epithelium (Javed *et al.*, 1992). Further, it was shown that submucosal neurons in sensitized animals were hyper-excitable, and that histamine antagonists reversed this hyper-excitability. It seems that signalling from mucosal mast cells to the enteric nervous system, using histamine as a paracrine messenger, is important in colonic responses to specific antigens (Frieling *et al.*, 1994).

Enkephalins and other opioid derivates have antisecretory effects in the small intestine of many species through interactions with δ-opioid receptors (Quock *et al.*, 1999). For example, the selective δ-opioid agonist [D-Pen^2,D-Pen^5]enkephalin (DPDPE) increased net salt absorption by tetrodotoxin- and naloxone-sensitive mechanisms, and blunted I_{sc} elevations evoked by transmural electrical stimulation of submucosal nerves (Quito and Brown, 1991). It was recently found that DPDPE inhibited the rise in I_{sc} evoked by histamine, compound 48/80 and β-lactoglobulin in pigs sensitized to milk proteins (Poonyachoti and Brown, 2001), an effect that was prevented by naltrindole, a highly selective δ-opioid receptor blocker. This functional relationship between mast cell products and opioid-sensitive neural pathways was underscored by histochemical results demonstrating that mast cells are in close proximity to nerve fibres expressing δ-opioid receptors (Poonyachoti and Brown, 2001). A possible mechanism for these opioid nerve–mast cell interactions is via tryptase-induced activation of protease-activated receptor-2 (PAR-2). Green *et al.* (2000) showed that PAR-2 activation by trypsin or a receptor-activating peptide, SLIGRL-NH2, caused I_{sc} elevation in the porcine ileum. This I_{sc} rise was inhibited by saxotoxin (neuronal conduction blocker) and by DPDPE, and PAR-2-like immunoreactivity was seen on both cholinergic and non-cholinergic submucosal nerve fibres. As mast cells containing PAR-2-activating proteases are closely associated with nerve fibres in the normal and inflamed intestine, it was suggested that mast cell-degranulation results in cleavage of PAR-2 on submucosal neurons to trigger the release of neurotransmitters that stimulate fluid and electrolyte secretion from enterocytes.

It seems plausible that mast cell mediators of different types interact with various types of enteric neurons, which modulate epithelial responses to antigen challenge. Interactions between the central nervous system (CNS) and mucosal mast cells have also been implicated in the regulation of ion secretion in food hypersensitivity. MacQueen et al. (1989) showed that Pavlovian conditioning was able to activate mucosal mast cells in rats via neuroimmune signals from the CNS. Antigen challenge, which stimulates mast cells (indicated by release of RMCP II, a highly specific protease of mucosal mast cells), was paired with an audiovisual cue once per week for 3 weeks. In the fifth week, animals re-exposed to only the audiovisual cue released a significantly greater quantity of RMCP II than negative control animals (previously exposed to cue and antigen in a non-contingent manner), and not different from animals re-exposed to both the cue and the antigen. In a later study of intestinal tissues in Ussing chambers, rats subjected to the same conditioning protocol demonstrated increased jejunal ion secretion in response to the antigen-associated audiovisual cue (Djuric et al., 1994), supporting a role for the CNS as a functional effector of mast cell-induced changes in gut mucosal function, at least in the allergic state. Moreover, central vagal activation in rats induced release of mast cell mediators, paralleled by increased protein leakage to the intestinal lumen (Santos et al., 1996).

It has also been shown that information on immune status in the gut is rapidly relayed to the CNS. Castex et al. (1995) demonstrated neuronal activation in vagal brain stem nuclei, by Fos immunohistochemistry, after intestinal anaphylaxis. The mechanism responsible for this finding was mast cell-induced stimulation of mesenteric afferent nerves via activation of serotonin 5-HT$_3$ and histamine H$_1$ receptors (Jiang et al., 2000). This specific information on the inflammatory state of the bowel enables timely responses to the presence of intraluminal antigen. Moreover, it was recently shown that mast cells can be a major source of substance P when the vanilloid receptor 1 is stimulated by capsaicin (Moriarty et al., 2001), providing another method for the mucosa to respond to changes in the luminal environment independently from afferent nerve activation.

MUCOSAL BARRIER FUNCTION

The involvement of nerve–mast cell interactions in antigen-induced enhanced intestinal permeability was first suggested by Crowe et al. (1993). Sensitized but not control rats demonstrated a 15-fold increase in ^{51}Cr-EDTA uptake after intraluminal antigen challenge, whereas no change occurred with addition of an unrelated protein. Moreover, uptake from the lumen of the macromolecular protein antigen (OVA) itself was enhanced 14-fold in sensitized animals. Antigen challenge was accompanied by release of RMCP II and morphological evidence of mast cell degranulation in sensitized rats. The neurotoxin tetrodotoxin (applied directly to the serosal surface of ligated jejunal segments) inhibited OVA-induced uptake of ^{51}Cr-EDTA and antigen. In addition, in jejunal mucosa mounted in Ussing chambers, tetrodotoxin (Crowe et al., 1993; McKay et al., 1996) and neuropeptide Y (McKay et al., 1996) inhibited the I_{sc} response to luminal, but not serosal, antigen challenge. These results indicate that neural factors may influence the uptake of macromolecules from the gut lumen during intestinal anaphylaxis.

The rise in I_{sc} upon antigen challenge to the serosal side was immediate (within 30 s), whereas addition of antigen to the luminal side involved a lag phase of approximately 3 min before an I_{sc} rise was recorded (Baird et al., 1985; Crowe et al., 1990). In addition, the I_{sc} response to luminally applied antigen was reduced by 50% compared with the response to serosal challenge, and the response to luminal antigen was abolished if the antigen was first applied to the serosal side. These findings suggested a final common effector mast cell, with the lag phase after luminal antigen challenge reflecting the time required for transepithelial antigen transport before submaximal activation of these cells. In addition, they indicated the presence of a novel system for rapid transepithelial antigen transport in sensitized animals, since in normal epithelial cells 20–30 min was required for luminal protein to reach the basolateral surface (Keljo and Hamilton, 1983; Bomsel et al., 1989).

To study mechanisms and routes of this rapid antigen transport , rodents were sensitized to horseradish peroxidase (HRP) (Berin et al., 1997, 1998). HRP was used since it is similar in size to common food protein antigens, the concentration of the intact molecule can be accurately determined by measuring enzymatic activity, and its location within cells and tissue can be visualized by electron microscopy. In HRP-sensitized rats, enhanced uptake of HRP within the endosomal compartment of jejunal enterocytes was documented even at 2 min after luminal antigen challenge (before mast cell activation) (Berin et al., 1997). Both the area of HRP-containing endosomes within enterocytes and their rate of transcellular transport were increased in sensitized animals compared with controls, with HRP already in the lamina propria. This enhanced uptake was antigen specific, and the transport pathway was exclusively transcellular at this time point. At 30 min post-challenge (well after mast cell activation and the I_{sc} response), HRP was visualized not only in endosomes, but also within the tight junctions and paracellular spaces between the enterocytes in sensitized rats. At this phase, an increase in tissue conductance and transmucosal flux of HRP was also recorded. These studies suggested two phases of transepithelial transport, with the first phase (before the hypersensitivity reaction) being transcellular and endosomal, and the second phase (after mast cell activation) involving the paracellular pathway. Studies using sensitized mast cell-deficient rats (Ws/Ws rats, with a similar spontaneous mutation in the c-kit gene as W/W^V mice) confirmed the requirement for mast cells in the second phase of transepithelial transport, since the increase in conductance and flux were totally absent in these rats (Berin et al., 1998). On the other hand, the first phase of rapid transcellular uptake of antigen was not affected by absence of mast cells.

Recently, Yang et al. (2000) showed that the specific and rapid antigen transport system during the first phase was mediated by IgE antibodies bound to low-affinity receptors on epithelial cells. Immunohistochemistry demonstrated that sensitization induced expression of CD23, the low-affinity IgE receptor (also denoted FcεRII), on epithelial cells, and whole serum but not IgE-depleted serum from sensitized rats was able to transfer the enhanced antigen transport phenomenon. The number of immunogold-labelled CD23 receptors on the enterocyte microvillous membrane was significantly increased in sensitized rats and was subsequently reduced after antigen challenge, when CD23 and HRP were co-localized within the same endosomes. Yu et al. (2001) have extended these studies in gene deficient mice, showing that IL-4 is required for IgE/CD23-mediated enhanced antigen uptake in enterocytes of sensitized mice, that the role of IL-4 is not solely to promote IgE synthesis, but also involves upregulation of CD23 expression on enterocytes.

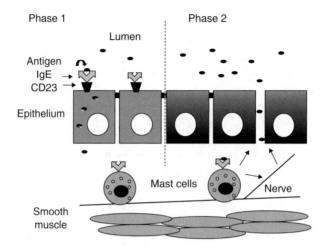

Figure 11.2   Current understanding of the various steps in intestinal mucosal pathophysiology in food allergy. Phase 1 involves enhanced transepithelial uptake of specific antigen via increased expression of CD23. Binding of IgE and antigen to this receptor allows rapid transcytosis through the enterocytes. Subsequently, antigen activates mast cells in the lamina propria to release mediators that induce ion secretion and phase 2 antigen transport by increasing the permeability of the tight junctions, thus allowing more antigen through the paracellular spaces.

The current understanding of the functional consequences of nerve–mast cell interactions in food hypersensitivity are summarized in Figure 11.2. Active sensitization induces a mast cell-independent, IgE/CD23-mediated enhanced antigen uptake resulting in rapid activation of mast cells in the lamina propria. Mast cell mediators (histamine, PGs, 5-HT, RMCP II, etc.) and enteric nerve signalling (acetylcholine, VIP, substance P, neuropeptide Y, etc.) then act together to alter epithelial function, inducing increased ion and water secretion, as well as increased permeability to small and large molecules.

MUCOSAL CHANGES IN THE LATE PHASE OF INTESTINAL ALLERGIC REACTIONS

Relatively little information exists concerning the late phase of the allergic reaction in the gastrointestinal tract. Yang *et al.* (2001) characterized jejunal mucosal pathophysiology and inflammation hours to days after oral antigen challenge of sensitized rats, and examined the role of mast cells in events after challenge. Intestine from sensitized Sprague–Dawley rats demonstrated enhanced ion secretion and permeability up to 72 h after challenge, and electron microscopy revealed abnormal mitochondria within enterocytes and disruption of the epithelial basement membrane associated with influx into the mucosa of mast cells, eosinophils, neutrophils, and mononuclear cells. In contrast, antigen-challenged mast cell-deficient Ws/Ws rats demonstrated no functional changes or inflammatory cell infiltrate. Thus, oral antigen challenge of sensitized rats induced sustained epithelial dysfunction, and mast cells mediated both epithelial pathophysiology and recruitment of additional inflammatory cells that may have contributed to persistent pathophysiology and symptoms.

NERVE–MAST CELL INTERACTIONS IN
PARASITE INFECTIONS

In addition to models of food hypersensitivity, models of parasitic infection have provided important insights in our understanding of neuroimmune regulation of intestinal epithelial function. These models include infections in rodents and guinea pigs with the nematodes, *Nippostongulus brasiliensis* and *Trichinella spiralis*. Both of these parasites infest the proximal small bowel in rats, causing an acute inflammation between day 7 and 10. After the worms have been expelled from the host (~35 days post-infection) mast cell hyperplasia develops in the intestinal mucosa (Perdue *et al.*, 1990), and these mast cells are in contact with peptidergic nerves (Stead *et al.*, 1987). Thereafter, rats become immune to re-infection, and an anaphylactic intestinal response occurs upon re-exposure to the parasite, leading to rapid expulsion.

Studies with *N. brasiliensis* showed that during acute inflammation (day 7–10 post-infection), when active worm expulsion began, changes in mucosal function were present and paralleled by extensive epithelial damage at the villus tips, decreased numbers of stained mast cells in the mucosa, and high serum levels of RMCP II (Perdue *et al.*, 1989). In jejunal mucosa, an increased basal I_{sc} and secretion of Na^+ and Cl^- were found (Perdue *et al.*, 1990), and the magnitude of the I_{sc} response to electrical transmural stimulation of enteric nerves was significantly reduced to 17–33% of control values, suggesting abnormalities of mucosal nerves. In addition, *in vivo* permeability to ^{51}Cr-EDTA and OVA (measured in blood and urine 5 h after injection into a ligated loop of jejunum) were increased approximately 20-fold at day 10 (Ramage *et al.*, 1988).

Following worm expulsion (day 35), baseline ion transport and serum levels of RMCP II returned to normal. However, a pronounced (8–10-fold) increase in mast cell numbers was apparent in the intestinal mucosa. At this time, I_{sc} responses to transmural nerve stimulation in the absence of antigen were enhanced, and challenge of the intestine with worm antigen induced increased ion and water secretion (Perdue *et al.*, 1989). Moreover, intravenous challenge with worm antigen in primed rats induced enhanced (approximately 10-fold compared with controls) intestinal permeability to ^{51}Cr-EDTA and OVA (Ramage *et al.*, 1988). Using *ex vivo* perfusion of the jejunum of *N. brasiliensis*-primed rats, Scudamore *et al.* (1995) demonstrated a time-dependent correlation and dose response between the release of RMCP II and increased permeability to macromolecules following intravascular challenge with worm antigen, whereas no morphological abnormalities were seen in the epithelium. Therefore, release of RMCP II from mast cells induced by worm antigen cross-linking IgE on the mast cell surface appeared to account for the increased gut permeability in this anaphylactic response.

Similar results were found in *T. spiralis*-infected rats. Worm antigen challenge in primed rats induced a significant decrease in net fluid absorption, an increase in net intestinal Cl^- secretion and rapid parasite rejection (Harari *et al.*, 1987). Antigenic challenge of sensitized jejunum caused local release of 5-HT, histamine, and PGE_2. The antigen-induced rise in I_{sc} was mimicked by exogenous 5-HT or histamine and blocked by pretreatment of tissue with 5-HT and histamine H1-antagonists. Atropine and tetrodotoxin significantly blunted the I_{sc} response, as well as the change in I_{sc} caused by exogenous 5-HT or histamine (Castro *et al.*, 1987).

Taken together, these findings suggest that changes in mast cells and enteric nerves occur during acute inflammation and antigen re-challenge of nematode infected animals and implicate nerve–mast cell interactions with the epithelium in producing the ion-transport and permeability abnormalities.

NERVE–MAST CELL INTERACTIONS IN STRESS

There is a large body of evidence that severe physical stress, for example, trauma, burns, and also major surgery, can cause gastrointestinal dysfunction and pathology, that is, stress ulcers, multiple organ dysfunction, bacterial translocation, increased intestinal permeability, etc. To study the possibility of an intestinal disease-promoting effect of life stressors, animal models of non-traumatic 'physiological' stress have been developed. Models of acute stress used in studies of intestinal mucosal function in humans include dichotomous hearing and cold-induced hand pain stress (Barclay and Turnberg, 1987, 1988; Santos et al., 1998). Animal models used in this field include components of both psychological and physical stressors. Even the stress of transport and handling of animals was shown to affect the intestinal barrier (Wilson and Baldwin 1999; Meddings and Swain, 2000).

The most widely used model for studies of intestinal function is acute (a single exposure) restraint stress (RS), or immobilization stress, in rodents and mice. A variant of RS is cold restraint stress (CRS), where the restrained animals are placed in a cold room, 4–8 °C for various periods of time. Water avoidance stress (WAS, placing animals on a small platform surrounded by room-tempered water) has been used as a model of pure psychological stress. Experiments with chronic (sequential exposure over several days) mild stress, such as WAS, has been widely used in psychological and psychiatric research, and has recently been introduced in studies of intestinal mucosal function (for review see Söderholm and Perdue (2001)). Social defeat and depravation have also been utilized as stress models in psychological research and are currently gaining interest from researchers studying intestinal function (Coutinho et al., 2000; Rosztoczy et al., 2001; Söderholm et al., 2002a).

ION SECRETION

Studies using intestinal perfusion techniques in the human jejunum revealed effects of acute stress on intestinal secretion. Barclay and Turnberg (1987, 1988) found that psychological stress, induced by dichotomous listening or cold-induced hand pain stress, reduced mean net water absorption and changed net sodium and chloride absorption to secretion. These stress effects were inhibited by i.v. atropine infusion, suggesting a cholinergic parasympathetic nervous mechanism, and studies in rodents have confirmed these findings. In acute CRS in rats, an elevated baseline I_{sc} was found in isolated jejunal segments (Saunders et al., 1994). Net Cl^- secretion was induced, and substitution of Cl^- in the bathing buffer eliminated the abnormality, indicating that stress stimulated chloride ion secretion. On the other hand, the magnitude of the I_{sc} response to electrical transmural stimulation of enteric nerves was significantly less in tissues from CRS than from control rats, suggesting an impaired responsiveness. The ability of the epithelium to secrete in response to exogenous stimulation with the neurotransmitter agonists, for example, bethanechol or VIP, was unimpaired, implicating a stress-induced neural change or depletion of neurotransmitter. The finding of an elevated baseline I_{sc} following stress was verified (Saunders et al., 1997), and shown to involve cholinergic nerves. Moreover, the magnitude of the stress response was inversely correlated to mucosal cholinesterase activity in the two rat strains used in the study (Wistar-Kyoto rats more stress susceptible but with low cholinesterase activity compared with Wistar rats). Similar findings regarding I_{sc} were found in the rat colon after CRS (Santos et al., 1999a). Taken together these results suggest that acetylcholine can mediate stress-induced abnormalities in chloride secretion via muscarinic receptors located in the gastrointestinal tract.

Corticotropin-releasing hormone (CRH) has been implicated in various stress-induced abnormalities, including gastrointestinal function. RS-induced functional changes in colonic epithelium were inhibited by the CRH antagonist α-helical CRH (Santos et al., 1999a). Moreover, peripheral (i.p.) injection of CRH mimicked stress-induced changes in colonic ion secretion, and these effects were inhibited by doxantrazole. The CRH-induced effects were not inhibited by blocking steroid synthesis with aminoglutethimide (Castagliuolo et al., 1996a; Million et al., 1999; Santos et al., 1999a). The involvement of neurons was indicated by modulation of the CRH response by atropine, hexamethonium, and bretylium (Santos et al., 1999a). Taken together the findings in these studies suggest that CRH is important for stress-induced changes in colonic ion secretion, and that its effects are mediated by peripherally located receptors, probably on enteric nerves, and involve activation of mast cells.

By extending the cold-induced pain stress model in humans, Santos et al. (1998) found that jejunal water secretion induced by stress was paralleled by luminal release of the mast cell mediators tryptase and histamine, suggesting an interaction between the CNS, mast cells, and the intestinal mucosa. Several studies from various groups have highlighted the importance of mast cells in stress-related changes in intestinal mucosal function in animal models. Moreover, studies of mucosal ultrastructure have found mucosal mast cell activation in combination with various signs of intestinal function disturbances during stress (Kiliaan et al., 1998; Theoharides et al., 1999; Wilson and Baldwin, 1999). These results confirm the results of mast cell activation by Pavlovian conditioning in food hypersensitivity showing that CNS has the ability to modulate intestinal mast cells, and suggest that mast cells play a role in stress-related gut dysfunction. In a study of intestinal barrier function during chronic stress (5 days of WAS) in mast cell-deficient and normal rats, Santos et al. (2000) showed that the stress-induced rise in jejunal I_{sc} was absent in mast cell-deficient animals. In normal rats subjected to 5-day stress, there was an increased number of activated mast cells in the colonic mucosa, some in contact with nerves which appeared depleted of neurotransmitter (Figure 11.3).

To summarize, acute CRS in rats increases baseline Cl⁻ in the jejunum and colon by a mechanism involving CRH and mast cells, and increased baseline activity of nerves. The ability to respond to noxious substances may, however, be impaired. A similar pattern is seen in chronic WAS and involves mucosal mast cell hyperplasia.

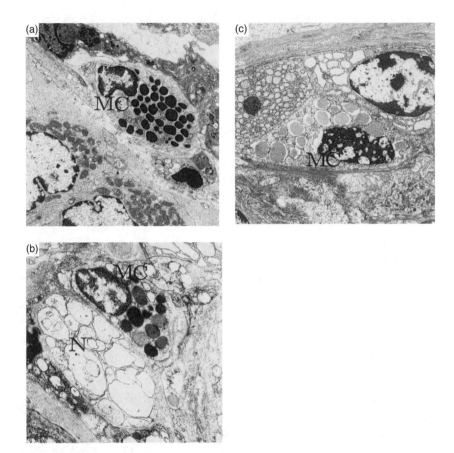

Figure 11.3   Effect of chronic stress on mast cells (MC) in the colonic mucosa. Rats ($+/+$) were subjected to water avoidance stress or sham stress (1 h/day) for 5 days. Electron photomicrographs show: (a) a mast cell in the mucosa of a sham stressed rat; (b) a mast cell and closely apposed nerve, both with signs of activation, in the mucosa of a stressed rat; (c) a mast cell with reduced granule density in the mucosa of a stressed rat. From Santos *et al.* (2001), with permission from *Gut*.

MUCUS SECRETION

In a series of experiments, Castagliuolo *et al.* (1996a,b, 1998) described the effects of RS on colonic mucin secretion. Thirty minutes of immobilization stress caused a significant increase in mucin release from colonic mucosal explants and goblet cell depletion by histological evaluation, paralleled by increased colonic mucosal levels of the mast cell mediators RMCP II, PGE_2, and cyclooxygenase-2 mRNA (Castagliuolo *et al.*, 1996a). The stress-associated changes were reproduced by injection of CRH in non-stressed rats, and pre-treatment of rats with a CRH antagonist or the mast cell stabilizer lodoxamide inhibited the stress-induced effects. In another study, Castagliuolo *et al.* (1996b) suggested the neuropeptide, neurotensin, as a candidate for transmitting the CRH-induced effects within the colonic wall. These findings suggested that RS stimulates colonic mucin secretion via release of CRH and activation of neurons and mast cells. To directly assess the contribution of mast cells, colonic responses to RS were compared in mast cell-deficient and normal mice (Castagliuolo *et al.*, 1998). RS-induced mucin and prostaglandin release in the colon did not occur in mast cell-deficient mice, in contrast to normal animals. However, mast cell-deficient mice that had their mast cell population reconstituted by injection of bone marrow-derived mast cells from normal animals had the same colonic response to stress as normal mice. The findings of a stress-activated mucin release have recently been corroborated by findings of goblet cell activation by environmental stress in rats (Wilson and Baldwin, 1999). These studies further underline the importance of nerve–mast cell interaction in the regulation of intestinal transport.

MUCOSAL BARRIER FUNCTION

Saunders *et al.* (1994) found that rats subjected to 4 h of RS showed overt barrier dysfunction, as assessed by jejunal conductance and permeability to the inert marker molecules mannitol and ^{51}Cr-EDTA, despite normal light microscopy. A follow-up study showed an enhanced permeability response to RS in Wistar-Kyoto rats (with low cholinesterase activity) compared to the parent Wistar strain (Saunders *et al.*, 1997), and the possibility that cholinergic nerves were involved in modulating barrier function was further emphasized by inhibition of the stress response by atropine or atropine methyl nitrate (does not pass blood–brain barrier). To determine whether the stress-induced epithelial barrier defect extended to macromolecules with antigenic potential, Kiliaan *et al.* (1998) studied transport of HRP in isolated jejunal segments. Wistar-Kyoto rats exposed to RS for 2 h showed increased permeability to HRP (in this case, a bystander macromolecule) and to ^{51}Cr-EDTA. Moreover, electron microscopy revealed an increased number of HRP-containing endosomes in the enterocytes, and HRP within the tight junctions in stressed rats. The increase in transcellular endosomal uptake and flux of the macromolecule was also inhibited by atropine. In a separate study, the cholinergic agonist carbachol alone increased transcellular and paracellular flux of HRP in isolated small intestine from normal rats (Bijlsma *et al.*, 1996). The importance of cholinergic mechanisms for regulation of the response of the intestinal barrier to stress was also shown by Castagliuolo *et al.* (1996a): atropine inhibited colonic mucin and RMCPII release following RS. A stress-induced defect in the colonic barrier was recently reported by Santos *et al.* (1999a). Isolated colonic segments from Wistar-Kyoto rats showed increased conductance, as well as increased permeability to HRP and the bacterial chemotactic peptide f-MLP after 2 h of RS. In addition, electron microscopy showed an increased uptake of HRP via both the transcellular and paracellular pathways in stressed rats. Experiments involving i.p. treatment with CRH implicated a role for CRH in colonic hyper-permeability following stress (Santos *et al.*,1999a). The CRH-induced effects were inhibited by doxantrazole, suggesting that CRH released during stress increases permeability by activation of mast cells.

A role for opioids in the stress-induced increase in colonic permeability was found in a study of adaptation to acute stress in Wistar-Kyoto rats (Yates *et al.*, 2001). While continuous WAS for 60 min significantly increased I_{sc}, conductance, and HRP flux, rats subjected to stress 3×20 min (with 5 min resting periods in between) showed a rapid adaptation, with functional values similar to those in non-stressed controls. Pre-treatment with naloxone reversed this adaptive response, and all variables returned to the abnormal values obtained following continuous stress (Yates *et al.*, 2001).

With *in vivo* techniques for estimating permeability, Meddings and Swain (2000) recently confirmed that acute stress increases intestinal permeability. Using models involving either RS or swimming with a floatation device, signs of increased paracellular permeability were found throughout the gastrointestinal tract by assessing fractional absorption of sucrose, lactulose/mannitol, and sucralose. Stress-induced permeability changes were absent in adrenalectomized and glucocorticoid antagonist-treated animals, and were mimicked by dexamethasone. Increased permeability related to high-dose dexamethasone has been reported previously (Spitz *et al.*, 1996).

A recent study suggests that long-term perceived stress in humans is more important than acutely stressful life events for the risk of exacerbation of ulcerative colitis (Levenstein *et al.*, 2000). Two animal studies showed that subjecting rodents to stress lowered their threshold for induction of inflammation in hapten-induced colitis (Million *et al.*, 1999; Qiu *et al.*, 1999), corroborating the clinical observations. Qiu *et al.* (1999) showed that reactivation of colitis in mice that had recovered from hapten-induced colitis 8 weeks earlier, required the combination of chronic stress and CD4+ T cells, thus underlining the importance of neuroimmune interactions in intestinal inflammation.

The effect of chronic psychological stress on rat growth rate, jejunal epithelial physiology and the role of mast cells in mucosal responses was investigated in rats exposed to one-hour sessions of WAS on five consecutive days (Santos *et al.*, 2000). Stressed rats reduced their food intake and lost weight during the 5-day period. Following chronic stress, epithelial conductance, and HRP flux was increased in wild-type rats. Mast cell-deficient (Ws/Ws) rats exposed to stress did lose weight, but were normal in epithelial barrier function, suggesting an important role for mast cells in the pathophysiology of stress-mediated barrier disturbances (reconstitution of mast cells is not possible in the used strain of rats). In addition, the effect of chronic stress on colonic permeability was studied in Ws/Ws and control rats (Santos *et al.*, 2001). Chronic stress significantly increased conductance and HRP flux in the colon (Figure 11.4). Moreover, epithelial mitochondrial swelling was found in stressed control rats, but not in Ws/Ws rats. The epithelial barrier defect and mitochondrial damage was paralleled by mucosal mast cell hyperplasia and activation. Activated mast cells were found in contact with depleted nerve varicosities (Figure 11.3). This study provides further support for an important role for mast cells in stress-induced colonic mucosal pathophysiology. A more prolonged period of repetitive stress was shown to induce significant pathology in the intestinal mucosa (Söderholm *et al.*, 2002b). Rats were exposed to repetitive WAS for 10 days, and the outcome with respect to mucosal function, structure, and inflammation was studied. In control rats, chronic stress induced barrier dysfunction in the ileum and colon and ultrastructural changes in epithelial cells (enlarged mitochondria and presence of autophagosomes). Moreover, mast cell activation and infiltration of inflammatory cells was documented in the mucosa. In Ws/Ws rats, epithelial function and mucosal morphology were unchanged by chronic stress. These findings suggest that chronic psychological stress impairs epithelial barrier function and can initiate intestinal inflammation in a naïve host, and highlights the importance of mast cells in this process (Söderholm *et al.*, 2002b).

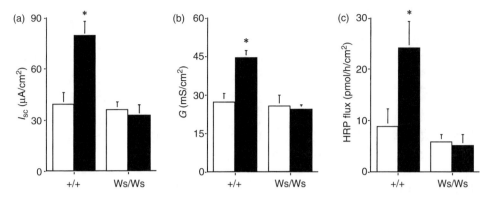

Figure 11.4   Effect of chronic stress on colonic physiology and macromolecule permeability. Mast cell-deficient rats (Ws/Ws) and their normal mast cell-containing littermates (+/+) were submitted to sham stress (white bars) or water avoidance stress (black bars) for five consecutive days. The figure shows: (a) baseline short circuit current (I_{sc}); (b) conductance (G); (c) horseradish peroxidase (HRP) flux studied in Ussing chambers 6 h after the final sham stress session. Bars represent the mean ± SEM; $n = 6$ rats/group, with two to four tissues averaged per rat; *$p < 0.01$ versus all other groups. This illustrates mast cell-dependent changes in colonic mucosal function induced by chronic stress. From Santos et al. (2001), with permission from Gut.

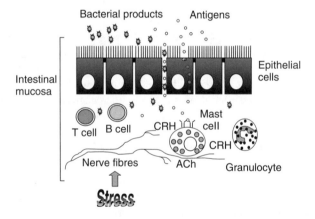

Figure 11.5   Schema of neuroimmune interactions involved in stress-induced functional changes of the intestinal mucosa (information mainly from studies conducted in the authors' laboratory). Our studies suggest activation of mucosal nerves, possibly releasing CRH and/or acetylcholine, to activated mast cells (CRH may be derived from nerves or immune/inflammatory cells). Mast cells (and possibly neurons) release bioactive chemicals (as yet not completely defined) that enhance epithelial ion secretion and permeability of both the transcellular and paracellular pathways. From Söderholm and Perdue (2001), with permission from Am. J. Physiol.

In summary, studies of intestinal mucosal function in stressed rats have demonstrated the ability of a combined physical and psychological stressor to modulate jejunal and colonic permeability, both by increasing transcellular endocytosis of macromolecules and by facilitating paracellular passage of ions and molecules of various sizes. Mucosal mast cells play a key role in the mechanism, possibly activated via neurons releasing CRH and acetylcholine, whereas various other factors affect the magnitude of the stress response. The accumulated evidence from these studies suggests that chronic stress may be implicated in the initiation, perpetuation, or exacerbation of intestinal inflammatory processes. A simplified model of the current view on mechanisms involved in stress-related barrier dysfunction is shown in Figure 11.5.

FUNCTIONAL CHANGES INDUCED BY NEUROIMMUNE INTERACTIONS IN OTHER SITUATIONS

NITRIC OXIDE (NO)

As discussed above, fluid secretion into the intestinal lumen is osmotically driven by the active transport of chloride ions, a process that is regulated by numerous soluble mediators and neurotransmitters (McKay and Baird, 1999). NO plays an important role in modulating epithelial fluid secretion. NO donors were shown to stimulate chloride secretion in the small intestine (MacNaughton, 1993; Tamai and Gaginella, 1993), and were also implicated in cholera toxin-induced intestinal secretion (Turvill *et al.*, 1999) and duodenal bicarbonate secretion (Holm *et al.*, 1998). During acute mucosal inflammation, epithelial secretion was impaired (Bell *et al.*, 1995). Recently, it was shown that colitis caused a prolonged (6 weeks) secretory hypo-responsiveness, and this chronic suppression of epithelial secretory function was caused by ongoing NO production through upregulation of iNOS (Asfaha *et al.*, 1999). These contradictory results were explained by differences in concentration of NO: low amounts being stimulatory, but higher amounts, as produced by iNOS upregulation, being inhibitory. Epithelial barrier function was also shown to be regulated by nitric oxide. NO donors resulted in an increase in the permeability of cultured epithelial cells (Menconi *et al.*, 1998). In addition, the permeability-increasing effect of interferon (IFN)-γ was suggested to be mediated via NO (Chavez *et al.*, 1999), since inhibitors of NOS prevented the increase in permeability induced by IFN-γ. NO is produced by mast cells by a constitutively expressed NOS, and this NO production appears to down-regulate the release of many other inflammatory mediators from mast cells, for example, histamine, PAF, and TNFα (Wallace and Miller, 2000). Kanwar *et al.* (1994) showed that administration of an inhibitor of NOS resulted in significant increases in intestinal permeability in the rat, which was inhibited by histamine receptors antagonists or mast cell stabilizers. This suggested that NOS inhibition caused degranulation of mucosal mast cell mediators producing alterations in epithelial permeability.

IgA SECRETION

To study the effect of direct nerve-stimulation on intestinal IgA secretion, isolated vascularly perfused porcine ileal segments with preserved extrinsic nerve supply were used (Schmidt *et al.*, 1999). IgA secretion in the luminal effluent was increased by somatostatin and inhibited by noradrenaline. Electrical stimulation of the extrinsic nerves significantly stimulated secretion of IgA only during blockade of α-adrenergic receptors, muscarinic receptors, or both, suggesting involvement of neuropeptides (Schmidt *et al.*, 1992). This stimulatory effect was abolished by blockade of nicotinic receptors. Possibly the released neuropeptides acted directly on receptors on plasma cells, which have been shown to express receptors for neuropeptides (Stanisz *et al.*, 1987), and are found close to somatostatin-immunoreactive nerve fibres (Feher *et al.*, 1992).

INTESTINAL INFLAMMATION

In addition to the relatively large literature in food allergy, nematode infections, and stress, the role of mast cells has also been studied in other animal models of inflammation associated with early mucosal barrier dysfunction. Two recent studies by Andoh *et al.* (1999, 2001) using mast cell-deficient rats, showed that ischemia/reperfusion injury was suppressed in Ws/Ws rats, and there was no increase in intestinal permeability during reperfusion in these rats. In controls rats, mast cell stabilizers inhibited the permeability increase. Work from the same group also showed that mast cells were important for the induction of colitis by dextran sodium sulphate (DSS) (Araki *et al.*, 2000). Following exposure of 3% DSS for 10 days, Ws/Ws rats had a significantly lower macroscopic and microscopic mucosal damage score. A correlation between mucosal histamine levels and mucosal damage was also demonstrated. A more complex involvement of mast cells was found in a model of radiation enteropathy. Mast cell hyperplasia was a characteristic feature of the late phase of intestinal radiation injury, whereas there was a rapid decrease in the number of mucosal mast cells early after radiation (Sedgwick and Ferguson, 1994). Zheng *et al.* (2000) subjected Ws/Ws rats to local irradiation of the ileum and studied early and late phase effects. Compared to irradiated control rats, Ws/Ws rats exhibited strikingly exacerbated mucosal injury at 2 weeks, but a minimal reactive intestinal wall fibrosis at 26 weeks. The authors concluded that mucosal mast cells play a critical role in protecting the intestine against radiation-induced mucosal injury, whereas connective tissue mast cells are involved in the pathogenesis of delayed intestinal wall fibrosis.

NEUROIMMUNE INTERACTION IN CHRONIC INTESTINAL DISEASES IN HUMANS

NORMAL HUMAN INTESTINE

As mentioned previously, studies using intestinal perfusion in normal human subjects have revealed effects of acute stress on intestinal secretion (Barclay and Turnberg, 1987, 1988). These effects were inhibited by atropine infusion (Barclay and Turnberg, 1987), and were paralleled by luminal release of the mast cell mediators tryptase and histamine (Santos *et al.*, 1998). This suggests that the CNS, mast cells, and the intestinal mucosa can interact in humans. These *in vivo* findings have been corroborated by studies of human intestine in Ussing chambers. In non-inflamed segments of surgically resected human bowel, anti-IgE containing serum (to stimulate mast cells) induced a rise in I_{sc} (Crowe and Perdue, 1993; Stack *et al.*, 1995) that was inhibited by the histamine1-receptor antagonists, cyclo-oxygenase inhibitors, and the neurotoxin, tetrodotoxin, suggesting that activation of mast cells releases mediators which stimulate intestinal ion transport through direct epithelial action and via nerves. Other examples of neurotransmitters shown to have the potential of mediating functional changes by neuroimmune interaction in human intestine are nitric oxide (Stack *et al.*, 1996; Riegler *et al.*, 1999), and neurotensin (Riegler *et al.*, 2000). These studies provide important evidence that immunoregulation of intestinal ion transport does occur in humans.

INFLAMMATORY BOWEL DISEASE

Inflammatory bowel diseases (IBD), including Crohn's disease and ulcerative colitis, are chronic inflammatory diseases of unknown etiology. An understanding of the pathogenesis of IBD is, however, evolving. The current paradigm is that IBD represents the outcome of the interaction between the predisposition of the host (due to genetic and environmental factors), the mucosal immune response, and the bacterial flora in the intestinal lumen. Mucosal homeostasis depends on the integrity of each of these interacting components. Studies in knockout mice have shown that induction of defects at various levels of host defence (e.g. immune regulation, neural function, barrier function) leads to chronic intestinal inflammation. In other words, there appears to be a multiplicity of pathways involving neuroimmune interactions which can lead to the outcome we call IBD.

Evidence for the role of mast cells in the pathogenesis of IBD includes increased numbers and degranulation of mast cells in IBD mucosa (Dvorak *et al.*, 1978), as well as elevated levels of histamine in the gut lumen of patients with IBD (Rampton *et al.*, 1980; Knutson *et al.*, 1990). Enhanced release of histamine and eicosanoids has also been described in actively inflamed IBD tissues after isolated intestinal mast cells were stimulated by cross-linking IgE receptors (Bischoff *et al.*, 1996). Moreover, mast cells have been shown to be a major producer of TNF-α in the intestine of CD patients (Bischoff *et al.*, 1999; Lilja *et al.*, 2000). Crowe *et al.* (1997) studied the significance of mast cells for mucosal functional disturbances in IBD, and found that the interactions between mast cells and the epithelium were altered. The I_{sc} response to anti-IgE and histamine was reduced in IBD mucosa, whereas mast cell numbers were increased. Histamine pre-treatment of mucosa from control patients reduced anti-IgE responses to levels seen in IBD, suggesting that prior activation of mast cells may account for the reduced secretory response to anti-IgE in IBD mucosa.

Altered intestinal permeability, indicating disturbances in mucosal barrier function, is a feature of intestinal inflammation in humans (Bjarnason *et al.*, 1995), and has been implicated as a pathogenic and perhaps etiologic factor in inflammatory bowel disease (Meddings, 1997; Peeters *et al.*, 1997; Söderholm *et al.*, 1999). Several neuroimmune mediators of importance in IBD were shown to have effects on intestinal epithelial permeability, for example, IFN-γ (Madara and Stafford, 1989), TNF-α (Mullin and Snock, 1990; Zareie *et al.*, 1998), IL-4 (Colgan *et al.*, 1994; Berin *et al.*, 1999b), and NO (Salzman *et al.*, 1995). Moreover it was shown that lamina propria-derived monocytes/macrophages from patients with Crohn's disease spontaneously secreted TNF-α, inducing changes in baseline I_{sc} and permeability when co-cultured with T84 human colonic monolayers (Zareie *et al.*, 2001). The importance of this finding is indicated by the efficacy of anti-TNFα treatment in Crohn's disease, and the recent finding by several independent groups of a susceptibility gene (dysfunction of the so-called NOD2 gene on chromosome 16), which could explain monocyte hyper-activation in these patients (Hampe et al., 2001; Hugot et al., 2001; Ogura et al., 2001).

Long-term psychological stress has been implicated as a factor in reactivation of inflammation in IBD (Levenstein et al., 2000; Collins, 2001). Moreover, electrical spinal cord stimulation was reported to precipitate relapse of ulcerative colitis (Kemler et al., 1999). Several physical stressors, for example, surgery and trauma, have been shown to increase intestinal permeability in humans. There are however, to date, no studies looking specifically at the influence of psychological stress on intestinal permeability in humans.

Structural, as well as functional, alterations of enteric nerves have been reported in both ulcerative colitis and Crohn's disease. The main histological findings include: ganglion cell hyperplasia, axonal necrosis, alterations in neuropeptide innervation, and neural damage in both inflamed and non-inflamed areas (Collins et al., 1997). Widespread, severe, axonal necrosis has been proposed as a histological marker discriminating Crohn's disease from other intestinal inflammation (Dvorak and Silen, 1985). MHC class II antigens are expressed by the enteroglial sheet surrounding the nerve fibres in the myenteric plexus in Crohn's disease patients (Geboes et al., 1992). Therefore, the axonal damage may be triggered by antigen presentation to CD4+ cells at these sites. An altered expression of the neuropeptides substance P (Koch et al., 1987; Mazumdar and Das, 1992) and VIP (O'Morain et al., 1984; Kubota et al., 1992) has been found in both Crohn's disease and ulcerative colitis. Nerve growth factor immunostaining is significantly increased in Crohn's disease tissue (Strobach et al., 1990), whereas an increased expression of CRH has been demonstrated in mucosal inflammatory cells in ulcerative colitis (Kawahito et al., 1995). Studies of functional changes of enteric nerves in IBD have mainly been focused on colonic contractility, showing a decreased rectal compliance in IBD (Collins et al., 1997). Also, a general autonomic nerve dysfunction has been shown in IBD (Lindgren et al., 1991, 1993). The consequences of these neuronal changes for mucosal function in IBD have yet to be investigated.

FOOD ALLERGY

Gastrointestinal symptoms occur in a large number of patients with food allergies. Immediate hypersensitivity mechanisms may give rise to the nausea, vomiting, abdominal pain, and diarrhoea experienced by these patients. As discussed above, animal models suggest that these symptoms can be explained by changes in ion secretion and permeability induced by mast cell activation. To investigate whether human intestinal hypersensitivity reactions are associated with detectable release of inflammatory mediators from activated cells, patients with food allergy and healthy volunteers were studied with a closed-segment perfusion technique. Intra-luminal administration of food antigens induced a rapid increase in intestinal release of the mast cell mediators, tryptase, histamine, and PGD_2 (Santos et al., 1999b). The increased release of these mediators was previously associated with a notable water secretory response, and leakage of albumin into the lumen (Bengtsson et al., 1997). These results demonstrate that intestinal hypersensitivity reactions also in humans are characterized by prompt activation of mast cells and other immune cells, with notable and immediate secretion of water and inflammatory mediators into the intestinal lumen. Data on normal human intestine in Ussing chambers (Crowe and Perdue, 1993) showing that chloride secretion induced by anti-IgE-mediated mast cell activation was inhibited by tetrodotoxin, and the similarity in jejunal mast cell mediator-release by cold pain stress and luminal antigen challenge in patients with food allergy (Santos *et al.*, 1998), strongly suggest neural involvement in mucosal anaphylaxis in humans. It was also shown that intraluminal challenge of birch antigen in atopic patients allergic to birch induced a net increase in albumin leakage and net secretion of ions and water. Thus, exposure to antigen normally acting on the respiratory tract induced increased permeability of the gastrointestinal mucosa, which would suggest a general allergic recognition shared by several immunocompetent tissues in the body (Knutson *et al.*, 1996).

An increased intestinal permeability in patients with food allergy has been shown *in vivo* with inert probes (Troncone *et al.*, 1994; Bjarnason *et al.*, 1995), as well as in biopsies studies in Ussing-type chambers (Heyman *et al.*, 1988). In jejunal biopsies from children with cow's milk allergy, transepithelial HRP fluxes were significantly higher, about eight-fold, together with increased I_{sc}. Subsequently, mechanisms involved in the permeability increase were studied (Heyman *et al.*, 1994). Peripheral blood mononuclear cells from infants with cow's milk allergy were cultured in the presence of milk proteins, the release of cytokines was measured, and the effects of culture supernatants were tested on intestinal cell monolayers. When stimulated by milk protein, mononuclear cells from infants with milk allergy released TNF-α, and culture supernatants induced a significant decrease in electrical resistance and increase in flux of HRP. These effects were reversed when culture supernatants were neutralized with anti-TNF-α antibodies. Thus, suggesting an important role for TNF-α in the intestinal barrier dysfunction in patients with food allergy.

IRRITABLE BOWEL SYNDROME

Irritable bowel syndrome (IBS) is a functional disorder characterized by abdominal pain and alterations in bowel habits in the absence of demonstrable pathology by endoscopy, radiology, histology, etc. A relationship between stress and symptom severity is well described in IBS, and changes in the stress response in the autonomic nervous system, the hypothalamo–pituitary–adrenal axis, and the visceral pain perception have been shown in patients with IBS, for review, see Mayer (2000). Stressful life events increased the risk of developing IBS after a bacterial gastroenteritis (Gwee *et al.*, 1996), and in these patients, immune activation and increased permeability persisted several months after the infection (Spiller *et al.*, 2000). An increased number of mast cells were found in the terminal ileum (Weston *et al.*, 1993) and caecum (O'Sullivan *et al.*, 2000) of patients with IBS. Patients with an increased number of mast cells in the mucosa also showed changes in anorectal function (Libel *et al.*, 1993). This finding suggested that the multiple effects of the intestinal mast cell alone, or as a participant of a persistent inflammatory response, might be of importance in the pathogenesis of IBS. CRH has also been implicated in IBS. Fukudo *et al.* (1998) studied the effects of i.v. injection of CRH in IBS patients and controls. The IBS patients had an increased colonic motility response to CRH, and produced significantly higher plasma levels of ACTH than in controls. This suggests that the brain–gut axis in IBS patients has an exaggerated response to CRH. Whether this also would imply exaggerated mucosal functional responses to stress in IBS patients is to date not known.

CONCLUSIONS AND PERSPECTIVES

To summarize, an extensive literature is consistent with an important role for neuroimmune regulation of intestinal mucosal function. From animal models, it is clear that mast cells activated by enteric nerves can release mediators (e.g. histamine, PGs, TNF-α, proteases) that affect epithelial ion secretion, mucus secretion, and permeability. Also, mast cells can affect enteric nerves to enhance nerve-mediated ion secretion. In addition, a link between intestinal mucosal mast cells and the CNS has been shown in studies of Pavlovian conditioning in a model of intestinal hypersensitivity and in studies of stress induced changes in mucosal function. Moreover, recent data suggest that a more chronic activation of mast cells, for example, by chronic stress and in the late phase of intestinal allergic reactions, can induce inflammatory cell infiltration into the lamina propria.

The role of other nerve–immune cell interactions in the regulation of epithelial cell physiology is just beginning to be delineated. The rapid growth in the field of intestinal immunology has provided a wealth of information regarding the potential of immunocytes and immune mediators to act as regulators of intestinal mucosal function. It would be surprising if the findings of neuroimmune interactions in hypersensitivity and stress were not also true for other immunocyte-mediated processes, for example, the rapid nerve-mediated increase in IgA secretion, or monocyte induced hyper-permeability. Explorations in this area are likely to reveal several intriguing avenues for future research.

There is also a growing literature on structural and functional neuroimmune abnormalities in human intestinal diseases, such as food allergy, IBD, and IBS. Patients in these groups demonstrate changes in intestinal mucosal function. The challenge now is to identify the critical pathophysiological steps that will lead to novel pharmacological therapeutic modalities to treat those suffering from chronic intestinal diseases. From a clinical perspective, it is also important to remember that there is now emerging evidence for a functional link between the CNS and the intestinal mucosa. Stress-related changes in disease activity may not just be something in the patient's mind, it may be in his mind–gut axis!

REFERENCES

Andoh, A., Kimura, T., Fukuda, M., Araki, Y., Fujiyama, Y. and Bamba, T. (1999) Rapid intestinal ischaemia-reperfusion injury is suppressed in genetically mast cell-deficient Ws/Ws rats. *Clin. Exp. Immunol.*, **116**, 90–93.

Andoh, A., Fujiyama, Y., Araki, Y., Kimura, T., Tsujikawa, T. and Bamba, T. (2001) Role of complement activation and mast cell degranulation in the pathogenesis of rapid intestinal ischemia/reperfusion injury in rats. *Digestion*, **63** (Suppl 1), 103–107.

Araki, Y., Andoh, A., Fujiyama, Y. and Bamba, T. (2000) Development of dextran sulphate sodium-induced experimental colitis is suppressed in genetically mast cell-deficient Ws/Ws rats. *Clin. Exp. Immunol.*, **119**, 264–269.

Armstrong, W.M. (1987) Cellular mechanisms of ion transport in the small intestine. In *Physiology of the Gastrointestinal Tract*, edited by L.R. Johnson, pp. 1251–1265. New York: Raven Press.

Asfaha, S., Bell, C.J., Wallace, J.L. and MacNaughton, W.K. (1999) Prolonged colonic epithelial hyporesponsiveness after colitis: role of inducible nitric oxide synthase. *Am. J. Physiol.*, **276**, G703–G710.

Baird, A.W., Cuthbert, A.W. and Pearce, F.L. (1985) Immediate hypersensitivity reactions in epithelia from rats infected with Nippostrongylus brasiliensis. *Br. J. Pharmacol.*, **85**, 787–795.

Baird, A.W., Cuthbert, A.W. and MacVinish, L.J. (1987) Type 1 hypersensitivity reactions in reconstructed tissues using syngeneic cell types. *Br. J. Pharmacol.*, **91**, 857–869.

Barclay, G.R. and Turnberg, L.A. (1987) Effect of psychological stress on salt and water transport in the human jejunum. *Gastroenterology*, **93**, 91–97.

Barclay, G.R. and Turnberg, L.A. (1988) Effect of cold-induced pain on salt and water transport in the human jejunum. *Gastroenterology*, **94**, 994–998.

Beagley, K.W. and Husband, A.J. (1998) Intraepithelial lymphocytes: origins, distribution, and function. *Crit. Rev. Immunol.*, **18**, 237–254.

Befus, A.D. (1994) Reciprocity of mast cell-nervous system interactions. In *Innervation of the Gut, Pathophysiological Implications*, edited by Y. Tache, D.L. Wingate and T.F. Burks, pp. 315–329. Boca Raton: CRC Press.

Bell, C.J., Gall, D.G. and Wallace, J.L. (1995) Disruption of colonic electrolyte transport in experimental colitis. *Am. J. Physiol.*, **268**, G622–G630.

Bengtsson, U., Knutson, T.W., Knutson, L., Dannaeus, A., Hallgren, R. and Ahlstedt, S. (1997) Eosinophil cationic protein and histamine after intestinal challenge in patients with cow's milk intolerance. *J. Allergy Clin. Immunol.*, **100**, 216–221.

Berin, M.C., Kiliaan, A.J., Yang, P.C., Groot, J.A., Taminiau, J.A. and Perdue, M.H. (1997) Rapid transepithelial antigen transport in rat jejunum: impact of sensitization and the hypersensitivity reaction. *Gastroenterology*, **113**, 856–864.

Berin, M.C., Kiliaan, A.J., Yang, P.C., Groot, J.A., Kitamura, Y. and Perdue, M.H. (1998) The influence of mast cells on pathways of transepithelial antigen transport in rat intestine. *J. Immunol.*, **161**, 2561–2566.

Berin, M.C., McKay, D.M. and Perdue, M.H. (1999a) Immune-epithelial interactions in host defense. *Am. J. Trop. Med. Hyg.*, **60**, 16–25.

Berin, M.C., Yang, P.C., Ciok, L., Waserman, S. and Perdue, M.H. (1999b) Role for IL-4 in macromolecular transport across human intestinal epithelium. *Am. J. Physiol.*, **276**, C1046–C1052.

Bienenstock, J., Tomioka, M., Stead, R., Ernst, P., Jordana, M., Gauldie, J. *et al.* (1987) Mast cell involvement in various inflammatory processes. *Am. Rev. Respir. Dis.*, **135**, S5–S8.

Bijlsma, P.B., Kiliaan, A.J., Scholten, G., Heyman, M., Groot, J.A. and Taminiau, J.A. (1996) Carbachol, but not forskolin, increases mucosal-to-serosal transport of intact protein in rat ileum in vitro. *Am. J. Physiol.*, **271**, G147–G155.

Bischoff, S.C., Schwengberg, S., Wordelmann, K., Weimann, A., Raab, R. and Manns, M.P. (1996) Effect of c-kit ligand, stem cell factor, on mediator release by human intestinal mast cells isolated from patients with inflammatory bowel disease and controls. *Gut*, **38**, 104–114.

Bischoff, S.C., Lorentz, A., Schwengberg, S., Weier, G., Raab, R. and Manns, M.P. (1999) Mast cells are an important cellular source of tumour necrosis factor alpha in human intestinal tissue. *Gut*, **44**, 643–652.

Bjarnason, I., MacPherson, A. and Hollander, D. (1995) Intestinal permeability: an overview. *Gastroenterology*, **108**, 1566–1581.

Bomsel, M., Prydz, K., Parton, R.G., Gruenberg, J. and Simons, K. (1989) Endocytosis in filter-grown Madin-Darby canine kidney cells. *J. Cell Biol.*, **109**, 3243–3258.

Castagliuolo, I., Lamont, J.T., Qiu, B., Fleming, S.M., Bhaskar, K.R., Nikulasson, S.T. *et al.* (1996a) Acute stress causes mucin release from rat colon: role of corticotropin releasing factor and mast cells. *Am. J. Physiol.*, **271**, G884–G892.

Castagliuolo, I., Leeman, S.E., Bartolak-Suki, E., Nikulasson, S., Qiu, B., Carraway, R.E. and Pothoulakis, C. (1996b) A neurotensin antagonist, SR 48692, inhibits colonic responses to immobilization stress in rats. *Proc. Natl. Acad. Sci. USA*, **93**, 12611–12615.

Castagliuolo, I., Wershil, B.K., Karalis, K., Pasha, A., Nikulasson, S.T. and Pothoulakis, C. (1998) Colonic mucin release in response to immobilization stress is mast cell dependent. *Am. J. Physiol.*, **274**, G1094–G1100.

Castex, N., Fioramonti, J., Fargeas, M.J. and Bueno, L. (1995) c-fos expression in specific rat brain nuclei after intestinal anaphylaxis: involvement of 5-HT3 receptors and vagal afferent fibers. *Brain Res.*, **688**, 149–160.

Castro, G.A., Harari, Y. and Russell, D. (1987) Mediators of anaphylaxis-induced ion transport changes in small intestine. *Am. J. Physiol.*, **253**, G540–G548.

Cereijido, M., Valdes, J., Shoshani, L. and Contreras, R.G. (1998) Role of tight junctions in establishing and maintaining cell polarity. *Annu. Rev. Physiol.*, **60**, 161–177.

Chavez, A.M., Menconi, M.J., Hodin, R.A. and Fink, M.P. (1999) Cytokine-induced intestinal epithelial hyper-permeability: role of nitric oxide. *Crit. Care Med.*, **27**, 2246–2251.

Clamp, J. R. and Creeth, J. M. (1984) Some non-mucin components of mucus and their possible biological roles. *Ciba Found. Symp.*, **109**, 121–136.

Colgan, S.P., Resnick, M.B., Parkos, C.A., Delp-Archer, C., McGuirk, D., Bacarra, A.E. *et al.* (1994) IL-4 directly modulates function of a model human intestinal epithelium. *J. Immunol.*, **153**, 2122–2129.

Collins, S.M. (2001) Stress and the gastrointestinal tract IV. Modulation of intestinal inflammation by stress: basic mechanisms and clinical relevance. *Am. J. Physiol.*, **280**, G315–G318.

Collins, S.M., Van Assche, G. and Hogaboam, C.M. (1997) Alterations in enteric nerve and smooth-muscle function in inflammatory bowel diseases. *Infl. Bowel Dis.*, **3**, 38–48.

Cooke, H.J. (1998) "Enteric Tears": chloride secretion and its neural regulation. *News Physiol. Sci.*, **13**, 269–274.

Coutinho, S. V., Sablad, M. R., Miller, J. C., Zhou, H., Lam, A., Bayati, A. I. *et al.* (2000) Neonatal maternal separation results in stress-induced visceral hyperalgesia in adult rats: a new model for IBS. *Gastroenterology*, **118** (Suppl 2), A637 (abstract).

Crowe, S.E., Sestini, P. and Perdue, M.H. (1990) Allergic reactions of rat jejunal mucosa. Ion transport responses to luminal antigen and inflammatory mediators. *Gastroenterology*, **99**, 74–82.

Crowe, S.E. and Perdue, M.H. (1992) Gastrointestinal food hypersensitivity: Basic mechanisms of pathophysiology. *Gastroenterology*, **103**, 1075–1095.

Crowe, S.E., Soda, K., Stanisz, A.M. and Perdue, M.H. (1993) Intestinal permeability in allergic rats: nerve involvement in antigen-induced changes. *Am. J. Physiol.*, **264**, G617–G623.

Crowe, S.E. and Perdue, M.H. (1993) Anti-immunoglobulin E-stimulated ion transport in human large and small intestine. *Gastroenterology*, **105**, 764–772.

Crowe, S.E., Luthra, G.K. and Perdue, M.H. (1997) Mast cell mediated ion transport in intestine from patients with and without inflammatory bowel disease. *Gut*, **41**, 785–792.

Djuric, V.J., Bienenstock, J. and Perdue, M.H. (1994) Psychological and neural regulation of intestinal hypersensitivity. In *Advances in Psychoneuroimmunology*, edited by I. Berczi and J. Szeleny, pp. 291–301. New York: Plenum Press.

Dvorak, A.M., Monahan, R.A., Osage, J.E. and Dickersin, G.R. (1978) Mast-cell degranulation in Crohn's disease. *Lancet*, **1**, 498.

Dvorak, A.M. and Silen, W. (1985) Differentiation between Crohn's disease and other inflammatory conditions by electron microscopy. *Ann. Surg.*, **201**, 53–63.

Dvorak, A.M., McLeod, R.S., Onderdonk, A.B., Monahan-Earley, R.A., Cullen, J.B., Antonioli, D.A. *et al.* (1992) Human gut mucosal mast cells: ultrastructural observations and anatomic variation in mast cell-nerve associations in vivo. *Int. Arch. Allergy Immunol.*, **98**, 158–168.

Farquhar, M.G. and Palade, G.E. (1963) Junctional complexes in various epithelia. *J. Cell Biol.*, **17**, 375–412.

Feher, E., Fodor, M. and Burnstock, G. (1992) Distribution of somatostatin-immunoreactive nerve fibres in Peyer's patches. *Gut*, **33**, 1195–1198.

Felten, D.L., Felten, S.Y., Bellinger, D.L., Carlson, S.L., Ackerman, K.D., Madden, K.S. *et al.* (1987) Noradrenergic sympathetic neural interactions with the immune system: structure and function. *Immunol.Rev.*, **100**, 225–260.

Fiocchi, C. (1997) Intestinal inflammation: a complex interplay of immune and nonimmune cell interactions. *Am. J. Physiol.*, **273**, G769–G775.

Forstner, G. (1995) Signal transduction, packaging and secretion of mucins. *Annu. Rev. Physiol.*, **57**, 585–605.

Frieling, T., Cooke, H.J. and Wood, J.D. (1994) Neuroimmune communication in the submucous plexus of guinea pig colon after sensitization to milk antigen. *Am. J. Physiol.*, **267**, G1087–G1093.

Fukudo, S., Nomura, T. and Hongo, M. (1998) Impact of corticotropin-releasing hormone on gastrointestinal motility and adrenocorticotropic hormone in normal controls and patients with irritable bowel syndrome. *Gut*, **42**, 845–849.

Gardner, M.L. (1988) Gastrointestinal absorption of intact proteins. *Annu. Rev. Nutr.*, **8**, 329–350.

Geboes, K., Rutgeerts, P., Ectors, N., Mebis, J., Penninckx, F., Vantrappen, G. and Desmet, V.J. (1992) Major histocompatibility class II expression on the small intestinal nervous system in Crohn's disease. *Gastroenterology*, **103**, 439–447.

Gordon, L.G., Kottra, G. and Fromter, E. (1989) Electrical impedance analysis of leaky epithelia: theory, techniques and leak artifact problems. *Methods Enzymol.*, **171**, 642–663.

Green, B.T., Bunnett, N.W., Kulkarni-Narla, A., Steinhoff, M. and Brown, D.R. (2000) Intestinal type 2 proteinase-activated receptors: expression in opioid-sensitive secretomotor neural circuits that mediate epithelial ion transport. *J. Pharmacol. Exp. Ther.*, **295**, 410–416.

Gwee, K.A., Graham, J.C., McKendrick, M.W., Collins, S.M., Marshall, J.S., Walters, S.J. and Read, N.W. (1996) Psychometric scores and persistence of irritable bowel after infectious diarrhoea. *Lancet*, **347**, 150–153.

Hampe, J., Cuthbert, A., Croucher, P.J., Mirza, M.M., Mascheretti, S., Fisher, S. *et al.* (2001) Association between insertion mutation in NOD2 gene and Crohn's disease in German and British populations. *Lancet*, **357**, 1925–1928.

Harari, Y., Russell, D.A. and Castro, G.A. (1987) Anaphylaxis-mediated epithelial Cl⁻ secretion and parasite rejection in rat intestine. *J. Immunol.*, **138**, 1250–1255.

Heyman, M., Grasset, E., Ducroc, R. and Desjeux, J.F. (1988) Antigen absorption by the jejunal epithelium of children with cow's milk allergy. *Pediatr. Res.*, **24**, 197–202.

Heyman, M., Darmon, N., Dupont, C., Dugas, B., Hirribaren, A., Blaton, M.A. and Desjeux, J.F. (1994) Mononuclear cells from infants allergic to cow's milk secrete tumor necrosis factor alpha, altering intestinal function. *Gastroenterology*, **106**, 1514–1523.

Holm, M., Johansson, B., Pettersson, A. and Fandriks, L. (1998) Acid-induced duodenal mucosal nitric oxide output parallels bicarbonate secretion in the anaesthetized pig. *Acta Physiol. Scand.*, **162**, 461–468.

Hugot, J.P., Chamaillard, M., Zouali, H., Lesage, S., Cezard, J.P., Belaiche, J. *et al.* (2001) Association of NOD2 leucine-rich repeat variants with susceptibility to Crohn's disease. *Nature*, **411**, 599–603.

Iwasaki, A. and Kelsall, B.L. (1999) Mucosal immunity and inflammation. I. Mucosal dendritic cells: their specialized role in initiating T cell responses. *Am. J. Physiol.*, **276**, G1074–G1078.

Javed, N.H., Wang, Y.Z. and Cooke, H.J. (1992) Neuroimmune interactions: role for cholinergic neurons in intestinal anaphylaxis. *Am. J. Physiol.*, **263**, G847–G852.

Jiang, W., Kreis, M.E., Eastwood, C., Kirkup, A.J., Humphrey, P.P. and Grundy, D. (2000) 5-HT(3) and histamine H(1) receptors mediate afferent nerve sensitivity to intestinal anaphylaxis in rats. *Gastroenterology*, **119**, 1267–1275.

Kagnoff, M.F. (1998) Current concepts in mucosal immunity. III. Ontogeny and function of gamma delta T cells in the intestine. *Am. J. Physiol.*, **274**, G455–G458.

Kanwar, S., Wallace, J.L., Befus, D. and Kubes, P. (1994) Nitric oxide synthesis inhibition increases epithelial permeability via mast cells. *Am. J. Physiol.*, **266**, G222–G229.

Karlsson, J. and Artursson, P. (1992) A new diffusion chamber system for the determination of drug permeability coefficients across the human intestinal epithelium that are independent of the unstirred water layer. *Biochim. Biophys. Acta*, **1111**, 204–210.

Kawahito, Y., Sano, H., Mukai, S., Asai, K., Kimura, S., Yamamura, Y. *et al.* (1995) Corticotropin releasing hormone in colonic mucosa in patients with ulcerative colitis. *Gut*, **37**, 544–551.

Keljo, D.J. and Hamilton, J.R. (1983) Quantitative determination of macromolecular transport rate across intestinal Peyer's patches. *Am. J. Physiol.*, **244**, G637–G644.

Kemler, M.A., Barendse, G.A. and Van Kleef, M. (1999) Relapsing ulcerative colitis associated with spinal cord stimulation. *Gastroenterology*, **117**, 215–217.

Kerneis, S., Bogdanova, A., Kraehenbuhl, J.P. and Pringault, E. (1997) Conversion by Peyer's patch lymphocytes of human enterocytes into M cells that transport bacteria. *Science*, **277**, 949–952.

Kiliaan, A.J., Saunders, P.R., Bijlsma, P.B., Berin, M.C., Taminiau, J.A., Groot, J.A. and Perdue, M.H. (1998) Stress stimulates transepithelial macromolecular uptake in rat jejunum. *Am. J. Physiol.*, **275**, G1037–G1044.

Knutson, L., Ahrenstedt, O., Odlind, B. and Hallgren, R. (1990) The jejunal secretion of histamine is increased in active Crohn's disease. *Gastroenterology*, **98**, 849–854.

Knutson, T.W., Bengtsson, U., Dannaeus, A., Ahlstedt, S. and Knutson, L. (1996) Effects of luminal antigen on intestinal albumin and hyaluronan permeability and ion transport in atopic patients. *J. Allergy Clin. Immunol.*, **97**, 1225–1232.

Koch, T.R., Carney, J.A. and Go, V.L. (1987) Distribution and quantitation of gut neuropeptides in normal intestine and inflammatory bowel diseases. *Dig. Dis. Sci.*, **32**, 369–376.

Kraehenbuhl, J.P. and Neutra, M.R. (2000) Epithelial M cells: differentiation and function. *Annu. Rev. Cell Dev. Biol.*, **16**, 301–332.

Kubota, Y., Petras, R.E., Ottaway, C.A., Tubbs, R.R., Farmer, R.G. and Fiocchi, C. (1992) Colonic vasoactive intestinal peptide nerves in inflammatory bowel disease. *Gastroenterology*, **102**, 1242–1251.

Levenstein, S., Prantera, C., Varvo, V., Scribano, M.L., andreoli, A., Luzi, C., Arca, M., Berto, E., Milite, G. and Marcheggiano, A. (2000) Stress and exacerbation in ulcerative colitis: a prospective study of patients enrolled in remission. *Am. J. Gastroenterol.*, **95**, 1213–1220.

Libel, R., Biddle, W.L. and Miner, P.B. Jr. (1993) Evaluation of anorectal physiology in patients with increased mast cells. *Dig. Dis. Sci.*, **38**, 877–881.

Lilja, I., Gustafson-Svard, C., Franzen, L. and Sjödahl, R. (2000) Tumor necrosis factor-alpha in ileal mast cells in patients with Crohn's disease. *Digestion*, **61**, 68–76.

Lin, T.J. and Befus, A.D. (1999) Mast cells and eosinophils in mucosal defenses and pathogenesis. In *Mucosal Immunology*, edited by Ogra, P., Mestecky, J., Lamm, M., Strober, W., Bienenstock, J. and McGhee, J.R., pp. 469–482. New York: Academic Press.

Lindgren, S., Lilja, B., Rosen, I. and Sundkvist, G. (1991) Disturbed autonomic nerve function in patients with Crohn's disease. *Scand. J. Gastroenterol.*, **26**, 361–366.

Lindgren, S., Stewenius, J., Sjolund, K., Lilja, B. and Sundkvist, G. (1993) Autonomic vagal nerve dysfunction in patients with ulcerative colitis. *Scand. J. Gastroenterol.*, **28**, 638–642.

Lorentz, A., Schwengberg, S., Sellge, G., Manns, M.P. and Bischoff, S.C. (2000) Human intestinal mast cells are capable of producing different cytokine profiles: role of IgE receptor cross-linking and IL-4. *J. Immunol.*, **164**, 43–48.

MacNaughton, W.K. (1993) Nitric oxide-donating compounds stimulate electrolyte transport in the guinea pig intestine in vitro. *Life Sci.*, **53**, 585–593.

MacQueen, G., Marshall, J., Perdue, M., Siegel, S. and Bienenstock, J. (1989) Pavlovian conditioning of rat mucosal mast cells to secrete rat mast cell protease II. *Science*, **243**, 83–85.

Madara, J.L. (1998) Regulation of the movement of solutes across tight junctions. *Annu. Rev. Physiol.*, **60**, 143–159.

Madara, J.L. and Stafford, J. (1989) Interferon-gamma directly affects barrier function of cultured intestinal epithelial monolayers. *J. Clin. Invest.*, **83**, 724–727.

Mandel, L.J., Bacallao, R. and Zampighi, G. (1993) Uncoupling of the molecular 'fence' and paracellular 'gate' functions in epithelial tight junctions. *Nature*, **361**, 552–555.

Marcon-Genty, D., Tome, D., Kheroua, O., Dumontier, A.M., Heyman, M. and Desjeux, J.F. (1989) Transport of beta-lactoglobulin across rabbit ileum in vitro. *Am. J. Physiol.*, **256**, G943–G948.

Maric, I., Holt, P.G., Perdue, M.H. and Bienenstock, J. (1996) Class II MHC antigen (Ia)-bearing dendritic cells in the epithelium of the rat intestine. *J. Immunol.*, **156**, 1408–1414.

Mayer, E.A. (2000) The neurobiology of stress and gastrointestinal disease. *Gut*, **47**, 861–869.

Mazumdar, S. and Das, K.M. (1992) Immunocytochemical localization of vasoactive intestinal peptide and substance P in the colon from normal subjects and patients with inflammatory bowel disease. *Am. J. Gastroenterol.*, **87**, 176–181.

McKay, D.M. and Baird, A.W. (1999) Cytokine regulation of epithelial permeability and ion transport. *Gut*, **44**, 283–289.

McKay, D.M., Berin, M.C., Fondacaro, J.D. and Perdue, M.H. (1996) Effects of neuropeptide Y and substance P on antigen-induced ion secretion in rat jejunum. *Am. J. Physiol.*, **271**, G987–G992.

Meddings, J.B. (1997) Review article: intestinal permeability in Crohn's disease. *Aliment. Pharmacol. Ther*, **11**(Suppl 3), 47–53.

Meddings, J.B. and Swain, M.G. (2000) Environmental stress-induced gastrointestinal permeability is mediated by endogenous glucocorticoids in the rat. *Gastroenterology*, **119**, 1019–1028.

Menconi, M.J., Unno, N., Smith, M., Aguirre, D.E. and Fink, M.P. (1998) Nitric oxide donor-induced hyperpermeability of cultured intestinal epithelial monolayers: role of superoxide radical, hydroxyl radical and peroxynitrite. *Biochim. Biophys. Acta*, **1425**, 189–203.

Metcalfe, D.D., Baram, D. and Mekori, Y.A. (1997) Mast cells. *Physiol. Rev.*, **77**, 1033–1079.

Million, M., Tache, Y. and Anton, P. (1999) Susceptibility of Lewis and Fischer rats to stress-induced worsening of TNB-colitis: protective role of brain CRF. *Am. J. Physiol.*, **276**, G1027–G1036.

Mitic, L.L., Van Itallie, C.M. and Anderson, J.M. (2000) Molecular physiology and pathophysiology of tight junctions I. Tight junction structure and function: lessons from mutant animals and proteins. *Am. J. Physiol.*, **279**, G250–G254.

Moriarty, D., Selve, N., Baird, A.W. and Goldhill, J. (2001) Potent NK1 antagonism by SR-140333 reduces rat colonic secretory response to immunocyte activation. *Am. J. Physiol.*, **280**, C852–C858.

Mullin, J.M. and Snock, K.V. (1990) Effect of tumor necrosis factor on epithelial tight junctions and transepithelial permeability. *Cancer Res.*, **50**, 2172–2176.

Neutra, M.R., Phillips, T.L. and Phillips, T.E. (1984) Regulation of intestinal goblet cells in situ, in mucosal explants and in the isolated epithelium. *Ciba Found. Symp.*, **109**, 20–39.

Nusrat, A., Turner, J.R. and Madara, J.L. (2000) Molecular physiology and pathophysiology of tight junctions. IV. Regulation of tight junctions by extracellular stimuli: nutrients, cytokines and immune cells. *Am. J. Physiol.*, **279**, G851–G857.

O'Morain, C., Bishop, A.E., McGregor, G.P., Levi, A.J., Bloom, S.R., Polak, J.M. and Peters, T.J. (1984) Vasoactive intestinal peptide concentrations and immunocytochemical studies in rectal biopsies from patients with inflammatory bowel disease. *Gut*, **25**, 57–61.

O'Sullivan, M., Clayton, N., Breslin, N.P., Harman, I., Bountra, C., McLaren, A. and O'Morain, C.A. (2000) Increased mast cells in the irritable bowel syndrome. *Neurogastroenterol. Motil.*, **12**, 449–457.

Ogura, Y., Bonen, D.K., Inohara, N., Nicolae, D.L., Chen, F.F., Ramos, R. *et al.* (2001) A frameshift mutation in NOD2 associated with susceptibility to Crohn's disease. *Nature*, **411**, 603–606.

Ottaway, C.A., Lewis, D.L. and Asa, S.L. (1987) Vasoactive intestinal peptide-containing nerves in Peyer's patches. *Brain Behav. Immun.*, **1**, 148–158.

Ouellette, A.J. IV. (1999) Paneth cell antimicrobial peptides and the biology of the mucosal barrier. *Am. J. Physiol.*, **277**, G257–G261.

Owen, R.L. (1999) Uptake and transport of intestinal macromolecules and microorganisms by M cells in Peyer's patches–a personal and historical perspective. *Semin. Immunol.*, **11**, 157–163.

Pascual, D.W., Stanisz, A.M., Bienenstock, J. and Bost, K.L. (1999) Neural intervention in mucosal immunity. In *Mucosal Immunology*, edited by Ogra, P., Mestecky, J., Lamm, M., Strober, W., Bienenstock, J. and McGhee, J.R., pp. 631–642. New York: Academic Press.

Peeters, M., Geypens, B., Claus, D., Nevens, H., Ghoos, Y., Verbeke, G. *et al.* (1997) Clustering of increased small intestinal permeability in families with Crohn's disease. *Gastroenterology*, **113**, 802–807.

Perdue, M.H., Chung, M. and Gall, D.G. (1984a) Effect of intestinal anaphylaxis on gut function in the rat. *Gastroenterology*, **86**, 391–397.

Perdue, M.H., Forstner, J.F., Roomi, N.W. and Gall, D.G. (1984b) Epithelial response to intestinal anaphylaxis in rats: goblet cell secretion and enterocyte damage. *Am. J. Physiol.*, **247**, G632–G637.

Perdue, M.H. and Gall, D.G. (1985) Transport abnormalities during intestinal anaphylaxis in the rat: effect of antiallergic agents. *J. Allergy Clin. Immunol.*, **76**, 498–503.

Perdue, M.H. and Gall, D.G. (1986) Rat jejunal mucosal response to histamine and anti-histamines in vitro. Comparison with antigen-induced changes during intestinal anaphylaxis. *Agents Actions*, **19**, 5–9.

Perdue, M.H., Marshall, J. and Masson, S. (1990) Ion transport abnormalities in inflamed rat jejunum. Involvement of mast cells and nerves. *Gastroenterology*, **98**, 561–567.

Perdue, M.H., Masson, S., Wershil, B.K. and Galli, S.J. (1991) Role of mast cells in ion transport abnormalities associated with intestinal anaphylaxis. Correction of the diminished secretory response in genetically mast cell-deficient W/Wv mice by bone marrow transplantation. *J. Clin. Invest.*, **87**, 687–693.

Perdue, M.H., Ramage, J.K., Burget, D., Marshall, J. and Masson, S. (1989) Intestinal mucosal injury is associated with mast cell activation and leukotriene generation during Nippostrongylus-induced inflammation in the rat. *Dig. Dis. Sci.*, **34**, 724–731.

Plaisancie, P., Bosshard, A., Meslin, J.C. and Cuber, J.C. (1997) Colonic mucin discharge by a cholinergic agonist, prostaglandins and peptide YY in the isolated vascularly perfused rat colon. *Digestion*, **58**, 168–175.

Poonyachoti, S. and Brown, D.R. (2001) Delta-opioid receptors inhibit neurogenic intestinal secretion evoked by mast cell degranulation and type I hypersensitivity. *J. Neuroimmunol.*, **112**, 89–96.

Qiu, B.S., Vallance, B.A., Blennerhassett, P.A. and Collins, S.M. (1999) The role of CD4+ lymphocytes in the susceptibility of mice to stress-induced reactivation of experimental colitis. *Nat. Med.*, **5**, 1178–1182.

Quito, F.L. and Brown, D.R. (1991) Neurohormonal regulation of ion transport in the porcine distal jejunum. Enhancement of sodium and chloride absorption by submucosal opiate receptors. *J. Pharmacol. Exp. Ther*, **256**, 833–840.

Quock, R.M., Burkey, T.H., Varga, E., Hosohata, Y., Hosohata, K., Cowell, S.M. *et al.* (1999) The delta-opioid receptor: molecular pharmacology, signal transduction and the determination of drug efficacy. *Pharmacol. Rev.*, **51**, 503–532.

Ramage, J.K., Stanisz, A., Scicchitano, R., Hunt, R.H. and Perdue, M.H. (1988) Effect of immunologic reactions on rat intestinal epithelium. Correlation of increased permeability to chromium 51-labeled ethylenedi-aminetetraacetic acid and ovalbumin during acute inflammation and anaphylaxis. *Gastroenterology*, **94**, 1368–1375.

Rampton, D.S., Murdoch, R.D. and Sladen, G.E. (1980) Rectal mucosal histamine release in ulcerative colitis. *Clin.Sci. (Colch.)*, **59**, 389–391.

Rescigno, M., Urbano, M., Valzasina, B., Francolini, M., Rotta, G., Bonasio, R. *et al.* (2001) Dendritic cells express tight junction proteins and penetrate gut epithelial monolayers to sample bacteria. *Nat. Immunol.*, **2**, 361–367.

Riegler, M., Castagliuolo, I., So, P.T., Lotz, M., Wang, C., Wlk, M. *et al.* (1999) Effects of substance P on human colonic mucosa in vitro. *Am. J. Physiol.*, **276**, G1473–G1483.

Riegler, M., Castagliuolo, I., Wang, C., Wlk, M., Sogukoglu, T., Wenzl, E. *et al.* (2000) Neurotensin stimulates Cl(−) secretion in human colonic mucosa in vitro: role of adenosine. *Gastroenterology*, **119**, 348–357.

Rosztoczy, A., Fioramonti, J., Lonovics, J., Wittmann, T. and Bueno, L. (2001) The influence of different neonatal maternal deprivation patterns on the development of visceral hypersensitivity in rats. *Gastroenterology*, **120** (Suppl. 1), A716 (abstract).

Salzman, A.L., Menconi, M.J., Unno, N., Ezzell, R.M., Casey, D.M., Gonzalez, P.K. and Fink, M.P. (1995) Nitric oxide dilates tight junctions and depletes ATP in cultured Caco-2BBe intestinal epithelial monolayers. *Am. J. Physiol.*, **268**, G361–G373.

Sanderson, I.R. and Walker, W.A. (1993) Uptake and transport of macromolecules by the intestine: possible role in clinical disorders (an update). *Gastroenterology*, **104**, 622–639.

Sansonetti, P.J. and Phalipon, A. (1999) M cells as ports of entry for enteroinvasive pathogens: mechanisms of interaction, consequences for the disease process. *Semin. Immunol.*, **11**, 193–203.

Santos, J., Saperas, E., Mourelle, M. Antolin, M. and Malagelada, J.R. (1996) Regulation of intestinal mast cells and luminal protein release by cerebral thyrotropin-releasing hormone in rats. *Gastroenterology*, **111**, 1465–1473.

Santos, J., Saperas, E., Nogueiras, C., Mourelle, M., Antolin, M., Cadahia, A. and Malagelada, J.R. (1998) Release of mast cell mediators into the jejunum by cold pain stress in humans. *Gastroenterology*, **114**, 640–648.

Santos, J., Saunders, P.R., Hanssen, N.P., Yang, P.C., Yates, D., Groot, J.A. and Perdue, M.H. (1999a) Corticotropin-releasing hormone mimics stress-induced colonic epithelial pathophysiology in the rat. *Am. J. Physiol.*, **277**, G391–G399.

Santos, J., Bayarri, C., Saperas, E., Nogueiras, C., Antolin, M., Mourelle, M. *et al.* (1999b) Characterisation of immune mediator release during the immediate response to segmental mucosal challenge in the jejunum of patients with food allergy. *Gut*, **45**, 553–558.

Santos, J., Benjamin, M., Yang, P.C., Prior, T. and Perdue, M.H. (2000) Chronic stress impairs rat growth and jejunal epithelial barrier function: role of mast cells. *Am. J. Physiol.*, **278**, G847–G854.

Santos, J., Yang, P.C., Soderholm, J.D., Benjamin, M. and Perdue, M.H. (2001) Role of mast cells in chronic stress-induced colonic epithelial barrier dysfunction in the rat. *Gut*, **48**, 630–636.

Saunders, P.R., Hanssen, N.P. and Perdue, M.H. (1997) Cholinergic nerves mediate stress-induced intestinal transport abnormalities in Wistar–Kyoto rats. *Am. J. Physiol.*, **273**, G486–G490.

Saunders, P.R., Kosecka, U., McKay, D.M. and Perdue, M.H. (1994) Acute stressors stimulate ion secretion and increase epithelial permeability in rat intestine. *Am. J. Physiol.*, **267**, G794–G799.

Schaerer, E., Neutra, M.R. and Kraehenbuhl, J.P. (1991) Molecular and cellular mechanisms involved in trans-epithelial transport. *J. Membr. Biol.*, **123**, 93–103.

Schmidt, P., Rasmussen, T.N. and Holst, J.J. (1992) Nervous control of the release of substance P and neurokinin A from the isolated perfused porcine ileum. *J. Auton. Nerv. Syst.*, **38**, 85–95.

Schmidt, P.T., Eriksen, L., Loftager, M., Rasmussen, T.N. and Holst, J.J. (1999) Fast acting nervous regulation of immunoglobulin A secretion from isolated perfused porcine ileum. *Gut*, **45**, 679–685.

Scudamore, C.L., Thornton, E.M., McMillan, L., Newlands, G.F. and Miller, H.R. (1995) Release of the mucosal mast cell granule chymase, rat mast cell protease-II, during anaphylaxis is associated with the rapid development of paracellular permeability to macromolecules in rat jejunum. *J. Exp. Med.*, **182**, 1871–1881.

Sedgwick, D.M. and Ferguson, A. (1994) Dose–response studies of depletion and repopulation of rat intestinal mucosal mast cells after irradiation. *Int. J. Radiat. Biol.*, **65**, 483–495.

Shao, L., Serrano, D. and Mayer, L. (2001) The role of epithelial cells in immune regulation in the gut. *Semin. Immunol.*, **13**, 163–176.

Smith, M.W., James, P.S. and Tivey, D.R. (1987) M cell numbers increase after transfer of SPF mice to a normal animal house environment. *Am. J. Pathol.*, **128**, 385–389.

Söderholm, J.D., Peterson, K.H., Olaison, G., Franzen, L.E., Westrom, B., Magnusson, K.E. and Sjodahl, R. (1999) Epithelial permeability to proteins in the noninflamed ileum of Crohn's disease? *Gastroenterology*, **117**, 65–72.

Söderholm, J.D. and Perdue, M.H. (2001) Stress and intestinal barrier function. *Am. J. Physiol.*, **280**, G7–G13.

Söderholm, J.D., Yates, D., Gareau, M., MacQueen, G. and Perdue, M.H. (2002a) Maternal separation predisposes adult rats to colonic barrier dysfunction in response to mild stress. *Am. J. Physiol.*, **283**, G1257–G1263.

Söderholm, J.D., Yang, P.C., Ceponis, P., Vohra, A., Riddell, R., Sherman, P. and Perdue, M.H. (2002b) Chronic stress induces mast cell-dependent bacterial adherence and initiates mucosal inflammation in rat intestine. *Gastroenterology*, **123**, 1099–1108.

Spiller, R.C., Jenkins, D., Thornley, J.P., Hebden, J.M., Wright, T., Skinner, M. and Neal, K.R. (2000) Increased rectal mucosal enteroendocrine cells, T lymphocytes and increased gut permeability following acute *Campylobacter enteritis* and in post-dysenteric irritable bowel syndrome. *Gut*, **47**, 804–811.

Spitz, J.C., Ghandi, S., Taveras, M., Aoys, E. and Alverdy, J.C. (1996) Characteristics of the intestinal epithelial barrier during dietary manipulation and glucocorticoid stress. *Crit. Care Med.*, **24**, 635–641.

Stack, W.A., Keely, S.J., O'Donoghue, D.P. and Baird, A.W. (1995) Immune regulation of human colonic electrolyte transport in vitro. *Gut*, **36**, 395–400.

Stack, W.A., Filipowicz, B. and Hawkey, C.J. (1996) Nitric oxide donating compounds stimulate human colonic ion transport in vitro. *Gut*, **39**, 93–99.

Stanisz, A.M., Scicchitano, R., Dazin, P., Bienenstock, J. and Payan, D.G. (1987) Distribution of substance P receptors on murine spleen and Peyer's patch T and B cells. *J. Immunol.*, **139**, 749–754.

Stead, R.H., Tomioka, M., Quinonez, G., Simon, G.T., Felten, S.Y. and Bienenstock, J. (1987) Intestinal mucosal mast cells in normal and nematode-infected rat intestines are in intimate contact with peptidergic nerves. *Proc. Natl. Acad. Sci. USA*, **84**, 2975–2979.

Stead, R.H. (1992) Nerve remodelling during intestinal inflammation. *Ann. N. Y. Acad. Sci.*, **664**, 443–455.

Strobach, R.S., Ross, A.H., Markin, R.S., Zetterman, R.K. and Linder, J. (1990) Neural patterns in inflammatory bowel disease: an immunohistochemical survey. *Mod. Pathol.*, **3**, 488–493.

Suzuki, R., Furuno, T., McKay, D.M., Wolvers, D., Teshima, R., Nakanishi, M. and Bienenstock, J. (1999) Direct neurite–mast cell communication in vitro occurs via the neuropeptide substance P. *J. Immunol.*, **163**, 2410–2415.

Swaan, P.W. (1998) Recent advances in intestinal macromolecular drug delivery via receptor-mediated transport pathways. *Pharm. Res.*, **15**, 826–834.

Tamai, H. and Gaginella, T.S. (1993) Direct evidence for nitric oxide stimulation of electrolyte secretion in the rat colon. *Free Radic. Res. Commun.*, **19**, 229–239.

Theoharides, T.C., Bondy, P.K., Tsakalos, N.D. and Askenase, P.W. (1982) Differential release of serotonin and histamine from mast cells. *Nature*, **297**, 229–231.

Theoharides, T.C., Letourneau, R., Patra, P., Hesse, L., Pang, X., Boucher, W. *et al.* (1999) Stress-induced rat intestinal mast cell intragranular activation and inhibitory effect of sulfated proteoglycans. *Dig. Dis. Sci.*, **44**, 87S–93S.

Troncone, R., Caputo, N., Florio, G. and Finelli, E. (1994) Increased intestinal sugar permeability after challenge in children with cow's milk allergy or intolerance. *Allergy*, **49**, 142–146.

Turvill, J.L., Mourad, F.H. and Farthing, M.J. (1999) Proabsorptive and prosecretory roles for nitric oxide in cholera toxin induced secretion. *Gut*, **44**, 33–39.

Walker, W.A. and Isselbacher, K.J. (1974) Uptake and transport of macromolecules by the intestine. Possible role in clinical disorders. *Gastroenterology*, **67**, 531–550.

Wallace, J.L. and Miller, M.J. (2000) Nitric oxide in mucosal defense: a little goes a long way. *Gastroenterology*, **119**, 512–520.

Wang, L., Stanisz, A.M., Wershil, B.K., Galli, S.J. and Perdue, M.H. (1995) Substance P induces ion secretion in mouse small intestine through effects on enteric nerves and mast cells. *Am. J. Physiol.*, **269**, G85–G92.

Weston, A.P., Biddle, W.L., Bhatia, P.S. and Miner, P.B., Jr. (1993) Terminal ileal mucosal mast cells in irritable bowel syndrome. *Dig. Dis. Sci.*, **38**, 1590–1595.

Wilson, L.M. and Baldwin, A.L. (1999) Environmental stress causes mast cell degranulation, endothelial and epithelial changes and edema in the rat intestinal mucosa. *Microcirculation*, **6**, 189–198.

Yang, P.C., Berin, M.C., Yu, L.C., Conrad, D.H. and Perdue, M.H. (2000) Enhanced intestinal transepithelial antigen transport in allergic rats is mediated by IgE and CD23 (FcepsilonRII). *J. Clin. Invest*, **106**, 879–886.

Yang, P.C., Berin, M.C., Yu, L. and Perdue, M.H. (2001) Mucosal pathophysiology and inflammatory changes in the late phase of the intestinal allergic reaction in the rat. *Am. J. Pathol.*, **158**, 681–690.

Yates, D.A., Santos, J., Soderholm, J.D. and Perdue, M.H. (2001) Adaptation of stress-induced mucosal pathophysiology in rat colon involves opioid pathways. *Am. J. Physiol.*, **281**, G124–G128.

Yu, L.C., Yang, P.C., Berin, M.C., Di, L., V, Conrad, D.H., McKay, D.M. *et al.* (2001) Enhanced transepithelial antigen transport in intestine of allergic mice is mediated by ige/cd23 and regulated by interleukin-4. *Gastroenterology*, **121**, 370–381.

Zareie, M., McKay, D.M., Kovarik, G.G. and Perdue, M.H. (1998) Monocyte/macrophages evoke epithelial dysfunction: indirect role of tumor necrosis factor-alpha. *Am. J. Physiol.*, **275**, C932–C939.

Zareie, M., Singh, P.K., Irvine, E.J., Sherman, P.M., McKay, D.M. and Perdue, M.H. (2001) Monocyte/macrophage activation by normal bacteria and bacterial products: implications for altered epithelial function in Crohn's disease. *Am. J. Pathol.*, **158**, 1101–1109.

Zheng, H., Wang, J. and Hauer-Jensen, M. (2000) Role of mast cells in early and delayed radiation injury in rat intestine. *Radiat. Res.*, **153**, 533–539.

12 Infectious Pathogens and the Neuroenteric System

Charalabos Pothoulakis[1] and Ignazio Castagliuolo[2]

[1]*Division of Gastroenterology, Beth Israel Deaconess Medical Center, Harvard Medical School, Boston, MA, USA*
[2]*Institute of Microbiology, University of Padua, Padua, Italy*

Interactions between nerves and immune and inflammatory cells of the intestine are critically involved in the pathogenesis of bacterial and viral infection and control important gastrointestinal functions, including intestinal motility changes and transport, permeability to ions and larger molecules, diarrhoea and intestinal inflammation. During the past few years, a large number of neuropeptides, including serotonin, substance P, calcitonin-gene related peptide, and more recently neurotensin and galanin have been identified as important mediators in the development and progress of intestinal infectious conditions. These and other peptides, released in response to infectious agents, exert their effects by interacting, directly or indirectly, with their receptors on nerves, epithelial cells, and immune and inflammatory cells such as mast cells, and macrophages. Activation of these cells leads to the release of immune cell mediators to initiate or augment diarrhoea and intestinal inflammation. Immune cell mediators can also activate nerve endings to release neuropeptides. This chapter summarizes our recent understanding on the identification of peptide hormones and immune cells and mediators participating in the pathophysiology of bacterial and viral infection with focus on the small and large intestine, and reviews the possible mechanism(s) of action involved in these processes.

KEY WORDS: colon; cytokines; diarrhoea; small intestine; enterotoxins; inflammation; neurons; neuropeptides; mast cells; parasites; secretion.

ABBREVIATIONS: Ach, acetylcholine; CGRP, calcitonin gene-related peptide; CT, cholera toxin; *C. difficile, Clostridium difficile*; *C. parvum, Cryptosporidium parvum*; cGMP, cyclic guanosine $3'$, $5'$-monophosphate; cAMP, cyclic adenosine monophosphate; EC, enterochromaffin cells; EAEC, enteroaggressive *Escherichia coli*; EPEC, enteropathogenic *Escherichia coli*; ETEC, enterotoxigenic *Escherichia coli*; ENS, enteric nervous system; 5-HT, 5-hydroxytryptamine; Gal-R, galanin receptor; H1, histamine 1; HIV, human immunodeficiency virus; IFN-γ, interferon gamma; IL, interleukin; Ig, immunoglobulin; LT, heat-labelled; MAP, mitogen-activating protein; NF-κB, nuclear factor-kappa B; NK, neurokinin; NPY, neuropeptide Y; NT, neurotensin; *N. brasiliensis, Nippostrongylus brasiliensis*; NO, nitric oxide; PAF, platelet activating factor; PYY, peptide YY; PGE2, prostaglandin E2; PMNs, polymorphonuclear leukocytes; *S. paratyphi, Salmonella paratyphi*; *S. typhi, Salmonella typhi*; *S. typhimurium, Salmonella typhimurium*; SNAC, S-nitroso-*N*-acetyl-L-cysteine; *S. mansoni, Schistostoma mansoni*; SOM, somatostatin; SP; substance P; ST; heat-stable; *T. spiralis, Trichinella spiralis*; TNF-α, tumor necrosis factor alpha; VIP, vasoactive intestinal polypeptide; *V. cholerae, Vibrio cholerae*.

SECRETORY DIARRHOEA DUE TO BACTERIAL
ENTEROTOXINS CHOLERA TOXIN

Vibrio cholerae represents a major cause of morbidity and mortality worldwide, still reaching from time to time endemic proportions. *V. cholerae*, a gram negative bacterium, colonizes the gastrointestinal tract and releases cholera toxin (CT), a potent enterotoxin which mediates in large part diarrhoea in patients infected with this pathogen. CT is an 84 kDa protein consisting of five binding subunits and one catalytic subunit representing the enzymatic part of the toxin molecule. The mechanism by which CT mediates fluid secretion has been an exciting and some times controversial research topic. Experiments with enterocytes *in vitro* indicated that CT-mediated electrolyte transport and fluid secretion involves binding to a GM1 ganglioside receptor on the intestinal brush border, ADP ribosylation of the G-protein Giα, leading to activation of adenyl cyclase, increased intracellular cyclic adenosine monophosphate (cAMP), and secretion of chloride from epithelial cells (Goyal and Hirano, 1996). Indeed, patients infected with *V. cholerae* have increased intestinal amounts of cAMP (Chen *et al.*, 1971), and cAMP has been closely associated with chloride secretion (Clarke *et al.*, 1992). Based on these considerations, direct activation of the cAMP system in intestinal epithelial cells by CT may be solely responsible for cholera diarrhoea.

ENS involvement in CT diarrhoea

Lundgren and coworkers presented strong pharmacologic evidence for an alternative mechanism for CT action that involves the enteric nervous system (ENS) via activation of neuronal reflexes and release of peptide mediators (Figure 12.1). Their early studies demonstrated that exposure of cat intestine to CT stimulated release of vasoactive intestinal polypeptide (VIP) (Cassuto *et al.*, 1981), and administration of the neuronal blocker tetrodotoxin or serosal application of lidocaine, attenuated cholera secretion in isolated intestinal segments of anesthetized rats (Cassuto *et al.*, 1983). Administration of the cholinergic nicotinic receptor antagonist hexamethonium also diminished CT secretion in animal intestine (Cassuto *et al.*, 1982, 1983). Moreover, the cholinergic agonist carbachol enhanced CT-induced secretion in both, human colonic cell lines and rat ileal mucosa placed in Ussing chambers (Mahmood *et al.*, 2000). These results indicate that participation of the ENS in CT-induced intestinal fluid secretion involves activation of cholinergic and non-cholinergic nervous reflexes (Lundgren, 1998). Jodal and Lundgren (1995) demonstrated that chemical ablation of the myenteric plexus in rat small intestine diminished CT-induced fluid secretion, indicating that the myenteric plexus plays a central role in the mediation of signals in the intramural secretory reflex activated by this toxin. In guinea pig intestine CT induces the appearance of activated, Fos immunoreactive, neurons localized in the submucosal and myenteric plexuses, and this effect was antagonized by tetrodotoxin (Ganser *et al.*, 1983; Kirchgessner *et al.*, 1992). Interestingly, fluid secretion in response to intraluminal administration of cAMP is also inhibited by nerve blockers (Eklund *et al.*, 1984), suggesting that secretion induced by cAMP may involve activation of the ENS. Further evidence for ENS involvement in CT-induced secretion was demonstrated by Nocerino *et al.* (1995). These investigators showed that CT administration to a jejunal segment not only induced a significant local secretory response, but also stimulated secretion in the colon, which was not exposed to cholera toxin (Nocerino *et al.*, 1995).

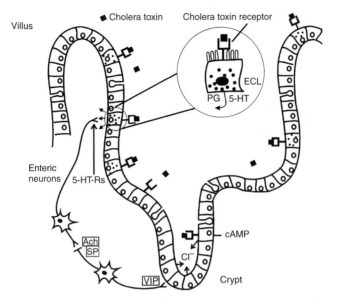

Figure 12.1 Neuroimmune regulation of CT-mediated intestinal secretion. CT binds to the GM1 ganglioside receptor on enterochromaffin cells (EC) leading to release of serotonin (5-HT) and prostaglandins that in turn activate enteric neurons in the lamina propria. Enteric nerves release acetylcholine (Ach), substance P (SP), and VIP which directly stimulates crypt cells to secrete Cl⁻ and water. CT can also stimulate Cl⁻ secretion by binding to GM1 ganglioside receptors on epithelial cells and stimulating cAMP (modified from Pothoulakis *et al.* (1998) *News Physiol Sci*, **13**, 58–63).

Moreover, colonic secretion following jejunal administration of CT was abolished by disruption of the anatomic continuity of enteric nerves, underlying the importance of the ENS in the mediation of this response (Nocerino *et al.*, 1995).

The mechanism of ENS activation by CT may involve direct interactions of the toxin with enteric nerves. Thus, CT binding sites are present in VIP-containing neurons in the submucosal plexus and these neurons could be activated by CT exposure (Jiang *et al.*, 1993). This is consistent with the notion that CT subunits can be delivered intact to the serosal surface of the basolateral membrane of colonocytes (Lencer *et al.*, 1995) and, in an *in vivo* scenario, possibly to supepithelial cells, including the submucosal plexus. Interestingly, VIP antagonists, dose dependently decreased fluid secretion induced by CT, indicating a role for this peptide in cholera toxin-induced secretion (Mourad and Nassar, 2000). Along these lines, intra-arterial infusion of neuropeptide Y (NPY), an established inhibitor of VIP, reduced cholera toxin-induced secretion and VIP release (Sjoqvist *et al.*, 1988), suggesting that cholera diarrhoea is mediated by inhibition of mucosal secretory VIP neurons.

5-hydroxytryptamine and CT secretion

The secretagogue 5-hydroxytryptamine (5-HT), a major peptide identified in enterochromaffin cells (EC) of the gastrointestinal mucosa, is stored in granules and released in response to several stimuli. Several intriguing early observations suggested that EC cells and 5-HT may be associated with the intestinal responses to CT. Bhide *et al.* (1970) noted increased 5-HT levels in the blood of rabbits infected with *V. cholerae* and Osaka *et al.* (1975) showed that CT administration into the duodenum causes degranulation of EC cells. 5-HT content of individual EC in intestinal segments exposed to CT was diminished, and this effect was correlated with increased net fluid transport in response to CT (Nilsson *et al.*, 1983). Moreover, experimental tachyphylaxis in animals inhibited CT-induced fluid secretion (Cassuto *et al.*, 1983).

Several groups confirmed and expanded these observations by demonstrating the functional significance of 5-HT and its receptors in the secretory responses to CT. Beubler *et al.* (1989) showed that CT, dose dependently, caused a significant rise in 5-HT and prostaglandin E2 (PGE2) levels in rat small intestine. Interestingly, fluid secretion and PGE2 release, but not 5-HT release, in response to cholera toxin, were inhibited by indomethacin and the 5-HT2 receptor blocker ketanserin (Beubler *et al.*, 1989). The 5-HT3 receptor antagonist, granisetron, also reduced intestinal fluid and electrolyte secretion induced by CT (Mourad *et al.*, 1995). Administration of 5-HT2 and 5-HT3 receptor blockers to rats abolished cholera toxin-induced secretion without influencing the expected rise in cAMP levels in response to this toxin (Beubler and Horina, 1990). These results provided evidence for an important role of 5-HT in CT-induced secretion. The sequence of events of this response involves an early release of 5-HT from EC cells that causes PGE2 formation via 5-HT2 receptors and activation of neurons via 5-HT1 and 5-HT3 receptors (Figure 12.1). These two effects may result in fluid secretion, which can be completely abolished by a combination of a 5-HT2 blocker and a 5-HT3 blocker. The 5-HT4 subtype also appears to participate in CT-induced secretion (Moore *et al.*, 1996). These studies support the notion that the secretory effects of CT, at least in part, may be independent of its effects on cAMP formation in enterocytes.

Studies in human volunteers demonstrated that CT administration to duodenal segments stimulated fluid secretion and increased 5-HT release into the intestinal lumen (Bearcroft *et al.*, 1996), indicating that 5-HT may play a role in mediating CT-induced diarrhoea in humans. Two clinical studies, however, failed to demonstrate a beneficial effect for 5-HT3 receptor antagonists in cholera toxin-induced secretion (Eherer *et al.*, 1994; Bearcroft *et al.*, 1997). Although the mechanism of 5-HT release from EC cells in response to CT is not known, calcium channels known to participate in exocytosis of many cell types may be involved. In support of this hypothesis, some studies showed that calcium channel blockers diminished CT-induced fluid secretion, as well as mucosal 5-HT in response to this toxin (Timar Peregrin *et al.*, 1997; Peregrin *et al.*, 1999), suggesting that calcium channel blockade may be beneficial in the treatment of cholera diarrhoea.

Role of prostaglandins

Prostaglandins are important mediators of many intestinal functions, including secretion of fluid and intestinal inflammation. Several groups have suggested a role for prostaglandins in the mechanism of cholera diarrhoea. Exposure of intestinal loops to CT results in enhanced PGE2 release (Bedwani and Okpako, 1975; Beubler *et al.*, 1989; Peterson and Ochoa, 1989) and patients with cholera-associated diarrhoea have enhanced intestinal PGE2 synthesis (Speelman *et al.*, 1985). Moreover, inhibitors of the cyclooxygenase pathway reduce cholera toxin-induced secretion (Gots *et al.*, 1974; Wald *et al.*, 1977) by a cAMP-independent mechanism (Wald *et al.*, 1977; Beubler *et al.*, 1986). Increased myo-electric activity in response to CT in rabbits was also attenuated after indomethacin pretreatment (Mathias *et al.*, 1977), suggesting that prostaglandins may be involved in motility changes during cholera infection.

Different pathways appear to be involved in CT-induced PGE2 formation. CT can directly induce synthesis of PGE2 in many cells, including intestinal epithelial cells (Peterson *et al.*, 1994), by mechanisms involving activation of the cyclooxygenase-2 gene (Beubler *et al.*, 2001), and phospholipase A2 (Peterson *et al.*, 1996). Platelet activating factor (PAF) is also involved in the effects of CT *in vitro*, and PAF, itself, stimulates PGE2 formation (Thielman *et al.*, 1997). CT-induced intestinal PGE2 formation has been also linked to 5-HT release from EC via neuronal and non-neuronal pathways (Beubler and Horina, 1990). Thus, a 5-HT-mediated neuronal pathway may be linked to PGE2 formation and fluid secretion during CT intestinal exposure. The results discussed above suggest that inhibition of prostaglandin synthesis may have a role in the treatment of cholera-induced diarrhoea. Using an intestinal perfusion system, Van Loon *et al.* (1992) demonstrated that administration of the cyclooxygenase inhibitor indomethacin reduced jejunal fluid secretion and PGE2 release in patients with acute cholera infection. However, a randomized, controlled clinical trial did not demonstrate a beneficial effect of indomethacin in 29 patients with cholera diarrhoea (Rabbani and Butler, 1985).

Substance P, nitric oxide and somatostatin

Several other neuropeptides have been also implicated in the intestinal effects of CT (Table 12.1). Studies with substance P (SP) antagonists indicated that the SP receptors neurokinin (NK)-1 and -2 are involved in CT-induced jejunal water and electrolyte secretion (Turvill *et al.*, 2000). Pre-treatment of animals with capsaicin, which desensitizes sensory neurons, did not influence cholera secretion, indicating that SP-containing sensory nerves may not participate in this CT response (Castagliuolo *et al.*, 1994; Turvill *et al.*, 2000). Participation of SP in these experiments may involve actions of this neuropeptide within intrinsic neurons of the myenteric plexus. Using different experimental approaches other groups, however, were unable to show a SP involvement in CT-induced secretion (Sjoqvist *et al.*, 1993; Pothoulakis *et al.*, 1994). Nitric oxide (NO) is a multifunctional molecule that modulates several intestinal functions, including blood flow and peristalsis. Janoff *et al.* (1997) demonstrated that levels of stable NO metabolites were significantly increased in sera from patients with active cholera infection, and CT administration to rabbits had a similar effect. Systemic administration of the NO synthase inhibitor NG-nitro-L-arginine methyl ester caused a dose dependent reduction in CT-induced secretion in the rat jejunum, suggesting involvement of NO in the intestinal responses to CT (Turvill *et al.*, 1999b). However, studies focusing on the rat terminal ileum failed to show a NO connection to the effects of CT (Qiu *et al.*, 1996).

Somatostatin (SOM) has been also implicated in cholera secretion. Administration of SOM or its analogues to animals *in vivo* suppressed CT-induced secretion in rats (Yoshioka *et al.*, 1987; Eklund *et al.*, 1988; Sjoqvist, 1992; Botella *et al.*, 1993). The mechanisms of SOM participation appear to involve a nervous secreto-motor reflex

TABLE 12.1
Neurotransmitters and neuropeptides involved
in enterotoxin-mediated secretory diarrhoea.

Acetylcholine
Enkephalins
Neuropeptide Y (NPY)
Nitric oxide
Peptide YY (PYY)
Serotonin (5-hydroxyltryptmine)
Somatostatin
Substance P
Vasoactive intestinal polypeptide

activated by CT (Sjoqvist, 1992), release of VIP (Eklund *et al.*, 1988), or a direct anti-secretory effect at the enterocyte level (Botella *et al.*, 1993). Clinical studies are required to study the usefulness of SOM analogues in cholera diarrhoea.

Role of opioid receptors

Opiod receptors appear to be involved in CT-mediated jejunal water and electrolyte secretion. Thus, opioid blockade by naloxon increased CT-induced secretion (Sjoqvist, 1991), while administration of morphine to rats inhibited this CT response (Sjoqvist, 1992). The mechanism of morphine inhibition appears to be indirect, via interactions with sympathetic nerve terminals in the intestine (Sjoqvist, 1992). Jonsdottir *et al.* (1999) reported that afferent stimulation of the sciatic nerve strongly inhibited CT-induced small intestinal fluid secretion primarily by a central opioid mechanism and to a lesser extent, by a peripheral mechanism.

Enkephalinases are enzymes that participate in the digestion of peptides, including enkephalins, which interact with opioid receptors. Acetorphan, a potent enkephalinase inhibitor, significantly decreased CT-induced water and electrolyte secretion in dogs (Primi *et al.*, 1999). Acetorphan prevented jejunal water and electrolyte secretion induced by CT in human subjects (Hinterleitner *et al.*, 1997). Interestingly, in cultured rat epithelial cells containing delta opioid receptors CT-stimulated cAMP synthesis was inhibited by a delta-agonist, while an enkephalinase inhibitor potentiated this effect (Nano *et al.*, 2000), indicating that involvement of opioid receptors in CT-induced secretion may be exerted at the enterocyte level. These results suggest a functional link between CT diarrhoea and opioid receptors and the possible use of enkephalinase inhibitors in cholera diarrhoea (Schwartz, 2000). Sigma receptors, previously thought to belong to the opioid receptor family, are present in the CNS and on the periphery, including the intestinal mucosa and submucosa. Activation of these receptors has been also linked to intestinal secretory processes (Pascaud *et al.*, 1993) and inhibition of Ach release. Turvill *et al.* (1999a) reported that CT-induced secretion in rat jejunum was inhibited by the sigma ligand igmesine. Igmesine was also successful in reducing established CT-mediated secretion (Turvill *et al.*, 1999a), indicating its therapeutic potential for treatment of CT diarrhoea.

Role of mast cells

Mast cells, by releasing soluble mediators and interacting with many different cell populations in the intestinal mucosa, participate in important intestinal functions, including fluid secretion. A limited number of studies examined the possible participation of mast cells in CT-induced diarrhoea. Mucosal endoscopy biopsies from the rectum of patients with cholera demonstrated an increased number of mast cells around blood vessels and nerves (Pulimood *et al.*, 1998). The same group also reported increased degranulation of mucosal mast cells in the small intestine of cholera patients (Mathan *et al.*, 1995). *In vitro* evidence also indicates that CT can have several immunomudolatory effects on mast cells, including histamine release (Saito *et al.*, 1988), potentiation of immunoglobulin E (IgE)-mediated serotonin secretion (McCloskey, 1988), increased interleukin (IL)-6 synthesis and decreased tumor necrosis alpha (TNF-α) production (Leal-Berumen *et al.*, 1996). However, evidence that mast cells are functionally linked to cholera-induced diarrhoea is lacking (Castagliuolo *et al.*, 1994; Klimpel *et al.*, 1995).

ESCHERICHIA COLI TOXINS

Enterotoxigenic strains of *Escherichia coli* (ETEC) represent the most important aetiologic agent of travellersí diarrhoea worldwide. ETEC exert its secretory effects by producing two different protein exotoxins, the heat-labelled (LT), and heat-stable (ST) enterotoxins. LT possess many structural and functional similarities to CT. Like CT, it binds to GM1 ganglioside receptors on enterocytes, activates adenylate cyclase and increases cAMP, leading to a chloride secretory state (Spangler, 1992). Some ETEC strains produce ST, an enterotoxic with a different mode of action. The A subunit of ST binds to guanylyl cyclase on the enterocyte brush border, leading to increased cyclic guanosine $3'$, $5'$-monophosphate (cGMP) and stimulation of chloride transport, and fluid and electrolyte secretion (Wedel and Garbers, 1997; Lucas *et al.*, 2000). Several pieces of evidence demonstrate that, as in CT diarrhoea, enteric nerves are also involved in the secretory responses to ST and LT. Thus, administration of tetrodotoxin, lidocaine, or hexamethonium significantly reduced ST- as well as LT-induced fluid secretion (Lundgren *et al.*, 1989; Turvill *et al.*, 1998). The secreto-motor neurons involved in the secretory effects of ST and LT appear to involve the neurotransmitters Ach and VIP (Lundgren *et al.*, 1989; Mourad and Nassar, 2000). However, despite these common neuronal pathways, there are also differences in the mechanism of neuronal stimulation between LT-, ST- and CT-induced secretion. The most intriguing difference is the lack of involvement of 5-HT in the effects of LT toxin, despite their structural and functional similarities. Thus, in contrast to CT, LT-induced secretion is not associated with 5-HT release, and it is not affected by 5-HT receptor antagonism (Turvill *et al.*, 1998). Differences in structure in specific regions of these two toxin molecules, binding to different receptors, or activation of different secretory neuronal reflexes by LT and CT may account for these differences (Lundgren, 1998). The latter hypothesis is supported by the inability of SP receptor antagonists to inhibit LT-induced fluid secretion, in contrast to their anti-secretory effects in CT-induced diarrhoea (Turvill *et al.*, 2000).

E. coli *enterotoxins and sigma ligands*

In animal studies intravenous administration of the sigma ligand igmesin inhibited ST- and LT-induced intestinal water and electrolyte secretion, indicating involvement of sigma receptors in neuronally mediated secretion (Turvill *et al.*, 1999a). The mechanism of this response appears to involve the neuropeptides neuropeptide Y (NPY) and peptide YY (PYY) which possess potent anti-secretory activities *in vivo* (Saria and Beubler, 1985), via neuronal activation of sigma receptors (Riviere *et al.*, 1993). Interestingly, igmesin also reduces intestinal secretion in human volunteers (Roze *et al.*, 1998), suggesting its potential application for the treatment of enterotoxigenic diarrhoea.

VIRAL SECRETORY DIARRHOEA

Although solid evidence exists for the involvement of enteric nerves and neuroimmune mediators in diarrhoea due to bacterial enterotoxins, only few studies identified similar interactions in viral infections. Sharkey *et al.* (1992) reported that human immunodeficiency virus (HIV) patients had increased abnormal patterns of immunoreactivity for SP, VIP and SOM in enteric nerves and EC cells, and altered patterns of immunoreactivity in rectal nerves. Extensive damage of enteric autonomic nerve fibres in the jejunum was also observed in biopsy specimens from asymptomatic as well as symptomatic HIV patients (Griffin *et al.*, 1988). Thus, gastrointestinal symptoms frequently seen in HIV patients may be associated with altered function of intestinal nerves or EC. Moreover, in isolated intestine, IgA from HIV-infected patients bound to the muscarinic acetylcholine receptor on villous epithelium and stimulated cellular responses similar to that seen after ligand binding to this receptor (Sales *et al.*, 1997), indicating a cholinergic response to HIV.

A recent study examined whether enteric nerves participate in rotavirus diarrhoea, one of the most common aetiologic agents of gastroenteritis in children. As shown by Ussing chamber experiments, rotavirus increased fluid secretion and this effect was significantly attenuated by pretreatment of the intestinal segments with tetrodotoxin, lidocaine, or mesalamine, a ganglionic blocker (Lundgren *et al.*, 2000). The results suggested that the ENS mediated two-thirds of the rotavirus-induced secretion. Although the mechanism of this response has not been elucidated, it is possible that the rotavirus non-structural protein, NSP4, a novel enterotoxin with secretory properties (Morris and Estes, 2001), mediates this effect.

INFLAMMATORY INFECTIONS DUE TO BACTERIA

ENTEROADHERENT BACTERIA

Enteroadherent bacteria, including enteropathogenic *E. coli* (EPEC), enteroaggressive *E. coli* (EAEC), diffusely adherent *E. coli* (DAEC), enterohaemorrhagic *E. coli* (EHEC) and *Salmonella typhimurium*, colonize the small and/or large intestine and do not appear to be invasive to any appreciable extent *in vivo*. Enteroadherent bacteria, with the exception of shiga-like toxins of EHEC, are not known to elaborate any of the classical enterotoxins although they cause diarrhoeal illness. The pathogenesis of gastrointestinal lesions caused by enteroadherent pathogens, such as the adherence and attaching phenotype, has been studied in detail and are induced by pathogen-derived proteins (Donnenberg and Whittam, 2001). However, the mechanisms by which these organisms induce shortening and necrosis of the villi, oedema with an inflammatory infiltrate of the submucosa and diarrhoea have only recently started to emerge. For example, EAEC release a serine protease (Pet) that causes disruption of the membrane skeleton leading to cytotoxic and enterotoxic effects (Villaseca et al., 2000). A variety of in vitro studies have clearly demonstrated that intestinal epithelial cells mount a coordinated response to enteroadherent bacterial pathogens by secreting water, salt, mucin and anti-microbial peptides, releasing cytokines and expressing adhesion molecules (Kagnoff and Eckmann, 1997).

Recent exciting studies indicate that intestinal epithelial cells may respond to enteroadhesive pathogens by expressing specific neuropeptide receptors, which modify the responsiveness of the intestinal mucosa to secretory peptides. Thus, noninvasive enteroadherent E. coli pathogens, but not normal commensal E. coli, induce mRNA and protein synthesis of the receptor for the neuropeptide galanin (Gal1-R) in intestinal epithelial cells in vitro (Hecht et al., 1999). Galanin is a neuropeptide found in enteric nerve terminals lining the gastrointestinal tract and in immune cells during the inflammatory process (King et al., 1989). When secreted in the gastrointestinal tract, galanin binds to its receptors modulating intestinal transit and stimulating Cl^- secretion (Bauer et al., 1989; Benya et al., 1999). Up-regulation of Gal1-R expression in colonocytes is evident also after oral administration of EHEC (Hecht et al., 1999). This observation may be relevant to EHEC infection since normal mouse colon is unresponsive to galanin, whereas the colon from EHEC-infected mice rapidly respond to galanin with increased chloride secretion (Hecht et al., 1999). Furthermore, mice genetically lacking Gal1 receptors are resistant to EHEC-mediated fluid secretion, underscoring the importance of these neuropeptide receptors in the pathophysiology of EHEC infection (Matkowskyj et al., 2000). Gal1-R gene up-regulation in colonocytes is mediated by activation and nuclear translocation of the nuclear factor kappa B (NF-κB) (Hecht et al., 1999), a key event in expression of several genes involved in the inflammatory response to pathogens (Naumann, 2000). Thus, the expression of specific neuropeptide receptors is part of the orchestrated response in the intestinal epithelial cells to enteroadherent pathogens. De novo expression of a specific neuropeptide receptor (Gal1-R) will modify the responsiveness of the mucosa to potent secretagogues and may account for a portion of the diarrhoea triggered by non-invading enteric pathogens (Figure 12.2).

Migration of polymorphonuclear leukocytes (PMNs) into the intestinal mucosa, a hallmark of inflammatory infectious diarrhoea, is triggered by cytokines released from epithelial cells. Several bacterial species like Shigella, S. typhimurium and EAEC trigger IL-8 production from intestinal epithelial cells. However, the severity of the inflammatory infiltrate may range from a massive PMNs recruitment with epithelial cell destruction, to a subtler inflammatory infiltrate. Thus, IL-8 release from epithelial cells may have different effects depending on the activation of inflammatory cells and neurons in the intestinal lamina propria. Neuropeptides, such as SOM have been shown to exert an active role in the regulation of mucosal inflammatory responses (McIntosh, 1985). SOM released from neuronal and non-neuronal cells distributed throughout the length of the gastrointestinal tract may

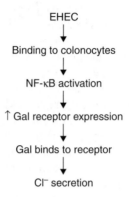

EHEC
↓
Binding to colonocytes
↓
NF-κB activation
↓
↑ Gal receptor expression
↓
Gal binds to receptor
↓
Cl⁻ secretion

Figure 12.2   Involvement of galanin (Gal) in fluid secretion mediated by enteroadherent *E. coli*.

TABLE 12.2
Neuropeptides and immune mediators involved in
bacteria-induced inflammatory diarrhoea.

Neuropeptides	Immune mediators
Galanin	Prostaglandin E2
Serotonin (5-hydroxyltryptamine)	Interleukin-8
Somatostatin	Interleukin-10
Substance P	Interleukin-12
	Interferon-gamma

modulate the inflammatory response to Salmonella infection by inhibiting the release of pro-inflammatory cytokine such as IL-10 and IL-8 from intestinal epithelial cells (Chowers *et al.*, 2000). Diarrhoea mediated by *Salmonella* and *Shigella* has been linked to the activation of the ENS. Pretreatment with nerve blockers inhibited fluid accumulation in the jejunum and ileum of rats caused by inoculation with *S. typhimurium* (Brunsson, 1987). Based on the type of the pharmacologic inhibitors used, this study suggested that this *Salmonella* strain activates a prostaglandin-dependent nerve reflex that contains a nicotinergic transmission (Brunsson, 1987). Consistent with this observation, Grondahl *et al.* (1998) demonstrated that the secretory pathways stimulated by *S. typhimurium* infection involve release of 5-HT and PGE2, and activation of 5-HT3 receptors. *S. typhimurium* and *Shigella flexerii* infection in mice also results in a dramatic increase in the expression of Gal1 receptors and binding of galanin to these receptors leads to significant colonic fluid secretion (Matkowskyj *et al.*, 2000). Interestingly, genetic deletion of the Gal1 receptor gene completely diminished fluid secretion, although it did not change the intestinal inflammatory response to these pathogens (Matkowskyj *et al.*, 2000) (Table 12.2).

ENTEROINVASIVE PATHOGENS: ROLE OF SP

Salmonella serovars are an important cause of gastrointestinal disorders characterized by mucosal and submucosal invasion followed by intrahost dissemination (*S. typhi and S. paratyphi*) (Buchwald and Blaser, 1984). Following oral administration, *Salmonella* species are able to survive the gastric acidic environment, reach the intestinal lumen and then adhere to and invade the ileal mucosa. The mediators released in response to Salmonella infection and the mucosal damage induced is strictly dependent upon the specific Salmonella serovar involved. Thus, *S. typhi* will invade and translocate across the epithelium and can be isolated from extra-intestinal sites. Indeed, these pathogens enter and apparently survive for many hours in submucosal monocytes, which facilitate their transport to distal sites in the host (spleen, liver and bone marrow). The role of neuropeptides and their receptors in the inflammatory response to these invasive pathogens has received limited attention. Enteroinvasive strains of *Salmonella dublin* invade the intestinal mucosa and they disseminate into peripheral tissues as intracellular pathogens of macrophages (Hsu, 1989). *S. typhi* survives in intestinal macrophages and activates them to secrete IL-12 that in turn triggers interferon gamma (IFN-γ) production (Kincy-Cain *et al.*, 1996). As well, the initiation of an efficient mucosal immune response against *Salmonella* requires increased expression of neuropeptides. Bost (1995) reported that following oral administration of *S. typhi*, SP and neurokinin A mRNA precursors are increased in Peyer's patches, lymph nodes and spleen. Up-regulation of preprotachykinin mRNA occurred mainly in mononuclear leukocytes (Bost, 1995). Furthermore, following oral administration of *Salmonella*, SP receptor mRNA is up-regulated in Peyer's patches, and mesenteric lymph nodes (Kincy-Cain and Bost, 1996). SP and SP receptor appear to play a key role in mediating macrophage activation following oral administration of *Salmonella*, since administration of a SP receptor antagonist enhances the severity of this infection (Kincy-Cain and Bost, 1996). It is therefore possible that Salmonella infection triggers SP and SP receptor up-regulation in macrophages and that SP, working in a paracrine fashion, further activates macrophages inducing IL-12 and IFN-γ expression (Kincy-Cain and Bost, 1997). Thus, these studies suggest that SP and his receptor may be involved in the early immune response against Salmonella infection, contributing to mount a coordinated response to kill the endocellular pathogens (Table 12.2).

INFLAMMATORY DIARRHOEA DUE TO BACTERIAL ENTEROTOXIS

CLOSTRIDIUM DIFFICILE

Clostridium difficile, a gram negative anaerobic bacteria, is the major aetiologic factor of antibiotic-associated diarrhoea and colitis in animals and humans (Kelly *et al.*, 1994). Diarrhoea and enterocolitis in response to this pathogen is mediated by release of two large molecular weight protein exotoxins, toxin A and B (Pothoulakis and Lamont, 2001). The primary cellular mechanism of these toxins is related to monoglucosylation of the small GTP binding proteins belonging to Rho family of proteins (Aktories *et al.*, 2000), leading to dissagregation of stress actin fibres and opening of the tight junctions in colonocytes

(Riegler *et al.*, 1995). However, in addition to actin-related alterations, *C. difficile* infection is associated with acute inflammatory changes present in patients with *C. difficile*-mediated colitis (Kelly *et al.*, 1994), as well in ileal or colonic loops exposed to purified *C. difficile* toxin A (Triadafilopoulos *et al.*, 1989). The cellular proinflammatory pathway triggered by toxin A involves binding to bush border glycoprotein receptors (Pothoulakis *et al.*, 1991), early mitochondrial damage in target cells (He *et al.*, 2000), mitogen-activating protein (MAP) kinase activation (Warny *et al.*, 2000), and nuclear translocation of NF-κB (Jefferson *et al.*, 1999). The end product of the cellular pathway is release of various cytokines with potent pro-inflammatory properties (Castagliuolo *et al.*, 1998a; Jefferson *et al.*, 1999; Warny *et al.*, 2000). In animal models, toxin A causes fluid accumulation and, in contrast to cholera and *E. coli* toxins, intestinal inflammation accompanied by epithelial cell destruction and necrosis, and transmigration of neutrophils to the intestinal mucosa. Communication between enterocytes, sensory neurons, endothelial cells and inflammatory cells of the lamina propria leads to the development of inflammatory enterocolitis in response to *C. difficile* toxin A (Pothoulakis, 1996).

ROLE OF SENSORY NEUROPEPTIDES

Studies with anesthetized animals demonstrated that either local application of lidocaine or i.v. administration of the ganglionic blocker hexamethonium dramatically inhibited *C. difficile* toxin A-induced fluid accumulation, and intestinal permeability and neutrophil transmigration (Castagliuolo *et al.*, 1994; Sorensson *et al.*, 2001). Interestingly, administration of the neurotoxin capsaicin, that depletes the nerve endings of sensory neurons from neuropeptides, or extrinsic surgical denervation, abrogated these toxin A responses (Castagliuolo *et al.*, 1994; Mantyh *et al.*, 1996b, 2000), suggesting that sensory neurons mediate toxin A-mediated enteritis. Since SP and calcitonin gene-related peptide (CGRP) are the primary neuropeptides synthesized by primary sensory neurons, the possibility that these peptides mediate the effects of toxin A was examined using specific peptide and peptide receptor antagonists. Injection of SP (NK-1) receptor antagonists or a CGRP antagonist diminished toxin A-induced inflammation and secretion, and secretion of proinflammatory mediators from rat intestine (Pothoulakis *et al.*, 1994; Mantyh *et al.*, 1996b; Keates *et al.*, 1998). Increased SP and CGRP expression in the cell bodies of dorsal root ganglia was also demonstrated followed by increased release of these peptide in the intestinal mucosa during the course of toxin A enteritis (Castagliuolo *et al.*, 1997; Keates *et al.*, 1998). Following toxin A administration lamina propria macrophages also express NK-1 receptors and ligand binding to these receptors caused TNF-α secretion (Castagliuolo *et al.*, 1997). Consistent with these observations, mice with genetic deletion of NK-1 receptors have diminished toxin A-induced responses (Castagliuolo *et al.*, 1998b), while, in contrast, mice genetically lacking neural endopeptidase, an enzyme that degrades SP, have significantly enhanced toxin A-mediated intestinal inflammation (Kirkwood *et al.*, 2001). These studies indicate that primary sensory afferent nerves and binding of sensory peptides to their activated receptors on mucosal cells are major determinants in the expression of toxin A-induced inflammatory diarrhoea *in vivo* (Figure 12.3). In line with the observations in animal models of toxin A enterocolitis, immunohistochemical studies also showed increased expression of SP receptors in small blood vessels and lymphoid

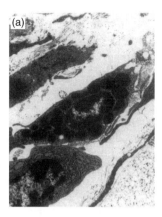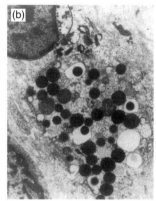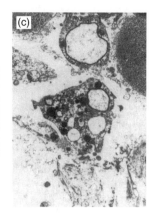

Figure 12.3   Mast cell degranulation during *C. difficile* toxin A-induced enteritis. Rat ileal loops were injected with either buffer (a) or purified *C. difficile* toxin A (B, C) and ileal explants were prepared and placed in medium. After 15 min (b), and 60 min (c), ileal samples were processed for electron microscopy. Note the round homogeneous dense granules in buffer exposed intestine (a) as opposed to many granules that lost their electron density or some granules that are entirely empty 15 min after toxin A administration (b). Sixty minutes after toxin A administration almost complete degranulation of mast cells is observed (c) (from Castagliuolo *et al.* (1994) *Gastroenterology*, **107**, 657–665, with permission).

aggregates in biopsies from patients with *C. difficile*-associated pseudomembranous colitis (Mantyh *et al.*, 1996a). These results suggest that SP receptor antagonists may have a place in therapy of *C. difficile* or other forms of colitis.

Although sensory neurons via release of SP and CGRP represent a major amplification system in acute inflammation caused by toxin A, the pathways by which signals from the intestinal epithelium are transmitted to the sensory nerves are not known. It is possible that release of pro-inflammatory cytokines (Castagliuolo *et al.*, 1998a), linked to the toxin A-induced activation of NF-κB system (Jefferson *et al.*, 1999; Warny *et al.*, 2000), can stimulate the nerve endings of the sensory neurons to initiate a neuronally dependent inflammatory response. This hypothesis is consistent with the notion that cytokines can affect the function of sensory neurons (Vasko *et al.*, 1994; Nicol *et al.*, 1997). Recent results with isolated guinea pig submucosal neurons suggest that toxin A has direct effects on enteric neurons *in vitro* (Xia *et al.*, 2000), but the importance of this interaction in the pathophysiology of toxin A enteritis remains to be elucidated.

NEUROTENSIN AND THE INTESTINAL INFLAMMATORY RESPONSE

The brain–gut peptide neurotensin (NT) has been implicated in several important colonic responses, including chloride secretion and stress-mediated colonic mucin and prostaglandin release (Castagliuolo *et al.*, 1996; Riegler *et al.*, 2000). Castagliuolo *et al.* (1999) reported that intraluminal toxin A administration caused a significant early rise in NT and NT receptor mRNA and protein in rat colon. Pretreatment with a specific NT type 1 receptor antagonist resulted in reduced toxin A-associated colonic responses (Castagliuolo *et al.*, 1999), indicating that secretion and inflammation mediated by toxin A involves NT. Participation of NT in the toxin A model appears to involve SP receptors

(Castagliuolo *et al*., 1999), suggesting a functional communication between NT and SP in the development of enterotoxin-induced colonic inflammation (Figure 12.3).

The mechanism of up-regulation of receptors for both SP and NT on intestinal epithelial cells soon after toxin A administration (Mantyh *et al*., 1996b; Pothoulakis *et al*., 1998), and of the increased expression of colonic SP receptors in patients with *C. difficile* colitis (Mantyh *et al*., 1996a) has not been elucidated. Interestingly, SP and NT receptors are G-protein-coupled receptors and the promoter region of these receptors contain NF-κB binding sites. As noted above, toxin A also stimulates increased NF-κB activity which in turn leads to the release of several inflammatory cytokines (Pothoulakis and Lamont, 2001). Thus, the NF-κB system may represent a pro-inflammatory mechanism of neuropeptide receptor up-regulation during microbial–epithelial cell interactions.

INVOLVEMENT OF MAST CELLS

Intestinal mast cells are critically involved in the pathophysiology of intestinal inflammation by release of potent mediators and communication with inflammatory cells of the lamina propria, as well as with neurons in the subepithelium. Mast cell-derived mediators can also influence epithelial pathophysiology by interacting directly with colonocytes and enterocytes. Several lines of evidence underscored the importance of these cells in *C. difficile* toxin A-mediated fluid secretion and neutrophil activation in toxin A-mediated inflammation. In animal models toxin A triggered acute mucosal mast cell degranulation (Castagliuolo *et al*., 1994) (Figure 12.4), and increased blood levels of the rat mast cell protease II and other potent mast cell products (Pothoulakis *et al*., 1993; Castagliuolo *et al*., 1994). Administration of the mast cell inhibitor ketotifen attenuated toxin A-associated intestinal secretion, mucosal permeability and neutrophil infiltration, and reduced levels of mast cell derived mediators (Pothoulakis *et al*., 1993). In addition, toxin A-induced increased albumin permeability in rat mesenteric venules was diminished after treatment with the mast cell blocker lodoxamide, or the histamine 1 (H1) receptor blocker hydroxyzine (Kurose *et al*., 1994), indicating that histamine secreted from mast cells is involved in leukocyte–endothelial cell interactions by engaging H1-receptors on endothelial cells.

Direct evidence for mast cell participation in the intestinal responses to toxin A was provided by mice genetically deficient in mast cells. Wershil *et al*. (1998) demonstrated that mast cell-deficient mice had lower toxin A-induced intestinal secretion and inflammation compared to controls. In addition, this effect could be restored by reconstitution of mast cell-deficient mice with mast cells derived from their normal counterparts (Wershil *et al*., 1998), indicating the importance of mast cells in the expression of toxin A-associated inflammatory diarrhoea. Mast cell activation in this model is controlled by sensory neurons, since chronic pre-treatment with capsaicin abolished toxin A-mediated mucosal mast activation *in vivo* (Castagliuolo *et al*., 1994). Experiments using SP and NT receptor antagonists *in vivo* also suggested that both neuropeptides are involved in the mast cell-dependent pro-inflammatory signalling mechanism following toxin A exposure (Pothoulakis *et al*., 1994; Castagliuolo *et al*., 1996; Wershil *et al*., 1998). Toxin A at low doses was also reported to stimulate release of TNF-α from peritoneal mast cells *in vitro* (Calderon *et al*., 1998). This may represent another mechanism by which mast cells, like intestinal macrophages (Castagliuolo *et al*., 1997), may be activated after the inflammatory response is established, and toxin A has access to mast cells in the lamina propria (Figure 12.3).

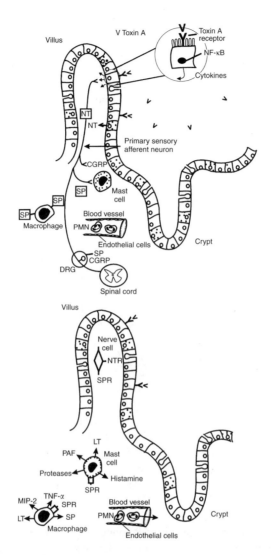

Figure 12.4 Pathogenesis of inflammatory diarrhoea caused by *C. difficile* toxin A. Top panel: *C. difficile* toxin A binds to epithelial cell receptor(s) causing activation and nuclear translocation of the transcription factor κB, leading to the release of pro-inflammatory cytokines from these cells. Cytokines released into subepithelial tissues activate lamina propria immune cells and primary sensory afferent neurons whose cell bodies are localized in the dorsal root ganglia (DRG). Activation of primary sensory neurons causes early release of SP and calcitonin gene related peptide (CGRP) that stimulates mucosal mast cells and other resident immune cells, such as macrophages and mast cells. Neurotensin (NT) is also released from specialized neuroendocrine cells (N cells) and intestinal neurons and SP can also be released form lamina propria macrophages. Bottom panel: The next step involves activation of SP receptors (SPR) and NT receptors (NTR) on enteric nerves, macrophages and possibly mast cells. Following stimulation by SP, CGRP and NT, significant mucosal mast cell degranulation and macrophage activation occurs leading to the release of several pro-inflammatory mediators such as histamine, platelet activating factor (PAF), leukotrienes C4 (LT) and proteases. Activated intestinal lamina propria macrophages also release potent inflammatory mediators such as macrophage inflammatory protein-2 (MIP-2), leukotrienes, tumor necrosis factor-α (TNF-α) and SP. These mediators may stimulate fluid secretion from epithelial cells and also upregulate expression of adhesion molecules on endothelial cells and neutrophils, causing entry of neutrophils into the tissue (modified from Pothoulakis *et al.* (1998) *News Physiol Sci,* **13**, 58–63).

Nitric oxide (NO) plays an important role in the communication between mast cells and nerves in enterotoxin-mediated diarrhoea and epithelial cell injury. Treatment with the neuronal NO synthase inhibitor 7-nitroindazole substantially enhanced, while the NO donor *S*-nitroso-*N*-acetyl-L-cysteine (SNAC) inhibited toxin A-associated ileal secretion and inflammation (Qiu *et al.*, 1996). Mast cell and neutrophil activation in response to *C. difficile* toxin A were also blocked by the NO donor SNAC (Qiu *et al.*, 1996). Thus, neuronal NO protects the intestinal mucosa from the effects of toxin A by reducing mast cell degranulation and neutrophil transmigration induced by this toxin.

PARASITIC INFECTIONS

TRICHINELLA SPIRALIS INFECTION

Intestinal helminths are common in many areas of the world, but they are more prevalent in non-industrialized countries with poor sanitary conditions. Animal models of parasitic infections have been extremely valuable to study intestinal pathophysiology of gut inflammation and the importance of neuroimmune–endocrine interactions in gastrointestinal functions. One of the most studied diseases is *Trichinella spiralis* infection that causes intestinal symptoms upon ingestion of meat contaminated with *T. spiralis* lavrae. The lavrae remain in the intestinal mucosa where they molt and cause intestinal inflammatory changes. The use of animal models of *T. spiralis* infection enabled investigators to identify the type of cells and mediators involved in inflammatory, secretory and neuromuscular changes in the gut during, as well after the parasites have been expelled from the intestine (Palmer *et al.*, 1998; Venkova *et al.*, 1999). Addition of a *T. spiralis* antigen to colonic strips of guinea pigs infected for 6–8 weeks with the nematode resulted in increased chloride secretion, that was mediated by submucosal cholinergic nerves and mast cells through histamine release (Wang *et al.*, 1991). Moreover, changes in net ion transport in rat jejunum in response to a *T. spiralis* antigen in animals primed with the nematode involves 5-HT, histamine, and prostaglandin acting through intrinsic nerves (Castro and Harari, 1991). Longitudinal muscle-myenteric plexus preparations from *T. spiralis*-infected animals showed suppression of electrical field stimulation and release of norepinephrine several weeks after infection (Swain *et al.*, 1991). Acetylcholine (Ach) release from myenteric plexus-muscle preparations of *T. spiralis*-infected rats was reduced, immunoreactive SP concentrations were increased and corticosteroid pretreatment reversed both these responses (Collins *et al.*, 1992; Swain *et al.*, 1992). Thus, alterations of neuropeptide levels may result from intestinal inflammatory changes during nematode infection. Interestingly, the cytokine IL-1β appears to mediate both, increased SP and reduced Ach levels in the myenteric plexus-muscle preparations of *T. spiralis* infected animals (Hurst *et al.*, 1993; Main *et al.*, 1993), while IL-6 appears to also contribute to the Ach response (Ruhl *et al.*, 1994). Increased SP immunoreactivity in muscle-myenteric plexus preparations during *T. spiralis* infection was also diminished by chronic capsaicin pretreatment and was absent in athymic rats infected with the nematode (Swain *et al.*, 1992), suggesting that sensory neurons and T cells are important for this SP response.

Macrophages are instrumental in the *T. spiralis*-mediated intestinal inflammatory and muscle contractile responses. Galeazzi *et al.* (2000) reported macrophage infiltration in the jejunum of mice infected with this parasite. Moreover, *T. spiralis*-induced macrophage infiltration was associated with reduced Ach release from the myenteric plexus, and pre-treatment with liposomes containing dichloromethylene diphosphate, an apoptotic agent for macrophages, attenuated this response (Galeazzi *et al.*, 2000). Further studies showed that macrophage colony-stimulating factor recruits macrophages that suppress enteric neuronal function in this model (Galeazzi *et al.*, 2001). Thus, intestinal macrophages play an important role in the functional impairment of cholinergic nerves in the intestinal myenteric plexus during parasitic infection. Using IL-5-deficient mice Vallance *et al.* (1999) recently demonstrated that this cytokine mediates muscle hypercontractility associated with *T. spiralis* infection.

NIPPOSTRONGYLUS BRASILIENSIS

Infection with *Nippostrongylus brasiliensis* has been widely used as a model to study intestinal pathophysiology. It is well established that infection with this parasite causes acute pathological and biochemical changes in the small intestine, including villus atrophy, crypt hyperplasia, and inflammation (Ramage *et al.*, 1988), motility changes (Vermillion *et al.*, 1988; Crosthwaite *et al.*, 1990), and alterations in intestinal permeability (Ramage *et al.*, 1988). Mast cells have been also recognized as an important cell population associated with the pathophysiology of inflammation and colonic motility changes during *N. brasiliensis* infection (Woodbury *et al.*, 1984; King *et al.*, 1985; Marzio *et al.*, 1992).

Immunologically activated rat intestinal mast cells obtained from *N. brasiliensis*-infected animals release more prostaglandins and leukotrienes (Heavey *et al.*, 1988). Moreover, intestinal mucosal injury in response to the nematode was temporally associated with mast cell activation and increased levels of leukotrienes in the jejunum (Perdue *et al.*, 1989). In *N. brasiliensis*-sensitized rats, mucosal mast cell degranulation is associated with jejunal paracellular permeability changes (Scudamore *et al.*, 1995). As well, intestinal hypersensitivity and motility changes following infection with *N. brasiliensis* in rats depend on mast cell activation (McLean *et al.*, 1998; Gay *et al.*, 2000a) (Figure 12.5).

Several reports provided evidence for the pathophysiologic importance of intestinal nerves in the development of *N. brasiliensis* infection. Using Ussing chambers, Masson *et al.* (1996) demonstrated that neuronally mediated ion secretion was diminished in rat jejunum 10 and 35 days after *N. brasiliensis* infection, and this effect was associated with increased SP levels in nerve fibres of infected rats. Consistent with this observation, Jodal *et al.* (1993) reported that net fluid transport changes during *N. brasiliensis* infection are mediated in large part by the enteric nervous system. Infection with this nematode also resulted in early activation of

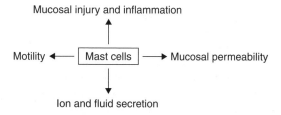

Figure 12.5   Mast cell-mediated effects during parasitic infections.

the brain Fos expression (a parameter of nerve cell activation) that was associated with intestinal inflammatory and motility changes (Castex *et al.*, 1998), indicating activation of neuronal afferents that communicate signals to the brain during gut inflammation.

Important anatomical observations also underscored the existence of close mast cells-nerve interactions in the intestinal lamina propria and opened the field for further studies to identify the functional significance of this interaction. In the *N. brasiliensis* nematode animal model, 67% of intestinal mucosal mast cells were found to be in contact with nerves in the small intestinal subepithelium (Bienenstock *et al.*, 1987; Stead *et al.*, 1987). Nerve endings associated with mast cells in these studies contained the neuropeptides SP and CGRP. Electron microscopy experiments identified a population of mast cells having membrane to membrane contact with unmyelinated axons, while another mast cell population embraced nerve bundles through lamellopodia formation (Stead *et al.*, 1987). Substantial remodelling of intestinal nerves also takes place during the different phases of *N. brasiliensis* infection (Stead *et al.*, 1991). These results provided substantial evidence for a possible cross-talk between the immune and nervous systems in the gut. Mast cell–nerve interactions play a functional role in the intestinal responses to *N. brasiliensis*. Perdue *et al.* (1990) demonstrated activation of mucosal mast cells during *N. brasiliensis*-induced inflammation, and implicated mast cell–nerve interactions in the modulation of ion secretion from the intestinal epithelium. Along these lines, mast cell stabilization with ketotifen decreases jejunal inflammation in the acute stage of the infection, but it results in prolonged intestinal inflammation (Gay *et al.*, 2000b). Ablation of extrinsic sensory neurons with capsaicin worsened intestinal inflammation in *N. brasiliensis*-infected rats, but it did not affect the duration of the infection (Gay *et al.*, 2000b), indicating that sensory neurons may play a protective role in this model via a pathway that does not involve mast cells.

SCHISTOSOMA MANSONI INFECTION

Schistosomiasis mansoni infection is a parasitic disease characterized by the formation of granulomas around schistosome eggs in the liver and small and large intestine. Several studies indicate that neuroimmunoregulatory circuits are involved in the pathogenesis of this infection. Immunohistochemical experiments demonstrated destruction of enteric nerves in the myenteric plexus in the ileum, and the mucosal and submucosal plexuses of the colon of *Schistosoma mansoni*-infected mice (Varilek *et al.*, 1991). Moreover, granulomas isolated from *S. mansoni*-infected animals produce measurable amounts of SP and its mRNA (Weinstock *et al.*, 1988), and SP release can be stimulated by calcium ionophore or histamine (Weinstock and Blum, 1990b). The neuropeptide VIP is also secreted from *S. mansoni* granulomas (Weinstock and Blum, 1990a) and behaves as an immune modulator. Thus, VIP enhances IL-5 production from ganuloma T cells, while it suppresses T cell proliferation and IL-2 release, by acting on specific T cell VIP receptors (Weinstock, 1996), whose expression can be modulated by IL-4 (Metwali *et al.*, 2000). Interactions between SP and SOM also exert an important role in the regulation of the inflammatory response within the *S. mansoni* granulomas. Thus, SP, acting via authentic NK-1 receptors (Cook *et al.*, 1994), stimulates antigen-induced IFN-γ production from *S. mansoni*-infected granulomas and spleen cells (Blum *et al.*, 1993). Granuloma development and expression of IFN-γ is reduced in animals genetically lacking the SP (NK-1) receptor, confirming the importance of SP in *S. mansoni* granuloma formation (Blum *et al.*, 1999). Macrophages obtained from granulomas from mice infected with *S. mansoni* secrete SOM which down-regulates the IFN-γ response (Blum *et al.*, 1992). The observation that SP, via NK-1 receptors, inhibits SOM expression in splenocytes and granulomas in *S. mansoni*-infected animals (Blum *et al.*, 1998), further underscores the pathophysiologic importance of SP-SOM interactions in this model of parasitic inflammatory infection (Weinstock and Elliott, 1998).

NEURONAL INVOLVEMENT IN DIARRHOEA DUE TO CRYPTOSPORIDIUM

Cryptosporidium is a protozoal parasite that causes chronic watery diarrhoea which can be life threatening. The extent of the disease is mostly dependent on the immune status of the host and *Cryptosporidium*-related diarrhoea became prominent because of its high incidence in patients with AIDS. Among *Cryptosporidium* species, *Cryptosporidium parvum* (*C. parvum*) represents the most common parasite affecting animals as well as humans. Work with neonate piglets infected with *C. parvum* demonstrate substantial intestinal pathology in response to this parasite, including villus atrophy, crypt hyperplasia and inflammatory infiltration in the intestinal mucosa (Argenzio *et al.*, 1990). These pathological changes were also associated with increased electrolyte and fluid secretion and reduced NaCl absorption, consistent with a diarrhoeal malabsorptive state (Argenzio *et al.*, 1990). Interestingly, prostaglandins released during the course of the infection from intestinal epithelial cells (Laurent *et al.*, 1998), mediate part of the diarrhoea seen in *C. parvum*-infected piglets (Argenzio *et al.*, 1993). The mechanism of prostaglandin mediation of NaCl transport changes in infected animals involve cholinergic, VIP containing neurons (Argenzio *et al.*, 1996), indicating an interplay between epithelial-derived prostaglandins and the enteric nervous system in this response. The potent pro-inflammatory mediator TNF-α also participates in Cl⁻ secretion in response to this parasite, and its release probably involves a prostaglandin-dependent mechanism (Kandil *et al.*, 1994). Exposure of colonic epithelial cell monolayers and human intestinal xenografts to *C. parvum* also resulted in secretion of proinflammatory cytokines that may contribute to intestinal inflammatory changes during *C. parvum* infection (Laurent *et al.*, 1997).

Peptide YY (PYY), released from neuroendocrine cells in the colon and small intestine, has been shown to attenuate prostaglandin and VIP-mediated fluid and electrolyte secretion in animal intestine (Saria and Beubler, 1985; Chariot *et al.*, 2000). Argenzio *et al.* (1997) found that PYY dramatically attenuated Cl⁻ secretion in intestinal tissues from piglets infected with *C. parvum*. Pharmacological experiments indicated that this PYY effect is mediated by inhibition of prostaglandin induction of enteric nerves (Argenzio *et al.*, 1997), underscoring the importance of neuroimmune interactions mediated by PYY in cryptosporidial infection.

CONCLUSIONS

The identification of complex neuroimmune interactions in the development of infectious diarrhoea and inflammation of multiple aetiologies opens up an exciting and difficult research area, that is, the specific *in vivo* mechanism(s), and the cellular pathways involved in these pathophysiologic processes. This is not an easy task since many types of neurons, neuropeptides, cytokines, and intestinal cell types have been implicated in the modulation of infectious conditions in the intestine. Moreover, these different molecules and cell types also participate in the normal gut physiology. The existence of surface receptors for several neuropeptides such as SP, NT and galanin on enterocytes and colonocytes, and their modulation during secretory and inflammatory diarrhoea also raises the possibility that these, and possibly other neuropeptides act directly on intestinal epithelial cells to modulate inflammatory and secretory responses. Mast cells and macrophages residing in the intestinal lamina propria are also key participants in the pathogenesis of inflammatory diarrhoea since they interact with neuropeptides and cytokines and, in response, release potent inflammatory mediators. Based on the available evidence from animal experiments, targeting of neuropeptide and cytokines and their specific receptors may be a novel and promising approach to treat infectious or enterotoxin-mediated diarrhoea in humans.

REFERENCES

Aktories, K., Schmidt, G. and Just, I. (2000) Rho GTPases as targets of bacterial protein toxins. *Biol Chem*, **381**, 421–6.

Argenzio, R. A., Armstrong, M., Blikslager, A. and Rhoads, J. M. (1997) Peptide YY inhibits intestinal Cl⁻ secretion in experimental porcine cryptosporidiosis through a prostaglandin-activated neural pathway. *J Pharmacol Exp Ther*, **283**, 692–7.

Argenzio, R. A., Armstrong, M. and Rhoads, J. M. (1996) Role of the enteric nervous system in piglet cryptosporidiosis. *J Pharmacol Exp Ther*, **279**, 1109–15.

Argenzio, R. A., Lecce, J. and Powell, D. W. (1993) Prostanoids inhibit intestinal NaCl absorption in experimental porcine cryptosporidiosis. *Gastroenterology*, **104**, 440–7.

Argenzio, R. A., Liacos, J. A., Levy, M. L., Meuten, D. J., Lecce, J. G. and Powell, D. W. (1990) Villous atrophy, crypt hyperplasia, cellular infiltration, and impaired glucose-Na absorption in enteric cryptosporidiosis of pigs. *Gastroenterology*, **98**, 1129–40.

Bauer, F. E., Zintel, A., Kenny, M. J., Calder, D., Ghatei, M. A. and Bloom, S. R. (1989) Inhibitory effect of galanin on postprandial gastrointestinal motility and gut hormone release in humans. *Gastroenterology*, **97**, 260–4.

Bearcroft, C. P., Andre, E. A. and Farthing, M. J. (1997) In vivo effects of the 5-HT3 antagonist alosetron on basal and cholera toxin-induced secretion in the human jejunum: a segmental perfusion study. *Aliment Pharmacol Ther*, **11**, 1109–14.

Bearcroft, C. P., Perrett, D. and Farthing, M. J. (1996) 5-hydroxytryptamine release into human jejunum by cholera toxin. *Gut*, **39**, 528–31.

Bedwani, J. R. and Okpako, D. T. (1975) Effects of crude and pure cholera toxin on prostaglandin. *Prostaglandins*, **10**, 117–27.

Benya, R. V., Marrero, J. A., Ostrovskiy, D. A., Koutsouris, A. and Hecht, G. (1999) Human colonic epithelial cells express galanin-1 receptors, which when activated cause Cl⁻ secretion. *Am J Physiol*, **276**, G64–G72.

Beubler, E., Bukhave, K. and Rask-Madsen, J. (1986) Significance of calcium for the prostaglandin E2-mediated secretory response to 5-hydroxytryptamine in the small intestine of the rat in vivo. *Gastroenterology*, **90**, 1972–7.

Beubler, E. and Horina, G. (1990) 5-HT2 and 5-HT3 receptor subtypes mediate cholera toxin-induced intestinal fluid secretion in the rat. *Gastroenterology*, **99**, 83–9.

Beubler, E., Kollar, G., Saria, A., Bukhave, K. and Rask-Madsen, J. (1989) Involvement of 5-hydroxytryptamine, prostaglandin E2, and cyclic adenosine monophosphate in cholera toxin-induced fluid secretion in the small intestine of the rat in vivo. *Gastroenterology*, **96**, 368–76.

Beubler, E., Schuligoi, R., Chopra, A. K., Ribardo, D. A. and Peskar, B. A. (2001) Cholera toxin induces prostaglandin synthesis via post-transcriptional activation of cyclooxygenase-2 in the rat jejunum. *J Pharmacol Exp Ther*, **297**, 940–5.

Bhide, M. B., Aroskar, V. A. and Dutta, N. K. (1970) Release of active substances by cholera toxin. *Indian J Med Res*, **58**, 548–50.

Bienenstock, J., Tomioka, M., Matsuda, H., Stead, R. H., Quinonez, G., Simon, G. T., Coughlin, M. D. and Denburg, J. A. (1987) The role of mast cells in inflammatory processes: evidence for nerve/mast cell interactions. *Int Arch Allergy Appl Immunol*, **82**, 238–43.

Blum, A. M., Elliott, D. E., Metwali, A., Li, J., Qadir, K. and Weinstock, J. V. (1998) Substance P regulates somatostatin expression in inflammation. *J Immunol*, **161**, 6316–22.

Blum, A. M., Metwali, A., Cook, G., Mathew, R. C., Elliott, D. and Weinstock, J. V. (1993) Substance P modulates antigen-induced, IFN-gamma production in murine Schistosomiasis mansoni. *J Immunol*, **151**, 225–33.

Blum, A. M., Metwali, A., Kim-Miller, M., Li, J., Qadir, K., Elliott, D. E., Lu, B., Fabry, Z., Gerard, N. and Weinstock, J. V. (1999) The substance P receptor is necessary for a normal granulomatous response in murine schistosomiasis mansoni. *J Immunol*, **162**, 6080–5.

Blum, A. M., Metwali, A., Mathew, R. C., Cook, G., Elliott, D. and Weinstock, J. V. (1992) Granuloma T lymphocytes in murine schistosomiasis mansoni have somatostatin receptors and respond to somatostatin with decreased IFN- gamma secretion. *J Immunol*, **149**, 3621–6.

Bost, K. L. (1995) Inducible preprotachykinin mRNA expression in mucosal lymphoid organs following oral immunization with *Salmonella*. *J Neuroimmunol*, **62**, 59–67.

Botella, A., Vabre, F., Fioramonti, J., Thomas, F. and Bueno, L. (1993) In vivo inhibitory effect of lanreotide (BIM 23014), a new somatostatin analog, on prostaglandin- and cholera toxin-stimulated intestinal fluid in the rat. *Peptides*, **14**, 297–301.

Brunsson, I. (1987) Enteric nerves mediate the fluid secretory response due to *Salmonella typhimurium* R5 infection in the rat small intestine. *Acta Physiol Scand*, **131**, 609–17.

Buchwald, D. S. and Blaser, M. J. (1984) A review of human salmonellosis: II. Duration of excretion following infection with nontyphi *Salmonella*. *Rev Infect Dis*, **6**, 345–56.

Calderon, G. M., Torres-Lopez, J., Lin, T. J., Chavez, B., Hernandez, M., Munoz, O., Befus, A. D. and Enciso, J. A. (1998) Effects of toxin A from *Clostridium difficile* on mast cell activation and survival. *Infect Immun*, **66**, 2755–61.

Cassuto, J., Fahrenkrug, J., Jodal, M., Tuttle, R. and Lundgren, O. (1981) Release of vasoactive intestinal polypeptide from the cat small intestine exposed to cholera toxin. *Gut*, **22**, 958–63.

Cassuto, J., Jodal, M. and Lundgren, O. (1982) The effect of nicotinic and muscarinic receptor blockade on cholera toxin induced intestinal secretion in rats and cats. *Acta Physiol Scand*, **114**, 573–7.

Cassuto, J., Siewert, A., Jodal, M. and Lundgren, O. (1983) The involvement of intramural nerves in cholera toxin induced intestinal secretion. *Acta Physiol Scand*, **117**, 195–202.

Castagliuolo, I., Keates, A. C., Qiu, B., Kelly, C. P., Nikulasson, S., Leeman, S. E. and Pothoulakis, C. (1997) Increased substance P responses in dorsal root ganglia and intestinal macrophages during *Clostridium difficile* toxin A enteritis in rats. *Proc Natl Acad Sci USA*, **94**, 4788–93.

Castagliuolo, I., Keates, A. C., Wang, C. C., Pasha, A., Valenick, L., Kelly, C. P., Nikulasson, S. T., LaMont, J. T. and Pothoulakis, C. (1998a) *Clostridium difficile* toxin A stimulates macrophage-inflammatory protein-2 production in rat intestinal epithelial cells. *J Immunol*, **160**, 6039–45.

Castagliuolo, I., Riegler, M., Pasha, A., Nikulasson, S., Lu, B., Gerard, C., Gerard, N. P. and Pothoulakis, C. (1998b) Neurokinin-1 (NK-1) receptor is required in *Clostridium difficile*-induced enteritis. *J Clin Invest*, **101**, 1547–50.

Castagliuolo, I., LaMont, J. T., Letourneau, R., Kelly, C., O'Keane, J. C., Jaffer, A., Theoharides, T. C. and Pothoulakis, C. (1994) Neuronal involvement in the intestinal effects of *Clostridium difficile* toxin A and *Vibrio cholerae* enterotoxin in rat ileum. *Gastroenterology*, **107**, 657–65.

Castagliuolo, I., Leeman, S. E., Bartolak-Suki, E., Nikulasson, S., Qiu, B., Carraway, R. E. and Pothoulakis, C. (1996) A neurotensin antagonist, SR 48692, inhibits colonic responses to immobilization stress in rats. *Proc Natl Acad Sci USA*, **93**, 12611–15.

Castagliuolo, I., Wang, C. C., Valenick, L., Pasha, A., Nikulasson, S., Carraway, R. E. and Pothoulakis, C. (1999) Neurotensin is a proinflammatory neuropeptide in colonic inflammation. *J Clin Invest*, **103**, 843–9.

Castex, N., Fioramonti, J., Ducos de Lahitte, J., Luffau, G., More, J. and Bueno, L. (1998) Brain Fos expression and intestinal motor alterations during nematode-induced inflammation in the rat. *Am J Physiol*, **274**, G210–16.

Castro, G. A. and Harari, Y. (1991) Immunoregulation of endometrial and jejunal epithelia sensitized by infection. *Int Arch Allergy Appl Immunol*, **95**, 184–90.

Chariot, J., Tsocas, A., Souli, A., Presset, O. and Roze, C. (2000) Neural mechanism of the antisecretory effect of peptide YY in the rat colon in vivo. *Peptides*, **21**, 59–63.

Chen, L. C., Rohde, J. E. and Sharp, G. W. (1971) Intestinal adenyl-cyclase activity in human cholera. *Lancet*, **1**, 939–41.

Chowers, Y., Cahalon, L., Lahav, M., Schor, H., Tal, R., Bar-Meir, S. and Levite, M. (2000) Somatostatin through its specific receptor inhibits spontaneous and TNF-alpha- and bacteria-induced IL-8 and IL-1 beta secretion from intestinal epithelial cells. *J Immunol*, **165**, 2955–61.

Clarke, L. L., Grubb, B. R., Gabriel, S. E., Smithies, O., Koller, B. H. and Boucher, R. C. (1992) Defective epithelial chloride transport in a gene-targeted mouse model of cystic fibrosis. *Science*, **257**, 1125–8.

Collins, S. M., Blennerhassett, P., Vermillion, D. L., Davis, K., Langer, J. and Ernst, P. B. (1992) Impaired acetylcholine release in the inflamed rat intestine is T cell independent. *Am J Physiol*, **263**, G198–201.

Cook, G. A., Elliott, D., Metwali, A., Blum, A. M., Sandor, M., Lynch, R. and Weinstock, J. V. (1994) Molecular evidence that granuloma T lymphocytes in murine schistosomiasis mansoni express an authentic substance P (NK-1) receptor. *J Immunol*, **152**, 1830–5.

Crosthwaite, A. I., Huizinga, J. D. and Fox, J. A. (1990) Jejunal circular muscle motility is decreased in nematode-infected rat. *Gastroenterology*, **98**, 59–65.

Donnenberg, M. S. and Whittam, T. S. (2001) Pathogenesis and evolution of virulence in enteropathogenic and enterohemorrhagic *Escherichia coli. J Clin Invest*, **107**, 539–48.

Eherer, A. J., Hinterleitner, T. A., Petritsch, W., Holzer-Petsche, U., Beubler, E. and Krejs, G. J. (1994) Effect of 5-hydroxytryptamine antagonists on cholera toxin-induced secretion in the human jejunum.*Eur J Clin Invest*, **24**, 664–8.

Eklund, S., Cassuto, J., Jodal, M. and Lundgren, O. (1984) The involvement of the enteric nervous system in the intestinal secretion evoked by cyclic adenosine 3′5′-monophosphate. *Acta Physiol Scand*, **120**, 311–16.

Eklund, S., Sjoqvist, A., Fahrenkrug, J., Jodal, M. and Lundgren, O. (1988) Somatostatin and methionine-enkephalin inhibit cholera toxin-induced ejunal net fluid secretion and release of vasoactive intestinal polypeptide in the cat in vivo. *Acta Physiol Scand*, **133**, 551–7.

Galeazzi, F., Haapala, E. M., van Rooijen, N., Vallance, B. A. and Collins, S. M. (2000) Inflammation-induced impairment of enteric nerve function in nematode-infected mice is macrophage dependent. *Am J Physiol Gastrointest Liver Physiol*, **278**, G259–65.

Galeazzi, F., Lovato, P., Blennerhassett, P. A., Haapala, E. M., Vallance, B. A. and Collins, S. M. (2001) Neural change in *Trichinella*-infected mice is MHC II independent and involves M-CSF-derived macrophages. *Am J Physiol Gastrointest Liver Physiol*, **281**, G151–8.

Ganser, A. L., Kirschner, D. A. and Willinger, M. (1983) Ganglioside localization on myelinated nerve fibres by cholera toxin binding. *J Neurocytol*, **12**, 921–38.

Gay, J., Fioramonti, J., Garcia-Villar, R. and Bueno, L. (2000a) Alterations of intestinal motor responses to various stimuli after *Nippostrongylus brasiliensis* infection in rats: role of mast cells. *Neurogastroenterol Motil*, **12**, 207–14.

Gay, J., Fioramonti, J., Garcia-Villar, R. and Bueno, L. (2000b) Development and sequels of intestinal inflammation in nematode-infected rats: role of mast cells and capsaicin-sensitive afferents. *Neuroimmunomodulation*, **8**, 171–8.

Gots, R. E., Formal, S. B. and Giannella, R. A. (1974) Indomethacin inhibition of *Salmonella typhimurium, Shigella flexneri*, and cholera-mediated rabbit ileal secretion. *J Infect Dis*, **130**, 280–4.

Goyal, R. K. and Hirano, I. (1996) The enteric nervous system. *N Engl J Med*, **334**, 1106–15.

Griffin, G. E., Miller, A., Batman, P., Forster, S. M., Pinching, A. J., Harris, J. R. and Mathan, M. M. (1988) Damage to jejunal intrinsic autonomic nerves in HIV infection. *AIDS*, **2**, 379–82.

Grondahl, M. L., Jensen, G. M., Nielsen, C. G., Skadhauge, E., Olsen, J. E. and Hansen, M. B. (1998) Secretory pathways in *Salmonella typhimurium*-induced fluid accumulation in the porcine small intestine. *J Med Microbiol*, **47**, 151–7.

He, D., Hagen, S. J., Pothoulakis, C., Chen, M., Medina, N. D., Warny, M. and LaMont, J. T. (2000) *Clostridium difficile* toxin A causes early damage to mitochondria in cultured cells. *Gastroenterology*, **119**, 139–50.

Heavey, D. J., Ernst, P. B., Stevens, R. L., Befus, A. D., Bienenstock, J. and Austen, K. F. (1988) Generation of leukotriene C4, leukotriene B4, and prostaglandin D2 by immunologically activated rat intestinal mucosa mast cells. *J Immunol*, **140**, 1953–7.

Hecht, G., Marrero, J. A., Danilkovich, A., Matkowskyj, K. A., Savkovic, S. D., Koutsouris, A. and Benya, R. V. (1999) Pathogenic *Escherichia coli* increase Cl⁻ secretion from intestinal epithelia by upregulating galanin-1 receptor expression. *J Clin Invest*, **104**, 253–62.

Hinterleitner, T. A., Petritsch, W., Dimsity, G., Berard, H., Lecomte, J. M. and Krejs, G. J. (1997) Acetorphan prevents cholera-toxin-induced water and electrolyte secretion in the human jejunum. *Eur J Gastroenterol Hepatol*, **9**, 887–91.

Hsu, H. S. (1989) Pathogenesis and immunity in murine salmonellosis. *Microbiol Rev*, **53**, 390–409.

Hurst, S. M., Stanisz, A. M., Sharkey, K. A. and Collins, S. M. (1993) Interleukin 1 beta-induced increase in substance P in rat myenteric plexus. *Gastroenterology*, **105**, 1754–60.

Janoff, E. N., Hayakawa, H., Taylor, D. N., Fasching, C. E., Kenner, J. R., Jaimes, E. and Raij, L. (1997) Nitric oxide production during *Vibrio cholerae* infection. *Am J Physiol*, **273**, G1160–7.

Jefferson, K. K., Smith, M. F., Jr. and Bobak, D. A. (1999) Roles of intracellular calcium and NF-kappa B in the *Clostridium difficile* toxin A-induced up-regulation and secretion of IL-8 from human monocytes. *J Immunol*, **163**, 5183–91.

Jiang, M. M., Kirchgessner, A., Gershon, M. D. and Surprenant, A. (1993) Cholera toxin-sensitive neurons in guinea pig submucosal plexus. *Am J Physiol*, **264**, G86–94.

Jodal, M. and Lundgren, O. (1995) Nerves and cholera secretion. *Gastroenterology*, **108**, 287–8.

Jodal, M., Wingren, U., Jansson, M., Heidemann, M. and Lundgren, O. (1993) Nerve involvement in fluid transport in the inflamed rat jejunum. *Gut*, **34**, 1526–30.

Jonsdottir, I. H., Sjoqvist, A., Lundgren, O. and Thoren, P. (1999) Somatic nerve stimulation and cholera-induced net fluid secretion in the small intestine of the rat: evidence for an opioid effect. *J Auton Nerv Syst*, **78**, 18–23.

Kagnoff, M. F. and Eckmann, L. (1997) Epithelial cells as sensors for microbial infection. *J Clin Invest*, **100**, 6–10.

Kandil, H. M., Berschneider, H. M. and Argenzio, R. A. (1994) Tumour necrosis factor alpha changes porcine intestinal ion transport through a paracrine mechanism involving prostaglandins. *Gut*, **35**, 934–40.

Keates, A. C., Castagliuolo, I., Qiu, B., Nikulasson, S., Sengupta, A. and Pothoulakis, C. (1998) CGRP upregulation in dorsal root ganglia and ileal mucosa during *Clostridium difficile* toxin A-induced enteritis. *Am J Physiol*, **274**, G196–202.

Kelly, C. P., Pothoulakis, C. and LaMont, J. T. (1994) *Clostridium difficile* colitis. *N Engl J Med*, **330**, 257–62.

Kincy-Cain, T. and Bost, K. L. (1996) Increased susceptibility of mice to *Salmonella* infection following in vivo treatment with the substance P antagonist, spantide II. *J Immunol*, **157**, 255–64.

Kincy-Cain, T. and Bost, K. L. (1997) Substance P-induced IL-12 production by murine macrophages. *J Immunol*, **158**, 2334–9.

Kincy-Cain, T., Clements, J. D. and Bost, K. L. (1996) Endogenous and exogenous interleukin-12 augment the protective immune response in mice orally challenged with *Salmonella dublin*. *Infect Immun*, **64**, 1437–40.

King, S. C., Slater, P. and Turnberg, L. A. (1989) Autoradiographic localization of binding sites for galanin and VIP in small intestine. *Peptides*, **10**, 313–7.

King, S. J., Miller, H. R., Newlands, G. F. and Woodbury, R. G. (1985) Depletion of mucosal mast cell protease by corticosteroids: effect on intestinal anaphylaxis in the rat. *Proc Natl Acad Sci USA*, **82**, 1214–18.

Kirchgessner, A. L., Liu, M. T., Tamir, H. and Gershon, M. D. (1992) Identification and localization of 5-HT1P receptors in the guinea pig pancreas. *Am J Physiol*, **262**, G553–66.

Kirkwood, K. S., Bunnett, N. W., Maa, J., Castagliolo, I., Liu, B., Gerard, N., Zacks, J., Pothoulakis, C. and Grady, E. F. (2001) Deletion of neutral endopeptidase exacerbates intestinal inflammation induced by *Clostridium difficile* toxin A. *Am J Physiol Gastrointest Liver Physiol*, **281**, G544–51.

Klimpel, G. R., Chopra, A. K., Langley, K. E., Wypych, J., Annable, C. A., Kaiserlian, D., Ernst, P. B. and Peterson, J. W. (1995) A role for stem cell factor and c-kit in the murine intestinal tract secretory response to cholera toxin. *J Exp Med*, **182**, 1931–42.

Kurose, I., Pothoulakis, C., LaMont, J. T., Anderson, D. C., Paulson, J. C., Miyasaka, M., Wolf, R. and Granger, D. N. (1994) *Clostridium difficile* toxin A-induced microvascular dysfunction. Role of histamine. *J Clin Invest*, **94**, 1919–26.

Laurent, F., Eckmann, L., Savidge, T. C., Morgan, G., Theodos, C., Naciri, M. and Kagnoff, M. F. (1997) *Cryptosporidium parvum* infection of human intestinal epithelial cells induces the polarized secretion of C-X-C chemokines. *Infect Immun*, **65**, 5067–73.

Laurent, F., Kagnoff, M. F., Savidge, T. C., Naciri, M. and Eckmann, L. (1998) Human intestinal epithelial cells respond to *Cryptosporidium parvum* infection with increased prostaglandin H synthase 2 expression and prostaglandin E2 and F2alpha production. *Infect Immun*, **66**, 1787–90.

Leal-Berumen, I., Snider, D. P., Barajas-Lopez, C. and Marshall, J. S. (1996) Cholera toxin increases IL-6 synthesis and decreases TNF-alpha production by rat peritoneal mast cells. *J Immunol*, **156**, 316–21.

Lencer, W. I., Moe, S., Rufo, P. A. and Madara, J. L. (1995) Transcytosis of cholera toxin subunits across model human intestinal epithelia. *Proc Natl Acad Sci USA*, **92**, 10094–8.

Lucas, K. A., Pitari, G. M., Kazerounian, S., Ruiz-Stewart, I., Park, J., Schulz, S., Chepenik, K. P. and Waldman, S. A. (2000) Guanylyl cyclases and signaling by cyclic GMP. *Pharmacol Rev*, **52**, 375–414.

Lundgren, O. (1998) 5-Hydroxytryptamine, enterotoxins, and intestinal fluid secretion. *Gastroenterology*, **115**, 1009–12.

Lundgren, O., Peregrin, A. T., Persson, K., Kordasti, S., Uhnoo, I. and Svensson, L. (2000) Role of the enteric nervous system in the fluid and electrolyte secretion of rotavirus diarrhea. *Science*, **287**, 491–5.

Lundgren, O., Svanvik, J. and Jivegard, L. (1989) Enteric nervous system. I. Physiology and pathophysiology of the intestinal tract. *Dig Dis Sci*, **34**, 264–83.

Mahmood, B., Warhurst, G., Higgs, N. and Turnberg, L. A. (2000) Carbachol potentiates cholera toxin-induced secretion in a colonic epithelial cell line (HT29-19A) and rat ileal mucosa in vitro. *J Health Popul Nutr*, **18**, 49–53.

Main, C., Blennerhassett, P. and Collins, S. M. (1993) Human recombinant interleukin 1 beta suppresses acetylcholine release from rat myenteric plexus. *Gastroenterology*, **104**, 1648–54.

Mantyh, C. R., Maggio, J. E., Mantyh, P. W., Vigna, S. R. and Pappas, T. N. (1996a) Increased substance P receptor expression by blood vessels and lymphoid aggregates in *Clostridium difficile*-induced pseudomembranous colitis. *Dig Dis Sci*, **41**, 614–20.

Mantyh, C. R., Pappas, T. N., Lapp, J. A., Washington, M. K., Neville, L. M., Ghilardi, J. R., Rogers, S. D., Mantyh, P. W. and Vigna, S. R. (1996b) Substance P activation of enteric neurons in response to intraluminal *Clostridium difficile* toxin A in the rat ileum. *Gastroenterology*, **111**, 1272–80.

Mantyh, C. R., McVey, D. C. and Vigna, S. R. (2000) Extrinsic surgical denervation inhibits *Clostridium difficile* toxin A-induced enteritis in rats. *Neurosci Lett*, **292**, 95–8.

Marzio, L., Blennerhassett, P., Vermillion, D., Chiverton, S. and Collins, S. (1992) Distribution of mast cells in intestinal muscle of nematode-sensitized rats. *Am J Physiol*, **262**, G477–82.

Masson, S. D., McKay, D. M., Stead, R. H., Agro, A., Stanisz, A. and Perdue, M. H. (1996) *Nippostrongylus brasiliensis* infection evokes neuronal abnormalities and alterations in neurally regulated electrolyte transport in rat jejunum. *Parasitology*, **113**, 173–82.

Mathan, M. M., Chandy, G. and Mathan, V. I. (1995) Ultrastructural changes in the upper small intestinal mucosa in patients with cholera. *Gastroenterology*, **109**, 422–30.

Mathias, J. R., Carlson, G. M., Bertiger, G., Martin, J. L. and Cohen, S. (1977) Migrating action potential complex of cholera: a possible prostaglandin-induced response. *Am J Physiol*, **232**, E529–34.

Matkowskyj, K. A., Danilkovich, A., Marrero, J., Savkovic, S. D., Hecht, G. and Benya, R. V. (2000) Galanin-1 receptor up-regulation mediates the excess colonic fluid production caused by infection with enteric pathogens. *Nat Med*, **6**, 1048–51.

McCloskey, M. A. (1988) Cholera toxin potentiates IgE-coupled inositol phospholipid hydrolysis and mediator secretion by RBL-2H3 cells. *Proc Natl Acad Sci USA*, **85**, 7260–4.

McIntosh, C. H. (1985) Gastrointestinal somatostatin: distribution, secretion and physiological significance. *Life Sci*, **37**, 2043–58.

McLean, P. G., Picard, C., Garcia-Villar, R., Ducos de Lahitte, R., More, J., Fioramonti, J. and Bueno, L. (1998) Role of kinin B1 and B2 receptors and mast cells in post intestinal infection-induced hypersensitivity to distension. *Neurogastroenterol Motil*, **10**, 499–508.

Metwali, A., Blum, A. M., Li, J., Elliott, D. E. and Weinstock, J. V. (2000) IL-4 regulates VIP receptor subtype 2 mRNA (VPAC2) expression in T cells in murine schistosomiasis. *FASEB J*, **14**, 948–54.

Moore, B. A., Sharkey, K. A. and Mantle, M. (1996) Role of 5-HT in cholera toxin-induced mucin secretion in the rat small intestine. *Am J Physiol*, **270**, G1001–9.

Morris, A. P. and Estes, M. K. (2001) Microbes and microbial toxins: paradigms for microbial-mucosal interactions. VIII. Pathological consequences of rotavirus infection and its enterotoxin. *Am J Physiol Gastrointest Liver Physiol*, **281**, G303–10.

Mourad, F. H. and Nassar, C. F. (2000) Effect of vasoactive intestinal polypeptide (VIP) antagonism on rat jejunal fluid and electrolyte secretion induced by cholera and *Escherichia coli* enterotoxins. *Gut*, **47**, 382–6.

Mourad, F. H., O'Donnell, L. J., Dias, J. A., Ogutu, E., Andre, E. A., Turvill, J. L. and Farthing, M. J. (1995) Role of 5-hydroxytryptamine type 3 receptors in rat intestinal fluid and electrolyte secretion induced by cholera and *Escherichia coli* enterotoxins. *Gut*, **37**, 340–5.

Nano, J. L., Fournel, S. and Rampal, P. (2000) Characterization of delta-opioid receptors and effect of enkephalins on IRD 98 rat epithelial intestinal cell line. *Pflugers Arch*, **439**, 547–54.

Naumann, M. (2000) Nuclear factor-kappa B activation and innate immune response in microbial pathogen infection. *Biochem Pharmacol*, **60**, 1109–14.

Nicol, G. D., Lopshire, J. C. and Pafford, C. M. (1997) Tumor necrosis factor enhances the capsaicin sensitivity of rat sensory neurons. *J Neurosci*, **17**, 975–82.

Nilsson, O., Cassuto, J., Larsson, P. A., Jodal, M., Lidberg, P., Ahlman, H., Dahlstrom, A. and Lundgren, O. (1983) 5-Hydroxytryptamine and cholera secretion: a histochemical and physiological study in cats. *Gut*, **24**, 542–8.

Nocerino, A., Iafusco, M. and Guandalini, S. (1995) Cholera toxin-induced small intestinal secretion has a secretory effect on the colon of the rat. *Gastroenterology*, **108**, 34–9.

Osaka, M., Fujita, T. and Yanatori, Y. (1975) On the possible role of intestinal hormones as the diarrhoeagenic messenger in cholera. *Virchows Arch B Cell Pathol*, **18**, 287–96.

Palmer, J. M., Wong-Riley, M. and Sharkey, K. A. (1998) Functional alterations in jejunal myenteric neurons during inflammation in nematode-infected guinea pigs. *Am J Physiol*, **275**, G922–35.

Pascaud, X. B., Chovet, M., Roze, C. and Junien, J. L. (1993) Neuropeptide Y and sigma receptor agonists act through a common pathway to stimulate duodenal alkaline secretion in rats. *Eur J Pharmacol*, **231**, 389–94.

Perdue, M. H., Marshall, J. and Masson, S. (1990) Ion transport abnormalities in inflamed rat jejunum. Involvement of mast cells and nerves. *Gastroenterology*, **98**, 561–7.

Perdue, M. H., Ramage, J. K., Burget, D., Marshall, J. and Masson, S. (1989) Intestinal mucosal injury is associated with mast cell activation and leukotriene generation during *Nippostrongylus*-induced inflammation in the rat. *Dig Dis Sci*, **34**, 724–31.

Peregrin, A. T., Ahlman, H., Jodal, M. and Lundgren, O. (1999) Involvement of serotonin and calcium channels in the intestinal fluid secretion evoked by bile salt and cholera toxin. *Br J Pharmacol*, **127**, 887–94.

Peterson, J. W., Lu, Y., Duncan, S., Cantu, J. and Chopra, A. K. (1994) Interactions of intestinal mediators in the mode of action of cholera toxin. *J Med Microbiol*, **41**, 3–9.

Peterson, J. W. and Ochoa, L. G. (1989) Role of prostaglandins and cAMP in the secretory effects of cholera toxin. *Science*, **245**, 857–9.

Peterson, J. W., Saini, S. S., Dickey, W. D., Klimpel, G. R., Bomalaski, J. S., Clark, M. A., Xu, X. J. and Chopra, A. K. (1996) Cholera toxin induces synthesis of phospholipase A2-activating protein. *Infect Immun*, **64**, 2137–43.

Pothoulakis, C. (1996) Pathogenesis of *Clostridium difficile*-associated diarrhoea. *Eur J Gastroenterol Hepatol*, **8**, 1041–7.

Pothoulakis, C., Castagliuolo, I., LaMont, J. T., Jaffer, A., O'Keane, J. C., Snider, R. M. and Leeman, S. E. (1994) CP-96,345, a substance P antagonist, inhibits rat intestinal responses to *Clostridium difficile* toxin A but not cholera toxin. *Proc Natl Acad Sci USA*, **91**, 947–51.

Pothoulakis, C., Castagliuolo, I., Leeman, S. E., Wang, C. C., Li, H., Hoffman, B. J. and Mezey, E. (1998) Substance P receptor expression in intestinal epithelium in *Clostridium difficile* toxin A enteritis in rats. *Am J Physiol*, **275**, G68–75.

Pothoulakis, C., Karmeli, F., Kelly, C. P., Eliakim, R., Joshi, M. A., O'Keane, C. J., Castagliuolo, I., LaMont, J. T. and Rachmilewitz, D. (1993) Ketotifen inhibits *Clostridium difficile* toxin A-induced enteritis in rat ileum. *Gastroenterology*, **105**, 701–7.

Pothoulakis, C. and Lamont, J. T. (2001) Microbes and microbial toxins: paradigms for microbial-mucosal interactions II. The integrated response of the intestine to *Clostridium difficile* toxins. *Am J Physiol Gastrointest Liver Physiol*, **280**, G178–83.

Pothoulakis, C., LaMont, J. T., Eglow, R., Gao, N., Rubins, J. B., Theoharides, T. C. and Dickey, B. F. (1991) Characterization of rabbit ileal receptors for *Clostridium difficile* toxin A. Evidence for a receptor-coupled G protein. *J Clin Invest*, **88**, 119–25.

Primi, M. P., Bueno, L., Baumer, P., Berard, H. and Lecomte, J. M. (1999) Racecadotril demonstrates intestinal antisecretory activity in vivo. *Aliment Pharmacol Ther*, **13**(Suppl 6), 3–7.

Pulimood, A. B., Mathan, M. M. and Mathan, V. I. (1998) Quantitative and ultrastructural analysis of rectal mucosal mast cells in acute infectious diarrhea. *Dig Dis Sci*, **43**, 2111–16.

Qiu, B., Pothoulakis, C., Castagliuolo, I., Nikulasson, Z. and LaMont, J. T. (1996) Nitric oxide inhibits rat intestinal secretion by *Clostridium difficile* toxin A but not *Vibrio cholerae* enterotoxin. *Gastroenterology*, **111**, 409–18.

Rabbani, G. H. and Butler, T. (1985) Indomethacin and chloroquine fail to inhibit fluid loss in cholera. *Gastroenterology*, **89**, 1035–7.

Ramage, J. K., Hunt, R. H. and Perdue, M. H. (1988) Changes in intestinal permeability and epithelial differentiation during inflammation in the rat. *Gut*, **29**, 57–61.

Riegler, M., Castagliuolo, I., Wang, C., Wlk, M., Sogukoglu, T., Wenzl, E., Matthews, J. B. and Pothoulakis, C. (2000) Neurotensin stimulates Cl(−) secretion in human colonic mucosa in vitro: role of adenosine. *Gastroenterology*, **119**, 348–57.

Riegler, M., Sedivy, R., Pothoulakis, C., Hamilton, G., Zacherl, J., Bischof, G., Cosentini, E., Feil, W., Schiessel, R., LaMont, and *et al.* (1995) *Clostridium difficile* toxin B is more potent than toxin A in damaging human colonic epithelium in vitro. *J Clin Invest*, **95**, 2004–11.

Riviere, P. J., Rao, R. K., Pascaud, X., Junien, J. L. and Porreca, F. (1993) Effects of neuropeptide Y, peptide YY and sigma ligands on ion transport in mouse jejunum. *J Pharmacol Exp Ther*, **264**, 1268–74.

Roze, C., Bruley Des Varannes, S., Shi, G., Geneve, J. and Galmiche, J. P. (1998) Inhibition of prostaglandin-induced intestinal secretion by igmesine in healthy volunteers. *Gastroenterology*, **115**, 591–6.

Ruhl, A., Hurst, S. and Collins, S. M. (1994) Synergism between interleukins 1 beta and 6 on noradrenergic nerves in rat myenteric plexus. *Gastroenterology*, **107**, 993–1001.

Saito, H., Okajima, F., Molski, T. F., Sha'afi, R. I., Ui, M. and Ishizaka, T. (1988) Effect of cholera toxin on histamine release from bone marrow-derived mouse mast cells. *Proc Natl Acad Sci USA*, **85**, 2504–8.

Sales, M. E., Sterin-Borda, L., de Bracco, M. M., Rodriguez, M., Narbaitz, M. and Borda, E. (1997) IgA from HIV haemophilic patients triggers intracellular signals coupled to the cholinergic system of the intestine. *Clin Exp Immunol*, **110**, 189–95.

Saria, A. and Beubler, E. (1985) Neuropeptide Y (NPY) and peptide YY (PYY) inhibit prostaglandin E2-induced intestinal fluid and electrolyte secretion in the rat jejunum in vivo. *Eur J Pharmacol*, **119**, 47–52.

Schwartz, J. C. (2000) Racecadotril: a new approach to the treatment of diarrhoea. *Int J Antimicrob Agents*, **14**, 75–9.

Scudamore, C. L., Thornton, E. M., McMillan, L., Newlands, G. F. and Miller, H. R. (1995) Release of the mucosal mast cell granule chymase, rat mast cell protease-II, during anaphylaxis is associated with the rapid development of paracellular permeability to macromolecules in rat jejunum. *J Exp Med*, **182**, 1871–81.

Sharkey, K. A., Sutherland, L. R., Davison, J. S., Zwiers, H., Gill, M. J. and Church, D. L. (1992) Peptides in the gastrointestinal tract in human immunodeficiency virus infection. The GI/HIV Study Group of the University of Calgary. *Gastroenterology*, **103**, 18–28.

Sjoqvist, A. (1991) Interaction between antisecretory opioid and sympathetic mechanisms in the rat small intestine. *Acta Physiol Scand*, **142**, 127–32.

Sjoqvist, A. (1992) Difference between the antisecretory mechanisms of opioids and the somatostatin analogue octreotide in cholera toxin-induced small intestinal secretion in the rat. *Regul Pept*, **40**, 339–49.

Sjoqvist, A., Brunsson, I., Theodorson, E., Brodin, E., Jodal, M. and Lundgren, O. (1993) On the involvement of tachykinin neurons in the secretory nervous reflex elicited by cholera toxin in the small intestine. *Acta Physiol Scand*, **148**, 387–92.

Sjoqvist, A., Fahrenkrug, J., Jodal, M. and Lundgren, O. (1988) The effect of splanchnic nerve stimulation and neuropeptide Y on cholera secretion and release of vasoactive intestinal polypeptide in the feline small intestine. *Acta Physiol Scand*, **133**, 289–95.

Sorensson, J., Jodal, M. and Lundgren, O. (2001) Involvement of nerves and calcium channels in the intestinal response to *Clostridium difficile* toxin A: an experimental study in rats in vivo. *Gut*, **49**, 56–65.

Spangler, B. D. (1992) Structure and function of cholera toxin and the related *Escherichia coli* heat-labile enterotoxin. *Microbiol Rev*, **56**, 622–47.

Speelman, P., Rabbani, G. H., Bukhave, K. and Rask-Madsen, J. (1985) Increased jejunal prostaglandin E2 concentrations in patients with acute cholera. *Gut*, **26**, 188–93.

Stead, R. H., Kosecka-Janiszewska, U., Oestreicher, A. B., Dixon, M. F. and Bienenstock, J. (1991) Remodeling of B-50 (GAP-43)- and NSE-immunoreactive mucosal nerves in the intestines of rats infected with *Nippostrongylus brasiliensis*. *J Neurosci*, **11**, 3809–21.

Stead, R. H., Tomioka, M., Quinonez, G., Simon, G. T., Felten, S. Y. and Bienenstock, J. (1987) Intestinal mucosal mast cells in normal and nematode-infected rat intestines are in intimate contact with peptidergic nerves. *Proc Natl Acad Sci USA*, **84**, 2975–9.

Swain, M. G., Agro, A., Blennerhassett, P., Stanisz, A. and Collins, S. M. (1992) Increased levels of substance P in the myenteric plexus of Trichinella-infected rats. *Gastroenterology*, **102**, 1913–19.

Swain, M. G., Blennerhassett, P. A. and Collins, S. M. (1991) Impaired sympathetic nerve function in the inflamed rat intestine. *Gastroenterology*, **100**, 675–82.

Thielman, N. M., Marcinkiewicz, M., Sarosiek, J., Fang, G. D. and Guerrant, R. L. (1997) Role of platelet-activating factor in Chinese hamster ovary cell responses to cholera toxin. *J Clin Invest*, **99**, 1999–2004.

Timar Peregrin, A., Ahlman, H., Jodal, M. and Lundgren, O. (1997) Effects of calcium channel blockade on intestinal fluid secretion: sites of action. *Acta Physiol Scand*, **160**, 379–86.

Triadafilopoulos, G., Pothoulakis, C., Weiss, R., Giampaolo, C. and Lamont, J. T. (1989) Comparative study of *Clostridium difficile* toxin A and cholera toxin in rabbit ileum. *Gastroenterology*, **97**, 1186–92.

Turvill, J. L., Connor, P. and Farthing, M. J. (2000) Neurokinin 1 and 2 receptors mediate cholera toxin secretion in rat jejunum. *Gastroenterology*, **119**, 1037–44.

Turvill, J. L., Kasapidis, P. and Farthing, M. J. (1999a) The sigma ligand, igmesine, inhibits cholera toxin and *Escherichia coli* enterotoxin induced jejunal secretion in the rat. *Gut*, **45**, 564–9.

Turvill, J. L., Mourad, F. H. and Farthing, M. J. (1999b) Proabsorptive and prosecretory roles for nitric oxide in cholera toxin induced secretion. *Gut*, **44**, 33–9.

Turvill, J. L., Mourad, F. H. and Farthing, M. J. (1998) Crucial role for 5-HT in cholera toxin but not *Escherichia coli* heat-labile enterotoxin-intestinal secretion in rats. *Gastroenterology*, **115**, 883–90.

Vallance, B. A., Blennerhassett, P. A., Deng, Y., Matthaei, K. I., Young, I. G. and Collins, S. M. (1999) IL-5 contributes to worm expulsion and muscle hypercontractility in a primary *T. spiralis* infection. *Am J Physiol*, **277**, G400–8.

Van Loon, F. P., Rabbani, G. H., Bukhave, K. and Rask-Madsen, J. (1992) Indomethacin decreases jejunal fluid secretion in addition to luminal release of prostaglandin E2 in patients with acute cholera. *Gut*, **33**, 643–5.

Varilek, G. W., Weinstock, J. V., Williams, T. H. and Jew, J. (1991) Alterations of the intestinal innervation in mice infected with *Schistosoma mansoni*. *J Parasitol*, **77**, 472–8.

Vasko, M. R., Campbell, W. B. and Waite, K. J. (1994) Prostaglandin E2 enhances bradykinin-stimulated release of neuropeptides from rat sensory neurons in culture. *J Neurosci*, **14**, 4987–97.

Venkova, K., Palmer, J. M. and Greenwood-Van Meerveld, B. (1999) Nematode-induced jejunal inflammation in the ferret causes long-term changes in excitatory neuromuscular responses. *J Pharmacol Exp Ther*, **290**, 96–103.

Vermillion, D. L., Ernst, P. B., Scicchitano, R. and Collins, S. M. (1988) Antigen-induced contraction of jejunal smooth muscle in the sensitized rat. *Am J Physiol*, **255**, G701–8.

Villaseca, J. M., Navarro-Garcia, F., Mendoza-Hernandez, G., Nataro, J. P., Cravioto, A. and Eslava, C. (2000) Pet toxin from enteroaggregative *Escherichia coli* produces cellular damage associated with fodrin disruption. *Infect Immun*, **68**, 5920–7.

Wald, A., Gotterer, G. S., Rajendra, G. R., Turjman, N. A. and Hendrix, T. R. (1977) Effect of indomethacin on cholera-induced fluid movement, unidirectional sodium fluxes, and intestinal cAMP. *Gastroenterology*, **72**, 106–10.

Wang, Y. Z., Palmer, J. M. and Cooke, H. J. (1991) Neuroimmune regulation of colonic secretion in guinea pigs. *Am J Physiol*, **260**, G307–14.

Warny, M., Keates, A. C., Keates, S., Castagliuolo, I., Zacks, J. K., Aboudola, S., Qamar, A., Pothoulakis, C., LaMont, J. T. and Kelly, C. P. (2000) p38 MAP kinase activation by *Clostridium difficile* toxin A mediates monocyte necrosis, IL-8 production, and enteritis. *J Clin Invest*, **105**, 1147–56.

Wedel, B. J. and Garbers, D. L. (1997) New insights on the functions of the guanylyl cyclase receptors. *FEBS Lett*, **410**, 29–33.

Weinstock, J. V. (1996) Vasoactive intestinal peptide regulation of granulomatous inflammation in murine *Schistosomiasis mansoni*. *Adv Neuroimmunol*, **6**, 95–105.

Weinstock, J. V., Blum, A., Walder, J. and Walder, R. (1988) Eosinophils from granulomas in murine schistosomiasis mansoni produce substance P. *J Immunol*, **141**, 961–6.

Weinstock, J. V. and Blum, A. M. (1990a) Detection of vasoactive intestinal peptide and localization of its mRNA within granulomas of murine schistosomiasis. *Cell Immunol*, **125**, 291–300.

Weinstock, J. V. and Blum, A. M. (1990b) Release of substance P by granuloma eosinophils in response to secretagogues in murine schistosomiasis mansoni. *Cell Immunol*, **125**, 380–5.

Weinstock, J. V. and Elliott, D. (1998) The substance P and somatostatin interferon-gamma immunoregulatory circuit. *Ann N Y Acad Sci*, **840**, 532–9.

Wershil, B. K., Castagliuolo, I. and Pothoulakis, C. (1998) Direct evidence of mast cell involvement in *Clostridium difficile* toxin A-induced enteritis in mice. *Gastroenterology*, **114**, 956–64.

Woodbury, R. G., Miller, H. R., Huntley, J. F., Newlands, G. F., Palliser, A. C. and Wakelin, D. (1984) Mucosal mast cells are functionally active during spontaneous expulsion of intestinal nematode infections in rat. *Nature*, **312**, 450–2.

Xia, Y., Hu, H. Z., Liu, S., Pothoulakis, C. and Wood, J. D. (2000) *Clostridium difficile* toxin A excites enteric neurons and suppresses sympathetic neurotransmission in the guinea pig. *Gut*, **46**, 481–6.

Yoshioka, M., Asakura, H., Hamada, Y., Miura, S., Kobayashi, K., Morishita, T., Morita, A. and Tsuchiya, M. (1987) Inhibitory effect of somatostatin on cholera toxin-induced diarrhea and glycoenzyme secretion in rat intestine. *Digestion*, **36**, 141–7.

13 Neuroimmune Interactions in the Lung

Bradley J. Undem[1] and Daniel Weinreich[2]

[1]*Department of Medicine, Johns Hopkins School of Medicine,
Baltimore, MD 21205, USA*
[2]*Department of Pharmacology and Experimental Therapeutics,
University of Maryland School of Medicine, Baltimore, MD 21201, USA*

The nervous system and the immune system serve a vital role in protecting the airspace from inhaled irritants and pathogens. The mechanisms by which these disparate systems defend the airways are dissimilar. Activating the nervous system leads to changes in breathing patterns, coughing, increases in airway mucus secretion and bronchospasm. The immune system defends the host against inhaled pathogens by the cellular responses involved in innate and adaptive immunity. We review here the many ways in which mediators released as a consequence of immune cell activation can lead to changes in of neuronal reflex physiology, and conversely, how neurotransmitters released from airway nerves can modulate immune cell function. The lack of redundancy in defence mechanisms between the two system makes it likely that nerve–immune interactions in the airways increase the efficacy of host defence against inhaled ambient air. On the other hand, in those cases where immune stimulation is detrimental to the host, as in allergic asthma, the interactions between the immune and nervous system may contribute to disease. A better understanding of the molecular processes involved in nerve–immune interactions in the lungs would therefore increase our understanding of host defense mechanisms as well as the pathophysiology of certain airway diseases.

KEY WORDS: neuronal reflex; airway nerves; asthma.

INTRODUCTION

The interactions between the immune and the nervous systems in the lung, as in other organs, are bi-directional. The immune system can initiate processes leading to the release of mediators and cytokines that affect the neurophysiology of the airways. Conversely, transmitters released from nerves innervating the airways can act on cells of the immune system to affect their function. As can be expected, the majority of investigations on nerve–immune interactions in the lung have been carried out *in vitro* at the cellular level. There is relatively little information that addresses the role of the nervous system in pulmonary immunity. Likewise, the information pertaining to the influence of the immune system on the neurophysiology of the lung is, for the most part, of a speculative nature. In this chapter we have attempted to provide an overview of the research that supports the hypothesis that nerve and immune cells interact in the airway, and provide some information on the potential consequence of this interaction in health and disease.

NEURAL REGULATION OF THE IMMUNE SYSTEM IN THE LUNGS

The nervous system can modulate the function of the immune system in the airways by both direct and indirect mechanisms. We have assigned direct mechanisms to those that involve transmitters released from nerve fibres in the airway wall interacting with surrounding immunocytes. Indirect mechanisms, are defined within the context of this review, as those processes by which the nervous system affects the recruitment of immune cell to the airways or the function of immunocytes before they reach the airways.

DIRECT MECHANISMS

The airways are densely innervated from the nose to the alveolar walls (van der Velden and Hulsmann, 1999). The majority of nerve fibres reach the lungs via the vagus nerves. The vagal innervation is divided into vagal afferent (sensory) nerves and autonomic, (parasympathetic) nerves. Approximately 80% of the vagal fibres innervating the airways are afferent nerves (Agostoni et al., 1957). The afferent nerve fibres form a dense plexus just beneath the epithelium. Many afferent fibres contain sensory neuropeptides such as substance P (SP), neurokinin A (NKA), and calcitonin gene related peptide (CGRP), that theoretically may be released as a consequence of axon reflexes. Autonomic preganglionic fibres leave the vagus and synapse in airway parasympathetic ganglia where they communicate with postganglionic fibres. The postganglionic fibres innervate airway smooth muscle, submucosal blood vessels and mucus glands (van der Velden and Hulsmann, 1999). Postganglionic parasympathetic fibres can be further subdivided based on their transmitter content. In several species, including humans, there are both cholinergic and noncholinergic parasympathetic fibres in the airways (Ellis and Undem, 1994). The noncholinergic fibres use nitric oxide and vasoactive intestinal peptide as their transmitters (Belvisi et al., 1992; Ellis and Undem, 1992b; Canning et al., 1996). In addition to vagal innervation, airways receive sensory innervation from neurons in the dorsal root ganglia (Kummer et al., 1992) and sympathetic innervation from superior cervical ganglia and stellate ganglia. The sympathetic nerves use catecholamines as well as neuropeptides such as neuropeptide Y as their transmitters.

Inactivation of afferent and autonomic neuropeptides occurs by relatively slow enzymatic degredation rather than re-uptake by nerve terminals. This can be seen in the vagally innervated guinea pig bronchus preparation where vagus nerve stimulation leads to a cholinergic and a noncholinergic contraction of the bronchial smooth muscle (Undem et al., 1990). The noncholinergic component of the response is due to antidromic stimulation of tachykinergic afferent fibres. Upon termination of the stimulus, the cholinergic component of the response rapidly decays with a half-time of 5 s, whereas the tachykinergic phase of the response decays with a half time of over 5 min (Undem et al., 1990). The relatively slow inactivation of neuropeptide transmitters, considered with the dense peptidergic innervation of the airway wall, sets up the condition in which peptide transmitters may diffuse from the immediate target of innervation and affect the function of bystander cells. Based on this, it can be argued that virtually all cells types in the airway wall may come in contact with reasonable concentrations of neurotransmitter molecules.

Some immune cells in the airways are in intimate contact with nerves. This is best exemplified by the innervation of bronchial associated lymphoid tissue (BALT) (Nohr and Weihe, 1991). BALT is thought to play an important role in seeding the airways with IgA producing B lymphocytes, thereby playing an important role in airway mucosal immunity (Bienenstock et al., 1973a,b). A common feature of BALT is its direct innervation by various types of nerve fibres including those that contain tachykinins, CGRP, catecholamines, NPY and various opioids (Nohr and Weihe, 1991). There is little known about the effect of nerve stimulation on IgA production in the airways, but in the gastrointesitnal system, evidence supports a role for positive and negative neuroregulation of IgA production. Substance P and somatostatin enhance IgA production, whereas catecholamines inhibit production (Stanisz et al., 1986; Schmidt et al., 1999). The airway epithelium contains numerous dendritic cells as well as a rich supply of afferent nerve fibres. The juxtaposition of nerve fibres and dendritic cells suggests the possibility that the nervous system may regulate antigen-presentation or other aspects of dendritic cell function. This hypothesis, as far as we know, has not yet been experimentally addressed.

If immune cells in the airways come in contact neurotransmitters their function will be affected only if they contain neurotransmitter receptors on their plasma membranes. In fact, not only IgA producing B lymphocytes, but all immune cells in the lung, as in all tissues, possess cell-surface receptors for a variety of neurotransmitters. The function of monocytes and macrophages, for example, can be modified by adrenergic and cholinergic receptor activation (Hjemdahl et al., 1990; Shen et al., 1994). In addition, neuropeptides including tachykinins and vasoactive intestinal polypeptide (VIP) can interact with receptors on macrophages to modify their function (Brunelleschi et al., 1990; Litwin et al., 1992). Similar findings have been made with T and B lymphocytes, neutrophils, eosinophils, mast cells and basophils. The vast literature pertaining to the modulation of immune cells by various neurotransmitters has been reviewed elsewhere (Undem and Myers, 2000). In all cases, the cells possess receptors for various neurotransmitters, and in vitro studies have revealed that activation of the receptors can profoundly enhance or inhibit cellular function.

INDIRECT MECHANISMS

The nervous system can indirectly affect the immunology of the airways via modulation of immune cell function in the spleen, bone marrow and other lymphoid organs. The innervation of these structures is discussed elsewhere in this volume. Airway nerves can also indirectly affect the immunological response in the lungs by inducing a process referred to as 'neurogenic inflammation'. Neurogenic inflammation in the airways is caused by activation of neuropeptide containing afferent fibres that innervate the post-capilliary venules in the airways. Neurokinins released from these nerves lead to vasodilatation, an increase in leukocyte adherence to the venules, and an increase in plasma extravasation through the venular endothelium (McDonald, 1988; McDonald *et al.*, 1988). Thus, activation of neurokinin containing afferent nerves in the airways may therefore affect the immune system by increasing the numbers of leukocytes and plasma proteins within the airway wall (McDonald, 1988). Neurogenic inflammation has been thoroughly investigated in guinea pig and rats, but relatively little is known about neurogenic inflammation in the lower airways of humans. The number of neurokinin containing nerves in adult healthy human airways is much less than that seen in guinea pig and rat airways. The neurogenic inflammatory response, therefore, is likely to be less in healthy human airways than that observed in guinea pigs and rats. The extent of neurokinin innervation, however, appears to be increased in infant airways and in airways associated with inflammatory diseases such as asthma (Hislop *et al.*, 1990; Ollerenshaw *et al.*, 1991; Chu *et al.*, 2000). Thus, neurogenic inflammation may play more of a role in immune modulation in certain subsets of humans. In the human nose, activation of neuropeptide containing afferent nerves with capsaicin has been shown to increase leukocyte infiltration and plasma extravasation; an effect that is most prominent in inflamed noses (Sanico *et al.*, 1997, 1998).

Although the nervous system can affect immune cell function by both direct and indirect mechanisms, the question of how pulmonary immunity is modified by the nervous system remains unanswered. The lack of insight in this area is due in large part to the inability to design experiments that directly address meaningful questions pertaining to neuronal regulation of immunological responses within the airways.

To specifically address the question how the nervous system regulates the immunology of the airways, denervation studies are required. The natural experiment therefore may be to evaluate the immune response in the lungs of subjects that have survived lung transplantation. Although these lungs contain the intrinsic innervation, all nerve fibres that functionally connect the lungs to the central nervous system (CNS) are severed. Unfortunately, patients with transplanted lungs are often required to take potent immunosupressing drugs render them less than ideal population for the study of nerve-immune interactions in the lungs. The effect of various neurotransmitter receptor antagonists on airway immunology is another approach that can be used to experimentally address question pertaining to the effect of the nervous system on the immune system in the lungs. The disadvantage with this approach is that neurotransmitter molecules are found in numerous nonneuronal cell types. For example, substance P has been localized in dendritic cells (Lambrecht *et al.*, 1999) and acetylcholine may be synthesized in airway epithelial cells (Reinheimer *et al.*, 1996). Interestingly, neurokinin receptor antagonists had significant effects on the dendritic cell dependent lymphocyte proliferation in an *in vitro* design that was devoid of neurons (Lambrecht *et al.*, 1999). Chemical and surgical denervation studies in animal model provides the most direct approach to questions pertaining to the consequence neural regulation of the immune system in the airways. Some of these studies are discussed later in this review.

IMMUNE-MODULATION OF AIRWAY NERVE FUNCTION

From a physiological perspective, airway nerves function within the context of a reflex arc (Widdicombe, 1986). The reflex is initiated by mechanically or chemically induced action potential discharge in primary afferent nerve fibres within the airway wall. The action potentials are conducted along the afferent fibre to the CNS where they activate or inhibit, via synaptic transmission, secondary neurons in the reflex pathway. The information is integrated in the CNS and ultimately leads to changes in activity of preganglionic autonomic neurons, neurons involved in various sensations, and neurons involved in regulating breathing pattern. The preganglionic autonomic fibres activate postganglionic nerve fibres, again via synaptic transmission, located within parasympathetic ganglia within the airway wall, or sympathetic ganglia situated in the prevertebral column. The parasympathetic cholinergic nerves provide the dominant mechanism for bronchial constriction in the airways, whereas the NANC parasympathetic innervation provides the only relaxant innervation to human airway smooth muscle (Canning and Fischer, 2001). Both cholinergic and NANC nerves also regulate mucus secretion, and airway vascular function (Widdicombe, 1986; Laitinen *et al.*, 1987). The sympathetic nerves innervating airways contain catecholamines and neuropeptide Y and serve primarily as regulators of vascular function. The sympathetic fibres do not innervate the human bronchial smooth muscle (Canning and Fischer, 2001).

The types of immunological modulation of airway nerve function most commonly investigated are those associated with allergic inflammation. When the allergic subject inhales allergens, the direct activation of IgE-bearing cells in the airways (most notably mast cells and basophils) sets in motion a cascade of events that impact all aspects of the neuronal reflex arc. Thus, allergen challenge can directly increase the activity of primary afferent nerves, synaptic efficacy at the autonomic ganglia and transmitter secretion at the autonomic effector junction, and indirectly can lead to changes in CNS integration (Figure 13.1).

Sites of immune modulation of airway reflexes

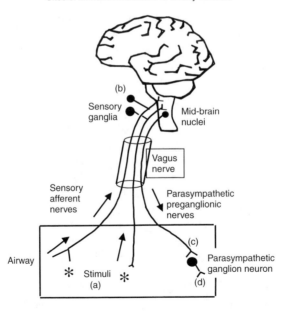

Figure 13.1 Potential mechanisms that underlie allergen-induced modulation of airway neuronal reflex activity. (a) Increase nerve ending stimulation, and neuropeptide secretion; (b) Induction of neuropeptide gene expression, affecting CNS integration; (c) Increase synaptic efficacy in autonomic ganglia; (d) Increase transmitter secretion at neuro-effector junction.

AFFERENT NERVES

The afferent nerves innervating the airways arise primary from the vagus nerves. They can be subdivided based on conduction velocity with the slowly conducting unmyelinated fibres referred to as C-fibres and the faster conducting myelinated fibres referred to as A fibres (Widdicombe, 1982). Many C-fibres are thought to serve a nociceptive-like function in that they are difficult to stimulate mechanically, but respond vigorously to various noxi-ous stimuli and chemicals such as capsaicin and bradykinin. Most A fibres in the airways are low threshold mechanosensors that can be further subdivided based on their adaptation to prolonged suprathreshold mechanical stimulus. Afferent fibres that have rapidly adapting receptors (RARs) lead to increases in reflexes such as cough and increases in parasympathetic activity in the airways (Sant'Ambrogio and Widdicombe, 2001). In this regard they are similar to C-fibres (Roberts et al., 1981). Activation of A fibres in the airways that have slowly adapting receptors (SARs) typically lead to decreases in parasympathetic tone and changes in breathing pattern (e.g. Herring Breuer reflex) (Widdicombe, 1982).

The allergic reaction in the airways has been shown to cause action potential discharge in RAR fibres within the airways (Mills and Widdicombe, 1970; Bergren et al., 1985). It is unlikely that the antigen interacts with antibodies directly on the nerve fibres, rather the activation of mast cells likely led to the production of mediators that in turn stimulated the afferent nerves. The mediators could interact directly with cell surface receptors on the afferent fibres to affect their electrophysiological properties. Alternatively, since most of the afferent fibres in the airways are mechanically sensitive, antigen challenge can indirectly lead to activation of afferent fibres secondary to bronchoconstrictrion or changes in vascular dynamics. This is especially the case for low-threshold mechanosensors. The indirect effect of mediators on afferent nerve activity is exemplified by histamine in guinea pig airways. Histamine does not appear to directly induce action potential discharge in afferent nerves in the airways wall when studied in vitro (Fox et al., 1993; Riccio et al., 1996). The histamine-induced bronchoconstriction observed in vivo, however, leads to activation of low threshold mechanosensors in the airways (both SARs and RARs), an effect that is inhibited by pretreating the animal with a bronchodilator (Matsumoto and Shimizu, 1994; Bergren, 1997).

An example of antigen inducing the release of mediators that interact directly with the afferent fibres has been noted in the guinea pig isolated airway preparation (Riccio et al., 1996). In this case, antigen challenge affected the electrophysiological properties of afferent nerve endings in the airway wall without overtly causing action potential discharge. Thus, antigen challenge failed to induce action potential discharge, but lead to a four-fold decrease in the amount of force required to activate RAR fibres. This increase in afferent nerve 'excitability' is likely to be secondary to inflammatory mediators interacting with cell surface receptors on the nerves resulting in alterations in the activity of various ion channels (Undem et al., 1993). In these in vitro studies, the mast cell likely plays a key role in initiating the antigen-induced change in afferent nerve excitability. In more chronic inflammatory states, however, other cell types are likely to participate in allergen–afferent nerve interactions. For example, eosinophils associated with airway allergic reactions are found in close proximity to submucosal nerve fibres in human and guinea pig airways (Costello *et al.*, 2000). Moreover, eosinophils can release cationic proteins that have been shown to cause impressive increases in excitability of pulmonary C-fibres (Lee *et al.*, 2001).

Allergic inflammation may also qualitatively change the responsivity of primary afferent nerves in the airways. In the somatosensory system, inflammation is associated with so-called silent fibres (respond to no known stimuli) becoming mechanically sensitive fibres (Handwerker *et al.*, 1991). Hypotheses relating to unmasking of silent receptors have not yet been addressed within the airways. Allergic inflammation of the airways has been shown, however, to lead to unmasking of chemical reception in vagal afferent nerves. Electrophysiological studies have shown that neurokinins have no effect on the resting membrane potential of guinea pig vagal sensory neurons. If, however, the sensory ganglia are obtained from guinea pigs that were previously subjected to allergen aerosol, substance P and neurokinin A cause robust depolarizations of the resident neurons (Moore *et al.*, 2000). The neurokinin-induced membrane depolarization observed subsequent to allergen challenge is secondary to neurokinin 2 receptor activation and a consequent opening of non-selective cation channels (Moore *et al.*, 1999).

The vagal afferent C-fibres innervating the airway wall synthesize neurokinins and calcitonin gene-related peptide. These neuropeptides are found in both the central terminals in the midbrain and peripheral terminals within the airway wall. The sensory neuropeptides can be released into the airway as a consequence of action potentials conducted down collateral branches of the afferent fibre (i.e. axon-reflex pathways). Alternatively, agonists such as capsaicin and trypsin can lead to neuropeptide release independently of action potentials (Carr *et al.*, 2000). Once release in the airway, sensory neuropeptides can participate in the inflammatory response via various mechanisms collectively referred to as neurogenic inflammation (discussed above). Numerous studies have shown that allergen challenge in the airways enhances neuropeptide release (Nieber *et al.*, 1991; Ellis and Undem, 1992a; Kageyama *et al.*, 1996; Kaltreider *et al.*, 1997; Lindstrom and Andersson, 1997). The mechanism for allergen-induced neuropeptide secretion has not been studied in detail, but mediators such as cysteinyl leukotrienes, lipoxins, histamine and trypsin are effective in causing, or potentiating, neurokinin secretion in airway tissues (Meini *et al.*, 1992; Ellis and Undem, 1992a; McAlexander *et al.*, 1998; Carr *et al.*, 2000). In various animal models of allergic airway disease, the increased neuropeptide release in the airway caused by allergen is thought to contribute to antigen-induced leukocyte recruitment, changes in epithelial ion transport, and airways hyperreactivity (Lundberg *et al.*, 1991; Bertrand and Geppetti, 1996).

Immunological challenge can increase neuropeptides in the airway by increasing their synthesis or by decreasing their metabolism. One day following allergen inhalation, there is a substantial increase in the preprotachykinin mRNA expressed in guinea pig vagal sensory ganglia, and a doubling in sensory neuropeptide content in the lungs (Fischer *et al.*, 1996). The major pathway for inactivation of neurokinins in the airway wall is via metabolism by neutral endopeptidase. Allergic inflammation or viral infection can lead to increases in the local concentration of sensory neuropeptides by decreasing the amount of neutral endopeptidase activity in the airway wall (Lazarus *et al.*, 1987; Kudlacz *et al.*, 1994; Lilly *et al.*, 1994).

CNS INTEGRATION

As discussed above, immunological processes in the airways can affect the intensity and pattern of action potential discharge of primary afferent nerve fibres. This will ultimately, depending on the afferent fibres involved, cause changes in sensations and reflex output from the CNS. Immunological stimulation can also change the neurochemistry of the afferent neurons in a fashion that may affect synaptic transmission between the central terminal of the airway afferent fibre and the secondary neurons within the CNS. The low threshold vagal afferent mechanosensors in the airways (SARs and RARs) are thought to use glutamate as an excitatory neurotransmitter (Haxhiu *et al.*, 1997, 2000). Release of glutamate at synapses in the nucleus tractus solitarious (NTS) causes fast excitatory post-synaptic potentials (EPSPs). The central terminals of nociceptive-like airway C-fibres contain sensory neuropeptides such as neurokinins. Neurokinins typically cause slow EPSPs that can act in a synergistic fashion with glutamate, causing a substantive augmentation of synaptic activity (Urban and Gebhart, 1999; Mutoh *et al.*, 2000). With this in mind it is interesting that in preliminary studies we have noted that immunological stimuli such as allergen inhalation or respiratory viral infection can lead to a phenotypic change in neurokinin innervation such that neurokinin synthesis occurs in low threshold mechanosensitive airway afferent fibres of the RAR subtype (Carr and Undem, 2001). This could set up a scenario in which neuropeptide-induced potentiation of synaptic transmission in the NTS will occur as a result of RAR activation (e.g. during respiration) without the requirement of nociceptive nerve activation. The induction by inflammatory stimuli of neurokinin synthesis in non-nociceptive neurons has also been noted in the somato-sensory system where it is thought to contribute to hyperalgesia and allodynic type disorders (Neumann *et al.*, 1996).

The mediators involved in the allergen- or viral-mediated phenotypic switch in airway sensory neuropeptide expressing neurons are not known. Neurotrophins such as nerve growth factor (NGF), however, are likely candidates. NGF is known to lead to activatation of preprotachykinin gene expression in neurons. Moreover, NGF was able to mimic the effect of allergen or virus challenge in inducing neurokinin expression in guinea pig airway RAR neurons (Hunter *et al.*, 2000).

In addition to NGF, other neurotrophins such as brain derived neurotrophic factor (BDNF) may regulate nerve function in the airways. Many immune cell types including mast cells, lymphocytes and macrophages express and release neurotrophins. Neurotrophin molecules are released in human upper and lower airways upon allergen challenge (Virchow *et al.*, 1998; Sanico *et al.*, 2000), and may contribute to both the cellular immune responses within the airways as well as changes in neuronal function (Braun *et al.*, 1999; Carr *et al.*, 2001). More specific assertions regarding the role of neurotrophins in immune–nerve interactions in the airways await additional investigation.

AUTONOMIC MODULATION

The changes in CNS integration brought about by immunological modulation of afferent nerve function and chemistry will ultimately result in major changes in activity of preganglionic nerve fibres. In addition, immunological process can lead to changes in autonomic neurophysiology by modulating synaptic transmission in autonomic ganglia, and by modulating neurotransmitter release at the nerve–effector junction within the airway wall.

The sympathetic innervation of the airways is derived from cell bodies located in the superior cervical, stellate and thoracic sympathetic ganglia (Kummer et al., 1992). Parasympathetic preganglionic nerve fibres by contrast synapse on neurons situated in small ganglia located within or near the airways (Myers, 2001).

A prototypical example of immune modulation of autonomic synaptic neurotransmission is the antigen-induced potentiation of EPSP amplitude in the superior cervical ganglion (SCG) (Weinreich and Undem, 1987; Weinreich et al., 1995; Cavalcante de Albuquerque et al., 1996). In these studies, the SCG was isolated from guinea pigs actively sensitized to ovalbumin, or passively sensitized with anti-ovalbumin antibodies. Recordings of either the postganglionic compound action potential or individual EPSPs was used to monitor the efficacy of synaptic transmission (Figure 13.2). The EPSP magnitude and the postganglionic compound action potential increased substantially within a few minutes of exposure of the isolated SCG to the sensitizing antigen. This coincided with activation of resident mast cells and the release of various inflammatory mediators. An intriguing aspect of the antigen-induced increase in synaptic transmission was its time-course. A 5-min exposure to the sensitizing antigen resulted in synaptic potentiation that did not reverse for the duration of the experiments (at least 3 h). This phenomenon was termed antigen-induced long-term potentiation (antigen-induced LTP) (Weinreich and Undem, 1987). The mediators and biophysical mechanisms underlying antigen-induced LTP have not been elucidated. Both presynaptic (increase quantal release of acetylcholine) and postsynaptic (changes in membrane potential and resistance) mechanisms appear to contribute to the antigen-induced LTP. The nature of the mediators involved in the response is also unclear. The LTP cannot be explained solely the actions of histamine, platelet activating factor, leukotrienes or prostaglandins. Other potential mast cell mediators including mast cell tryptase, certain cytokines, chemokines and neurotrophins have not yet been studied. In any event the increase in autonomic synaptic transmission initially studied in the SCG appears to be a common feature of the immediate hypersensitivity response. Thus, antigen challenge has been found to increase synaptic transmission in other sympathetic ganglia, parasympathetic and enteric ganglia (Undem et al., 1991; Frieling et al., 1994; Cavalcante de Albuquerque et al., 1996; Palmer and Greenwood-Van Meerveld, 2001).

Particularly relevant to this chapter is the finding that allergen challenge affects synaptic transmission in bronchial parasympathetic ganglia (Myers et al., 1991; Undem et al., 1991; Myers and Undem, 1995). There have been relatively few studies on the electrophysiological properties of parasympathetic ganglia within the airways (Baker et al., 1983; Coburn and Kalia, 1986). These ganglia typically contain 5–20 neurons, and project either cholinergic or NANC nerves to effectors cells in the airway wall (e.g. bronchial smooth muscle, glands, vasculature). Exposing bronchi isolated from actively or passively sensitized guinea pigs to a sensitizing antigen, results in mast cell activation and several consequent changes in the electrical properties of bronchial ganglion neurons. The accommodative properties of the neurons are diminished, the membrane potential is depolarized, and the membrane conductance is decreased (Myers et al., 1991). Importantly, as was noted in the sympathetic ganglia, antigen challenge is associated with an increase in synaptic efficacy in bronchial parasympathetic ganglia as evidenced by increased EPSP

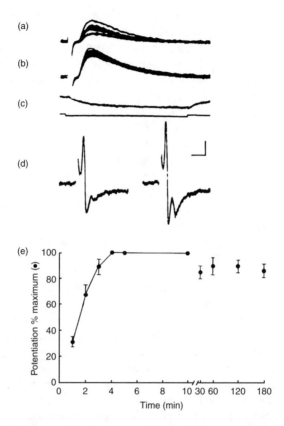

Figure 13.2   Effects of antigen application on synaptic transmission recorded in a sympathetic ganglion (superior cervical ganglion) removed from an actively sensitized guinea pig. Data recorded ~15 min after a 5-min application of the sensitizing antigen (10 μg/ml chick ovalbumin). Antigen application provokes a transient membrane depolarization due to histamine release (not shown) that lasted 4 min in this neuron. (a–c) Intracellular recording showing the effects of antigen treatment on action potential-evoked fast nicotinic EPSPs (0.3 Hz preganglionic nerve stimulation) and on electrotonic voltage transients. (a) Seven consecutive superimposed EPSPs recorded before antigen application. (b) Seven consecutive superimposed EPSPs recorded after antigen challenge; note that the fluctuations in EPSP amplitudes following antigen challenge is significantly less than control fluctuations indicating that the larger EPSPs after antigen treatment are likely due to enhanced acetylcholine release. (c) Two superimposed electrotonic voltage transients produced by 100 pA constant-current pulses shown below EPSPs, one before and one 15 min after antigen treatment showing that the larger EPSP are not due to change in resting membrane resistance. Resting membrane potential held at −62 mV. Calibration: 5 mV, 20 ms. (d) Extracellular recording of postganglionic compound action potentials (CAPs) from the same ganglion. Left CAP recorded before antigen application; right CAP recorded 15 min after antigen challenge. Calibration: 0.5 mV, 10 ms. (e) Time-course of antigen-induced potentiation of synaptic transmission monitored by measuring the amplitude of the CAP. Synaptic potentiation is long-lived, lasting >3 h following antigen application (applied for 5 min beginning at time zero).

amplitude. This can have a major consequence on airway physiology inasmuch as under normal conditions much of the preganglionic input to bronchial parasympathetic ganglia is filtered, that is, the EPSPs are subthreshold for action potential generation (Myers and Undem, 1991).

By increasing synaptic efficacy, allergen challenge can theoretically lead to a decrease in the filtering capacity of the ganglia and a consequent increase in the frequency of action potential discharge in postganglionic parasympathetic nerves within the airways (Undem et al., 2000). Immunological processes can also lead to elevations in the amount of neuro-transmitter released per action potential at the neuro-effector junction. A classical example of this can be found in the increase in vagal nerve stimulation-induced cholinergic contractions of guinea pig bronchi observed following allergic or viral inflammation (Fryer and Jacoby, 1991; Fryer and Wills-Karp, 1991; Sorkness et al., 1994; Costello et al., 1999). This has been found to be due in part to a decrease in muscarinic M2 receptor function (Fryer and Jacoby, 1991; Fryer and Wills-Karp, 1991; Ten Berge et al., 1996). Stimulation of muscarinic M2 receptors located on the postganglionic cholinergic nerve fibres by released acetylcholine leads to autoinhibition of further acetylcholine release (Blaber et al., 1985). Allergen- and viral-induced airway inflammation decreases this auto-inhibitory activity of M2 receptors thereby amplifying acetylcholine release. Inhibition of muscarinic M2 receptor function is not the only mechanism by which inflammation can lead to increased cholinergic responses in the airways (Sorkness et al., 1994). Certain autacoids released during inflammatory reactions may also lead to elevation in acetycholine release by interacting with autacoid receptors on the postganglionic cholinergic nerve (Barnes, 1992).

The effect of immunological stimulation on release of NANC autonomic transmitters in the airways has not received much attention. The metabolic degradation of one such transmitter, VIP, has been shown to be enhanced following allergen challenge (Miura et al., 1992; Lilly et al., 1994). This may lead to increases in bronchomotor tone, because VIP (along with nitric oxide) is thought to contribute to autonomic nerve mediated relaxations of airway smooth muscle.

CONSEQUENCES OF NERVE–IMMUNE INTERACTIONS IN THE AIRWAYS

With each inhalation the airways are exposed to myriad irritants and pathogens in the ambient air. The nervous system and the immune system share a common purpose in keeping the environment in the distal air spaces sterile. The effector mechanisms by which these two systems defend the airways, however, are dissimilar. The accessory cells of the immune system exposed to potential pathogens marshal forward a cellular, antibody and cytokine driven response to rid the host of the invading pathogens. Activation of the afferent limb of the nervous system by inhaled irritants, by contrast, triggers various physical mechanisms such as mucus secretion, bronchospasm, and cough to defend the air passages. Because the effector systems are not redundant, it may be beneficial for the immune system to recruit neuronal defense mechanisms in its defensive strategy. Likewise, it would be beneficial to the host for the nervous system to recruit processes of the immune system when exposed to inhaled irritants. As discussed in the sections above, numerous mechanisms are in place by which these positive interactions may take place.

Rats infected with mycoplasma pulmonis provide an experimental example of the beneficial consequence of nerve–immune interactions in the airways. This infection leads to a long-lasting potentiation of afferent nerve-induced plasma extravasation (neurogenic inflammation) (McDonald *et al.*, 1991). This inflammation, in turn, may assist the immune system in defending the airways from the mycoplasma organisms. Indeed, chemical denervated of afferent C-fibres with neonatal capsaicin treatment, prevented neurogenic inflammation in this model, and lead to a substantial increase in severity of infection and resulted in exaggerated airway pathology in response to subsequent mycoplasma infections (Bowden *et al.*, 1996). Mycoplasma infection may also lead to recruitment of neuronal defense mechanisms in human airways. In human asthmatic airways infected with mycoplasma pneumonia, there is an increase in substance P immunoreactivity and neurokinin receptor expression (Chu *et al.*, 2000).

Nerve–immune interactions in the airways may not always be beneficial to the host. In allergic airway disease the immune response to inhaled antigen is inappropriate and can have a negative reaction in the host. By recruiting neuronal mechanisms to the cause, the negative effects may be amplified. When the inhaled antigen (allergen) interacts with mast cell and basophil bound anti-IgE it initiates the classical immediate hypersensitivity reactions. From an immunological perspective this leads to a lymphocyte (Th-2) driven eosinophilic bronchitis (Busse and Lemanske, 2001). Antigenic activation of mast cells also augments airway neuronal activity. As discussed above this occurs as a result of multiple interactions at the level of afferent, CNS and autonomic nerves. The relatively mild eosinophilic bronchitis may then be transduced into excessive non-productive coughing, rapid and reversible bronchospasm, increased mucus secretion, and sensations of dyspnea, that is, attack of bronchial asthma. Designing strategies to thwart the transduction of allergic inflammation to altered neuronal activity may lead to novel therapeutic strategies aimed at alleviating the symptoms of allergic airway disease.

REFERENCES

Agostoni, E., Chinnock, J. E., De Burgh, D. M. and Murray, J. G. (1957) Functional and histological studies of the vagus nerve and its branches to the heart, lungs, and abdominal viscera in the cat. *J Physiol*, **135**, 182–205.

Baker, D. G., Basbaum, C. B., Herbert, D. A. and Mitchell, R. A. (1983) Transmission in airway ganglia of ferrets: inhibition by norepinephrine. *Neurosci Lett*, **41**, 139–43.

Barnes, P. J. (1992) Modulation of neurotransmission in airways. *Physiol Rev*, **72**, 699–729.

Belvisi, M. G., Stretton, C. D., Yacoub, M. and Barnes, P. J. (1992) Nitric oxide is the endogenous neurotransmitter of bronchodilator nerves in humans. *Eur J Pharmacol*, **210**, 221–2.

Bergren, D. R. (1997) Sensory receptor activation by mediators of defense reflexes in guinea-pig lungs. *Respir Physiol*, **108**, 195–204.

Bergren, D. R., Myers, D. L. and Mohrman, M. (1985) Activity of rapidly-adapting receptors to histamine and antigen challenge before and after sodium cromoglycate. *Arch Int Pharmacodyn Ther*, **273**, 88–99.

Bertrand, C. and Geppetti, P. (1996) Tachykinin and kinin receptor antagonists: therapeutic perspectives in allergic airway disease. *Trends Pharmacol Sci*, **17**, 255–9.

Bienenstock, J., Johnston, N. and Perey, D. Y. (1973a) Bronchial lymphoid tissue. I. Morphologic characteristics. *Lab Invest*, **28**, 686–92.

Bienenstock, J., Johnston, N. and Perey, D. Y. (1973b) Bronchial lymphoid tissue. II. Functional characterisitics. *Lab Invest*, **28**, 693–8.

Blaber, L. C., Fryer, A. D. and Maclagan, J. (1985) Neuronal muscarinic receptors attenuate vagally-induced contraction of feline bronchial smooth muscle. *Br J Pharmacol*, **86**, 723–8.

Bowden, J. J., Baluk, P., Lefevre, P. M., Schoeb, T. R., Lindsey, J. R. and McDonald, D. M. (1996) Sensory denervation by neonatal capsaicin treatment exacerbates Mycoplasma pulmonis infection in rat airways. *Am J Physiol*, **270**, L393–403.

Braun, A., Lommatzsch, M., Lewin, G. R., Virchow, J. C. and Renz, H. (1999) Neurotrophins: a link between airway inflammation and airway smooth muscle contractility in asthma? *Int Arch Allergy Immunol*, **118**, 163–5.

Brunelleschi, S., Vanni, L., Ledda, F., Giotti, A., Maggi, C. A. and Fantozzi, R. (1990) Tachykinins activate guinea-pig alveolar macrophages: involvement of NK2 and NK1 receptors. *Br J Pharmacol*, **100**, 417–20.

Busse, W. W. and Lemanske, R. F., Jr (2001) Asthma. *N Engl J Med*, **344**, 350–62.

Canning, B. J. and Fischer, A. (2001) Neural regulation of airway smooth muscle tone. *Respir Physiol*, **125**, 113–27.

Canning, B. J., Undem, B. J., Karakousis, P. C. and Dey, R. D. (1996) Effects of organotypic culture on parasympathetic innervation of guinea pig trachealis. *Am J Physiol*, **271**, L698–706.

Carr, M. J., Hunter, D. D. and Undem, B. J. (2001) Neurotrophins and asthma. *Curr Opin Pulm Med*, **7**, 1–7.

Carr, M. J., Schechter, N. M. and Undem, B. J. (2000) Trypsin-induced, neurokinin-mediated contraction of guinea pig bronchus. *Am J Respir Crit Care Med*, **162**, 1662–7.

Carr, M. J. and Undem, B. J. (2001) Inflammation-induced plasticity of the afferent innervation of the airways. *Environ Health Perspect*, **109** (Suppl 4), 567–71.

Cavalcante de Albuquerque, A. A., Leal-Cardoso, J. H. and Weinreich, D. (1996) Antigen-induced synaptic plasticity in sympathetic ganglia from actively and passively sensitized guinea-pigs. *J Auton Nerv Syst*, **61**, 139–44.

Chu, H. W., Kraft, M., Krause, J. E., Rex, M. D. and Martin, R. J. (2000) Substance P and its receptor neurokinin 1 expression in asthmatic airways. *J Allergy Clin Immunol*, **106**, 713–22.

Coburn, R. F. and Kalia, M. P. (1986) Morphological features of spiking and nonspiking cells in the paratracheal ganglion of the ferret. *J Comp Neurol*, **254**, 341–51.

Costello, R. W., Evans, C. M., Yost, B. L., Belmonte, K. E., Gleich, G. J., Jacoby, D. B. and Fryer, A. D. (1999) Antigen-induced hyperreactivity to histamine: role of the vagus nerves and eosinophils. *Am J Physiol*, **276**, L709–14.

Costello, R. W., Jacoby, D. B., Gleich, G. J. and Fryer, A. D. (2000) Eosinophils and airway nerves in asthma. *Histol Histopathol*, **15**, 861–8.

Ellis, J. L. and Undem, B. J. (1992a) Antigen-induced enhancement of noncholinergic contractile responses to vagus nerve and electrical field stimulation in guinea pig isolated trachea. *J Pharmacol Exp Ther*, **262**, 646–53.

Ellis, J. L. and Undem, B. J. (1992b) Inhibition by L-NG-nitro-L-arginine of nonadrenergic–noncholinergic-mediated relaxations of human isolated central and peripheral airway. *Am Rev Respir Dis*, **146**, 1543–7.

Ellis, J. L. and Undem, B. J. (1994) Pharmacology of non-adrenergic, non-cholinergic nerves in airway smooth muscle. *Pulm Pharmacol*, **7**, 205–23.

Fischer, A., McGregor, G. P., Saria, A., Philippin, B. and Kummer, W. (1996) Induction of tachykinin gene and peptide expression in guinea pig nodose primary afferent neurons by allergic airway inflammation. *J Clin Invest*, **98**, 2284–91.

Fox, A. J., Barnes, P. J., Urban, L. and Dray, A. (1993) An in vitro study of the properties of single vagal afferents innervating guinea-pig airways. *J Physiol*, **469**, 21–35.

Frieling, T., Palmer, J. M., Cooke, H. J. and Wood, J. D. (1994) Neuroimmune communication in the submucous plexus of guinea pig colon after infection with Trichinella spiralis. *Gastroenterology*, **107**, 1602–9.

Fryer, A. D. and Jacoby, D. B. (1991) Parainfluenza virus infection damages inhibitory M2 muscarinic receptors on pulmonary parasympathetic nerves in the guinea-pig. *Br J Pharmacol*, **102**, 267–71.

Fryer, A. D. and Wills-Karp, M. (1991) Dysfunction of M2-muscarinic receptors in pulmonary parasympathetic nerves after antigen challenge. *J Appl Physiol*, **71**, 2255–61.

Handwerker, H. O., Kilo, S. and Reeh, P. W. (1991) Unresponsive afferent nerve fibres in the sural nerve of the rat. *J Physiol*, **435**, 229–42.

Haxhiu, M. A., Erokwu, B. and Dreshaj, I. A. (1997) The role of excitatory amino acids in airway reflex responses in anesthetized dogs. *J Auton Nerv Syst*, **67**, 192–9.

Haxhiu, M. A., Yamamoto, B., Dreshaj, I. A., Bedol, D. and Ferguson, D. G. (2000) Involvement of glutamate in transmission of afferent constrictive inputs from the airways to the nucleus tractus solitarius in ferrets. *J Auton Nerv Syst*, **80**, 22–30.

Hislop, A. A., Wharton, J., Allen, K. M., Polak, J. M. and Haworth, S. G. (1990) Immunohistochemical localization of peptide-containing nerves in human airways: age-related changes. *Am J Respir Cell Mol Biol*, **3**, 191–8.

Hjemdahl, P., Larsson, K., Johansson, M. C., Zetterlund, A. and Eklund, A. (1990) Beta-adrenoceptors in human alveolar macrophages isolated by elutriation. *Br J Clin Pharmacol*, **30**, 673–82.

Hunter, D. D., Myers, A. C. and Undem, B. J. (2000) Nerve growth factor-induced phenotypic switch in guinea pig airway sensory neurons. *Am J Respir Crit Care Med*, **161**, 1985–90.

Kageyama, N., Ichinose, M., Igarashi, A., Miura, M., Yamauchi, H., Sasaki, Y., Ishikawa, J., Tomaki, M. and Shirato, K. (1996) Repeated allergen exposure enhances excitatory nonadrenergic noncholinergic nerve-mediated bronchoconstriction in sensitized guinea-pigs. *Eur Respir J*, **9**, 1439–44.

Kaltreider, H. B., Ichikawa, S., Byrd, P. K., Ingram, D. A., Kishiyama, J. L., Sreedharan, S. P., Warnock, M. L., Beck, J. M. and Goetzl, E. J. (1997) Upregulation of neuropeptides and neuropeptide receptors in a murine model of immune inflammation in lung parenchyma. *Am J Respir Cell Mol Biol*, **16**, 133–44.

Kudlacz, E. M., Shatzer, S. A., Farrell, A. M. and Baugh, L. E. (1994) Parainfluenza virus type 3 induced alter-ations in tachykinin NK1 receptors, substance P levels and respiratory functions in guinea pig airways. *Eur J Pharmacol*, **270**, 291–300.

Kummer, W., Fischer, A., Kurkowski, R. and Heym, C. (1992) The sensory and sympathetic innervation of guinea-pig lung and trachea as studied by retrograde neuronal tracing and double-labelling immunohisto-chemistry. *Neuroscience*, **49**, 715–37.

Laitinen, L. A., Laitinen, M. V. and Widdicombe, J. G. (1987) Parasympathetic nervous control of tracheal vascu-lar resistance in the dog. *J Physiol*, **385**, 135–46.

Lambrecht, B. N., Germonpre, P. R., Everaert, E. G., Carro-Muino, I., De Veerman, M., de Felipe, C., Hunt, S. P., Thielemans, K., Joos, G. F. and Pauwels, R. A. (1999) Endogenously produced substance P contributes to lymphocyte proliferation induced by dendritic cells and direct TCR ligation. *Eur J Immunol*, **29**, 3815–25.

Lazarus, S. C., Borson, D. B., Gold, W. M. and Nadel, J. A. (1987) Inflammatory mediators, tachykinins and enkephalinase in airways. *Int Arch Allergy Appl Immunol*, **82**, 372–6.

Lee, L. Y., Gu, Q. and Gleich, G. J. (2001) Effects of human eosinophil granule-derived cationic proteins on C-fibre afferents in the rat lung. *J Appl Physiol*, **91**, 1318–26.

Lilly, C. M., Kobzik, L., Hall, A. E. and Drazen, J. M. (1994) Effects of chronic airway inflammation on the acti-vity and enzymatic inactivation of neuropeptides in guinea pig lungs. *J Clin Invest*, **93**, 2667–74.

Lindstrom, E. G. and Andersson, R. G. (1997) Neurokinin A-LI release after antigen challenge in guinea-pig bronchial tubes: influence of histamine and bradykinin. *Br J Pharmacol*, **122**, 417–22.

Litwin, D. K., Wilson, A. K. and Said, S. I. (1992) Vasoactive intestinal polypeptide (VIP) inhibits rat alveolar macrophage phagocytosis and chemotaxis in vitro. *Regul Pept*, **40**, 63–74.

Lundberg, J. M., Alving, K., Karlsson, J. A., Matran, R. and Nilsson, G. (1991) Sensory neuropeptide involve-ment in animal models of airway irritation and of allergen-evoked asthma. *Am Rev Respir Dis*, **143**, 1429–30; discussion 1430–1.

Matsumoto, S. and Shimizu, T. (1994) Effects of isoprenaline on the responses of slowly adapting pulmonary stretch receptors to reduced lung compliance and to administered histamine. *Neurosci Lett*, **172**, 47–50.

McAlexander, M. A., Myers, A. C. and Undem, B. J. (1998) Inhibition of 5-lipoxygenase diminishes neurally evoked tachykinergic contraction of guinea pig isolated airway. *J Pharmacol Exp Ther*, **285**, 602–7.

McDonald, D. M. (1988) Neurogenic inflammation in the rat trachea. I. Changes in venules, leucocytes and epithelial cells. *J Neurocytol*, **17**, 583–603.

McDonald, D. M., Mitchell, R. A., Gabella, G. and Haskell, A. (1988) Neurogenic inflammation in the rat trachea. II. Identity and distribution of nerves mediating the increase in vascular permeability. *J Neurocytol*, **17**, 605–28.

McDonald, D. M., Schoeb, T. R. and Lindsey, J. R. (1991) Mycoplasma pulmonis infections cause long-lasting potentiation of neurogenic inflammation in the respiratory tract of the rat. *J Clin Invest*, **87**, 787–99.

Meini, S., Evangelista, S., Geppetti, P., Szallasi, A., Blumberg, P. M. and Manzini, S. (1992) Pharmacologic and neurochemical evidence for the activation of capsaicin-sensitive sensory nerves by lipoxin A4 in guinea pig bronchus. *Am Rev Respir Dis*, **146**, 930–4.

Mills, J. E. and Widdicombe, J. G. (1970) Role of the vagus nerves in anaphylaxis and histamine-induced bron-choconstrictions in guinea-pigs. *Br J Pharmacol*, **39**, 724–31.

Miura, M., Ichinose, M., Kimura, K., Katsumata, U., Takahashi, T., Inoue, H. and Takishima, T. (1992) Dysfunction of nonadrenergic noncholinergic inhibitory system after antigen inhalation in actively sensi-tized cat airways. *Am Rev Respir Dis*, **145**, 70–4.

Moore, K. A., Taylor, G. E. and Weinreich, D. (1999) Serotonin unmasks functional NK-2 receptors in vagal sen-sory neurones of the guinea-pig. *J Physiol*, **514**, 111–24.

Moore, K. A., Undem, B. J. and Weinreich, D. (2000) Antigen inhalation unmasks NK-2 tachykinin receptor-mediated responses in vagal afferents. *Am J Respir Crit Care Med*, **161**, 232–6.

Mutoh, T., Bonham, A. C. and Joad, J. P. (2000) Substance P in the nucleus of the solitary tract augments broncho-pulmonary C fibre reflex output. *Am J Physiol Regul Integr Comp Physiol*, **279**, R1215–23.

Myers, A. C. (2001) Transmission in autonomic ganglia. *Respir Physiol*, **125**, 99–111.

Myers, A. C. and Undem, B. J. (1991) Analysis of preganglionic nerve evoked cholinergic contractions of the guinea pig bronchus. *J Auton Nerv Syst*, **35**, 175–84.

Myers, A. C. and Undem, B. J. (1995) Antigen depolarizes guinea pig bronchial parasympathetic ganglion neurons by activation of histamine H1 receptors. *Am J Physiol*, **268**, L879–84.

Myers, A. C., Undem, B. J. and Weinreich, D. (1991) Influence of antigen on membrane properties of guinea pig bronchial ganglion neurons. *J Appl Physiol*, **71**, 970–6.

Neumann, S., Doubell, T. P., Leslie, T. and Woolf, C. J. (1996) Inflammatory pain hypersensitivity mediated by phenotypic switch in myelinated primary sensory neurons. *Nature*, **384**, 360–4.

Nieber, K., Baumgarten, C., Witzel, A., Rathsack, R., Oehme, P., Brunnee, T., Kleine-Tebbe, J. and Kunkel, G. (1991) The possible role of substance P in the allergic reaction, based on two different provocation models. *Int Arch Allergy Appl Immunol*, **94**, 334–8.

Nohr, D. and Weihe, E. (1991) The neuroimmune link in the bronchus-associated lymphoid tissue (BALT) of cat and rat: peptides and neural markers. *Brain Behav Immun*, **5**, 84–101.

Ollerenshaw, S. L., Jarvis, D., Sullivan, C. E. and Woolcock, A. J. (1991) Substance P immunoreactive nerves in airways from asthmatics and nonasthmatics. *Eur Respir J*, **4**, 673–82.

Palmer, J. M. and Greenwood-Van Meerveld, B. (2001) Integrative neuroimmunomodulation of gastrointestinal function during enteric parasitism. *J Parasitol*, **87**, 483–504.

Reinheimer, T., Bernedo, P., Klapproth, H., Oelert, H., Zeiske, B., Racke, K. and Wessler, I. (1996) Acetylcholine in isolated airways of rat, guinea pig, and human: species differences in role of airway mucosa. *Am J Physiol*, **270**, L722–8.

Riccio, M. M., Myers, A. C. and Undem, B. J. (1996) Immunomodulation of afferent neurons in guinea-pig isolated airway. *J Physiol*, **491**, 499–509.

Roberts, A. M., Kaufman, M. P., Baker, D. G., Brown, J. K., Coleridge, H. M. and Coleridge, J. C. (1981) Reflex tracheal contraction induced by stimulation of bronchial C-fibres in dogs. *J Appl Physiol*, **51**, 485–93.

Sanico, A. M., Atsuta, S., Proud, D. and Togias, A. (1997) Dose-dependent effects of capsaicin nasal challenge: in vivo evidence of human airway neurogenic inflammation. *J Allergy Clin Immunol*, **100**, 632–41.

Sanico, A. M., Atsuta, S., Proud, D. and Togias, A. (1998) Plasma extravasation through neuronal stimulation in human nasal mucosa in the setting of allergic rhinitis. *J Appl Physiol*, **84**, 537–43.

Sanico, A. M., Stanisz, A. M., Gleeson, T. D., Bora, S., Proud, D., Bienenstock, J., Koliatsos, V. E. and Togias, A. (2000) Nerve growth factor expression and release in allergic inflammatory disease of the upper airways. *Am J Respir Crit Care Med*, **161**, 1631–5.

Sant'Ambrogio, G. and Widdicombe, J. (2001) Reflexes from airway rapidly adapting receptors. *Respir Physiol*, **125**, 33–45.

Schmidt, P. T., Eriksen, L., Loftager, M., Rasmussen, T. N. and Holst, J. J. (1999) Fast acting nervous regulation of immunoglobulin A secretion from isolated perfused porcine ileum. *Gut*, **45**, 679–85.

Shen, H. M., Sha, L. X., Kennedy, J. L. and Ou, D. W. (1994) Adrenergic receptors regulate macrophage secretion. *Int J Immunopharmacol*, **16**, 905–10.

Sorkness, R., Clough, J. J., Castleman, W. L. and Lemanske, R. F., Jr (1994) Virus-induced airway obstruction and parasympathetic hyperresponsiveness in adult rats. *Am J Respir Crit Care Med*, **150**, 28–34.

Stanisz, A. M., Befus, D. and Bienenstock, J. (1986) Differential effects of vasoactive intestinal peptide, substance P, and somatostatin on immunoglobulin synthesis and proliferations by lymphocytes from Peyer's patches, mesenteric lymph nodes, and spleen. *J Immunol*, **136**, 152–6.

Ten Berge, R. E., Krikke, M., Teisman, A. C., Roffel, A. F. and Zaagsma, J. (1996) Dysfunctional muscarinic M2 autoreceptors in vagally induced bronchoconstriction of conscious guinea pigs after the early allergic reaction. *Eur J Pharmacol*, **318**, 131–9.

Undem, B. J., Hubbard, W. and Weinreich, D. (1993) Immunologically induced neuromodulation of guinea pig nodose ganglion neurons. *J Auton Nerv Syst*, **44**, 35–44.

Undem, B. J., Kajekar, R., Hunter, D. D. and Myers, A. C. (2000) Neural integration and allergic disease. *J Allergy Clin Immunol*, **106**, S213–20.

Undem, B. J. and Myers, A. C. (2000) *Neural Regulation of the Immune Reponse*, Blackwell Science Ltd, Oxford, England.

Undem, B. J., Myers, A. C., Barthlow, H. and Weinreich, D. (1990) Vagal innervation of guinea pig bronchial smooth muscle. *J Appl Physiol*, **69**, 1336–46.

Undem, B. J., Myers, A. C. and Weinreich, D. (1991) Antigen-induced modulation of autonomic and sensory neurons in vitro. *Int Arch Allergy Appl Immunol*, **94**, 319–24.

Urban, M. O. and Gebhart, G. F. (1999) Central mechanisms in pain. *Med Clin North Am*, **83**, 585–96.

van der Velden, V. H. and Hulsmann, A. R. (1999) Autonomic innervation of human airways: structure, function, and pathophysiology in asthma. *Neuroimmunomodulation*, **6**, 145–59.

Virchow, J. C., Julius, P., Lommatzsch, M., Luttmann, W., Renz, H. and Braun, A. (1998) Neurotrophins are increased in bronchoalveolar lavage fluid after segmental allergen provocation. *Am J Respir Crit Care Med*, **158**, 2002–5.

Weinreich, D., Undem, B. J., Taylor, G. and Barry, M. F. (1995) Antigen-induced long-term potentiation of nicotinic synaptic transmission in the superior cervical ganglion of the guinea pig. *J Neurophysiol*, **73**, 2004–16.

Weinreich, D. and Undem, B. J. (1987) Immunological regulation of synaptic transmission in isolated guinea pig autonomic ganglia. *J Clin Invest*, **79**, 1529–32.

Widdicombe, J. (1986) The neural reflexes in the airways. *Eur J Respir Dis Suppl*, **144**, 1–33.

Widdicombe, J. G. (1982) Pulmonary and respiratory tract receptors. *J Exp Biol*, **100**, 41–57.

14 Neuroimmune Control of the Pulmonary Immune Response

Armin Braun* and Harald Renz

Department of Clinical Chemistry and Molecular Diagnostics, Philipps-University Marburg, 35033 Marburg, Germany

Allergic bronchial asthma (BA) is charcterized by chronic inflammation of the airways, development of airway hyper-reactivity (AHR) and recurrent reversible airway obstruction. Target and effector cells responsible for AHR and airway obstruction include sensory and motor neurons as well as epithelial and smooth muscle cells. Although it is well established that the inflammatory process is controlled by T-helper (Th) 2 cells and the Th2-derived cytokines interleukin (IL)-4, IL-5 and IL-13, the mechanisms by which immune cells interact with neurons, epithelial cells or smooth muscle cells still remain uncertain. Since there is growing evidence for extensive communication between neurons and immune cells, the mechanisms of this neuro-immune crosstalk in lung and airways of asthmatic patients are recently becoming the focus of asthma research. This chapter will review the current literature on this subject and develop a hypothetical concept of bi-directional pathways connecting the nervous and the immune systems.

KEY WORDS: allergy; airway hyperreactivity; asthma; inflammation; neuron; NGF; neuroimmunology; neurotrophin; T cell.

T CELLS AND AIRWAY INFLAMMATION IN ALLERGIC ASTHMA

Allergic bronchial asthma (BA) is characterized by chronic airway inflammation that has been implied to play an important role in the development of airway hyper-reactivity (AHR) and recurrent reversible airway obstruction. There is overwhelming evidence that T cells play a central role in allergic BA (Kay, 1996). Strong evidence supports the notion that Th2 cells orchestrate allergic inflammation driven by effector functions of B cells, mast cells and eosinophils. Th2 cells produce a cytokine profile that predominantly includes IL-4, IL-5 and IL-13. In B cells, IL-4 is involved in isotype switching towards IgE and propagation of Th2 cells, while IL-5 possesses pro-inflammatory properties due to its

---

* Present address: Fraunhofer Institute of Toxicology and Aerosol Research, Drug Research and Clinical Inhibition, 30625 Hannover, Germany.

role in the development, differentiation, recruitment and survival of eosinophils. The importance of T cells or T cell-driven processes is further underlined by the effectiveness of anti-inflammatory therapies including glucocorticoids and several other mediations (Wong and Koh, 2000). This concept is currently leading to the development of novel drugs including anti-IL-5 or IgE inhibitors which allow testing of this concept under *in vivo* conditions (Kay, 2001a,b).

AIRWAY HYPER-REACTIVITY (AHR)

The success of anti-inflammatory therapy is not followed by a complete disappearance of symptoms (Milgrom *et al.*, 1999) indicating long lasting inflammation-induced effects. To date, no satisfying concept linking inflammation with persistent symptoms of asthma is available, but it is known that chronic inflammation is associated with clinical consequences including non-specific bronchial hyper-responsiveness, which may be defined as an increase in the ease in degree of airway narrowing in response to a wide range of bronchoconstrictor stimuli (Bousquet *et al.*, 2000). The development of AHR in response to allergic inflammation is mediated by multiple independent and additive pathways working in concert (Herz *et al.*, 1998; Wilder *et al.*, 1999; Wills-Karp, 1999). Several mechanisms have been identified to be involved in the occurrence of AHR. The airway changes leading to airway narrowing mainly include: (a) altered neuronal regulation of airway tone; (b) increases in muscle content or function; and (c) increased epithelial mucus production and airway oedema (Wills-Karp, 1999; Bousquet *et al.*, 2000). Therefore, the mode of measuring AHR is critical for identifying the underlying mechanisms (Table 14.1). Dependent on the method chosen for AHR measurement, different pathways can be distinguished. It has been previously shown that *in vitro*, electrical field stimulation of tracheal smooth muscle segments reflects specific neuronal airway dysfunction since the addition of both atropin (disruption of cholinergic pathways) and capsaicin (depletion of sensory neurons) completely blocks any reaction of the airway to electric field stimulation (Andersson and Grundstrom, 1983; Ellis and Undem, 1992). By using specific stimuli,

TABLE 14.1
Assessment of different pathways of airway responsiveness in allergic bronchial asthma.

Stimulus	Effector cells	Acting through
Electrical field stimulation (EFS)	Sensory neurons Motor neurons	Unspecific depolarization
Methacholine	Smooth muscle cells	M_3 receptors
Histamine	Smooth muscle cells Sensory neurons Motor neurons	H_1 receptors
Serotonin	Sensory neurons Motor neurons	$5\,HT_1$ receptors $5\,HT_3$ receptors
Capsaicin	Sensory neurons	Vanilloid receptor
Hypotonic H_2O	Sensory neurons	Unspecific

the measuring of lung function in animal models allows to distinguish between broncho-constriction due to direct stimulation of airway smooth muscle or due to alteration of sensory nerves. Capsaicin is a potent stimulant of sensory nerves and induces a characteristic modification of the normal breathing pattern. In contrast, methacholine acts via direct stimulation of airway smooth muscle cells. Other stimuli including histamine can affect both nerve and smooth muscle cells.

INNERVATION OF THE LUNG

The human airways are innervated via efferent and afferent autonomic nerves regulating many aspects of airway function, including airway smooth muscle tone, mucus secretion, bronchial micro-circulation, microvascular permeability as well as recruitment and subsequent activation of inflammatory cells (van der Velden and Hulsmann, 1999a). Innervation of the lung can be functionally divided into cholinergic, adrenergic and non-adrenergic non-cholinergic (NANC) pathways, which are not strictly anatomically separated (Table 14.2). At least certain NANC effects are mediated by the release of neuropeptides from classical cholinergic or adrenergic nerves (van der Velden and Hulsmann, 1999a). The NANC system has been subdivided into the e-NANC and i-NANC system. The e-NANC system exhibits excitatory, bronchoconstrictory, C-fibre-mediated and tachykinin-dependent functions. In contrast, the i-NANC system is an bronchodilatory pathway located within

TABLE 14.2
Innervation of the lung.

	Cholinergic	*Adrenergic*	*e-NANC*	*i-NANC*
Neurotransmitter	Acetylcholine	(a) Noradrenaline (b) Adrenaline	SP, NKA, NKB	VIP, NO
Receptors	Muscarinic (M_1, M_2, M_3)	(a) α-Adrenoreceptor (b) β_2-Adrenoreceptor	NK-1, NK-2, NK-3	VIP-R, guanylyl cyclase
Expressed on	Lymphocytes, smooth muscle (M_3), submucosal glands postganglionic cholinergic nerves (M_1, M_2)	(a) Cholinergic nerves, blood vessel (b) Smooth muscle, cholinergic nerves	Cholinergic nerves, blood vessel, smooth muscle, lymphocytes, granulocytes monocytes	Blood vessel, submucosal glands
Effects	Smooth muscle constriction	(a) Constriction (b) Dilation	Smooth muscle constriction	Smooth muscle dilation
Therapeutic intervention	M_3 antagonist	β_2 Agonist	NK-antagonists	

parasympathetic nerves, mediating its effects mainly by nitric oxide (NO) and vasoactive intestinal peptide (VIP) (van der Velden and Hulsmann, 1999a).

NEURONAL CHANGES IN ALLERGIC ASTHMA

In the last decade, growing evidence indicates neuronal dysregulation on several levels in allergic BA (Joos *et al.*, 2000a) (Table 14.3). Since cholinergic nerves represent the dominant bronchoconstrictory pathway and anticholinergic drugs are very effective bronchodilators in acute severe asthma, cholinergic mechanisms must be considered in the development of AHR. These possible mechanisms include enhanced cholinergic reflex activity, increased acetylcholine (ACh) release, enhanced sensitivity of smooth muscle to ACh and increased density of muscarinic receptors on airway smooth muscle. In addition, sensory nerves are able to modulate cholinergic function. Cholinergic activity was shown to be increased by tachykinins (Delaunois *et al.*, 1996; Mackay *et al.*, 1998). Contraction of airway smooth muscle is mediated by M3 muscarinic receptors on airway smooth muscle. There is no evidence, however, suggesting that hyperresponsiveness results from any alteration in the function of these M3 muscarinic receptors. In contrast, there is clearly increased release of the neurotransmitter ACh in animal models of hyper-reactivity and asthma (Larsen *et al.*, 1994). In addition, the loss of function of inhibitory M2 muscarinic receptors on the airway parasympathetic nerves enhances vagally mediated bronchoconstriction and hyper-responsiveness following allergen challenges (Fryer *et al.*, 1999). The M2 muscarinic receptors on the parasympathetic nerves in the lungs normally inhibit release of acetylcholine. When the receptors are blocked, the inhibitory effect is lost and, then, acetylcholine release is increased. Loss of M2 receptor function has been shown as a result of eosinophilic mediations, resulting in airway hyper-responsiveness (Jacoby *et al.*, 2001).

The sympathetic nervous system is less prominent than the parasympathetic nervous system within the human airways. It should be highlighted that there is a lack of

TABLE 14.3
Neuronal alterations in human BA and animal models of BA.

Neuronal pathway	Mediators	Changes observed in BA	Species
Sensory innervation	Tachykinins	Increased levels in BAL	Human, guinea pig
		Upregulation in cell bodies of ganglion nodosum	Guinea pig
		Increased mechano-sensitivity in sensory Aδ fibres	Guinea pig
		Increased excitability of sensory neurons	Guinea pig
		Increased response of sensory neurons to SP	Guinea pig
Cholinergic innervation	Acetylcholine	Increased acetylcholine release	Guinea pig, mouse
		Loss of function of inhibitory M2 muscarinic receptors	Guinea pig
Adrenergic innervation	NA, NPY	Increased sympathetic efficiacy	Guinea pig

sympathetic innervation of the human airway smooth muscle compared to other species (Barnes, 1995; Casale, 1996; Joos *et al.*, 2000b). Its main neurotransmitters are noradrenaline (NA) and neuropeptide Y (NPY). Guinea pig models indicate that antigen challenge induces a mast-cell mediated long-lasting increase in synaptic efficacy (Weinreich *et al.*, 1995). In asthmatic airways, no difference in the number of NPY-immunoreactive nerves has been found as compared to healthy controls (Howarth *et al.*, 1995).

The e-NANC system exhibits a high degree of plasticity in inflammatory conditions. SP and NKA, preferentially released by sensory C-fibres, are closely related members of the neuropeptide family termed tachykinins. They are synthesized preferentially in cell bodies of the sensory ganglia by a complex biosynthetic pathway. These neuropeptides are transferred via axonal transport not only to presynaptic axon endings in the spinal cord and the nucleus of the solitary tract, but also to peripheral sensory nerve endings (Brimijoin *et al.*, 1980). Upon exposure to mechanical, thermal, chemical (capsaicin, nicotine) or inflammatory stimuli (bradykinin, histamine, prostaglandins), tachykinins are released from nerve cells through a local (axon) reflex mechanism (Barnes, 1986). Tachykinins act in a dual fashion, as afferent neurotransmitters to the central nervous system as well as efferent neurosecretory mediators diffusing into the peripheral tissue. Increased levels of the neuropeptide substance P (SP) have been detected in the airways of asthmatic patients (Baumgarten *et al.*, 1996).

Additionally, allergen challenge increased neurokinin A (NKA) levels in bronchoalveolar lavage fluid (BALF) of asthmatic patients (Heaney *et al.*, 1998). Nerve fibres containing SP have been described in and around bronchi, bronchioles, the more distal airways and occasionally extend into the alveolar wall. The fibres are located beneath and within the airway epithelium, around blood vessels and submucosal glands, within the bronchial smooth muscle layer and around the local tracheo-bronchial ganglion cells (Lundberg *et al.*, 1984). Although there is evidence for an increase in both the number and length of SP immunoreactive nerve fibres in airways from subjects with bronchial asthma as compared to airways from healthy subjects (Ollerenshaw *et al.*, 1991), the study from Lilli *et al.* detected no difference in SP-like immunoreactivity (Lilly *et al.*, 1995). One reason for the latter finding may be that the nerves release SP due to continued stimulation. In a guinea pig model, it has been shown that sensory innervation of the airways is altered during allergic inflammation (Undem *et al.*, 1999). The increase of SP and NKA in the lung in response to allergen challenges has been related to an increased production of these neuropeptides in neurons of the nodose ganglion (Fischer *et al.*, 1996). Impaired degradation of tachykinins could further enhance their local activity (van der Velden and Hulsmann, 1999b). Tachykinins are degraded and inactivated by neutral endopeptidase (NEP), a membrane-bound metallopeptidase located mainly on the surface of airway epithelial cells and also present in airway smooth muscle cells, submucosal glands and fibroblasts. As a key role NEP limits the biological actions of mediators like tachykinins via enzymatic degradation. Allergen exposure, inhalation of cigarette smoke and other respiratory irritants are associated with a reduced NEP activity, thus enhancing the effects of tachykinins within the airways (Sont *et al.*, 1997; Di Maria *et al.*, 1998; Tudoric *et al.*, 2000).

In addition, antigen-induced functional changes in sensory neurons, including the depolarization of the resting membrane potential, changes in membrane resistance, increases in mechanosensitivity (of Aδ vagal afferent airway nerves) and enhanced responses to SP, have been described (Undem et al., 1993; Weinrich et al., 1997). Since these neuronal alterations in asthma are associated with local inflammation, the concept that inflammatory mediators could be responsible for neuronal dysfunction was developed.

EFFECTS OF INFLAMMATORY MEDIATORS ON NEURON FUNCTIONS

Numerous studies provide evidence for upregulation and/or increased expression of T cell derived cytokines in the airways of asthmatic patients. Increased levels of IL-2, IL-4, IL-5, IL-9, IL-13 and IL-18 could be determind in BAL-fluid and mucosal biopsies on both transcriptional and translational levels (Virchow et al., 1995; Kay, 1996; Hamid and Minshall, 2000). In addition, macrophage-derived pro-inflammatory mediators including IL-1β, IL-6 and TNF-α have also been detected during allergic responses (Hamid and Minshall, 2000). Many of these mediators are implied to regulate and affect airway responsiveness in BA. Particularly the role of IL-4, IL-5 and IL-13 has been demonstrated in this respect by utilizing genetically manipulated mice or antibody treatment strategies. However, whether T cell or macrophage-derived inflammatory mediators modulate neuronal smooth muscle innervation in a direct or indirect fashion, or whether they directly affect the smooth muscle cell itself still remains unclear. In addition, there are no reports available concerning the direct effects of the cytokines normally thought to be responsible for the development of AHR like IL-5 and IL-13 on nerve cells innervating the lung.

In contrast, other cytokines, especially the neurotrophic cytokines including ciliary neurotrophic factor (CNTF), leukaemia inhibitory factor (LIF), oncostatin-M (OSM), IL-6 and cardiotrophin-1 (CT-1), were shown to influence sensory neurons (Horton et al., 1998). They are predominantly produced by macrophages. Particulary IL-6 is known to affect peripheral sensory neurons and to support neuronal, SP production and hypersensitivity (Murphy et al., 1999; Thier et al., 1999) (Table 14.4). In addition, IL-1 and TNF-α induce changes in sensory neurons and are involved in the inflammation associated hyperalgesia (Woolf et al., 1997; Ek et al., 1998). Furthermore, other mediators of allergic inflammation like bradykinin are able to augment the response of airway sensory nerves in response to capsaicin or citric acid (Fox et al., 1996).

TABLE 14.4
Cytokine effects on PNS neurons.

Cytokine	Susceptible neuron	Effects
CNTF, LIF, OSM and CT-1	Enteroceptive neurons of the nodose ganglion	Supported throughout their development
CNTF, LIF, OSM and CT-1	Sympathetic neurons	Acetylcholine production
IL-6	Sensory neurons	SP-production, survival and hypersensitivity, hyperalgesia, neuroprotection
IL-1, TNF-α	Sensory neurons	Neuroprotection, hyperalgesia

THE PUTATIVE ROLE OF NEUROTROPHINS IN THE PATHOGENESIS OF BRONCHIAL ASTHMA

Some of the most effective mediators involved in inflammatory hyperalgesia are the neurotrophins nerve growth factor (NGF) and the brain derived neurotrophic factor (BDNF) (Donnerer et al., 1992; Dmitrieva et al., 1997; Mannion et al., 1999). NGF is a mediator with functions on both immune- and nerve cells (Levi Montalcini et al., 1996). It is a well-studied example of a target-derived neurotrophic factor that is essential for development, differentiation, maintenance and survival of peripheral sympathetic and neural crest-derived sensory nerve cells (Levi Montakine et al., 1995). NGF upregulates expression of neuropeptides in sensory neurons (Lindsay and Harmar, 1989) and contributes to inflammatory sensory hypersensitivity (Donnerer et al., 1992). In the central nervous system, NGF is a trophic factor for basal forebrain cholinergic neurons. The biological effects of neurotrophins are mediated by binding either to the specific high affinity (Kd -10^{-11}) glycoprotein receptors trkA (for NGF), trkB (for BDNF) and trkC (for NT-3) or the low affinity (Kd -10^{-9}) pan-neurotrophin receptor p75 (NTR). Neurotrophin receptors are widely expressed in the peripheral and the central nervous system as well as on cells of the immune system (Lewin and Barde, 1996).

SOURCES OF NEUROTROPHINS IN ALLERGIC DISEASE

Based on their expression profile, neurotrophins are excellent candidates for mediating immune–nerve cell interactions. During the inflammatory processes, NGF is produced by a wide range of immune cells including mast cells, macrophages, T cells and B cells (for review see Braun, 2000). Analysis of a murine model of allergic airway inflammation revealed that T cells, B cells and macrophages represent sources of enhanced NGF production (Table 14.5). In vitro, allergen stimulation of mononuclear cells from sensitized mice resulted in enhanced NGF synthesis (Braun et al., 1998). In addition, NGF production was enhanced by antigen stimulation in murine and human Th2 cell clones (Otten et al., 1994; Otten and Gradient, 1995; Lambiase et al., 1997). BDNF synthesis has been detected in activated human T cells, B cells, macrophages, mast cells and in platelets

TABLE 14.5
Neurotrophin production during allergic airway
inflammation in the mouse.

Cellular source	NGF	BDNF
Airway epithelium	−	+
Airway lumen		
Macrophages	+	+
T cells	−	−
Airway mucosa		
T cells	+	+
B cells	+	−

(Radka *et al.*, 1996; Braun *et al.*, 1999a; Kerschensteiner *et al.*, 1999). In addition to a constitutive production of BDNF by respiratory epithelial cells, we have demonstrated that activated murine macrophages and T cells, but not B cells, produce BDNF during allergic inflammation (Braun *et al.*, 1999a). Histological analysis of the inflamed lung revealed strong NGF and BDNF production by cells within the peribronchial inflammatory infiltrate (Braun *et al.*, 1998, 1999a). As of recently, there is also evidence for enhanced neurotrophin production in allergic patients.

Patients with allergic BA display increased levels of NGF in serum and BAL (Bonini *et al.*, 1996, 1999; Undem *et al.*, 1999). Increased neurotrophin production in response to allergen provocation was demonstrated in airways of subjects with allergic rhinitis and mild allergic asthma (Virchow *et al.*, 1998; Sanico *et al.*, 2000). After segmental allergen provocation in mildly asthmatic patients, the neurotrophin content in BAL increased markedly in allergen exposed lung segments as opposed to in saline exposed control segments. Notably, this upregulation was seen during the allergic late phase response but not in the early phase (Virchow *et al.*, 1998).

NEURONAL PLASTICITY AND AHR IN RESPONSE TO NEUROTROPHINS

Neuronal plasticity in the peripheral nervous system is as of yet not well characterized. For sensory neurons, the lowering of the activation threshold, changes in the processing of information and altered neurotransmitter synthesis are possible mechanisms. To some extent inflammation-induced hyperalgesia shows remarkable similarities to airway hyper-responsiveness, particularly with respect to the effects of neurotrophins. Hyperalgesia can be defined by a decrease in the threshold for painful stimuli and heightened reflex pathways in sensory neurons (Carr *et al.*, 2001). It is well established that neurotrophins play a central role in inflammation induced hyperalgesia (Safieh-Garabedian *et al.*, 1995; Woolf *et al.*, 1997).

There is some evidence suggesting that sensory neurons innervating the lung are also responsive to neurotrophins since local increase of neurotrophins in the lung could mediate similar neuronal changes in animal models as seen during allergic inflammation (Undem *et al.*, 1999; Hunter *et al.*, 2000). It has been well established that visceral sensory neurons localized in the nodose and dorsal root ganglia require neurotrophins for survival during development (Snider, 1994). In adults, functional properties of neurons are also affected by neurotrophins (Chalazonitis *et al.*, 1987). NGF was shown to upregulate neuropeptide production in sensory neurons and to contribute to inflammatory hypersensitivity (Donnerer *et al.*, 1992). Though cultured nodose ganglion neurons do not require NGF for survival, their SP production is regulated by NGF (MacLean *et al.*, 1988). In transgenic mice over-expressing NGF in airway restricted Clara-cells, a marked sensory and sympathetic hyper-innervation and increased neuropeptide content was observed in projecting sensory neurons (Hoyle *et al.*, 1998). In addition, these mice demonstrated AHR in response to capsaicin. In a guinea pig model, tracheal injection of NGF induced SP production in mechanically sensitive 'Aδ' fibres that do not produce SP under physiological conditions (Hunter *et al.*, 2000). These NGF mediated effects are comparable to neuronal changes observable during allergic inflammation (Undem *et al.*, 1999). The induction of neuropeptides in mechanically sensitive neurons may lead to exaggerated reflex responses to innocuous stimuli (Hunter *et al.*, 2000). In a murine model of allergic airway inflammation, we were able to demonstrate that blocking of NGF by local treatment with anti-NGF antibodies prevented the development of AHR (Braun *et al.*, 1998). Therefore, it is not surprising that NGF treatment induces AHR in the guinea pig. Along this line, de Vries and colleagues could block NGF induced AHR to histamine with the NK-1 specific tachykinin antagonist SR 140333, thus pointing again to the central role played by tachykinins in this condition (de Vries *et al.*, 1999). Taken together, these data provide further evidence that neurotrophins are central signalling molecules in immune-nerve cell communication as it occurs in pathophysiological conditions including BA (Braun *et al.*, 2000).

IMMUNOLOGICAL PLASTICITY IN RESPONSE TO NEUROTROPHINS

In addition to the effects of neurotrophins on neuronal plasticity, there is growing evidence as well for a sustained action of neurotrophins on immune cells involved in allergic inflammation, including mast cells, eosinophils, B cells and T cells. One of the first reports demonstrating that neurotrophins can modulate immune cell activities came from Aloe *et al.* After injection of NGF into neonatal rats, they observed an increased number of mast cells in these animals (Aloe and Levi Montalcini, 1977). Further studies characterized NGF as an important growth and differentiation factor for mast cells and basophils (Matsuda *et al.*, 1991; Kannan *et al.*, 1993; Bürgi *et al.*, 1996; Tam *et al.*, 1997). Since NGF is the best characterized neurotrophin, most of the available data pertains to it. NGF stimulates rapid degranulation of mast cells and basophils (Bischoff and Dahinden, 1992; Horigome *et al.*, 1993, 1994), promotes differentiation, activation and cytokine production of mast cells, granulocytes and macrophages (Matsuda *et al.*, 1991; Kannan *et al.*, 1993; Susaki *et al.*, 1996; Welker *et al.*, 1998), activates eosinophils (Hamada *et al.*, 1996), promotes proliferation of B- and T-cell subsets (Thorpe and Perez Polo, 1987; Otten *et al.*, 1989), enhances Th-2 cytokine production and IgE synthesis in a murine asthma model (Braun *et al.*, 1998) and induces differentiation of activated B cells in Ig-secreting plasma cells (Brodie and Gelfand, 1994). It needs to be emphasized that the majority of NGF effects have been observed in pre-activated cells. NGF by itself does not appear to activate the immune cells in physiologically relevant concentrations, but rather modulates their threshold to other triggering stimuli (Levi Montalcini *et al.*, 1996). Most of these investigations, however, were performed *in vitro*. Therefore, the physiological function of NGF *in vivo* remains to be elucidated. Compared to NGF, there are few data available about possible functions of other neurotrophins in the immune system. The presence of a variety of neurotrophin receptors on developing and mature immune cells, however, suggests that the effects of other neurotrophins have to be considered as well (Table 14.6).

Recent studies demonstrated the expression of neurotrophins and their receptors in bone marrow and thymus cells. Since bone marrow and thymus are the preferential organs for immune cell maturation and differentiation, this suggests neurotrophins play a role

TABLE 14.6
NGF-receptor expression and NGF effects on human immune cells.

Cell type	trkA	p75NTR	Effects
T cells	+	−	c-fos transcription ↑, activation ↑, proliferation ↑, cytokine production↑
B cells	+	+	Survival ↑, proliferation ↑, immunoglobulin production ↑
Monocytes/macrophages	+	+/−	Respiratory burst ↑
Eosinophils	+	+	Viability ↑, chemotaxis ↑, cytotoxity ↑
Mast cells	+	−	Histamine release ↑, proliferation ↑, differentiation, cytokine production↑
Basophils	+	−	Priming ↑

−, not detected; +, detected; +/−, conflicting data.

in immune cell differentiation as well (Laurenzi *et al.*, 1998; Labouyrie *et al.*, 1999). Findings of higher neurotrophin transcript levels at fetal as compared to at adult stages further support such a notion (Labouyrie *et al.*, 1999). Similar results were demonstrated in the thymus, where trkB on T cells inversely correlated with stages of maturation (Maroder *et al.*, 1996). Therefore, Maroder *et al.* hypothesized that stroma cell derived neurotrophins have a direct influence on developing thymocytes (Maroder *et al.*, 1996).

Taken together, these data provide evidence that NGF is involved in the development of 'immunological plasticity'. Immunological plasticity may be defined as a long-lasting change in immunological functions including cell differentiation, mediator production and release, or sensitivity to activating stimuli. Since the immune and nervous systems share many features including recognition of and reaction to unknown stimuli, it is not surprising that both systems use similar mediators and mechanisms to perform their tasks.

NEURONAL EFFECTS OF EFFECTOR CELL DERIVED MEDIATORS

In addition to the neuroactive mediators released by lymphocytes, many mediators of effector cells, including mast cells, macrophages and eosinophils, were shown to influence neuronal functions. Mast cells are localized at the interface of the internal and external environment and are often closely approximated to nerve endings within the lung where they respond to allergens and other exogenous stimuli. Upon stimulation, mast cells release: (a) neuroactive mediators including histamine; (b) arachidonic acid metabolites such as prostaglandin (PG)D2 and cysteinyl leukotrienes (LT), for example, LTC4; (c) thromboxane (Tx)A2; and (d) an array of lipid mediators like platelet activating factor (PAF) as well as growth factors, cytokines, and chemokines (Holgate, 2000) (Table 14.7).

TABLE 14.7
Effects of effector cell mediators on sensory neurons.

Mediator	Receptor	Effect
Histamine	H1 (upregulation following nerve damage) H3	Stimulation of acetylcholine release (H1), low dose:inhibition of the release of neuropeptides (H3)
Leukotriens	LT-receptors	Release of neuropeptides from sensory nerves
Prostaglandins	EP receptor	Sensitize C fibres, lowering threshold of firing, release of tachykinins
Platelet activating factor (PAF)	PAF receptor	Activates C fibres, upregulation of H1 receptors
Major basic protein (MBP)		Demyelination, inhibition of muscarinic M2 receptor, stimulation of sensory neurons, tachykinin release
Eosinophilic cationic protein (ECP)		Demyelination
Eosinophil peroxidase (EPO)		Demyelination, inhibition of muscarinic M2 receptor
Eosinophil derived neurotoxin (EDN)		Demyelination

In addition to its well-established effects on blood vessels and smooth muscle layers, histamine was also described to act on neurons. Three types of histamine receptors, H1–H3, have so far been recognized. The effect of histamine on cholinergic nerves is mediated, in part, by stimulation of acetylchloline release from postganglionic nerve terminals, suggesting that histamine acts on prejunctional H1 receptors (Barnes, 1991). Surprisingly, histamine H3 receptors inhibit the release of neuropeptides from NANC nerves (Ichinose and Barnes, 1989). In human airways, a low dose of histamine may act on H3 receptors expressed on cholinergic nerve terminals and ganglia to inhibit neurotransmission and, thus, prevents the activation of bronchoconstrictory reflexes. In contrast, massive release of histamine by allergen-induced IgE-crosslinking results in the activation of H1 receptors on endothelial and smooth muscle cells and ultimately leads to bronchoconstriction (Ichinose et al., 1990).

Leukotrienes are known to have profound biochemical and physiological effects in inflammation, including development of tissue edema, secretion of bronchial mucus from submucosal glands and direct smooth muscle constriction, all together contributing to airway obstruction. Leukotrienes act via two major classes of LT receptors, the BLT receptors – activated by LTB4 – and the cysteinyl-LT1–2 receptors activated by cys-LT. At least in guinea pigs, LTD4 stimulates the release of neuropeptides from sensory nerves (Ayala et al., 1988).

Prostaglandins such as PGD2, PGF2a and TxA2 are potent bronchoconstrictors as well. PGD2 functions as a mast cell-derived mediator to trigger asthmatic responses and elicits its biological action through interaction with the PGD receptor (DP). Presumably mediated via an EP receptor, PGE2 inhibits the cholinergic nerve activation, suggesting an inhibitory effect on ACh release. PGE2 is known to sensitize C-afferent sensory nerve fibres and lower their threshold of firing (Ellis and Conanan, 1996). Furthermore, PGE2 stimulates renal pelvic sensory nerves to release tachykinins like SP (Kopp and Cicha, 1999).

PAF has long been implied to take part in the pathophysiology of BA. PAF activates airway sensory C-fibres and is therefore able to increase AR (Perretti and Manzini, 1993). Moreover, PAF upregulates the expression of H1 receptor mRNA in trigeminal ganglia, thus pointing to a role as a neuroactive mediator (Nakasaki et al., 1999). In an animal model, PAF was shown to be an extremely potent mediator that elicits smooth muscle contraction at least in parts by postganglionic activation of parasympathetic nerves (Leff et al., 1987).

Nitric oxide (NO) is one of the major neurotransmitters of the i-NANC system (Belvisi et al., 1992, 1995). NO is formed during the conversion of L-arginine and oxygen to L-citrulline by the constitutively expressed form of the enzyme nitrite oxide synthase (cNOS). After production, NO is released by simple diffusion. NOS-containing nerves were identified in tracheal and bronchial smooth muscle, around submucosal glands and blood vessels (Ward et al., 1995). Allergen stimulation induces the synthesis of NO in alveolar macrophages. NO may be co-released with acetylcholine and VIP. NO possesses potent smooth muscle relaxing properties and is a neurotransmitter of bronchodilatation in human airways. NO may, however, also play a role in causing broncho-obstruction by increasing plasma exudation and amplifying the asthmatic inflammatory response. Pro-inflammatory cytokines and oxidants increase the expression of an inducible form of NOS in airway epithelial cells in asthma and this may be the reason for increased levels of NO found in the exhaled air of asthmatic patients (Barnes, 1996). Therefore, NO plays an important physiologic role in the airways by regulating bronchoconstriction and vascular tone (Belvisi et al., 1992, 1993).

There is considerable evidence suggesting that lymphocyte-derived cytokines control the eosinophilic infiltrate within the lungs of asthmatic individuals. A crucial element of allergic inflammation is the activation of eosinophils and the subsequent release of a variety of neuroactive mediators including cytokines, toxic basic proteins (such as major basic protein (MBP), eosinophil cationic protein (ECP), eosinophil derived neurotoxin (EDN), eosinophil peroxidase (EPO)) and lipid mediators including PGE2, PAF, as well as reactive oxygen species (Hamid and Minshall, 2000).

After allergen challenge, airway nerves are surrounded by infiltrating eosinophils (Costello et al., 1997). Considerable evidence supports the eosinophils' ability to inflict tissue damage in BA. MBP, ECP, EDN, and EPO damage airway epithelium, demyelinate nerve fibres and though contribute to development of airway hyper-responsiveness (Sunohara et al., 1989). There also is some evidence that eosinophils increase release of acetylcholine from the parasympathetic nerves. In a guinea pig model, after antigen challenge, eosinophils are actively recruited to the airway nerves, possibly through expression of chemotactic substances and adhesion molecules by the nerves. Activated eosinophils release eosinophil major basic protein (MBP) and eosinophil peroxidase (EPO), an endogenous antagonist for M(2) muscarinic receptors. The M(2) muscarinic receptors on the parasympathetic nerves in the lungs normally inhibit release of acetylcholine. When M(2) receptors are blocked by MBP. MBP inhibited specific binding of [3M]–*N*-methyl–scopolamine to M2 receptors, but not to M3 receptors. MBP also inhibited atropine-induced dissociation of scopolamine–receptor complex in a dose-dependent fashion, demonstrating that the interaction of MBP with M2 receptor is allosteric. EPO also inhibited specific ligand binding to M2 receptors, but not to M3, however, this effect was less potent on a molar basis. The M2 receptors on the parasympathetic nerves in the lung normally inhibit release of acetylcholine. When M2 receptors are blocked, natylcholine release is increased, resulting in airway hyper-responsiveness. Antibodies to MBP, as well as antibodies to the adhesion molecule very late antigen 4 (VLA-4) which presents eosinophils migration, block the develpment of airway hyper-responsiveness. Furthermore, MBP and EPO release by eosinophils is triggered by tachykinins. Treatment with NK-1 receptor antagonists which represents the receptor for tachykinins on eosinophils, also blocks development of AHR.

In response to stimulation, eosinophils undergo a respiratory burst accompanied with the release of super oxide ions and H_2O_2 (Nagata *et al.*, 1995). These reactive oxygen species induce an imbalance between muscarinic receptor-mediated contraction and the β-adrenergic-mediated relaxation of the pulmonary smooth muscle, thus resulting in bronchial hyper-responsiveness (Doelman and Bast, 1990).

EFFECT OF NEUROPEPTIDES ON IMMUNE CELLS

A substantial number of studies provide evidence that neuropeptides and neuromediators released from nerve endings of the parasympathetic, sympathetic, and sensory system directly influence immune cells and, thus, participate in immunomodulation (Table 14.8).

The major neurotransmitter of parasympathetic nerves is ACh, which exerts its biological activities via binding to the acetylcholine receptors. These receptors are classified as M1–M3 on the basis of functional inhibition by antagonists. The muscarinic receptors M1–M2 are expressed on T cells and their activation augments anti-CD3-induced mRNA expression of IL-2 and IL-2 receptors as well as T-cell proliferation (Fujino *et al.*, 1997). Macrophages express the M3 receptor and subsequent stimulation with Ach triggers the release of chemotactic substances by alveolar macrophages (Sato *et al.*, 1998).

Sympathetic nerves predominantly release NA and NPY. NA binds to α- and β-adrenergic receptor subtypes expressed on T cells, monocytes, and mast cells (Carstairs *et al.*, 1985; Kammer, 1988; Woiciechowsky *et al.*, 1998). In contrast to activation of muscarinic receptors, IL-2 production and thus, the proliferation of T cells are inhibited by β-adrenergic receptor activation (Kammer, 1988). Catecholamine-induced release of IL-10 from unstimulated monocytes appears to be rapid and direct, without involvement of immunological costimulation (Woiciechowsky *et al.*, 1998). Mast cells in the human lung bear the $β_2$-adrenergic receptor subtype (Carstairs *et al.*, 1985). NPY serum levels have been demonstrated to be increased during acute exacerbation of asthma (Cardell *et al.*, 1994). NPY is able to modulate immune cell functions such as T cell adhesion to the extracellular fibronectin matrix which is mediated by integrin expression via NPY receptors (Levite *et al.*, 1998). NPY was shown to induce Th2 cytokine release from a Th1 cell line and Th1 cytokines from a Th2 cell line breaking the commitment of T cell effector populations (Levite, 1998).

TABLE 14.8
Direct neurogenic effects on inflammation.

Neurotransmitter	Target	Effect
		Pro-inflammatory
ACh	T cell	IL-2/IL-2R expression and IL-2 release, proliferation
	Macrophage	Mediator release
CGRP	Lymphocyte	Chemotaxis and adhesion
	Eosinophil	Chemotaxis and adhesion
NO	T cell	Stimulation of Th 2 phenotype
		Inhibition of Th1 phenotype
SP	Lymphocyte	Stimulation
	T cell	Proliferation, chemotaxis, cytokine production, modulation of Th1/Th2 phenotype switching
	B cell	Differentiation, immunoglobulin switch
	Eosinophil	Migration, recruitment
	Neutrophil	Chemotaxis, superoxide production, adhesion
	Monocyte	TNF-α, IL-1, IL-6, and IL-10 release
		Anti-inflammatory
SOM	B cell	Inhibition of IgE production
	Basophil	Inhibition of mediator release
	T cell	Modulation of Th1/Th2 phenotype switching, adhesion
	Eosinophil	Increase adhesion and infiltration
NA	T cell	Inhibition of IL-2 release
	Monocyte	Stimulation IL-10 release
NPY	T cell	Th1/Th2 phenotype switching
VIP	T cell	Inhibition of IL-2, IL-4, and IL-10 production
	B cell	Inhibition of IgE secretion
	Mast cell	Inhibition of mediator release by mast cells
	Leukocyte	Chemotactic

Immunoactive secretory products from the i-NANC system include mediators such as VIP and NO, which appears to be the major neurotransmitter in this system (Belvisi *et al.*, 1992). Under normal conditions in mouse lungs, VIP receptors are localized on alveolar macrophages. Immunized and intratracheally challenged mice demonstrated elevated levels of VIP and VIP receptor expression on mononuclear cells and neutrophils in perivascular, peribronchiolar and alveolar inflammatory infiltrates (Kaltreider *et al.*, 1997). Direct immunologically effects of VIP (reviewed in Bellinger *et al.*, 1996) include inhibition of T cell proliferation, IL-2, IL-4 and IL-10 cytokine production, inhibition of IgE release by B cells and inhibition of mediator release from mast cells. Once produced, NO passes through membranes by simple diffusion and direct activation of soluble guanylate cyclase. A role of NO has been implied in skewing T lymphocytes towards a Th2 phenotype by inhibition of Th1 cells and their product IFN-γ (Taylor-Robinson *et al.*, 1994).

e-NANC system associated neuropeptides with immunomodulatory functions are somatostatin (SOM), calcitonin gene-related peptide (CGRP) and the members of the tachykinin family, SP and NKA, which all act via G protein-coupled receptors and thus share several effector functions. Specific receptors for CGRP and SOM have been demonstrated on monocytes, B cells, and T cells (McGillis *et al.*, 1991; van Hagen *et al.*, 1994). Similar to NPY, CGRP and SOM have the capacity to induce T cell adhesion to fibronectin and to drive distinct Th1 and Th2 populations to an atypical expression pattern of Th2 or Th1 cytokines respectively and, therefore, break the commitment to a distinct T helper phenotype (Levite, 1998; Levite *et al.*, 1998).

Somatostatin exerts various inhibitory functions on immune responses via specific receptor activation (reviewed in van Hagen *et al.*, 1994). SOM affects the suppression of Ig production in B cells, including IgE (Kimata *et al.*, 1992), modulation of lymphocyte proliferation (inhibitory effect at low concentrations or stimulatory effect at high concentrations) and reduction of (peritoneal) eosinophil infiltration in experimentally induced hypereosinophilia. Furthermore, SOM inhibits SP-induced mucus secretion in rats (Wagner *et al.*, 1995).

Calcitonin gene-related peptide inhibits SP-induced super oxide production in neutrophils and the proliferation as well as the antigen presentation by peripheral mononuclear cells (Tanabe *et al.*, 1996). It also stimulates chemotaxis and adhesion of lymphocytes (Levite *et al.*, 1998) and causes eosinophilia in the rat lung (Bellibas, 1996).

The tachykinin family includes neuropeptides such as SP, NKA, and NKB, which all bind to the three neurokinin receptors NK-1–NK-3 with different affinity. Although all tachykinins possess a high affinity and poor selectivity for their receptors, NK-1 shows highest affinity for SP, NK-2 for NKA, and NK-3 for NKB (Regoli *et al.*, 1994). In the airways, NK-1 receptors are primarily responsible for mediating the inflammatory effects of tachykinins, while NK-2 receptors represent the main mediators of bronchoconstriction (Joos *et al.*, 1987). This is in line with NK-1 receptor expression on immune cells such as B cells, T cells (Braun *et al.*, 1999b), monocytes, macrophages (Ho *et al.*, 1997), eosinophils and neutrophils (Iwamoto *et al.*, 1993). Like VIPR, NK-1R expression by lung-infiltrating leukocytes in systemically and subsequently intratracheally allergen challenged mice is strongly elevated (Kaltreider *et al.*, 1997). Since SP binds to NK-1 with the highest affinity, it is the predominant mediator of immunomodulatory effects among tachykinins. The actvities of SP on immune cells include a broad range of functional responses from neutrophils, eosinophils, mast cells, monocytes/macrophages and lymphocytes (reviewed in van der Velden and Hulsmann, 1999a). SP stimulates a number of neutrophil functions, including chemotaxis, superoxide production and adherence to epithelium and endothelium. Most of these effects require high concentrations of SP whereas at low doses, SP primes the response to other stimuli that otherwise would be ineffective. SP has a degranulating effect on eosinophils and induces human eosinophil migration *in vitro*. In an *in vivo* study with allergic rhinitis patients, it was shown that SP administered after repeated allergen challenge enhanced the recruitment of eosinophils. It has been demonstrated that SP can cause histamine release from human lung mast cells (Heaney *et al.*, 1995). This is underlined by an *in vitro* model using trachea from the SP-hyper-responsive Fisher 344 rat, in which SP stimulation of mast cells represented a major factor leading to bronchoconstriction (Joos *et al.*, 1997). Moreover, SP activates monocytes to release inflammatory cytokines, including TNF-α, IL-1, IL-6 and IL-10. In lymphocytes, SP inhibits glucocorticoid-induced thymocyte apoptosis (Dimri *et al.*, 2000), stimulates proliferation, cytokine production, chemotaxis (van der Velden and Hulsmann,

1999a) and a Th1/Th2 phenotype switch in T cells (Levite, 1998; Levite *et al.*, 1998) and induces differentiation and immunoglobulin switching in B cells (Braun *et al.*, 1999b).

Taken together, the accumulated evidence now points towards direct effects of neurotransmitters and neuropeptides on immune cells. A role of these neurogenic immunodulation in asthma is very likely, but since most of these data were obtained *in vitro* or in animal models, the clinical relevance of these observations requires further evaluation.

CONCLUSIONS

This review summarizes the current literature on immune–nerve cell interaction with respect to allergy and asthma. There is emerging evidence that many cell types and mediators are involved in this complex bi-directional network between immune and nervous system (Figure 14.1). Interaction between the two systems occurs on several levels: (1) Lung and airways are innervated via cholinergic, non-cholinergic, adrenergic and non-adrenergic pathways. (2) The neurotrophins NGF and BDNF are produced in increasing concentration by both immune and non immune cells in the asthmatic patient. (3) Neurotrophins affect the pathophysiology of asthma in several ways. The predominant effect on peripheral nerves is described by the term 'neuronal plasticity'. This is defined as qualitative and/or

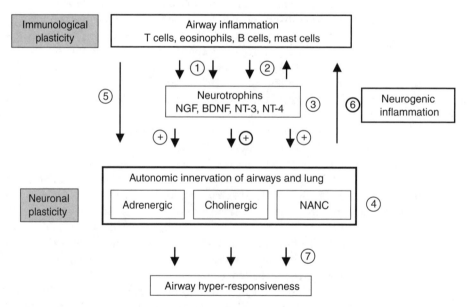

Figure 14.1   Tentative model of bi-directional interaction between neurons and immune cells in lung inflammation. (1) Neurotrophins are produced by immune cells during allergic immune response. (2) Neurotrophins amplify the immunological response; 'immunological plasticity'. (3) Neurotrophins mediate 'neuronal plasticity'. (4) Lung and airways are innervated by adrenergic, cholinergic and non-adrenergic non-cholinergic (NANC) neurons. (5) Inflammatory mediators released by T cells and effector cells (e.g. mast cells and eosinophils) modulate neuronal functions. (6) Direct immune modulatory effects of neurotransmitters (e.g. substance P and acetylcholine) on immune cells. (7) Altered neuronal function results in airway hyper-responsiveness.

quantitative changes in the functional activity and capacity of peripheral neurons. One result of these alterations is the development of airway hyper-responsiveness in bronchial asthma. (4) In parallel, neurotrophins also exhibit profound effects on immune cells residing in airways and lung tissue. These effects are described by the term 'immunological plasticity'. In this regard, neurotrophins act as amplifiers of the locally occurring immune dysbalance. It is important to note that these effects of neurotrophins are not immediate, but rather require some time to fully materialize. In this context, we propose the concept that aggravating and amplifying the pathology in bronchial asthma with all clinical consequences neurotrophins act as intermediate and long-term modulators of neuronal and immune functions.

In parallel to the above-described level of interaction, another level of bi-directional communication exists, which is mediated by cytokines and neuropeptides. Several pro-inflammatory mediators released by eosinophils, mast cells and macrophages during the allergic response have profound effects on many airway and lung innervating neurons. Vice versa, neuropeptides and tachykinins released by these nerves following activation also effect the functional activity of immunological effector cells. Since the immediate acting signals of immune- and neurotransmitters usually have a very short half-life time, they exhibit their functional capacity particularly in the local micro-environment at the site of release in a short-term fashion.

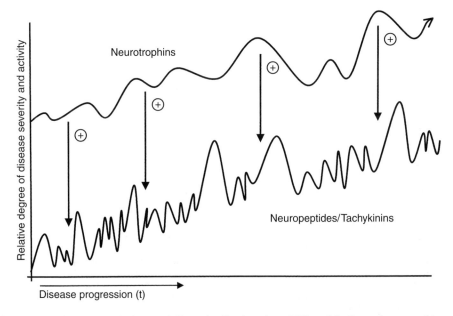

Figure 14.2   Time course of relevant mediators in allergic asthma. Differential effects of neurotrophins and neuropeptides (tachykinins) on disease activity, severity and progression. Upon stimulation, increased neurotrophin release can be detected over a relatively long period of time (weeks). They are not only detectable in the local environment of inflammation (e.g. BAL), but also systemically in the blood. They affect the level and degree of immune activation (immunological plasticity) and nerve cell activity (neuronal plasticity) in a long-term fashion. In contrast, tachykinins and neuropeptides have a relatively short half-life time upon release. They act immediately in the local environment and, therefore, contribute to the acute inflammatory response (neurogenic inflammation).

This complex situation leaves us with the concept of multi-dimensional interaction between the immune and the nervous system. Whereas the neurotrophins are conveyors of long-lasting and long-term signals between the two systems, many pro-inflammatory cytokines and the neuropeptides belong to another group of mediators which deliver short-term and immediate-type signals, acting in the local micro-environment of airway inflammation (Figure 14.2).

We are just at the beginning of unravelling the complex interaction between neural and immune system. The concepts developed above now await further exploration, particularly in suitable *in vivo* models. Ultimately, this concept requires to be proven using suitable intervention strategies.

ACKNOWLEDGEMENTS

This work was supported by the Volkswagen Foundation and the Deutsche Forschungsgemeinschaft DFG.

REFERENCES

Aloe L, Levi Montalcini R. Mast cells increase in tissues of neonatal rats injected with the nerve growth factor. *Brain Res* 1977; 133: 358–366.

Andersson RG, Grundstrom N. The excitatory non-cholinergic, non-adrenergic nervous system of the guinea-pig airways. *Eur J Respir Dis Suppl* 1983; 131: 141–157.

Ayala LE, Choudry NB, Fuller RW. LTD4-induced bronchoconstriction in patients with asthma: lack of a vagal reflex. *Br J Clin Pharmacol* 1988; 26: 110–112.

Barnes PJ. Asthma as an axon reflex. *Lancet* 1986; 1: 242–245.

Barnes PJ. Histamine receptors in the lung. *Agents Actions Suppl* 1991; 33: 103–122.

Barnes PJ. Overview of neural mechanisms in asthma. *Pulm Pharmacol* 1995; 8: 151–159.

Barnes PJ. NO or no NO in asthma? *Thorax* 1996; 51: 218–220.

Baumgarten CR, Witzel A, Kleine Tebbe J, Kunkel G. Substance P enhances antigen-evoked mediator release from human nasal mucosa. *Peptides* 1996; 17: 25–30.

Bellibas SE. The effect of human calcitonin gene-related peptide on eosinophil chemotaxis in the rat airway. *Peptides* 1996; 17: 563–564.

Bellinger DL, Lorton D, Brouxhon S, Felten S, Felten DL. The significance of vasoactive intestinal polypeptide (VIP) in immunomodulation. *Adv Neuroimmunol* 1996; 6: 5–27.

Belvisi MG, Stretton CD, Yacoub M, Barnes PJ. Nitric oxide is the endogenous neurotransmitter of bronchodila-tor nerves in humans. *Eur J Pharmacol* 1992; 210: 221–222.

Belvisi MG, Ward JK, Mitchell JA, Barnes PJ. Nitric oxide as a neurotransmitter in human airways. *Arch Int Pharmacodyn Ther* 1995; 329: 97–110.

Bischoff SC, Dahinden CA. Effect of nerve growth factor on the release of inflammatory mediators by mature human basophils. *Blood* 1992; 79: 2662–2669.

Bonini S, Lambiase A, Bonini S, Angelucci F, Magrini L, Manni L, Aloe L. Circulating nerve growth factor levels are increased in humans with allergic diseases and asthma. *Proc Natl Acad Sci USA* 1996; 93: 10955–10960.

Bonini S, Lambiase A, Bonini S, Levi-Schaffer F, Aloe L. Nerve growth factor: an important molecule in aller-gic inflammation and tissue remodelling. *Int Arch Allergy Immunol* 1999; 118: 159–162.

Bousquet J, Jeffery PK, Busse WW, Johnson M, Vignola AM. Asthma. From bronchoconstriction to airways inflammation and remodeling. *Am J Respir Crit Care Med* 2000; 161: 1720–1745.

Braun A, Appel E, Baruch R, Herz U, Botchkarev V, Paus R, Brodie C, Renz H. Role of nerve growth factor in a mouse model of allergic airway inflammation and asthma. *Eur J Immunol* 1998; 28: 3240–3251.

Braun A, Lommatzsch M, Mannsfeldt A, Neuhaus-Steinmetz U, Fischer A, Schnoy N, Lewin GR, Renz H. Cellular sources of enhanced Brain-Derived Neurotrophic Factor (BDNF) production in a mouse model of allergic inflammation. *Am J Respir Cell Mol Biol* 1999a; 21: 537–546.

Braun A, Wiebe P, Pfeufer A, Gessner R, Renz H. Differential modulation of human immunoglobulin isotype production by the neuropeptides substance P, NKA and NKB. *J Neuroimmunol* 1999b; 97: 43–50.

Braun A. Neurotrophins a new family of cytokines. *Mod Aspects Immunobiol* 2000; 1: 8–9.

Braun A, Lommatzsch M, Renz H. The role of neurotrophins in allergic bronchial asthma. *Clin Exp Allergy* 2000; 30: 178–186.

Brimijoin S, Lundberg JM, Brodin E, Hokfelt T, Nilsson G. Axonal transport of substance P in the vagus and sciatic nerves of the guinea pig. *Brain Res* 1980; 191: 443–457.

Brodie C, Gelfand EW. Regulation of immunoglobulin production by nerve growth factor: comparison with anti-CD40. *J Neuroimmunol* 1994; 52: 87–96.

Bürgi B, Otten UH, Ochensberger B, Rihs S, Heese K, Ehrhard PB, Ibanez CF, Dahinden CA. Basophil priming by neurotrophic factors. Activation through the trk receptor. *J Immunol* 1996; 157: 5582–5588.

Cardell LO, Uddman R, Edvinsson L. Low plasma concentrations of VIP and elevated levels of other neuropeptides during exacerbations of asthma. *Eur Respir J* 1994; 7: 2169–2173.

Carr MJ, Hunter DD, Undem BJ. Neurotrophins and asthma. *Curr Opin Pulm Med* 2001; 7: 1–7.

Carstairs JR, Nimmo AJ, Barnes PJ. Autoradiographic visualization of beta-adrenoceptor subtypes in human lung. *Am Rev Respir Dis* 1985; 132: 541–547.

Casale TB. *Neurogenic Control of Inflammation and Airway Function*, vol. 1. St. Louis, MO: Mosby, 1996.

Chalazonitis A, Peterson ER, Crain SM. Nerve growth factor regulates the action potential duration of mature sensory neurons. *Proc Natl Acad Sci USA* 1987; 84: 289–293.

Costello RW, Schofield BH, Kephart GM, Gleich GJ, Jacoby DB, Fryer AD. Localization of eosinophils to airway nerves and effect on neuronal M2 muscarinic receptor function. *Am J Physiol* 1997; 273: L93–L103.

Delaunois A, Gustin P, Segura P, Vargas M, Ansay M. Interactions between acetylcholine and substance P effects on lung mechanics in the rabbit. *Fundam Clin Pharmacol* 1996; 10: 278–288.

Di Maria GU, Bellofiore S, Geppetti P. Regulation of airway neurogenic inflammation by neutral endopeptidase. *Eur Respir J* 1998; 12: 1454–1462.

Dimri R, Sharabi Y, Shoham J. Specific inhibition of glucocorticoid-induced thymocyte apoptosis by substance P. *J Immunol* 2000; 164: 2479–2486.

Dmitrieva N, Shelton D, Rice AS, McMahon SB. The role of nerve growth factor in a model of visceral inflammation. *Neuroscience* 1997; 78: 449–459.

Doelman CJ, Bast A. Oxygen radicals in lung pathology. *Free Radic Biol Med* 1990; 9: 381–400.

Donnerer J, Schuligoi R, Stein C. Increased content and transport of substance P and calcitonin gene-related peptide in sensory nerves innervating inflamed tissue: evidence for a regulatory function of nerve growth factor in vivo. *Neuroscience* 1992; 49: 693–698.

Ek M, Kurosawa M, Lundeberg T, Ericsson A. Activation of vagal afferents after intravenous injection of interleukin-1beta: role of endogenous prostaglandins. *J Neurosci* 1998; 18: 9471–9479.

Ellis JL, Conanan ND. Prejunctional inhibition of cholinergic responses by prostaglandin E2 in human bronchi. *Am J Respir Crit Care Med* 1996; 154: 244–246.

Ellis JL, Undem BJ. Antigen-induced enhancement of noncholinergic contractile responses to vagus nerve and electrical field stimulation in guinea pig isolated trachea. *J Pharmacol Exp Ther* 1992; 262: 646–653.

Fischer A, McGregor GP, Saria A, Philippin B, Kummer W. Induction of tachykinin gene and peptide expression in guinea pig nodose primary afferent neurons by allergic airway inflammation. *J Clin Invest* 1996; 98: 2284–2291.

Fox AJ, Lalloo UG, Belvisi MG, Bernareggi M, Chung KF, Barnes PJ. Bradykinin-evoked sensitization of airway sensory nerves: a mechanism for ACE-inhibitor cough. *Nat Med* 1996; 2: 814–817.

Fryer AD, Adamko DJ, Yost BL, Jacoby DB. Effects of inflammatory cells on neuronal M2 muscarinic receptor function in the lung. *Life Sci* 1999; 64: 449–455.

Fujino H, Kitamura Y, Yada T, Uehara T, Nomura Y. Stimulatory roles of muscarinic acetylcholine receptors on T cell antigen receptor/CD3 complex-mediated interleukin-2 production in human peripheral blood lymphocytes. *Mol Pharmacol* 1997; 51: 1007–1014.

Hamada A, Watanabe N, Ohtomo H, Matsuda H. Nerve growth factor enhances survival and cytotoxic activity of human eosinophils. *Br J Haematol* 1996; 93: 299–302.

Hamid QA, Minshall EM. Molecular pathology of allergic disease: I: lower airway disease. *J Allergy Clin Immunol* 2000; 105: 20–36.

van Hagen PM, Krenning EP, Kwekkeboom DJ, Reubi JC, Anker-Lugtenburg PJ, Lowenberg B, Lamberts SW. Somatostatin and the immune and haematopoetic system; a review. *Eur J Clin Invest* 1994; 24: 91–99.

Heaney LG, Cross LJ, McGarvey LP, Buchanan KD, Ennis M, Shaw C. Neurokinin A is the predominant tachykinin in human bronchoalveolar lavage fluid in normal and asthmatic subjects. *Thorax* 1998; 53: 357–362.

Heaney LG, Cross LJ, Stanford CF, Ennis M. Substance P induces histamine release from human pulmonary mast cells. *Cin Exp Allergy* 1995; 25: 179–186.

Herz U, Braun A, Ruckert R, Renz H. Various immunological phenotypes are associated with increased airway responsiveness. *Clin Exp Allergy* 1998; 28: 625–634.

Ho WZ, Lai JP, Zhu XH, Uvaydova M, Douglas SD. Human monocytes and macrophages express substance P and neurokinin-1 receptor. *J Immunol* 1997; 159: 5654–5660.

Holgate ST. The role of mast cells and basophils in inflammation. *Clin Exp Allergy* 2000; 30 Suppl 1: 28–32.

Horigome K, Bullock ED, Johnson EM, Jr. Effects of nerve growth factor on rat peritoneal mast cells. Survival promotion and immediate-early gene induction. *J Biol Chem* 1994; 269: 2695–2702.

Horigome K, Pryor JC, Bullock ED, Johnson EM, Jr. Mediator release from mast cells by nerve growth factor. Neurotrophin specificity and receptor mediation. *J Biol Chem* 1993; 268: 14881–14887.

Horton AR, Bartlett PF, Pennica D, Davies AM. Cytokines promote the survival of mouse cranial sensory neurones at different developmental stages. *Eur J Neurosci* 1998; 10: 673–679.

Howarth PH, Springall DR, Redington AE, Djukanovic R, Holgate ST, Polak JM. Neuropeptide-containing nerves in endobronchial biopsies from asthmatic and nonasthmatic subjects. *Am J Respir Cell Mol Biol* 1995; 13: 288–296.

Hoyle GW, Graham RM, Finkelstein JB, Nguyen KPT, Gozal D, Friedman M. Hyperinnervation of the airways in transgenic mice overexpressing nerve growth factor. *Am J Respir Cell Molecular Biol* 1998; 18: 149–157.

Hunter DD, Myers AC, Undem BJ. Nerve growth factor-induced phenotypic switch in guinea pig airway sensory neurons. *Am J Respir Crit Care Med* 2000; 161: 1985–1990.

Ichinose M, Barnes PJ. Histamine H3-receptors modulate nonadrenergic noncholinergic neural bronchoconstriction in guinea-pig in vivo. *Eur J Pharmacol* 1989; 174: 49–55.

Ichinose M, Belvisi MG, Barnes PJ. Histamine H3-receptors inhibit neurogenic microvascular leakage in airways. *J Appl Physiol* 1990; 68: 21–25.

Iwamoto I, Nakagawa N, Yamazaki H, Kimura A, Tomioka H, Yoshida S. Mechanism for substance P-induced activation of human neutrophils and eosinophils. *Regul Pept* 1993; 46: 228–230.

Jacoby DB, Costello RM, Fryer AD. Eosinophil recruitment to the airway nerves. *J Allergy Clin Immunol* 2001; 107: 211–218.

Joos GF, Germonpre PR, Pauwels RA. Role of tachykinins in asthma. *Allergy* 2000a; 55: 321–337.

Joos GF, Germonpre PR, Pauwels RA. Neural mechanisms in asthma. *Clin Exp Allergy* 2000b; 30 (Suppl 1): 60–65.

Joos GF, Lefebvre RA, Bullock GR, Pauwels RA. Role of 5-hydroxytryptamine and mast cells in the tachykinin-induced contraction of rat trachea in vitro. *Eur J Pharmacol* 1997; 338: 259–268.

Joos G, Pauwels R, van der Straeten M. Effect of inhaled substance P and neurokinin A on the airways of normal and asthmatic subjects. *Thorax* 1987; 42: 779–783.

Kaltreider HB, Ichikawa S, Byrd PK, Ingram DA, Kishiyama JL, Sreedharan SP, Warnock ML, Beck JM, Goetzl EJ. Upregulation of neuropeptides and neuropeptide receptors in a murine model of immune inflammation in lung parenchyma. *Am J Respir Cell Mol Biol* 1997; 16: 133–144.

Kammer GM. The adenylate cyclase–cAMP–protein kinase A pathway and regulation of the immune response. *Immunol Today* 1988; 9: 222–229.

Kannan Y, Matsuda H, Ushio H, Kawamoto K, Shimada Y. Murine granulocyte-macrophage and mast cell colony formation promoted by nerve growth factor. *Int Arch Allergy Immunol* 1993; 102: 362–367.

Kay AB. Pathology of mild, severe, and fatal asthma. *Am J Respir Crit Care Med* 1996; 154: S66–S69.

Kay AB. Allergy and allergic diseases. Second of two parts. *N Engl J Med* 2001a; 344: 109–113.

Kay AB. Allergy and allergic diseases. First of two parts. *N Engl J Med* 2001b; 344: 30–37.

Kerschensteiner M, Gallmeier E, Behrens L, Leal VV, Misgeld T, Klinkert WE, Kolbeck R, Hoppe E, Oropeza-Wekerle RL, Bartke I et al. Activated human T cells, B cells, and monocytes produce brain-derived neurotrophic factor in vitro and in inflammatory brain lesions: a neuroprotective role of inflammation? *J Exp Med* 1999; 189: 865–870.

Kimata H, Yoshida A, Ishioka C, Mikawa H. Differential effect of vasoactive intestinal peptide, somatostatin, and substance P on human IgE and IgG subclass production [published erratum appears in *Cell Immunol* 1993 Jul; 149(2): 450]. *Cell Immunol* 1992; 144: 429–442.

Kopp UC, Cicha MZ. PGE2 increases substance P release from renal pelvic sensory nerves via activation of N-type calcium channels. *Am J Physiol* 1999; 276: R1241–R1248.

Labouyrie E, Dubus P, Groppi A, Mahon FX, Ferrer J, Parrens M, Reiffers J, de Mascarel A, Merlio JP. Expression of neurotrophins and their receptors in human bone marrow. *Am J Pathol* 1999; 154: 405–415.

Lambiase A, Bracci Laudiero L, Bonini S, Bonini S, Starace G, MM DE, De Carli M, Aloe L. Human CD4+ T cell clones produce and release nerve growth factor and express high-affinity nerve growth factor receptors. *J Allergy Clin Immunol* 1997; 100: 408–414.

Larsen GL, Fame TM, Renz H, Loader JE, Graves J, Hill M, Gelfand EW. Increased acetylcholine release in tracheas from allergen-exposed IgE-immune mice. *A J Physiol* 1994; 266: L263–L270.

Laurenzi MA, Beccari T, Stenke L, Sjolinder M, Stinchi S, Lindgren JA. Expression of mRNA encoding neu-rotrophins and neurotrophin receptors in human granulocytes and bone marrow cells–enhanced neu-rotrophin-4 expression induced by LTB4. *J Leukoc Biol* 1998; 64: 228–234.

Leff AR, White SR, Munoz NM, Popovich KJ, Shioya T, Stimler-Gerard NP. Parasympathetic involvement in PAF-induced contraction in canine trachealis in vivo. *J Appl Physiol* 1987; 62: 599–605.

Levi Montalcini R, Dal Toso R, della Valle F, Skaper SD, Leon A. Update of the NGF saga. *J Neurol Sci* 1995; 130: 119–127.

Levi Montalcini R, Skaper SD, Dal Toso R, Petrelli L, Leon A. Nerve growth factor: from neurotrophin to neu-rokine. *Trends Neurosci* 1996; 19: 514–520.

Levite M. Neuropeptides, by direct interaction with T cells, induce cytokine secretion and break the commitment to a distinct T helper phenotype. *Proc Natl Acad Sci USA* 1998; 95: 12544–12549.

Levite M, Cahalon L, Hershkoviz R, Steinman L, Lider O. Neuropeptides, via specific receptors, regulate T cell adhesion to fibronectin. *J Immunol* 1998; 160: 993–1000.

Lewin GR, Barde YA. Physiology of the neurotrophins. *Annu Rev Neurosci* 1996; 19: 289–317.

Lilly CM, Bai TR, Shore SA, Hall AE, Drazen JM. Neuropeptide content of lungs from asthmatic and nonasth-matic patients. *Am J Respir Crit Care Med* 1995; 151: 548–553.

Lindsay RM, Harmar AJ. Nerve growth factor regulates expression of neuropeptide genes in adult sensory neu-rons. *Nature* 1989; 337: 362–364.

Lundberg JM, Hokfelt T, Martling CR, Saria A, Cuello C. Substance P-immunoreactive sensory nerves in the lower respiratory tract of various mammals including man. *Cell Tissue Res* 1984; 235: 251–261.

Mackay TW, Hulks G, Douglas NJ. Non-adrenergic, non-cholinergic function in the human airway. *Respir Med* 1998; 92: 461–466.

MacLean DB, Lewis SF, Wheeler FB. Substance P content in cultured neonatal rat vagal sensory neurons: the effect of nerve growth factor. *Brain Res* 1988; 457: 53–62.

Mannion RJ, Costigan M, Decosterd I, Amaya F, Ma QP, Holstege JC, Ji RR, Acheson A, Lindsay RM, Wilkinson GA *et al*. Neurotrophins: peripherally and centrally acting modulators of tactile stimulus-induced inflam-matory pain hypersensitivity. *Proc Natl Acad Sci USA* 1999; 96: 9385–9390.

Maroder M, Bellavia D, Meco D, Napolitano M, Stigliano A, Alesse E, Vacca A, Giannini G, Frati L, Gulino A *et al*. Expression of trKB neurotrophin receptor during T cell development. Role of brain derived neurotrophic factor in immature thymocyte survival. *J Immunol* 1996; 157: 2864–2872.

Matsuda H, Kannan Y, Ushio H, Kiso Y, Kanemoto T, Suzuki H, Kitamura Y. Nerve growth factor induces devel-opment of connective tissue-type mast cells in vitro from murine bone marrow cells. *J Exp Med* 1991; 174: 7–14.

McGillis JP, Humphreys S, Reid S. Characterization of functional calcitonin gene-related peptide receptors on rat lymphocytes. *J Immunol* 1991; 147: 3482–3489.

Milgrom H, Fick RBJ, Su JQ, Reimann JD, Bush RK, Watrous ML, Metzger WJ. Treatment of allergic asthma with monoclonal anti-IgE antibody. *N Engl J Med* 1999; 341: 1966–1973.

Murphy PG, Ramer MS, Borthwick L, Gauldie J, Richardson PM, Bisby MA. Endogenous interleukin-6 con-tributes to hypersensitivity to cutaneous stimuli and changes in neuropeptides associated with chronic nerve constriction in mice. *Eur J Neurosci* 1999; 11: 2243–2253.

Nagata M, Sedgwick JB, Busse WW. Differential effects of granulocyte-macrophage colony-stimulating factor on eosinophil and neutrophil superoxide anion generation. *J Immunol* 1995; 155: 4948–4954.

Nakasaki T, Masuyama K, Fukui H, Ogino S, Eura M, Samejima Y, Ishikawa T, Yumoto E. Effects of PAF on his-tamine H1 receptor mRNA expression in rat trigeminal ganglia. *Prostaglandins Other Lipid Mediat* 1999; 58: 29–41.

Otten U, Ehrhard P, Peck R. Nerve growth factor induces growth and differentiation of human B lymphocytes. *Proc Natl Acad Sci USA* 1989; 86: 10059–10063.

Otten U, Gadient RA. Neurotrophins and cytokines-intermediaries between the immune and nervous systems. *Int J Dev Neurosci* 1995; 13: 147–151.

Otten U, Scully JL, Ehrhard PB, Gadient RA. Neurotrophins: signals between the nervous and immune systems. *Prog Brain Res* 1994; 103: 293–305.

Ollerenshaw SL, Jarvis D, Sullivan CE, Woolcock AJ. Substance P immunoreactive nerves in airways from asth-matics and nonasthmatics. *Eur Respir J* 1991; 4: 673–682.

Perretti F, Manzini S. Activation of capsaicin-sensitive sensory fibers modulates PAF-induced bronchial hyperre-sponsiveness in anesthetized guinea pigs. *Am Rev Respir Dis* 1993; 148: 927–931.

Radka SF, Holst PA, Fritsche M, Altar CA. Presence of brain-derived neurotrophic factor in brain and human and rat but not mouse serum detected by a sensitive and specific immunoassay. *Brain Res* 1996; 709: 122–301.

Regoli D, Boudon A, Fauchere JL. Receptors and antagonists for substance P and related peptides. *Pharmacol Rev* 1994; 46: 551–599.

Safieh-Garabedian B, Poole S, Allchorne A, Winter J, Woolf CJ. Contribution of interleukin-1 beta to the inflammation-induced increase in nerve growth factor levels and inflammatory hyperalgesia. *Br J Pharmacol* 1995; 115: 1265–1275.

Sanico AM, Stanisz AM, Gleeson TD, Bora S, Proud D, Bienenstock J, Koliatsos VE, Togias A. Nerve growth factor expression and release in allergic inflammatory disease of the upper airways. *Am J Respir Crit Care Med* 2000; 161: 1631–1635.

Sato E, Koyama S, Okubo Y, Kubo K, Sekiguchi M. Acetylcholine stimulates alveolar macrophages to release inflammatory cell chemotactic activity. *Am J Physiol* 1998; 274: L970–L979.

Snider WD. Functions of the neurotrophins during nervous system development: what the knockouts are teaching us. *Cell* 1994; 77: 627–638.

Sont JK, van Krieken JH, van Klink HC, Roldaan AC, Apap CR, Willems LN, Sterk PJ. Enhanced expression of neutral endopeptidase (NEP) in airway epithelium in biopsies from steroid- versus nonsteroid-treated patients with atopic asthma. *Am J Respir Cell Mol Biol* 1997; 16: 549–556.

Sunohara N, Furukawa S, Nishio T, Mukoyama M, Satoyoshi E. Neurotoxicity of human eosinophils towards peripheral nerves. *J Neurol Sci* 1989; 92: 1–7.

Susaki Y, Shimizu S, Katakura K, Watanabe N, Kawamoto K, Matsumoto M, Tsudzuki M, Furusaka T, Kitamura Y, Matsuda H. Functional properties of murine macrophages promoted by nerve growth factor. *Blood* 1996; 88: 4630–4637.

Tam SY, Tsai M, Yamaguchi M, Yano K, Butterfield JH, Galli SJ. Expression of functional TrkA receptor tyrosine kinase in the HMC-1 human mast cell line and in human mast cells. *Blood* 1997; 90: 1807–1820.

Tanabe T, Otani H, Zeng XT, Mishima K, Ogawa R, Inagaki C. Inhibitory effects of calcitonin gene-related peptide on substance-P-induced superoxide production in human neutrophils [published erratum appears in *Eur J Pharmacol* 1997 Feb 19; 321(1): 137–141]. *Eur J Pharmacol* 1996; 314: 175–183.

Taylor-Robinson AW, Liew FY, Severn A, Xu D, McSorley SJ, Garside P, Padron J, Phillips RS. Regulation of the immune response by nitric oxide differentially produced by T helper type 1 and T helper type 2 cells. *Eur J Immunol* 1994; 24: 980–984.

Thier M, Marz P, Otten U, Weis J, Rose-John S. Interleukin-6 (IL-6) and its soluble receptor support survival of sensory neurons. *J Neurosci Res* 1999; 55: 411–422.

Thorpe LW, Perez Polo JR. The influence of nerve growth factor on the in vitro proliferative response of rat spleen lymphocytes. *J Neurosci Res* 1987; 18: 134–139.

Tudoric N, Zhang M, Kljajic-Turkalj M, Niehus J, Cvoriscec B, Jurgovsky K, Kunkel G. Allergen inhalation challenge induces decrease of serum neutral endopeptidase (NEP) in asthmatics. *Peptides* 2000; 21: 359–364.

Undem BJ, Hubbard W, Weinreich D. Immunologically induced neuromodulation of guinea pig nodose ganglion neurons. *J Autonom Nerv Syst* 1993; 44: 35–44.

Undem BJ, Hunter DD, Liu M, Haak-Frendscho M, Oakragly A, Fischer A. Allergen-induced sensory neuroplasticity in airways. *Int Arch Allergy Immunol* 1999; 118: 150–153.

van der Velden VH, Hulsmann AR. Autonomic innervation of human airways: structure, function, and pathophysiology in asthma. *Neuroimmunomodulation* 1999a; 6: 145–159.

van der Velden VHJ, Hulsmann AR. Peptidases: structure, function and modulation of peptide-mediated effects in the human lung. *Clin Exp Allergy* 1999b; 29: 445–456.

Virchow JC, Julius P, Lommatzsch M, Luttmann W, Renz H, Braun A. Neurotrophins are increased in bronchoalveolar lavage fluid after segmental allergen provocation. *Am J Respir Crit Care Med* 1998; 158: 2002–2005.

Virchow JC, Jr., Walker C, Hafner D, Kortsik C, Werner P, Matthys H, Kroegel C. T cells and cytokines in bronchoalveolar lavage fluid after segmental allergen provocation in atopic asthma. *Am J Respir Crit Care Med* 1995; 151: 960–968.

de Vries A, Dessing MC, Engels F, Henricks PA, Nijkamp FP. Nerve growth factor induces a neurokinin-1 receptor- mediated airway hyperresponsiveness in guinea pigs. *Am J Respir Crit Care Med* 1999; 159: 1541–1544.

Wagner U, Fehmann HC, Bredenbroker D, Yu F, Barth PJ, von Wichert P. Galanin and somatostatin inhibition of substance P-induced airway mucus secretion in the rat. *Neuropeptides* 1995; 28: 59–64.

Ward JK, Barnes PJ, Springall DR, Abelli L, Tadjkarimi S, Yacoub MH, Polak JM, Belvisi MG. Distribution of human i-NANC bronchodilator and nitric oxide-immunoreactive nerves. *Am J Respir Cell Mol Biol* 1995; 13: 175–184.

Weinreich D, Moore KA, Taylor GE. Allergic inflammation in isolated vagal sensory ganglia unmasks silent NK-2 tachykinin receptors. *J Neurosci* 1997; 17: 7683–7693.

Weinreich D, Undem BJ, Taylor G, Barry MF. Antigen-induced long-term potentiation of nicotinic synaptic transmission in the superior cervical ganglion of the guinea pig. *J Neurophysiol* 1995; 73: 2004–2016.

Welker P, Grabbe J, Grutzkau A, Henz BM. Effects of nerve growth factor (NGF) and other fibroblast-derived growth factors on immature human mast cells (HMC-1). *Immunology* 1998; 94: 310–317.

Wilder JA, Collie DD, Wilson BS, Bice DE, Lyons CR, Lipscomb MF. Dissociation of airway hyperresponsiveness from immunoglobulin E and airway eosinophilia in a murine model of allergic asthma. *Am J Resp Cell
Mol Biol* 1999; 20: 1326–1334.

Wills-Karp M. Immunologic basis of antigen-induced airway hyperresponsiveness. *Annu Rev Immunol* 1999; 17:
255–281.

Woiciechowsky C, Asadullah K, Nestler D, Eberhardt B, Platzer C, Schoning B, Glockner F, Lanksch WR, Volk
HD, Docke WD. Sympathetic activation triggers systemic interleukin-10 release in immunodepression
induced by brain injury. *Nat Med* 1998; 4: 808–813.

Woolf CJ, Allchorne A, Safieh-Garabedian B, Poole S. Cytokines, nerve growth factor and inflammatory hyperalgesia: the contribution of tumour necrosis factor alpha. *Br J Pharmacol* 1997; 121.
http://research.bmn.com/medline/search/record?uid=97322892: 417–424.

Wong WS, Koh DS. Advances in immunopharmacology of asthma. *Biochem Pharmacol* 2000; 59: 1323–1335.

15 Regulatory Aspects of Neuroimmunology in the Skin

Martin Steinhoff, Sonja Ständer and Thomas A. Luger

Department of Dermatology and Boltzmann Institute for Cell- and Immunobiology of the Skin, University of Münster, 48129 Münster, Germany

The cutaneous nervous system regulates a variety of physiological and pathophysiological biological functions such as cellular development, growth, differentiation, immunity, inflammation, pruritus and tissue repair. Several structures and cells are involved including cutaneous nerve fibres, which release neuromediators and activate specific receptors on resident target cells or transient immune cells in the skin. Cutaneous neuromediators include different biochemical entities. Classical neurotransmitters such as catecholamines and acetylcholine are released from the autonomic nervous system or cutaneous cells to modulate inflammatory or immune functions in the epidermis and dermis via high-affinity receptors. Neuropeptides such as substance P, calcitonin gene-related peptide (CGRP), vasoactive intestinal peptide (VIP), or proopio-melano-corticotropins (POMC) peptides, for example, can be released from both sensory or autonomic nerve fibres to activate a variety of cutaneous cells through high-affinity neuropeptide receptors or by direct activation of intracellular G protein signalling cascades. Proteinases such as tryptase or neutral endopeptidase, for example, inactivate neuropeptides in the extracellular space or at the cell surface thereby terminating neuropeptide-induced inflammatory or immune responses. Proteinase-activated receptors or vanilloid (capsaicin) receptors are recently described receptors which may have a high impact in regulating cutaneous neurogenic inflammation. Together, a close multidirectional interaction between neuromediators, high-affinity receptors and regulatory proteases on nerves, cutaneous cells and transient or permanent immunomodulatory cells are involved to maintain tissue integrity and regulate inflammatory responses in the skin.

KEY WORDS: neuroimmunology; neurotransmitter; autonomic nerve; neuropeptide; proteinase-activated receptor; skin disease.

ABBREVIATIONS: ACE, angiotensin converting enzyme; ACh, acetylcholine; nAChR, nicotinergic acetylcholine receptor; mAChR, muscarinergic acetylcholine receptor; AR, adrenergic receptor; 6-BH4, (6R) L-erythro-5,6,7,8-tetrahydrobiopterin; CGRP, calcitonin gene-related peptide; CNS, central nervous system; CRH, corticotropin releasing hormone; DTH, delayed type hypersensitivity; GPCR, G protein-coupled receptors; HDMEC, human dermal microvascular endothelial cells; ICAM-1, intercellular cell adhesion molecule-1; IL, interleukin; MCR, melanocortin receptor; MSH, melanocyte-stimulating hormone; NE, norepinephrine; NEP, neutral endopeptidase; NF, neurofilament; NGF, nerve growth factor; NK, neurokinin; NO, nitric oxide; NOS, nitric oxide synthase; NPY, neuropeptide Y; PACAP, pituitary adenylate cyclase activating polypeptide; PAR, proteinase-activated receptor; PG, prostaglandin; PNS, peripheral nervous system; POMC, proopio-melano-corticotropin; PTHrP, parathyroid hormone-related protein; SOM, somatostatin; SP, substance P; TNF-α, tumor necrosis factor-α; VCAM-1, vascular cell adhesion molecule-1; VIP, vasoactive intestinal polypeptide; VPACR, vasoactive intestinal polypeptide/pituitary adenylate cyclase activating polypeptide-receptor; VR1, vanilloid (capsaicin) receptor 1.

ARCHITECTURE OF PERIPHERAL NERVES IN THE SKIN

AUTONOMIC NERVES

In contrast to extensive studies about the role of sensory nerves in the skin, only a small amount of attention has been paid on the role of autonomic nerves in the skin under normal or pathophysiological conditions. Mainly animal studies but also observations in patients with neurological disorders have clearly demonstrated that autonomic nerve fibres from sympathetic and parasympathetic neurons contribute to skin homeostasis. Interestingly, autonomic nerves are restricted to the dermis where they innervate blood vessels, arterio-venous anastomoses, lymphatic vessels, erector pili muscles, eccrine glands, apocrine glands and hair follicles. However, acetylcholine can be also produced by epidermal keratinocytes thereby activating muscarinic and nicotinergic acetylcholine receptors (Grando, 1997). These results clearly indicate a receptor-mediated autocrine and/or paracrine loop of epidermally generated neurotransmitters in the skin (see Section titled 'Vasoactive intestinal peptide' for details).

Selective immunohistochemical markers to differentiate autonomic nerves from sensory nerves are neurofilament protein (NF) in hair follicles, atrial natriuretic peptide (ANP), high-affinity choline transporter or tyrosine hydroxylase (Bjorklund *et al.*, 1986; Tainio *et al.*, 1987; Vitadello *et al.*, 1987; Roth and Kummer, 1994; Haberberger *et al.*, 2002). However, the classical neurotransmitters in autonomic nerves are acetylcholine and catecholamines. Cholinergic nerves mainly innervate sweat glands and arteriovenous anastomoses while adrenergic nerves predominantly innervate large blood vessels and erector pili muscles of the hair follicle. Beside this, autonomic nerves are also capable of releasing neuropeptides such as calcitonin gene-related peptide (CGRP), ANP and vasoactive intestinal polypeptitde (VIP), neuropeptide (NPY), tyrosin hydroxylase, and galanin. ANP, for example, may serve a similar role in the skin, since it regulates water and electrolyte balance in different organs and its immunoreactivity is found predominantly in sympathetic cholinergic fibres around sweat glands. VIP triggers sweat secretion from eccrine sweat glands through a cAMP-dependent mechanism (Tainio, 1987) indicating that VIP may be involved in regulating sweating. Similarly, VIP from parasympathetic nerves stimulate adenylate cyclase activity in vascular cells suggesting a regulatory role for this peptide in blood vessel regulation.

Autonomic nerves appear to play an important role in vascular regulation (Michikami *et al.*, 2001). Arterial sections of arteriovenous anastomoses, precapillary sphincters of metarterioles, arteries and capillaries appear to be the most intensely innervated regions. While sensory nerves are important for vasodilatation, neuropeptides from sympathetic neurons such as NPY mediate vasoconstriction. Parasympathetic nerves release acetyl-choline (ACh) and VIP/PHM to mediate vasodilatation through activation of venous sinusoides and adrenergic sympathetic nerves release noradrenaline (NA) and/or NPY to innervate arterioles, arteriovenous anastomoses and venous sinusoides which resulting in vasoconstriction (Brain and Williams, 1985; Wallengren *et al.*, 1995).

Both autonomic as well as sensory nerve fibres appear to be involved in hair follicle cycling (Paus *et al.*, 1997). For example, subcutaneous injections of the NA-depleting agents (guanethidine or 6-hydroxydopamine), but not of the β_2-adrenergic receptor (β_2-AR) agonist isoproterenol induced a premature onset of anagen in a mouse model supporting the concept that sympathetic nerves are intimately involved in hair growth control. These results may lead to novel therapeutical strategies for the treatment of hair growth disorders (Botchkarev *et al.*, 1999; Peters *et al.*, 1999b). In contrast, intact hair follicle innervation does not seem to be essential for anagen induction and development in a mouse model, although a minor modulatory role in depilation-induced hair growth was observed (Maurer *et al.*, 1998). Taken together, autonomic nerve fibres play a major role in cutaneous biology, mainly by regulating vascular and glandular but also immunomodulatory responses.

SENSORY NERVES

Afferent somatic nerves with fine unmyelinated (C-) or myelinated (Aδ-) primary afferent nerve fibres deriving from dorsal root ganglia transmit physiological stimuli such as pain, heat, cold, mechanical stimulation, distension, trauma, or pH changes. These specialized nocieptor fibres can be activated by a small or wide range of stimuli. The exact anatomical–physiological interrelationship of these different nerve fibre subsets, however, has to await further clarification (Schmelz *et al.*, 2000). Sensory nerve fibres release factors which modulate the function of epidermis, blood vessels, hair follicles, sweat glands, or apocrine glands. (Jiang *et al.*, 1998; McArthur *et al.*, 1998; Pergolizzi *et al.*, 1998; Simone *et al.*, 1998; Steinhoff *et al.*, 1999b).

Polymodal C-fibres represent the majority of all cutaneous C-fibres in the skin followed by Aδ-fibres exerting their antidromic effects on several target cells such as keratinocytes, melanocytes, Langerhans cells, Merkel cells, mast cells, endothelial cells and probably transient immunocompetent cells suggesting a close interaction between neuropeptides and the skin. Vice versa, various skin cells are also capable of releasing neuropeptides under physiological circumstances in the skin. Finally, sensory nerves have been demonstrated to be involved in the pathophysiology of several skin diseases such as atopic dermatitis, psoriasis, urticaria, hair disorders and prurigo, for example see Section titled 'Role of autonomic and sensory nerves in pruritus' (reviewed by Scholzen *et al.*, 1998; Luger *et al.*, 2000; Slominski and Wortsman, 2000; Steinhoff *et al.*, 2002b).

BIOLOGY OF ACETYLCHOLINE AND CATECHOLAMINES IN THE SKIN

ACETYLCHOLINE AND ACETYLCHOLINE RECEPTORS

The classical non-peptide neurotransmitters in the skin constitute of acetylcholine, epinephrine and norepinephrine. In the skin, they are generated by both neurons and non-neuronal cells (reviewed by Slominski and Wortsman, 2000) binding to high-affinity receptors in various cell types. Degradation of acetylcholine by acetylcholinesterase can be also generated by several cutaneous cells leading to effective inactivation. While choline acetyltransferase was detected in all epidermal layers of human skin, acetylcholinesterase is more confined to basal keratinocytes (Grando et al., 1996; Ndoye et al., 1998). Acetylcholine has been detected in autonomic nerve fibres (Schmelz et al., 2000), melanocytes (Iyengar, 1989; Lammerding-Koppel et al., 1997) and keratinocytes of human skin (Grando et al., 1993) as well as lymphocytes (Bering et al., 1987) where it may regulate proliferation, adhesion, migration, differentiation, cell viability, cell adhesion and motility.

Acetylcholine and its derivatives effectively stimulate cutaneous cells by activating nicotinergic or muscarinergic cell surface receptors (Peralta et al., 1987). The nicotinic receptors for acetylcholine (nAChR) are transmembrane ion channels formed of $\alpha 1$, $\alpha 2$, β, γ, δ and ϵ subunits. In contrast, muscarinic receptors (mAChR) belong to a subfamily of G protein-coupled receptors (GPCR) with seven transmembrane domains, defined as m1, m2, m3, m4 and m5 receptors. These receptors have all been cloned and well characterized (Grando, 1997; Hulme et al., 1999). Muscarinergic receptors have been detected in melanoma cell lines in vitro (Noda et al., 1998), keratinocytes (Grando, 1997), fibroblasts (Buchli et al., 1999), and endothelial cells.

Human keratinocytes generate acetylcholine, choline acetyltransferase, acetylcholine esterase and both muscarinergic as well as nicotinergic receptors (Grando, 1997). While $\alpha 5$ nACh receptor was detected homogeneously throughout all epidermal layers, the $\alpha 3$-, $\beta 2$-, and $\beta 4$-subunits seem to be restricted to the upper epithelium (Grando et al., 1995). Interestingly, expression of these subtypes appears to be stage-dependent during keratinocyte differentiation (Grando et al., 1996; Grando, 1997; Ndoye et al., 1998). In vitro, nAChRs influence motility, differentiation and cell survival of keratinocytes (Grando, 1997). In contrast, mAChRs appear to be involved in proliferation (Grando, 1997). Nicotinergic and muscarinergic pathways stimulate keratinocyte cell adhesion (Grando, 1997) and influence keratinocyte migration via $\alpha 3$- and $\alpha 7$-nAChRs (Zia et al., 1997). Finally, a recently cloned cholinergic receptor contains a $\alpha 9$-subunit and shows both muscarinergic as well as nicotinergic characteristics (Nguyen et al., 2000).

Keratinocytes also generate various muscarinic receptor subtypes such as the m1, m3, m4, and m5 (Grando, 1997). Additionally, m2 receptor distribution was recently described in rat skin (Haberberger and Bodenbenner, 2000). Thus, ACh may exert various biologic effects in human keratinocytes at different stages of cell differentiation by activating specific subtypes of cholinergic receptors. Together, several observations are in favour of an important role of cholinergic nerves in physiological and pathopysiological processes in the skin, although their precise role of cholinergic mediated cell activation in the skin is far from complete.

CATECHOLAMINES AND ADRENERGIC RECEPTORS

Central and peripheral nerves generate and release catecholamines. This especially pertains to postganglionic sympathetic nerves principally innervating ganglia, blood vessels, and smooth muscle cells. In addition, the human skin is able to synthesize catecholamines, their degrading enzymes and high-affinity receptors.

Catecholamines and their regulating enzymes have been detected in nerve fibres (Katz et al., 1983), keratinocytes (Schallreuter et al., 1993), and melanocytes (Schallreuter et al., 1994). They regulate activity of natural killer cells (Oya et al., 2000), monocytes

(Rouppe van der Voort *et al.*, 2000; Woods, 2000), and induce apoptosis in lymphocytes (Cioca *et al.*, 2000). Vice versa, catecholamine release may also be induced by lymphocytes such as T cells and B cells (Kohm *et al.*, 2000). Norepinephrine (NE) may also serve as an immunomodulatory agent during delayed type hypersensitivity (Dhabhar and McEwen, 1999) supporting a role for this neurotransmitter during cutaneous inflammation.

Keratinocytes generate epinephrine, NE, the co-factor (6R)L-erythro-5,6,7,8-tetrahydro-biopterin (6-BH4) as well as β_2-ARs (Schallreuter, 1997). Thus, keratinocytes are able to synthesize the whole factory needed for catecholamine-induced responses and regulation in the skin. The highest expression of these molecules can be detected at early stages of differentiation correlating with increase of intracellular $[Ca^{2+}]$ in keratinocytes in response to catecholamines (Schallreuter *et al.*, 1996a). These results suggest an important role of this signalling system in epidermal differentiation.

Melanocytes also produce the whole repertoire of the catecholamine system including transmitters, enzymes and receptors and may regulate melanogenesis (Schallreuter *et al.*, 1994). In patients with vitiligo, upregulation of 6-BH4 and monoamine oxidase was observed in keratinocytes which may lead to increased NE levels and β_2-AR density (Schallreuter *et al.*, 1996b).

Adrenergic receptors belong to a subfamily of GPCR, and comprise some of the most intensively studied members of this family of receptor molecules. Although originally identified as α or β-receptors, multiple subtypes are known to exist (ARs). For instance, α_1- and β_2-ARs were detected in keratinocytes and melanocytes of human skin (Schallreuter *et al.*, 1992; Drummond and Lipnicki, 1999). Additionally, β_2-AR agonists inhibited TNF-α release from mast cells (Bissonnette and Befus, 1997) and are potent inhibitors of mast cell function *in vitro* (Suzuki *et al.*, 2000). α_1- and β_2-ARs may also regulate important vascular responses in the skin such as vasoconstriction (Harada *et al.*, 1996).

The pathophysiology of different skin diseases may be caused by variations in AR receptor density or function. Decreased levels of β_2-ARs were observed in lesional and non-lesional skin of psoriasis patients, for example (Steinkraus *et al.*, 1993), while enhanced levels of α_2-AR were found in arterioles of patients with scleroderma (Flavahan *et al.*, 2000). Finally, LPS-induced IL-6 production in human microvascular endothelial cells could be increased by epinephrine as well as NE via β_1- and β_2-ARs (Gornikiewicz *et al.*, 2000).

Together, these reports clearly argue for an important role of autonomic nerves for the regulation of cutaneous inflammatory responses and immunomodulation.

FUNCTION AND PATHOPHYSIOLOGY OF AUTONOMIC NERVES IN THE SKIN

Sympathetic neurons are involved in pain transmission (Levine *et al.*, 1986). Noradrenaline hyperalgesia is mediated through the interaction with sympathetic postganglionic neuron terminals rather than activation of primary afferent nociceptors.

Cutaneous sympathetic neurons play an important role in regulating blood flow and body temperature or act as alert sensors. Different populations of sympathetic neurons innervate proximal arteries compared with more distal cutaneous arteries and can be activated selectively by specific physiological stimuli. These subpopulations of sympathetic neurons can contain characteristic combinations of cotransmitters. Mostly, sympathetic neurons utilize two or more cotransmitters including classical neurotransmitters as well as neuropeptides. As already described, NE, ATP, and NPY are effective vasoconstrictors released from sympathetic nerve fibres in many organs including the skin. Moreover, the distribution of postsynaptic receptors mediating vasoconstriction in response to exogenous agonists varies among different cutaneous vascular substructures. The high degree of heterogenity in innervation pattern and receptor distribution throughout the cutaneous vasculature suggests that sympathetic neurotransmission may also vary between vascular segments. Indeed, recent electrophysiological observations clearly demonstrated differences in the neurotransmission between the main ear artery and small distributing arteries and showed that NE and ATP mediate membrane potential changes in cutaneous arteries. Moreover, video microscopy studies suggest that NE and NPY act as co-transmitters to mediate sympathetic constriction of small arteries (Morris *et al.*, 1999).

There may be anatomical and functional variations of NE-mediated effects among arteries in the skin which may explain tissue- and region-specific responses of blood vessels. For example, nitric oxide synthase (NOS) inhibition was shown to potentiate NE- but not sympathetic nerve-mediated co-transmission in arteries (Smith *et al.*, 1999). In general, adrenergic as well as non-noradrenergic mechanisms appear to be involved in cutaneous vasoconstriction (Stephens *et al.*, 2001). Interestingly, the hormone status may influence these effects in women (Stephens *et al.*, 2002).

A close interaction of neuronal and non-neuronal derived transmitters in regulating inflammatory mediators has been reported. For example, NE reduces the release of neuropeptides from central terminals of primary afferent neurons by presynaptic inhibition. Moreover, skin-derived adrenoreceptors modulate CGRP and PGE_2 release from isolated rat skin (Averbeck and Reeh, 2001). These results support the idea that catecholamines affect PGE_2 release via 2-adrenoceptors from cutaneous cells despite sensory nerves while opioid receptors are expressed in primary afferent neurons and their activation reduces CGRP release and PGE_2 formation.

Neuropeptides from autonomic fibres are also involved in cutaneous vascular responses to sympathetic nerve stimulation. For example, NPY is involved in sympathetic contraction of the central ear arteries during cooling in rabbits. These effects were mediated via NPY Y1 receptors, are dependent of Ca^{2+} channels but are probably independent of endothelial NO (Garcia-Villalon *et al.*, 2000).

Adrenergic nerves may also exert functions during hair follicle development, growth, and/or cycling. For example, sympathetic nerve fibres modulate hair cycle-related plasticity (Botchkarev *et al.*, 1999) indicating a close functional bidirectional interaction of the hair follicle apparatus and autonomic nerves in the skin which may lead to new therapeutical tools for the treatment of hair diseases (Peters *et al.*, 1999a).

Autonomic cutaneous nerve fibres may be also involved in stress reactions (Gorbunova, 2000) and in regulating cutaneous inflammation. Recent results also show that α_2-ARs are predominately involved in the NE-sensitivity of inflamed rats during chronic inflammation (Banik *et al.*, 2001). However, the most important target cells involved in these processes during inflammation have to be defined in the future. Together, recent observations strongly suggest a role of cutaneous autonomic nerves during hyperalgesia, impairment of motor function, swelling, changes in sweating, vascular abnormalities and inflammation.

BIOLOGY AND FUNCTION OF NEUROPEPTIDES AND THEIR RECEPTORS IN THE SKIN

Neuropeptides are a relatively diverse group of regulatory peptides (Table 15.1) which exert their effects by activating high- or low-affinity receptors on cutaneous target cells. The definition is old-fashioned because several non-neuronal cells have been demonstrated to generate same peptides. Most of them bind to and activate GPCRs with seven transmembrane domains. However, some also directly activate G proteins in target cells without binding to transmembrane receptors. So far, more than 20 neuropeptides have been described in cutaneous nerve fibres and tissue cells within the skin (Table 15.1). In the following, a few well-characterized neuropeptides of the skin will be described. Other neuromediators such as cortiocotropin releasing hormone (CRH), urocortin, parathyroid hormone-related protein (PTHrP), growth hormone (GH), prolactin, galanin, dynorphin, neurotensin, gastrin-releasing peptide, NPY, glutamate, aspartate, endorphins, enkephalins, bradykinin, secretoneurin, serotonin, cholecystokinin, thyroid hormones, endothelins, ANP, NO, adenine or adenosine nucleotides and their purinergic receptors, adrenomedullin, L-DOPA, for example, are described in more detail elsewhere (reviewed by Luger *et al.*, 1997; Scholzen *et al.*, 1998; Slominski and Wortsman, 2000; Vaudry *et al.*, 2000; Steinhoff *et al.*, 2002b).

TACHYKININS

Substance P (SP), neurokinin A (NKA), neurokinin B (NKB), and the NKA-variants neuropeptide K (NP-K) and neuropeptide-γ (NP-γ) belong to a peptide family consisting of 10–13 amino acids (reviewed in Scholzen *et al.*, 1998). Often tachykinin-immunoreactive sensory nerves in the skin are associated with dermal blood vessels, mast cells, hair follicles or epidermal cells suggesting an important role of tachykinins for the regulation of the dermal as well as epidermal compartment (Eedy *et al.*, 1991, 1994; Reilly *et al.*, 1997). Interestingly, inflammatory mediators such as Interleukin-1, lipopolysarccharide (LPS) or nerve growth factor (NGF) may modulate the expression of tachykinins (Freidin and Kessler, 1991; Bost *et al.*, 1992; Vedder *et al.*, 1993).

In keratinocytes, NGF generation and release can be enhanced by SP and NKA (Burbach *et al.*, 2001). SP stimulates release of mediators such as IL-8, TNF-α, histamine, leukotriene B_4 or prostaglandin D_2, and provokes skin erythema, oedema and pruritus (Ansel *et al.*, 1993; Brown *et al.*, 1993; Kramp *et al.*, 1995; Columbo *et al.*, 1996; Furutani *et al.*, 1999; Okabe *et al.*, 2000). SP also induces upregulation of cell adhesion molecule expression on keratinocytes (Viac *et al.*, 1996).

In dermal microvascular endothelial cells, SP can directly upregulate expression of the cell adhesion molecules ICAM-1 and VCAM-1 (Wiedermann *et al.*, 1996; Quinlan *et al.*, 1998, 1999). Moreover, SP induces cutaneous neutrophilic and eosinophilic infiltration that is accompanied by translocation of P-selectin and upregulation of E-selectin

TABLE 15.1

Sources and target cells of neurotransmitters/neuropeptides in the skin (selected, as indicated in the text of this chapter).

Neuromediator	Receptor	Source	Target cells/ function	References
Acetylcholine	Nicotinergic and muscarinergic acetylcholine receptors	Autonomic cholinergic nerves, keratinocytes, lymphocytes, melanocytes	Innervation of sweat glands and arteriovenous anastomoses; keratinocyte and lymphocyte differentiation, proliferation, adhesion, migration	Grando, 1993, 1995, 1996, 1997; Schmelz, 2000; Iyengar, 1989; Bering, 1987; Ndoye, 1998; Zia, 1997; Lammerding-Koppel, 1997
Catecholamine, noradrenaline	Adrenergic receptors	Autonomic adrenergic nerves, keratinocytes, melanocytes	Innervation of blood vessels, erector pili muscles; pain transmission; regulates activity in natural killer cells and monocytes; induce apoptosis in lymphocytes	Michikami, 2001; Kaji et al., 1988; Advenier, 1993; Brain, 1985; Wallengren, 1995; Katz, 1983; Levine, 1986; Schallreuter, 1993, 1994; Oya, 2000; Woods, 2000; Cioca, 2000
Substance P	Tachykinin (neurokinin) receptor	Sensory nerve fibres	Mediates skin erythema, oedema, pruritus; upregulates cell adhesion molecule expession on keratinocytes and endothelial cells; release of IL-8, TNF-α, histamine, leukotriene B$_4$, prostaglandin D$_2$,	Brown, 1993; Ansel, 1993; Kramp, 1995; Columbo, 1996; Furutani, 1999; Viac, 1996; Wiederman, 1996; Quinlan, 1998; Smith, 1993; Regoli, 1997; Okabe, 2000

Neurokinin A	Tachykinin (neurokinin) receptor	Sensory nerve fibers	Upregulation of keratinocyte nerve growth factor expression	Regoli, 1997; Burbach, 2001
VIP	VPAC-receptors	Sensory nerve fibers Merkel cells	Sweat secretion, vasodilatation; proliferation, migration of keratinocytes; histamine release from mast cells	Tainio, 1987; Schulze, 1997; Hartschuh, 1983; Naukkarinen, 1993; Harvima, 1993; Haegerstrand, 1989; Wollina, 1997; Warren, 1993; Smith, 1992; Gonzalez, 1997; Eedy, 1994; Sato, 1987; Arimura, 1995; Vaudry, 2000
PACAP	VPAC-receptors	Autonomic and sensory nerve fibres, lymphocytes, dermal endothelial cells	Vasodilatation, immunomodulatory effect on T-cells and macrophages; modulates mast cell function, inhibits antigen-induced apoptosis on mature T lymphocytes, downregulates proinflammatory cytokines and chemokines in T cells,	Arimura, 1995; Vaudry, 2000; Moller, 1993; Odum, 1998; Steinhoff, 1999a; Warren, 1992; Delgado, 1998, 2000, 2001; Schmidt-Choudhury, 1999; Seebeck, 1998; Arimura, 1995; Vaudry, 2000; Steinhoff *et al.*, 2002a

TABLE 15.1
(Continued)

Neuromediator	Receptor	Source	Target cells/function	References
			Upregulates cytokines and cell adhesion molecules in dermal microvascular endothelial cells	Steinhoff, 2002
CGRP	CGRP-receptors	Sensory nerve fibres	Keratinocyte and endothelial cell proliferation, stimulates cytokine production	Wollina, 1997; Haegerstrand, 1989, 1990; Sung, 1992; Kiss, 1999; Kramp, 1995; Hara, 1996; Scholzen, 2000
POMC	Melanocortin receptors	Melanocytes, keratinocytes, endothelial cells, Langerhans cells, mast cells, fibroblasts, monocytes, macrophages	Antagonize effects of proinflammatory cytokines (IL-1α, 1β, 6,TNF-α, endotoxins); upregulates IL-10, releases histamine from mast cells	Luger, 2000; Catania, 2000; Hartmeyer, 1997; Grabbe, 1996; Adachi, 1999; Grutzkau, 2000; Teofoli, 1999; Cone, 1996; Brzoska, 1999

(Smith *et al.*, 1993), and directly regulates vasodilatation in the skin via NK1R *in vivo* (Habler *et al.*, 1999; Lofgren *et al.*, 1999).

In humans, increased numbers of SP-immunoreactive nerve fibres have been observed during inflammatory skin diseases (Misery, 1997). SP has been demonstrated to modulate immediate type skin hypersensitivity reactions (Streilein *et al.*, 1999) and promotes induction of CHS within normal skin (Niizeki *et al.*, 1999). In patients with atopic dermatitis, SP modulates proliferation and cytokine expression of peripheral blood mononuclear cells in response to house dust mice antigens (Yokote *et al.*, 1998). Thus, SP modulates various inflammatory and immune responses in the skin of patients with different forms of eczema. Acute immobilization stress induces mast cell degranulation via SP (Singh *et al.*, 1999) indicating that stress may trigger inflammatory skin responses and pruritus by release of SP (Theoharides *et al.*, 1998).

Three tachykinin (neurokinin) receptors (NKRs)can be differentiated because of their pharmacological characteristics (Grady *et al.*, 1996). They all belong to the GPCR family with seven transmembrane domains and bind SP, NKA and NKB with different affinities (Regoli *et al.*, 1997). Various cutaneous cells such as keratinocytes, endothelial cells, mast cells, fibroblasts, Merkel cells, and Langerhans cells have been demonstrated to express functional NKRs (reviewed by Scholzen *et al.*, 1998). Additionally, SP is capable of directly activating G proteins in a non-receptor mediated fashion (Bueb *et al.*, 1990). While NK2R expression is significantly higher in murine keratinocytes (Song *et al.*, 2000) human keratinocytes preferentially express NK1R (Quinlan *et al.*, 1997).

In summary, tachykinins are mainly released by sensory nerves in the skin which activate functional NKRs on various cell types thereby contributing to cutaneous inflammation and immune responses. Therefore, tachykinin receptor antagonists may be therapeutical tools to reduce inflammatory responses as well as pruritus in the skin. However, the precise role of tachykinins in mediating itch responses has to be clarified.

VASOACTIVE INTESTINAL PEPTIDE (VIP)

VIP is a peptide of 28 amino acids which belongs to the glucagon/secretin family. It is found in large amounts in the CNS and PNS. In the skin, immunoreactivity for VIP is predominantly observed in the dermis (Hartschuh *et al.*, 1983; Schulze *et al.*, 1997) associated with blood vessels, sweat or apocrine glands, hair follicles, Merkel cells or mast cells (Harvima *et al.*, 1993; Naukkarinen *et al.*, 1993).

VIP may be an important neuroimmunological molecule in the skin. VIP stimulates vasodilatation (Williams, 1982), proliferation (Haegerstrand *et al.*, 1989) and induces migration of keratinocytes which may be important in wound healing (Wollina *et al.*, 1997). VIP stimulates histamine release from mast cells and potentiates bradykinin-induced oedema (Warren *et al.*, 1993). VIP may be involved in neurogenic inflammation and increased plasma extravasation (Smith *et al.*, 1992), and may cause direct vasodilatation by inducing nitric oxide (NO) synthesis (Gonzalez *et al.*, 1997). Migration of monocytes into the inflammatory microenvironment is also promoted by VIP.

VIP has been reported to induce sweat production and may thus be involved in hyperhidrosis (Sato and Sato, 1987; Eedy *et al.*, 1994). Interestingly, a role of VIP is indicated in cutaneous inflammation since increased VIP levels were observed in patients with psoriasis as compared to controls (Eedy *et al.*, 1991). In contrast, no statistical differences of VIP-immunoreactivity and -concentrations were observed in patients with allergic contact dermatitis. However, VPAC2R expression was decreased in cutaneous monocytes of lesional as compared to non-lesional skin (Lundeberg and Nordlind, 1999). Moreover, application of VIP inhibited the challenge phase of allergic contact dermatitis of human skin, probably by enhanced production of γ-interferon (Lundeberg *et al.*, 1999).

VIP and PACAP analogues may serve as therapeutical tools for the treatment of inflammatory diseases. This is indicated by very recent results about an antiinflammatory role of VIP (Delgado *et al.*, 2001) and PACAP (Abad *et al.*, 2001) in experimentally induced arthritis of mice. Apparently, VIP and PACAP may exert pro- and anti-inflammatory responses dependent on the state of activation in these cells. For example, VIP downregulates pro-inflammatory cytokines in T-helper cells while IL-6 release was induced by VIP and PACAP in dermal microvascular endothelial cells *in vitro* (Steinhoff *et al.*, 2002b). These results also reflect that the neuropeptide function in cutaneous cells during inflammation is very complex. Thus, further studies in cutaneous disease models and humans are necessary to clarify the precise role of these peptides during cutaneous inflammation

PITUITARY ADENYLATE CYCLASE ACTIVATING POLYPEPTIDE (PACAP)

PACAP is a recently described member of the VIP/secretin-peptide family (Miyata *et al.*, 1989). PACAP is immunoreactive in nerve fibres of various tissues including the skin. Furthermore, PACAP positive fibres were found in lymphoid tissues and lymphocytes which indicates a role of this peptide in neuro-immunomodulation (reviewed by Arimura and Shioda, 1995; Vaudry *et al.*, 2000).

PACAP is present in cutaneous autonomic and capsaicin-sensitive sensory nerve fibres (Moller *et al.*, 1993). In human skin, immunoreactivity for PACAP was observed mainly in the dermis, around hair follicles and close to sweat glands (Odum *et al.*, 1998; Steinhoff *et al.*, 1999b) and its concentration was enhanced in patients with psoriasis (Steinhoff *et al.*, 1999b). PACAP produces a long-lasting depression of a C-fibre-evoked flexion reflex in rats (Zhang *et al.*, 1993), mediates plasma extravasation in rat skin (Cardell *et al.*, 1997), and is a potent vasodilatator and edema potentiator (Warren *et al.*, 1992). Thus, PACAP may play a role in cutaneous vascular regulation during inflammation.

PACAP has an immunomodulatory effect on several immunocompetent cells such as T-cells, macrophages, and inhibits the LPS-stimulated production of TNF-α (Delgado *et al.*, 1998). PACAP modulates mast cell function (Seebeck *et al.*, 1998; Schmidt-Choudhury *et al.*, 1999) and inhibits antigen-induced apoptosis of mature T lymphocytes (Delgado and Ganea, 2000). In a mouse arthritis model, PACAP was shown to inhibit release of pro-inflammatory mediators indicating immunosuppressive effects mediated by PACAP (Abad *et al.*, 2001). Thus, PACAP may have a vasoactive and immunoregulatory function that may also contribute to cutaneous inflammation.

Both VIP and PACAP partly share same receptors of the G protein-coupled seven transmembrane receptor family, recently defined as VPAC receptors. So far, three different VIP/PACAP receptors with additional splicing products have been cloned (Arimura and Shioda, 1995) which were recently defined as PVR1 (= PACAPR-1), VPAC1R (= VIP1R),

VPAC2R (= VIP2R) (Vaudry *et al.*, 2000). Since VIP and PACAP are able to binding same receptors but with different affinities, a differential interaction between these two peptides can be suggested. While VPAC1R and VPAC2R have been detected in human skin by several authors (Lundeberg *et al.*, 1999; Lundeberg and Nordlind, 1999; Steinhoff *et al.*, 2002b), the existence of PAC1R in the skin is still contradictory. Since various transient immuncompetent cells in different species release VIP as well as receptors for VIP it may be possible that RNA levels of inflammatory-associated transient immune cells were detected in certain tissues (Steinhoff *et al.*, 1999b). These results indicate a physiological and pathophysiological role of PACAP in normal and psoriatic human skin via activating one or more VPAC receptors.

Very recently, an important role of VIP and VPAC2R was demonstrated in allergic disease (Voice *et al.*, 2001). VPAC2R transgenic mice showed significant elevations of blood IgE, IgG1, and eosinophils as well as increased IgE antibody responses and depressed delayed-type hypersensitivity responses. Thus, VIP may enhance the ratio of Th2 to Th1 cell cytokines resulting in modified hypersensitivity. Another study showed enhanced delayed-type hypersensitivity and diminished immediate-type hypersensitivity in VPAC2R-deficient mice by creating a disbalance of the Th1 to Th2 helper cell ratio in these animals (Goetzl *et al.*, 2001). Thus, VIP as well as PACAP appear to significantly contribute to cutaneous inflammation and allergic or atopic diseases.

CALCITONIN GENE-RELATED PEPTIDE (CGRP)

CGRP consists of 37 amino acids. The molecular regulation of CGRP has been described elsewhere (Sternini *et al.*, 1992; Scholzen *et al.*, 1998). CGRP-immunoreactive nerve fibres are associated with mast cells (Botchkarev *et al.*, 1997), Merkel cells (Fantini *et al.*, 1992), melanocytes (Hara *et al.*, 1996), keratinocytes and Langerhans cells (Hosoi *et al.*, 1993; Asahina *et al.*, 1995a). CGRP-positive nerves were also detected close to smooth muscles cells and blood vessels (Wallengren *et al.*, 1987). In the epidermis, CGRP stimulates keratinocyte proliferation (Wollina *et al.*, 1997) (Haegerstrand *et al.*, 1989) and may regulate cytokine production in these cells (Kiss *et al.*, 1999). In endothelial cells, CGRP induced proliferation (Haegerstrand *et al.*, 1990) and adhesion of neutrophils and monocytes (Sung *et al.*, 1992). In dermal endothelial cells, CGRP stimulates IL-8 production (Kramp *et al.*, 1995; Scholzen *et al.*, 2000a) and NO release, and may be important for angiogenesis (Haegerstrand *et al.*, 1990). CGRP-α also enhanced phagocytosis in macrophages *in vitro* (Ichinose and Sawada, 1996) and stimulated mast cells to release TNF-α (Niizeki *et al.*, 1997).

In the skin, CGRP is regulated by UV light (Gillardon *et al.*, 1991) and suppresses DTH reactions by downregulating B7-2 and upregulating IL-10, for example (Hosoi *et al.*, 1993; Asahina *et al.*, 1995b; Torii *et al.*, 1997; Garssen *et al.*, 1998)

So far, CGRP receptors have been classified in two major classes, namely the CGRP1 and CGRP2 subtypes (Hara *et al.*, 1996). Both receptor subtypes can be differentiated by their pharmacological characteristics, although their molecular and pharmacological characteristics are not completely understood. CGRP1R responds to the antagonist CGRP8-37 whereas CGRP2R responses to other antagonists (Dumont *et al.*, 1997). However, the role of receptors for CGRP in the skin are not well characterized as yet.

PROOPIOMELANOCORTIN (POMC) PEPTIDES

Several bioactive peptides such as adrenocorticotrophic hormone (ACTH), β-lipotropin, α-, β-, γ-melanocyte-stimulating hormone (MSH), and β-endorphin are included in the POMC gene. Several POMC peptides are generated by melanocytes, keratinocytes, microvascular endothelial cells, Langerhans cells, mast cells and fibroblasts as well as immune cells such as monocytes and macrophages.

POMC peptides have a direct immunoregulatory and inflammatory effect in cutaneous cells *in vivo* and *in vitro*. α-MSH has been demonstrated as an important regulator of inflammation (Luger *et al.*, 2000) and host defence (Catania *et al.*, 2000). For example, α-MSH modulates IgE production and POMC-peptides are increased during atopic dermatitis (Rupprecht *et al.*, 1997) and antagonizes the effects of proinflammatory cytokines such as IL-1α, IL-1β, IL-6 and TNF-α or endotoxins (Hartmeyer *et al.*, 1997). α-MSH also upregulates the production of anti-inflammatory mediators such as IL-10 (Grabbe *et al.*, 1996), suggesting an anti-inflammatory role of α-MSH during cutaneous inflammation. Moreover, α-MSH induces release of histamine from mast cells, probably via MC1-R activation (Adachi *et al.*, 1999; Grutzkau *et al.*, 2000).

Human dermal fibroblasts express prohormone convertases 1 and 2 and produce POMC-derived peptides such as α-MSH (Schiller *et al.*, 2001) which upregulates matrix metalloproteinase-1 (MMP-1) expression (Kiss *et al.*, 1995) and IL-8 release (Kiss *et al.*, 1999). α-MSH itself can be upregulated by pro-inflammatory stimuli such as IL-1 (Schiller *et al.*, 2001). Moreover, β-endorphin from fibroblasts induced histamine release from mast cells demonstrating a possible role of POMC peptides in extracellular matrix regulation and tissue repair (Teofoli *et al.*, 1999).

Dermal endothelial cells have been shown to upregulate POMC after treatment with UV light or IL1-β (Scholzen *et al.*, 2000b). Moreover, HDMEC express MC1-R and α-MSH induces increased levels of IL-8 and modulates the production of chemokines such as IL-8 or Gro-α (Hartmeyer *et al.*, 1997; Bohm *et al.*, 1999).

In monocytes, α-MSH was found to downregulate MHC-class I expression and to inhibit the expression of CD 86 (B7-2) while MHC-class II and CD 80 (B7-1) expression were not significantly changed (Bhardwaj *et al.*, 1997). α-MSH also impaired immunologic functions *in vivo* (Hedley *et al.*, 2000; Slominski *et al.*, 2000) and downregulated the costimulatory molecule CD 86 on monocytes or macrophages (Bhardwaj *et al.*, 1996). Recent studies further indicate that some of the anti-inflammatory properties of α-MSH on cellular level are mediated by downregulation of NF-κB (Brzoska *et al.*, 1999; Ichiyama *et al.*, 2000).

Five subtypes of melanocortin receptors were characterized as of yet (MC1-R–MC5-R). MC-Rs differ in their tissue distribution and affinity for POMC peptides (Cone *et al.*, 1996). MC1-R is the predominant receptor in the skin and exhibits the highest affinity for α-MSH and ACTH, respectively (Luger *et al.*, 1998). Endothelial cells, fibroblasts, keratinocytes, monocytes, and melanocytes express MC1-R, while MC2-R and MC5-R are synthesized in muscle cells and adipocytes (Bohm *et al.*, 1998). Others have shown that MC5-R is expressed by keratinocytes, sebaceous and eccrine glands, and hair follicles (Thiboutot *et al.*, 2000). Furthermore, UVB treatment leads to upregulation of MC1-R in cultured keratinocytes (Jiang *et al.*, 1996), and dermal microvascular endothelial cells

(Scholzen *et al.*, 2000b). In conclusion, several observations are in favour of an important immunomodulatory role of POMC peptides such as α-MSH during cutaneous inflammation and immune response.

CUTANEOUS NEUROGENIC INFLAMMATION

REGULATORY ASPECTS OF CUTANEOUS NEUROGENIC INFLAMMATION

Capsaicin-sensitive C-fibres and to a lesser extent Aδ-fibres are not only capable of transporting impulses to the central nervous system (orthodromic signal) but also release neuropeptides (antidromic signal) which result in inflammatory activities in the skin. Neuropeptides released from cutaneous nerves are capable of acting on target cells via a paracrine, juxtacrine or endocrine pathway. These target cells express specific neuro-peptide receptors that are appropriately coupled to an intracellular signal transduction pathway or ion channels which, when activated, may result in activation of biological responses such as erythema, oedema, hyperthermia and pruritus. Since afferent sensory neurons express specific receptors for neuropeptides, prostaglandins, histamine, neurotrophins, proteases (Steinhoff *et al.*, 2000), vanilloids (Caterina *et al.*, 1997) and cytokines, an interactive communication network between sensory nerves and immune cells likely exists during cutaneous inflammation (Ansel *et al.*, 1996). Finally, cell-associated neuropeptide-degrading peptidases such as NEP or ACE have been shown to modulate neurogenic inflammation by limiting the effects of neuropeptides in the skin (Scholzen *et al.*, 2001). Thus, the interaction between sensory nerves releasing neuropeptides, target cells with functional receptors and neuropeptide-degrading peptidases appears to be critical for determining neurogenic inflammation.

REGULATORY ASPECTS OF PROTEINASE-ACTIVATED RECEPTORS IN CUTANEOUS NEUROGENIC INFLAMMATION

Recently, a new subfamily of G protein-coupled serpin receptors was described which is activated by serine proteases such as thrombin, cathepsin G, tryptase or trypsin, for exam-ple. So far, four proteinase-activated receptors (PARs) have been cloned and characterized (Dery *et al.*, 1998; Macfarlane *et al.*, 2001). PAR-2 is cleaved by trypsin, tryptase (Steinhoff *et al.*, 1999a) as well as proteases from house dust mite (Der p3 and Der p9) (Sun *et al.*, 2001) suggesting a role of PAR-2 in mediating inflammatory and allergic responses in the skin. Interestingly, PAR-2 mimics inflammatory responses observed by neuropep-tides such as SP or CGRP. This leads to the hypothesis that PAR-2 may be expressed by dorsal root ganglia and may contribute to neurogenic inflammation. Indeed, primary affer-ent neurons express functional PAR-2 which induces neuronal CGRP and SP release. Moreover, PAR-2 agonists caused marked oedema that was abrogated by antagonists of CGRP type 1- and NK1-receptors and by sensory denervation with capsaicin indicating a neuronal pathway after PAR-2 activation *in vivo*. Moreover, histological examination of paw skin indicated that PAR-2 caused a marked oedema and infiltration of granulocytes in

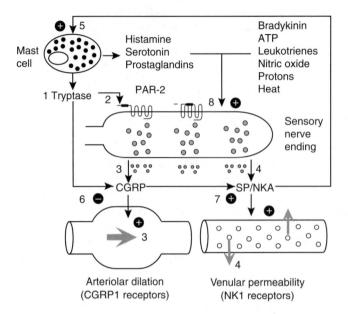

Figure 15.1   PAR-2 regulates inflammation by a neurogenic mechanism. (1) Tryptase released from degranulated mast cells cleaves PAR-2 at the plasma membrane of sensory nerve endings to expose a tethered ligand
domain that binds and activates the cleaved receptor. (2) Activation of PAR-2 stimulates the release of CGRP and
the tachykinins SP and neurokinin A (NKA) from sensory nerve endings. (3) CGRP interacts with the CGRP1
receptor to induce arteriolar dilation and hyperaemia. (4) SP interacts with the NK1R on endothelial cells of post-
capillary venules to cause gap formation and plasma extravasation. The hyperaemia and plasma extravasation
cause oedema. (5) SP may stimulate degranulation of mast cells, providing a positive feedback. (6) Tryptase
degrades CGRP and terminates its effects. (7) CGRP inhibits SP degradation by neutral endopeptidase and also
enhances SP release, thereby amplifying the effects. (8) Mediators from mast cells and other inflammatory cells
stimulate the release of vasoactive peptides from sensory nerves and also sensitize nerves (modified from
Steinhoff et al., 2000).

the dermis. Thus, PAR-2 agonists stimulate the release of SP and CGRP from spinal afferent C-fibres in the rat paw resulting in extravasation of plasma proteins and fluid but not on
infiltration of granulocytes (Figure 15.1) (Steinhoff et al., 2000). Additional observations
support the idea that PAR-2 regulates keratinocyte (Derian et al., 1997; Hou et al., 1998;
Steinhoff et al., 1999a) and endothelial cell function in the skin, for example, by upregulating cytokine release and cell adhesion molecule expression (Shpacovitch et al., 2001).
Together, serine protease may communicate with nerve fibres by activating neuronal
PAR-1 or PAR-2 leading to release of neuropeptides which contributes to inflammation
(Steinhoff et al., 2000; de Garavilla et al., 2001) and pain transmission (Vergnolle et al., 2001).

ROLE OF AUTONOMIC AND SENSORY NERVES
IN PRURITUS

Pruritus is one of the most frequent symptoms of skin diseases with a high impact on life
quality. Polymodal C-fibres and probably subtypes of Aδ-fibres appear to be crucial for

TABLE 15.2
Pathophysiological role of neuromediators in mammalian skin
(selected, as indicated in the text of this chapter).

Neuromediators	Pathophysiological role	References
Acetylcholine	Mediator of pruritus in atopic dermatitis	Heyer 1999, 1997
Noradrenaline	Psoriasis (decreased level of β_2-AR)	Steinkraus, 1993
	Skleroderma (enhanced levels of α_2-AR)	Flavahan, 2000
Substance P	Various inflammatory skin diseases	Al'Abadie et al., 1994
	(eczema, atopic dermatitis, prurigo),	Misery, 1997
	Pruritus (increased)	
VIP	Psoriasis (increased)	Eedy, 1991
PACAP	Psoriasis (increased)	Steinhoff, 1999b
POMC	Atopic dermatitis (increased)	Rupprecht, 1997

mediating stimuli to the spinal cord and CNS resulting in the symptom of itching. Recently it has been shown that certain areas of the CNS indeed participate in itch perception of humans (Hsieh et al., 1994; Darsow et al., 2000). In inflammatory dermatoses, several soluble mediators have been detected to provoke itch such as amines, neuropeptides, proteases, arachidonic acid derivates, cytokines or growth factors, respectively (Table 15.2). For example, intradermal injection of SP provokes itch along with wheal and flare. These responses were inhibited using antihistamines indicating an involvement of mast cell mediators in this process. On the other hand, there is evidence that SP-induced itch responses can be mediated by NK1R activation in mice (Andoh et al., 1998) supporting also a direct effect of SP in mediating pruritus in vivo. Intracutaneous injection of VIP and ACh both induce a dose-dependent pruritus in healthy skin and patients with atopic dermatitis. However, the subjective pruritus score did not differ between combined injections of VIP and ACh from ACh injections alone (Rukwied and Heyer, 1998) suggesting a predominant role of ACh over VIP involved in the pathophysiology of pruritus in patients with atopic dermatitis. Finally, opioids may induce itch response (Fjellner and Hagermark, 1982) and opioid μ-receptors seem to play an important role for central and peripheral (Ständer et al., 2002) neural mechanisms of itch sensation (Bergasa et al., 1992; Thomas et al., 1992).

CONCLUSIONS AND FUTURE DIRECTIONS

A substantial body of evidence clearly supports the idea that neurotransmitters and nerve fibres interact with cutaneous cells via high-affinity receptors to modulate physiological and pathophysiological responses in the skin. The skin is also equipped with a complete factory of enzymes which inactivate neurotransmitter function to maintain skin homeostasis under physiological and pathophysiological conditions. Transmembrane-coupled or extracellular proteases degrade cutaneous neuropeptides and other inflammatory mediators, or activate PARs resulting in a complex and effective neuroimmune network in the skin. Various neurotransmitters and neuropeptides have been demonstrated to play an important role for the pathophysiology of various cutaneous diseases such as inflammatory dermatoses, hypersensitivity reactions, vitiligo, wound healing and itching.

Understanding of the molecular mechanisms of neuropeptide/neurotransmitter and neuropeptide/neurotransmitter receptor regulation, advances in the understanding of mechanisms of neurogenic inflammation, cloning and characterization of further important neuroreceptors, proteinase-activated receptors as well as endopeptidases have brought new insights into the neuroimmune network. Future studies using modern techniques such as available knockout mice, genomic and proteomic approaches will help to clarify the role of the automomic and sensory nerves in the skin and may help to develop new therapeutical tools for the treatment of several skin diseases.

REFERENCES

Abad, C., Martinez, C., Leceta, J. *et al.* (2001) Pituitary adenylate cyclase-activating polypeptide inhibits collagen-induced arthritis: an experimental. immunomodulatory therapy. *J Immunol*, **167**, 3182–9.

Adachi, S., Nakano, T., Vliagoftis, H. *et al.* (1999) Receptor-mediated modulation of murine mast cell function by alpha-melanocyte stimulating hormone. *J Immunol*, **163**, 3363–8.

Al'Abadie, M. S., Senior, H. J., Bleehen, S. S., Gawkrodger, D. J. (1994) Neuropeptide and neuronal marker studies in vitiligo. *Br J Dermatol*, **131**, 160–5.

Andoh, T., Nagasawa, T., Satoh, M. *et al.* (1998) Substance P induction of itch-associated response mediated by cutaneous NK1 tachykinin receptors in mice. *J Pharmacol Exp Ther*, **286**, 1140–5.

Ansel, J. C., Brown, J. R., Payan, D. G. *et al.* (1993) Substance P selectively activates TNF-alpha gene expression in murine mast cells. *J Immunol*, **150**, 4478–85.

Ansel, J. C., Kaynard, A. H., Armstrong, C. A. *et al.* (1996) Skin–nervous system interactions. *J Invest Dermatol*, **106**, 198–204.

Arimura, A., Shioda, S. (1995) Pituitary adenylate cyclase activating polypeptide (PACAP) and its receptors: neuroendocrine and endocrine interaction. *Front Neuroendocrinol*, **16**, 53–88.

Asahina, A., Hosoi, J., Grabbe, S. *et al.* (1995a) Modulation of Langerhans cell function by epidermal nerves. *J Allergy Clin Immunol*, **96**, 1178–82.

Asahina, A., Moro, O., Hosoi, J. *et al.* (1995b) Specific induction of cAMP in Langerhans cells by calcitonin gene-related peptide: relevance to functional effects. *Proc Natl Acad Sci USA*, **92**, 8323–7.

Averbeck, B., Reeh, P. W. (2001) Interactions of inflammatory mediators stimulating release of calcitonin gene-related peptide, substance P and prostaglandin E(2) from isolated rat skin. *Neuropharmacology*, **40**, 416–23.

Banik, R. K., Kozaki, Y., Sato, J. *et al.* (2001) B2 receptor-mediated enhanced bradykinin sensitivity of rat cutaneous C-fiber nociceptors during persistent inflammation. *J Neurophysiol*, **86**, 2727–35.

Bergasa, N. V., Talbot, T. L., Alling, D. W. *et al.* (1992) A controlled trial of naloxone infusions for the pruritus of chronic cholestasis. *Gastroenterology*, **102**, 544–9.

Bering, B., Moises, H. W., Muller, W. E. (1987) Muscarinic cholinergic receptors on intact human lymphocytes. Properties and subclass characterization. *Biol Psychiatry*, **22**, 1451–8.

Bhardwaj, R., Becher, E., Mahnke, K. *et al.* (1997) Evidence for the differential expression of the functional alpha-melanocyte-stimulating hormone receptor MC-1 on human monocytes. *J Immunol*, **158**, 3378–84.

Bhardwaj, R. S., Schwarz, A., Becher, E. *et al.* (1996) Pro-opiomelanocortin-derived peptides induce IL-10 production in human monocytes. *J Immunol*, **156**, 2517–21.

Bissonnette, E. Y., Befus, A. D. (1997) Anti-inflammatory effect of beta 2-agonists: inhibition of TNF-alpha release from human mast cells. *J Allergy Clin Immunol*, **100**, 825–31.

Bjorklund, H., Dalsgaard, C. J., Jonsson, C. E. *et al.* (1986) Sensory and autonomic innervation of non-hairy and hairy human skin. An immunohistochemical study. *Cell Tissue Res*, **243**, 51–7.

Bohm, M., Schulte, U., Goez, R. *et al.* (1998) Human dermal fibroblasts express melanocortin-1 receptors and respond to α-melanocyte stimulating hormone with increased secretion of IL-8. *Arch Dermatol Res*, **290**, 104 (abstract).

Bohm, M., Schulte, U., Kalden, H. *et al.* (1999) Alpha-melanocyte-stimulating hormone modulates activation of NF-kappa B and AP-1 and secretion of interleukin-8 in human dermal fibroblasts. *Ann N Y Acad Sci*, **885**, 277–86.

Bost, K. L., Breeding, S. A., Pascual, D. W. (1992) Modulation of the mRNAs encoding substance P and its receptor in rat macrophages by LPS. *Reg Immunol*, **4**, 105–12.

Botchkarev, V. A., Eichmuller, S., Peters, E. M. *et al.* (1997) A simple immunofluorescence technique for simultaneous visualization of mast cells and nerve fibers reveals selectivity and hair cycle-dependent changes in mast cell–nerve fiber contacts in murine skin. *Arch Dermatol Res*, **289**, 292–302.

Botchkarev, V. A., Peters, E. M., Botchkareva, N. V. *et al.* (1999) Hair cycle-dependent changes in adrenergic skin innervation, and hair growth modulation by adrenergic drugs. *J Invest Dermatol*, **113**, 878–87.

Brain, S. D., Williams, T. J. (1985) Inflammatory oedema induced by synergism between calcitonin gene-related peptide (CGRP) and mediators of increased vascular permeability. *Br J Pharmacol*, **86**, 855–60.

Brown JR, Perry P, Hefeneider S *et al.* (1993) Neuropeptide modulation of keratinocyte cytokine production. In *Molecular and Cellular Biology of Cytokines*. (eds Oppenheim, Powanda, Kluger *et al.*), pp. 451–456: Wiley-Liss, Inc.

Brzoska, T., Kalden, D. H., Scholzen, T. *et al.* (1999) Molecular basis of the alpha-MSH/IL-1 antagonism. *Ann N Y Acad Sci*, **885**, 230–8.

Buchli, R., Ndoye, A., Rodriguez, J. G. *et al.* (1999) Human skin fibroblasts express m2, m4, and m5 subtypes of muscarinic acetylcholine receptors. *J Cell Biochem*, **74**, 264–77.

Bucb, J. L., Mousli, M., Landry, Y. *et al.* (1990) A pertussis toxin-sensitive G protein is required to induce histamine release from rat peritoneal mast cells by bradykinin. *Agents Actions*, **30**, 98–101.

Burbach, G. J., Kim, K. H., Zivony, A. S. *et al.* (2001) The neurosensory tachykinins substance P and neurokinin A directly induce keratinocyte nerve growth factor. *J Invest Dermatol*, **117**, 1075–82.

Cardell, L. O., Stjarne, P., Wagstaff, S. J. *et al.* (1997) PACAP-induced plasma extravasation in rat skin. *Regul Pept*, **71**, 67–71.

Catania, A., Airaghi, L., Colombo, G. *et al.* (2000) Alpha-melanocyte-stimulating hormone in normal human physiology. *Trends Endocrinol Metab*, **11**, 304–308.

Caterina, M. J., Schumacher, M. A., Tominaga, M. *et al.* (1997) The capsaicin receptor: a heat-activated ion channel in the pain pathway [see comments]. *Nature*, **389**, 816–24.

Cioca, D. P., Watanabe, N., Isobe, M. (2000) Apoptosis of peripheral blood lymphocytes is induced by catecholamines. *Jpn Heart J*, **41**, 385–98.

Columbo, M., Horowitz, E. M., Kagey-Sobotka, A. *et al.* (1996) Substance P activates the release of histamine from human skin mast cells through a pertussis toxin-sensitive and protein kinase C-dependent mechanism. *Clin Immunol Immunopathol*, **81**, 68–73.

Cone, R. D., Lu, D., Koppula, S. *et al.* (1996) The melanocortin receptors: agonists, antagonists, and the hormonal control of pigmentation. *Recent Prog Horm Res*, **51**, 287–317.

Darsow, U., Drzezga, A., Frisch, M., Munz, F., Weilke, F., Bartenstein, P., Schwaiger, M., Ring, J. (2000) Processing of histamine-induced itch in the human cerebral cortex: a correlation analysis with dermal reactions. *J Invest Dermatol*, **115**, 1029–33.

Delgado, M., Abad, C., Martinez, C. *et al.* (2001) Vasoactive intestinal peptide prevents experimental arthritis by downregulating both autoimmune and inflammatory components of the disease. *Nat Med*, **7**, 563–8.

Delgado, M., Ganea, D. (2000) Vasoactive intestinal peptide and pituitary adenylate cyclase-activating polypeptide inhibit antigen-induced apoptosis of mature T lymphocytes by inhibiting Fas ligand expression. *J Immunol*, **164**, 1200–10.

Delgado, M., Munoz-Elias, E. J., Kan, Y. *et al.* (1998) Vasoactive intestinal peptide and pituitary adenylate cyclase-activating polypeptide inhibit tumor necrosis factor alpha transcriptional activation by regulating nuclear factor-kB and cAMP response element-binding protein/c-Jun. *J Biol Chem*, **273**, 31427–36.

Derian, C. K., Eckardt, A. J., Andrade-Gordon, P. (1997) Differential regulation of human keratinocyte growth and differentiation by a novel family of protease-activated receptors. *Cell Growth Differ*, **8**, 743–9.

Dery, O., Corvera, C. U., Steinhoff, M. *et al.* (1998) Proteinase-activated receptors: novel mechanisms of signaling by serine proteases. *Am J Physiol*, **274**, C1429–52.

Dhabhar, F. S., McEwen, B. S. (1999) Enhancing versus suppressive effects of stress hormones on skin immune function. *Proc Natl Acad Sci USA*, **96**, 1059–64.

Drummond, P. D., Lipnicki, D. M. (1999) Noradrenaline provokes axon reflex hyperaemia in the skin of the human forearm. *J Auton Nerv Syst*, **77**, 39–44.

Dumont, Y., Fournier, A., St-Pierre, S. *et al.* (1997) A potent and selective CGRP2 agonist, [Cys(Et)2,7]hCGRP alpha: comparison in prototypical CGRP1 and CGRP2 in vitro bioassays. *Can J Physiol Pharmacol*, **75**, 671–6.

Eedy, D. J., Johnston, C. F., Shaw, C. *et al.* (1991) Neuropeptides in psoriasis: an immunocytochemical and radioimmunoassay study. *J Invest Dermatol*, **96**, 434–8.

Eedy, D. J., Shaw, C., Johnston, C. F. *et al.* (1994) The regional distribution of neuropeptides in human skin as assessed by radioimmunoassay and high-performance liquid chromatography. *Clin Exp Dermatol*, **19**, 463–72.

Fantini, F., Baraldi, A., Sevignani, C. *et al.* (1992) Cutaneous innervation in chronic renal failure patients. An immunohistochemical study. *Acta Derm Venereol*, **72**, 102–5.

Fjellner, B., Hagermark, O. (1982) Potentiation of histamine-induced itch and flare responses in human skin by the enkephalin analogue FK-33-824, beta-endorphin and morphine. *Arch Dermatol Res*, **274**, 29–37.

Flavahan, N. A., Flavahan, S., Liu, Q. *et al*. (2000) Increased alpha2-adrenergic constriction of isolated arterioles in diffuse scleroderma. *Arthritis Rheum*, **43**, 1886–90.

Freidin, M., Kessler, J. A. (1991) Cytokine regulation of substance P expression in sympathetic neurons. *Proc Natl Acad Sci USA*, **88**, 3200–3.

Furutani, K., Koro, O., Hide, M. *et al*. (1999) Substance P- and antigen-induced release of leukotriene B4, prostaglandin D2 and histamine from guinea pig skin by different mechanisms *in vitro*. *Arch Dermatol Res*, **291**, 466–73.

Garcia-Villalon, A. L., Fernandez, N., Monge, L. *et al*. (2000) Insulin effects on the sympathetic contraction of rabbit ear arteries. *Gen Pharmacol*, **34**, 221–6.

Garssen, J., Buckley, T. L., Van Loveren, H. (1998) A role for neuropeptides in UVB-induced systemic immuno-suppression. *Photochem Photobiol*, **68**, 205–10.

de Garavilla, L., Vergnolle, N., Young, S. H. *et al*. (2001) Agonists of proteinase-activated receptor 1 induce plasma extravasation by a neurogenic mechanism. *Br J Pharmacol*, **133**, 975–87.

Gillardon, F., Morano, I., Zimmermann, M. (1991) Ultraviolet irradiation of the skin attenuates calcitonin gene-related peptide mRNA expression in rat dorsal root ganglion cells. *Neurosci Lett*, **124**, 144–7.

Goetzl, E. J., Voice, J. K., Shen, S. *et al*. (2001) Enhanced delayed-type hypersensitivity and diminished immediate-type hypersensitivity in mice lacking the inducible VPAC2 receptor for vasoactive intestinal peptide. *Proc Natl Acad Sci USA*, **98**, 13854–9.

Gonzalez, C., Barroso, C., Martin, C. *et al*. (1997) Neuronal nitric oxide synthase activation by vasoactive intestinal peptide in bovine cerebral arteries. *J Cereb Blood Flow Metab*, **17**, 977–84.

Gorbunova, A. V. (2000) Autonomic ganglionic neurones in rabbits with differing resistance to emotional stress. *Stress*, **3**, 309–18.

Gornikiewicz, A., Sautner, T., Brostjan, C. *et al*. (2000) Catecholamines up-regulate lipopolysaccharide-induced IL-6 production in human microvascular endothelial cells. *FASEB J*, **14**, 1093–100.

Grabbe, S., Bhardwaj, R. S., Mahnke, K. *et al*. (1996) Alpha-melanocyte-stimulating hormone induces hapten-specific tolerance in mice. *J Immunol*, **156**, 473–8.

Grady, E. F., Baluk, P., Bohm, S. *et al*. (1996) Characterization of antisera specific to NK1, NK2, and NK3 neuro-kinin receptors and their utilization to localize receptors in the rat gastrointestinal tract. *J Neurosci*, **16**, 6975–86.

Grando, S. A. (1997) Biological functions of keratinocyte cholinergic receptors. *J Investig Dermatol Symp Proc*, **2**, 41–8.

Grando, S. A., Horton, R. M., Mauro, T. M. *et al*. (1996) Activation of keratinocyte nicotinic cholinergic receptors stimulates calcium influx and enhances cell differentiation. *J Invest Dermatol*, **107**, 412–8.

Grando, S. A., Horton, R. M., Pereira, E. F. *et al*. (1995) A nicotinic acetylcholine receptor regulating cell adhesion and motility is expressed in human keratinocytes. *J Invest Dermatol*, **105**, 774–81.

Grando, S. A., Kist, D. A., Qi, M. *et al*. (1993) Human keratinocytes synthesize, secrete, and degrade acetyl-choline. *J Invest Dermatol*, **101**, 32–6.

Grutzkau, A., Henz, B. M., Kirchhof, L. *et al*. (2000) Alpha-melanocyte stimulating hormone acts as a selective inducer of secretory functions in human mast cells. *Biochem Biophys Res Commun*, **278**, 14–9.

Haberberger, R. V., Bodenbenner, M. (2000) Immunohistochemical localization of muscarinic receptors (M2) in the rat skin. *Cell Tissue Res*, **300**, 389–96.

Haberberger, R. V., Pfeil, U., Lips, K. S., Kummer, W. (2002) Expression of the high-affinity choline transporter, CHT1, in the neuronal and non-neuronal cholinergic system of human and rat skin. *J Invest Dermatol*, **119**, 943–8.

Habler, H. J., Timmermann, L., Stegmann, J. U. *et al*. (1999) Involvement of neurokinins in antidromic vasodi-latation in hairy and hairless skin of the rat hindlimb. *Neuroscience*, **89**, 1259–68.

Haegerstrand, A., Dalsgaard, C. J., Jonzon, B. *et al*. (1990) Calcitonin gene-related peptide stimulates proliferation of human endothelial cells. *Proc Natl Acad Sci USA*, **87**, 3299–303.

Haegerstrand, A., Jonzon, B., Dalsgaard, C. J. *et al*. (1989) Vasoactive intestinal polypeptide stimulates cell proliferation and adenylate cyclase activity of cultured human keratinocytes. *Proc Natl Acad Sci USA*, **86**, 5993–6.

Hara, M., Toyoda, M., Yaar, M. *et al*. (1996) Innervation of melanocytes in human skin. *J Exp Med*, **184**, 1385–95.

Harada, K., Ohashi, K., Fujimura, A. *et al*. (1996) Effect of alpha 1-adrenoceptor antagonists, prazosin and urapidil, on a finger skin vasoconstrictor response to cold stimulation. *Eur J Clin Pharmacol*, **49**, 371–5.

Hartmeyer, M., Scholzen, T., Becher, E. *et al*. (1997) Human dermal microvascular endothelial cells express the melanocortin receptor type 1 and produce increased levels of IL-8 upon stimulation with alpha-melanocyte-stimulating hormone. *J Immunol*, **159**, 1930–7.

Hartschuh, W., Weihe, E., Reinecke, M. (1983) Peptidergic (neurotensin, VIP, substance P) nerve fibres in the skin. Immunohistochemical evidence of an involvement of neuropeptides in nociception, pruritus and inflammation. *Br J Dermatol*, **109**, 14–7.

Harvima, I. T., Viinamaki, H., Naukkarinen, A. *et al.* (1993) Association of cutaneous mast cells and sensory nerves with psychic stress in psoriasis. *Psychother Psychosom*, **60**, 168–76.

Hedley, S. J., Murray, A., Sisley, K. *et al.* (2000) Alpha-melanocyte stimulating hormone can reduce T-cell interaction with melanoma cells *in vitro*. *Melanoma Res*, **10**, 323–30.

Hosoi, J., Murphy, G. F., Egan, C. L. *et al.* (1993) Regulation of Langerhans cell function by nerves containing calcitonin gene-related peptide. *Nature*, **363**, 159–63.

Hou, L., Kapas, S., Cruchley, A. T. *et al.* (1998) Immunolocalization of protease-activated receptor-2 in skin: receptor activation stimulates interleukin-8 secretion by keratinocytes *in vitro*. *Immunology*, **94**, 356–62.

Hsieh, J. C., Hagermark, O., Stahle-Backdahl, M. *et al.* (1994) Urge to scratch represented in the human cerebral cortex during itch. *J Neurophysiol*, **72**, 3004–8.

Hulme, E. C., Lu, Z. L., Ward, S. D. *et al.* (1999) The conformational switch in 7-transmembrane receptors: the muscarinic receptor paradigm. *Eur J Pharmacol*, **375**, 247–60.

Ichinose, M., Sawada, M. (1996) Enhancement of phagocytosis by calcitonin gene-related peptide (CGRP) in cultured mouse peritoneal macrophages. *Peptides*, **17**, 1405–14.

Ichiyama, T., Okada, K., Campbell, I. L. *et al.* (2000) NF-kappaB activation is inhibited in human pulmonary epithelial cells transfected with alpha-melanocyte-stimulating hormone vector. *Peptides*, **21**, 1473–7.

Iyengar, B. (1989) Modulation of melanocytic activity by acetylcholine. *Acta Anat (Basel)*, **136**, 139–41.

Jiang, J., Sharma, S. D., Fink, J. L. *et al.* (1996) Melanotropic peptide receptors: membrane markers of human melanoma cells. *Exp Dermatol*, **5**, 325–33.

Jiang, W. Y., Raychaudhuri, S. P., Farber, E. M. (1998) Double-labeled immunofluorescence study of cutaneous nerves in psoriasis. *Int J Dermatol*, **37**, 572–4.

Kaji, A., Shigematsu, H., Fujita, K., Maeda, T., Watanabe, S. (1988) Parasympathetic innervation of cutaneous blood vessels by vasoactive intestinal polypeptide-immunoreactive and acetylcholinesterase-positive nerves: histochemical and experimental study on rat lower lip. *Neuroscience*, **25**, 353–62.

Katz, D. M., Markey, K. A., Goldstein, M. *et al.* (1983) Expression of catecholaminergic characteristics by primary sensory neurons in the normal adult rat *in vivo*. *Proc Natl Acad Sci USA*, **80**, 3526–30.

Kiss, M., Kemeny, L., Gyulai, R. *et al.* (1999) Effects of the neuropeptides substance P, calcitonin gene-related peptide and alpha-melanocyte-stimulating hormone on the IL-8/IL-8 receptor system in a cultured human keratinocyte cell line and dermal fibroblasts. *Inflammation*, **23**, 557–67.

Kiss, M., Wlaschek, M., Brenneisen, P. *et al.* (1995) Alpha-melanocyte stimulating hormone induces collagenase/matrix metalloproteinase-1 in human dermal fibroblasts. *Biol Chem Hoppe Seyler*, **376**, 425–30.

Kohm, A. P., Tang, Y., Sanders, V. M. *et al.* (2000) Activation of antigen-specific CD4+ Th2 cells and B cells *in vivo* increases norepinephrine release in the spleen and bone marrow. *J Immunol*, **165**, 725–33.

Kramp, J., Brown, J., Cook, P. *et al.* (1995) Neuropeptide induction of human microvascular endothelial cell interleukin 8. *J Invest Dermatol*, **104**, 586 (abstract).

Lammerding-Koppel, M., Noda, S., Blum, A. *et al.* (1997) Immunohistochemical localization of muscarinic acetylcholine receptors in primary and metastatic malignant melanomas. *J Cutan Pathol*, **24**, 137–44.

Levine, J. D., Taiwo, Y. O., Collins, S. D. *et al.* (1986) Noradrenaline hyperalgesia is mediated through interaction with sympathetic postganglionic neurone terminals rather than activation of primary afferent nociceptors. *Nature*, **323**, 158–60.

Lofgren, O., Qi, Y., Lundeberg, T. (1999) Inhibitory effects of tachykinin receptor antagonists on thermally induced inflammatory reactions in a rat model. *Burns*, **25**, 125–9.

Luger, T. A., Brzoska, T., Scholzen, T. E. *et al.* (2000) The role of alpha-MSH as a modulator of cutaneous inflammation. *Ann N Y Acad Sci*, **917**, 232–8.

Luger, T. A., Scholzen, T., Brzoska, T. *et al.* (1998) Cutaneous immunomodulation and coordination of skin stress responses by alpha-melanocyte-stimulating hormone. *Ann N Y Acad Sci*, **840**, 381–94.

Luger, T. A., Scholzen, T., Grabbe, S. (1997) The role of alpha-melanocyte-stimulating hormone in cutaneous biology. *J Investig Dermatol Symp Proc*, **2**, 87–93.

Lundeberg, L., Mutt, V., Nordlind, K. (1999) Inhibitory effect of vasoactive intestinal peptide on the challenge phase of allergic contact dermatitis in humans. *Acta Derm Venereol*, **79**, 178–82.

Lundeberg, L., Nordlind, K. (1999) Vasoactive intestinal polypeptide in allergic contact dermatitis: an immunohistochemical and radioimmunoassay study. *Arch Dermatol Res*, **291**, 201–6.

Macfarlane, S. R., Seatter, M. J., Kanke, T. *et al.* (2001) Proteinase-activated receptors. *Pharmacol Rev*, **53**, 245–82.

Maurer, M., Peters, E. M., Botchkarev, V. A. *et al.* (1998) Intact hair follicle innervation is not essential for anagen induction and development. *Arch Dermatol Res*, **290**, 574–8.

McArthur, J. C., Stocks, E. A., Hauer, P. *et al.* (1998) Epidermal nerve fiber density: normative reference range and diagnostic efficiency [see comments]. *Arch Neurol*, **55**, 1513–20.

Michikami, D., Iwase, S., Kamiya, A. *et al.* (2001) Interrelations of vasoconstrictor sympathetic outflow to skin core temperature during unilateral sole heating in humans. *Auton Neurosci*, **91**, 55–61.

Misery, L. (1997) Skin, immunity and the nervous system. *Br J Dermatol*, **137**, 843–50.

Miyata, A., Arimura, A., Dahl, R. R. *et al.* (1989) Isolation of a novel 38 residue-hypothalamic polypeptide which stimulates adenylate cyclase in pituitary cells. *Biochem Biophys Res Commun*, **164**, 567–74.

Moller, K., Zhang, Y. Z., Hakanson, R. *et al.* (1993) Pituitary adenylate cyclase activating peptide is a sensory neuropeptide: immunocytochemical and immunochemical evidence. *Neuroscience*, **57**, 725–32.

Morris, J. L., Zhu, B. S., Gibbins, I. L. *et al.* (1999) Subpopulations of sympathetic neurons project to specific vascular targets in the pinna of the rabbit ear. *J Comp Neurol*, **412**, 147–60.

Naukkarinen, A., Harvima, I., Paukkonen, K. *et al.* (1993) Immunohistochemical analysis of sensory nerves and neuropeptides, and their contacts with mast cells in developing and mature psoriatic lesions. *Arch Dermatol Res*, **285**, 341–6.

Ndoye, A., Buchli, R., Greenberg, B. *et al.* (1998) Identification and mapping of keratinocyte muscarinic acetylcholine receptor subtypes in human epidermis. *J Invest Dermatol*, **111**, 410–16.

Nguyen, V. T., Ndoye, A., Grando, S. A. (2000) Novel human alpha9 acetylcholine receptor regulating keratinocyte adhesion is targeted by pemphigus vulgaris autoimmunity. *Am J Pathol*, **157**, 1377–91.

Niizeki, H., Alard, P., Streilein, J. W. (1997) Calcitonin gene-related peptide is necessary for ultraviolet B-impaired induction of contact hypersensitivity. *J Immunol*, **159**, 5183–6.

Niizeki, H., Kurimoto, I., Streilein, J. W. (1999) A substance P agonist acts as an adjuvant to promote hapten-specific skin immunity. *J Invest Dermatol*, **112**, 437–42.

Noda, S., Lammerding-Koppel, M., Oettling, G. *et al.* (1998) Characterization of muscarinic receptors in the human melanoma cell line SK-Mel-28 via calcium mobilization. *Cancer Lett*, **133**, 107–14.

Odum, L., Petersen, L. J., Skov, P. S. *et al.* (1998) Pituitary adenylate cyclase activating polypeptide (PACAP) is localized in human dermal neurons and causes histamine release from skin mast cells. *Inflamm Res*, **47**, 488–92.

Okabe, T., Hide, M., Koro, O. *et al.* (2000) Substance P induces tumor necrosis factor-alpha release from human skin via mitogen-activated protein kinase. *Eur J Pharmacol*, **398**, 309–15.

Oya, H., Kawamura, T., Shimizu, T. *et al.* (2000) The differential effect of stress on natural killer T (NKT) and NK cell function. *Clin Exp Immunol*, **121**, 384–90.

Paus, R., Peters, E. M., Eichmuller, S. *et al.* (1997) Neural mechanisms of hair growth control. *J Invest Dermatol Symp Proc*, **2**, 61–8.

Peralta, E. G., Ashkenazi, A., Winslow, J. W. *et al.* (1987) Distinct primary structures, ligand-binding properties and tissue-specific expression of four human muscarinic acetylcholine receptors. *EMBO J*, **6**, 3923–9.

Pergolizzi, S., Vaccaro, M., Magaudda, L. *et al.* (1998) Immunohistochemical study of epidermal nerve fibres in involved and uninvolved psoriatic skin using confocal laser scanning microscopy. *Arch Dermatol Res*, **290**, 483–9.

Peters, E. M., Maurer, M., Botchkarev, V. A. *et al.* (1999) Hair growth-modulation by adrenergic drugs. *Exp Dermatol*, **8**, 274–81.

Quinlan, K. L., Olerud, J., Armstrong, C. A. *et al.* (1997) Neuropeptides upregulate expression of adhesion molecules on human keratinocytes and dermal microvascular endothelial cells. *J Invest Dermatol*, **108**, 551.

Quinlan, K. L., Song, I. S., Bunnett, N. W. *et al.* (1998) Neuropeptide regulation of human dermal microvascular endothelial cell ICAM-1 expression and function. *Am J Physiol*, **275**, C1580–90.

Quinlan, K. L., Song, I. S., Naik, S. M. *et al.* (1999) VCAM-1 Expression on human dermal microvascular endothelial cells is directly and specifically up-regulated by substance P. *J Immunol*, **162**, 1656–61.

Regoli, D., Nguyen, K., Calo, G. (1997) Neurokinin receptors. Comparison of data from classical pharmacology, binding, and molecular biology. *Ann N Y Acad Sci*, **812**, 144–6.

Reilly, D. M., Ferdinando, D., Johnston, C. *et al.* (1997) The epidermal nerve fibre network: characterization of nerve fibres in human skin by confocal microscopy and assessment of racial variations. *Br J Dermatol*, **137**, 163–70.

Roth, S., Kummer, W. (1994) A quantitative ultrastructural investigation of tyrosine hydroxylase-immunoreactive axons in the hairy skin of the guinea pig. *Anat Embryol*, **190**, 155–62.

Rouppe van der Voort, C., Kavelaars, A., van de Pol, M. *et al.* (2000) Noradrenaline induces phosphorylation of ERK-2 in human peripheral blood mononuclear cells after induction of alpha(1)-adrenergic receptors. *J Neuroimmunol*, **108**, 82–91.

Rukwied, R., Heyer, G. (1998) Cutaneous reactions and sensations after intracutaneous injection of vasoactive intestinal polypeptide and acetylcholine in atopic eczema patients and healthy controls. *Arch Dermatol Res*, **290**, 198–204.

Rupprecht, M., Salzer, B., Raum, B. *et al.* (1997) Physical stress-induced secretion of adrenal and pituitary hormones in patients with atopic eczema compared with normal controls. *Exp Clin Endocrinol Diabetes*, **105**, 39–45.

Sato, K., Sato, F. (1987) Effect of VIP on sweat secretion and cAMP accumulation in isolated simian eccrine glands. *Am J Physiol*, **253**, R935–41.

Schallreuter, K. U. (1997) Epidermal adrenergic signal transduction as part of the neuronal network in the human epidermis. *J Investig Dermatol Symp Proc*, **2**, 37–40.

Schallreuter, K. U., Korner, C., Pittelkow, M. R. *et al.* (1996a) The induction of the alpha-1-adrenoceptor signal transduction system on human melanocytes. *Exp Dermatol*, **5**, 20–3.

Schallreuter, K. U., Wood, J. M., Pittelkow, M. R. *et al.* (1996b) Increased monoamine oxidase A activity in the epidermis of patients with vitiligo. *Arch Dermatol Res*, **288**, 14–18.

Schallreuter, K. U., Wood, J. M., Lemke, R. *et al.* (1992) Production of catecholamines in the human epidermis. *Biochem Biophys Res Commun*, **189**, 72–8.

Schallreuter, K. U., Wood, J. M., Pittelkow, M. R. *et al.* (1993) Increased *in vitro* expression of beta 2-adrenoceptors in differentiating lesional keratinocytes of vitiligo patients. *Arch Dermatol Res*, **285**, 216–20.

Schallreuter, K. U., Wood, J. M., Pittelkow, M. R. *et al.* (1994) Regulation of melanin biosynthesis in the human epidermis by tetrahydrobiopterin. *Science*, **263**, 1444–6.

Schiller, M., Raghunath, M., Kubitscheck, U. *et al.* (2001) Human dermal fibroblasts express prohormone convertases 1 and 2 and produce proopiomelanocortin-derived peptides. *J Invest Dermatol*, **117**, 227–35.

Schmelz, M., Michael, K., Weidner, C. *et al.* (2000) Which nerve fibers mediate the axon reflex flare in human skin? *Neuroreport*, **11**, 645–8.

Schmidt-Choudhury, A., Furuta, G. T., Galli, S. J. *et al.* (1999) Mast cells contribute to PACAP-induced dermal oedema in mice. *Regul Pept*, **82**, 65–9.

Scholzen, T., Armstrong, C. A., Bunnett, N. W. *et al.* (1998) Neuropeptides in the skin: interactions between the neuroendocrine and the skin immune systems. *Exp Dermatol*, **7**, 81–96.

Scholzen, T., Steinhoff, M., Bonaccorsi, P. *et al.* (2001) Neutral endopeptidase terminates substance P-induced inflammation in allergic contact dermatitis. *J Immunol*, **166**, 1285–91.

Scholzen, T. E., Fastrich, M., Brzoska, T. *et al.* (2000a) Calcitonin gene-related peptide (CGRP) activation of human dermal microvascular endothelial cell (HDMEC) transcription factors NF-kappa B and CREB. *J Invest Dermatol.*, **115**, 534.

Scholzen, T. E., Kalden, D., Brzoska, T. *et al.* (2000b) Expression of proopiomelanocortin peptides in human dermal microvascular endothelial cells: evidence for a regulation by ultraviolet light and interleukin-1. *J Invest Dermatol.*, **115**.

Schulze, E., Witt, M., Fink, T. *et al.* (1997) Immunohistochemical detection of human skin nerve fibers. *Acta Histochem*, **99**, 301–9.

Seebeck, J., Kruse, M. L., Schmidt-Choudhury, A. *et al.* (1998) Pituitary adenylate cyclase activating polypeptide induces degranulation of rat peritoneal mast cells via high-affinity PACAP receptor-independent activation of G proteins. *Ann N Y Acad Sci*, **865**, 141–6.

Shpacovitch, V. M., Brzoska, T., Buddenkotte, J. *et al.* (2001) Agonists of proteinase-activated receptor-2 induce cytokine release and activation of nuclear transcription factor kappa B in human dermal microvascular endothelial cells. *J Invest Dermatol*, **118**.

Simone, D. A., Nolano, M., Johnson, T. *et al.* (1998) Intradermal injection of capsaicin in humans produces degeneration and subsequent reinnervation of epidermal nerve fibers: correlation with sensory function. *J Neurosci*, **18**, 8947–59.

Singh, L. K., Pang, X., Alexacos, N. *et al.* (1999) Acute immobilization stress triggers skin mast cell degranulation via corticotropin releasing hormone, neurotensin, and substance P: A link to neurogenic skin disorders. *Brain Behav Immun*, **13**, 225–39.

Slominski, A., Wortsman, J. (2000) Neuroendocrinology of the skin. *Endocr Rev*, **21**, 457–87.

Slominski, A., Wortsman, J., Luger, T. *et al.* (2000) Corticotropin releasing hormone and proopiomelanocortin involvement in the cutaneous response to stress. *Physiol Rev*, **80**, 979–1020.

Smith, C. H., Atkinson, B., Morris, R. W. *et al.* (1992) Cutaneous responses to vasoactive intestinal polypeptide in chronic idiopathic urticaria. *Lancet*, **339**, 91–3.

Smith, C. H., Barker, J. N., Morris, R. W. *et al.* (1993) Neuropeptides induce rapid expression of endothelial cell adhesion molecules and elicit granulocytic infiltration in human skin. *J Immunol*, **151**, 3274–82.

Smith, K. M., Macmillan, J. B., McCulloch, K. M. *et al.* (1999) NOS inhibition potentiates norepinephrine but not sympathetic nerve-mediated co-transmission in resistance arteries. *Cardiovasc Res*, **43**, 762–71.

Song, I. S., Bunnett, N. W., Olerud, J. E. *et al.* (2000) Substance P induction of murine keratinocyte PAM 212 interleukin 1 production is mediated by the neurokinin 2 receptor (NK-2R). *Exp Dermatol*, **9**, 42–52.

Ständer, S., Gunzer, M., Metze, D., Luger, T., Steinhoff, M. (2002) Detection of μ-opioid receptor 1A in human skin. *Regulat Peptides*, in press.

Steinhoff, A., Grevelhörster, A., Schmidt, W. E., Goetzl, E. J., Luger, T. A., Steinhoff, M. (2002a) The neuro-peptides VIP and PACAP upregulate cytokine and cell adhesion molecule expression and release in human dermal microvascular endothelial cells. *Arch Dermatol Res*, **294**, 116 (abstr).

Steinhoff, M., Bunnett, N. W., Scholzen, T. *et al.* (2002b) Neurocutaneous control of inflammation. In *Molecular Mechanisms of Cutaneous Disease*, (eds Kupper, T., Norris, D.), 2nd edn., in press.

Steinhoff, M., Corvera, C. U., Thoma, M. S. *et al.* (1999a) Proteinase-activated receptor-2 in human skin: tissue distribution and activation of keratinocytes by mast cell tryptase. *Exp Dermatol*, **8**, 282–94.

Steinhoff, M., McGregor, G. P., Radleff-Schlimme, A. *et al.* (1999b) Identification of pituitary adenylate cyclase activating polypeptide (PACAP) and PACAP type 1 receptor in human skin: expression of PACAP-38 is increased in patients with psoriasis. *Regul Pept*, **80**, 49–55.

Steinhoff, M., Vergnolle, N., Young, S. H. *et al.* (2000) Agonists of proteinase-activated receptor 2 induce inflam-mation by a neurogenic mechanism. *Nat Med*, **6**, 151–8.

Steinkraus, V., Steinfath, M., Stove, L. *et al.* (1993) Beta-adrenergic receptors in psoriasis: evidence for down-regulation in lesional skin. *Arch Dermatol Res*, **285**, 300–4.

Stephens, D. P., Aoki, K., Kosiba, W. A. *et al.* (2001) Nonnoradrenergic mechanism of reflex cutaneous vasoconstriction in men. *Am J Physiol Heart Circ Physiol*, **280**, H1496–504.

Stephens, D. P., Bennett, L. A., Aoki, K. *et al.* (2002) Sympathetic nonnoradrenergic cutaneous vasocon-striction in women is associated with reproductive hormone status. *Am J Physiol Heart Circ Physiol*, **282**, H264–72.

Sternini, C., De Giorgio, R., Furness, J. B. (1992) Calcitonin gene-related peptide neurons innervating the canine digestive system. *Regul Pept*, **42**, 15–26.

Streilein, J. W., Alard, P., Niizeki, H. (1999) Neural influences on induction of contact hypersensitivity. *Ann N Y Acad Sci*, **885**, 196–208.

Sun, G., Stacey, M. A., Schmidt, M. *et al.* (2001) Interaction of mite allergens Der p3 and Der p9 with protease-activated receptor-2 expressed by lung epithelial cells. *J Immunol*, **167**, 1014–21.

Sung, C. P., Arleth, A. J., Aiyar, N. *et al.* (1992) CGRP stimulates the adhesion of leukocytes to vascular endothe-lial cells. *Peptides*, **13**, 429–34.

Suzuki, H., Ueno, A., Takei, M. *et al.* (2000) The effects of S1319, a novel marine sponge-derived beta2-adreno-ceptor agonist, on IgE-mediated activation of human cultured mast cells. *Inflamm Res*, **49**, 86–94.

Tainio, H. (1987) Cytochemical localization of VIP-stimulated adenylate cyclase activity in human sweat glands. *Br J Dermatol*, **116**, 323–8.

Tainio, H., Vaalasti, A., Rechardt, L. (1987) The distribution of substance P-, CGRP-, galanin- and ANP-like immunoreactive nerves in human sweat glands. *Histochem J*, **19**, 375–80.

Teofoli, P., Frezzolini, A., Puddu, P. *et al.* (1999) The role of proopiomelanocortin-derived peptides in skin fibrob-last and mast cell functions. *Ann N Y Acad Sci*, **885**, 268–76.

Theoharides, T. C., Singh, L. K., Boucher, W. *et al.* (1998) Corticotropin-releasing hormone induces skin mast cell degranulation and increased vascular permeability, a possible explanation for its proinflammatory effects. *Endocrinology*, **139**, 403–13.

Thiboutot, D., Sivarajah, A., Gilliland, K. *et al.* (2000) The melanocortin 5 receptor is expressed in human seba-ceous glands and rat preputial cells. *J Invest Dermatol*, **115**, 614–19.

Thomas, D. A., Williams, G. M., Iwata, K. *et al.* (1992) Effects of central administration of opioids on facial scratching in monkeys. *Brain Res*, **585**, 315–17.

Torii, H., Hosoi, J., Beissert, S. *et al.* (1997) Regulation of cytokine expression in macrophages and the Langerhans cell-like line XS52 by calcitonin gene-related peptide. *J Leukoc Biol*, **61**, 216–23.

Vaudry, D., Gonzalez, B. J., Basille, M. *et al.* (2000) Pituitary adenylate cyclase-activating polypeptide and its receptors: from structure to functions. *Pharmacol Rev*, **52**, 269–324.

Vedder, H., Affolter, H. U., Otten, U. (1993) Nerve growth factor (NGF) regulates tachykinin gene expression and biosynthesis in rat sensory neurons during early postnatal development. *Neuropeptides*, **24**, 351–7.

Vergnolle, N., Bunnett, N. W., Sharkey, K. A. *et al.* (2001) Proteinase-activated receptor-2 and hyperalgesia: A novel pain pathway. *Nat Med*, **7**, 821–6.

Viac, J., Gueniche, A., Doutremepuich, J. D. *et al.* (1996) Substance P and keratinocyte activation markers: an *in vitro* approach. *Arch Dermatol Res*, **288**, 85–90.

Vitadello, M., Triban, C., Fabris, M., Dona, M., Gorio, A., Schiaffino, S.A. (1987) Developmentally regulated isoform of 150,000 molecular weight neurofilament protein specifically expressed in autonomic and small sensory neurons. *Neuroscience*, **23**, 931–41.

Voice, J. K., Dorsam, G., Lee, H. *et al.* (2001) Allergic diathesis in transgenic mice with constitutive T cell expres-sion of inducible vasoactive intestinal peptide receptor. *FASEB J*, **15**, 2489–96.

Wallengren, J., Badendick, K., Sundler, F. *et al.* (1995) Innervation of the skin of the forearm in diabetic patients: relation to nerve function. *Acta Derm Venereol*, **75**, 37–42.

Wallengren, J., Ekman, R., Sundler, F. (1987) Occurrence and distribution of neuropeptides in the human skin. An immunocytochemical and immunochemical study on normal skin and blister fluid from inflamed skin. *Acta Derm Venereol*, **67**, 185–92.

Warren, J. B., Larkin, S. W., Coughlan, M. *et al.* (1992) Pituitary adenylate cyclase activating polypeptide is a potent vasodilator and oedema potentiator in rabbit skin *in vivo. Br J Pharmacol*, **106**, 331–4.

Warren, J. B., Wilson, A. J., Loi, R. K. *et al.* (1993) Opposing roles of cyclic AMP in the vascular control of edema formation. *FASEB J*, **7**, 1394–400.

Wiedermann, C. J., Auer, B., Sitte, B. *et al.* (1996) Induction of endothelial cell differentiation into capillary-like structures by substance P. *Eur J Pharmacol*, **298**, 335–8.

Williams, T. J. (1982) Vasoactive intestinal polypeptide is more potent than prostaglandin E2 as a vasodilator and oedema potentiator in rabbit skin. *Br J Pharmacol*, **77**, 505–9.

Wollina, U., Huschenbeck, J., Knoll, B. *et al.* (1997) Vasoactive intestinal peptide supports induced migration of human keratinocytes and their colonization of an artificial polyurethane matrix. *Regul Pept*, **70**, 29–36.

Woods, J. A. (2000) Exercise and neuroendocrine modulation of macrophage function. *Int J Sports Med*, **21**(Suppl 1), S24–30.

Yokote, R., Yagi, H., Furukawa, F. *et al.* (1998) Regulation of peripheral blood mononuclear cell responses to Dermatophagoides farinae by substance P in patients with atopic dermatitis. *Arch Dermatol Res*, **290**, 191–7.

Zhang, Y. Z., Sjolund, B., Moller, K. *et al.* (1993) Pituitary adenylate cyclase activating peptide produces a marked and long-lasting depression of a C-fibre-evoked flexion reflex. *Neuroscience*, **57**, 733–7.

Zia, S., Ndoye, A., Nguyen, V. T. *et al.* (1997) Nicotine enhances expression of the alpha 3, alpha 4, alpha 5, and alpha 7 nicotinic receptors modulating calcium metabolism and regulating adhesion and motility of respiratory epithelial cells. *Res Commun Mol Pathol Pharmacol*, **97**, 243–62.

16 Neuroimmune Connections and Regulation of Function in the Urinary Bladder

Theoharides C. Theoharides[1] and Grannum R. Sant[2]

[1]*Department of Pharmacology and Experimental Therapeutics, Tufts University School of Medicine, Boston, MA 02111, USA*
[2]*Department of Urology, Tufts University, School of Medicine and New England Medical Center, Boston, MA 02111, USA*

Interactions between immune cells and neurons are increasingly implicated in normal and pathological processes in different organs and associated diseases. A cell that has emerged at the centre of these interactions is the mast cell, a cell more commonly recognized for its role in allergic reactions. Anatomical and functional associations are reviewed with respect to bladder function and the pathophysiology of a non-infectious painful bladder condition, interstitial cystitis. The availability of bladder biopsies has provided the unique opportunity to investigate these interactions in humans. Moreover, the existence of overlapping syndromes, such as irritable bowel syndrome, has shown that the underlying mechanisms involving mucosal permeability and sensory neuron activation depend on neuropeptide activation of mast cells. *In vitro* evidence is presented for anatomical and functional interactions between mucosal-like mast cells and neuron-like pheochromocytoma cells. Stress affects the mast cell–neuronal interactions by release of corticotropin-releasing hormone or the related peptide urocortin (directly, and via neurotensin and substance P), which lead to mast cell activation. Cytokines released from mast cells may then further activate the stress response. Interruption of stress-induced mast cell activation may offer new therapeutic alternatives.

KEY WORDS: bladder; chondroitin sulphate; corticotropin-releasing hormone; interstitial cystitis; mast cells; neurons; neuropeptides; quercetin; secretion.

ABBREVIATIONS: ACh, acetylcholine; CTMC, connective tissue mast cell; CNS, central nervous system; CRH, corticotropin-releasing hormone; GAG, glycosaminoglycan; CGRP, calcitonin gene related peptide; HPA, hypothalamic–pituitary–adrenal; IgE, immunoglobulin E; IMLC, intermediolateral column; INF, interferon; IL-6, interleukin 6; IC, interstitial cystitis; IBS, irritable bowel syndrome; LP, lamina propria; MMC, mucosal mast cells; NK, neurokinin; NPY, neuropeptide Y; NE, norepinephrine; NGF, nerve growth factor; NT, neurotensin; NO, nitric oxide; PGD_2, prostaglandin D_2; RBL, rat basophil leukaemia; SCF, stem cell factor; SP, substance P; SPN, sacral parasympathetic nucleus; TNF, tumour necrosis factor; Ucn, urocortin; VIP, vasoactive intestinal peptide.

NEUROPHYSIOLOGY OF BLADDER FUNCTION

Bladder function is regulated by complex and interrelated neural circuits (Figure 16.1) involving the peripheral and central nervous systems (CNS) (Elbadawi, 1991). The main

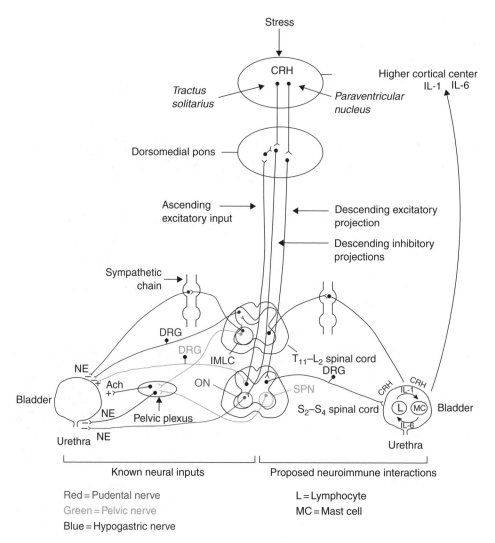

Figure 16.1   Diagram depicting neural and neuroimmune bladder circuits. (See Colour Plate I.)

bladder functions are: (a) urine storage, holding urine at low pressures by relaxing the detrusor muscle and constricting the external urethral sphincter; (b) voiding, releasing urine through contraction of the detrusor muscle and relaxation of the sphincter. These seemingly simple but inter-related functions involve the closely coordinated interactions of the autonomic, sympathetic and somatic divisions of the nervous system. Increasing evidence indicates that, in addition to the classic neurotransmitters acetylcholine (ACh) and norepinephrine (NE), neuropeptides released from the vagus, splanchnic, hypogastric, pelvic and pudendal nerves, as well as biogenic amines and cytokines liberated from mast cells, leukocytes and the urothelium profoundly affect bladder function (Kinder and

Mundy, 1991). Dysfunction of these interactions could lead to bladder hyperactivity with irritative voiding, urinary incontinence, neuropathic bladder and pelvic pain, symptoms present in conditions such as interstitial cystitis, chronic pelvic pain and non-bacterial prostatitis. Increased bladder permeability could permit penetration of toxic substances through the damaged urothelium and lead to systemic inflammatory symptoms.

The excitatory stimulus responsible for micturition is provided by the parasympathetic division of the autonomic nervous system and originates in the sacral parasympathetic nucleus (SPN) at the level of S2–4 of the spinal cord (Elbadawi, 1991; Chai and Steers, 1997). Cholinergic preganglionic efferent neurons form the pelvic nerve, which synapses either on postganglionic neurons in the pelvic plexus or within the bladder (Chai and Steers, 1997). The pelvic plexus also includes sympathetic efferent nerve fibres that originate in the intermediolateral column (IMLC) at the T11–12 level of the spinal cord; these synapse at the sympathetic chain from which postganglionic fibres contribute to the pelvic nerve. Other efferent fibres pass through the sympathetic ganglia and travel in the hypogastric nerve with synapses in the pelvic plexus. The smooth muscle of the bladder neck receives a rich nerve supply containing nonepinephrine (Gosling et al., 1999). It is generally thought from animal studies that sympathetic outflow can either inhibit or overcome the parasympathetic input and thus reduce urge incontinence in humans (Elbadawi, 1991).

The afferent innervation of the bladder and urethra synapse at the dorsal root ganglia (DRG) at the corresponding levels of the spinal cord (Chai and Steers, 1997). Low threshold myelinated afferents (Aσ) and unmyelinated C-fibres transmit noxious stimuli via the pelvic nerve to the S2–4 DRG; high threshold nociceptive afferents transmit to the T11~L2 DRG (Dixon and Gilpin, 1987). The afferent plexus is thickest around the neck of the bladder and contains numerous calcitonin gene related peptide (CGRP)-positive neurons (Kluck, 1980; Gabella and Davis, 1998). There is bilateral innervation with many neurons crossing the midline (Gabella and Davis, 1998). Activation of afferents carried by the vagus nerve appears to be important in bladder function, especially in the stress response (Peeker et al., 2000).

The somatic neurons innervating the external urethral sphincter and the musculature of the pelvic floor derive from the anterior horn at the level of S2–4 of the spinal cord and form the pudendal nerve. There is evidence that the external urethra also receives input from both the pelvic and hypogastric nerves (Chai and Steers, 1997; von Heyden et al., 1998). In general, the pudendal nerve is important for maintaining pelvic floor integrity and continence. Continence is achieved by activation of sympathetic and somatic afferents, inhibition of parasympathetic efferents, as well as input from sacral spinal cord levels (Kinder and Mundy, 1991). Local injury or inflammation, however, rapidly interfere with this local input. There is also the possibility that autoantibodies against muscarinic or other receptors may interfere with bladder function (Van de Merwe and Arendsen, 2000). Afferents in the pelvic nerve transmit mechanoceptic messages upon bladder filling and activate thoracolumbar sympathetic preganglionic firing, which increases outlet resistance through the hypogastric nerve (Gosling et al., 1999). However, sympathetic input may by required only for large volume, as mechanical or chemical sympathectomy in animals does not appear to affect urine storage (Chai and Steers, 1997). Afferent innervation from vagina, uterine cervix and rectum supplies inhibitory input to the parasympathetic preganglionic fibres and increases urethral resistance.

Unlike other visceral organs, the bladder receives neural input also from higher cortical centres. For instance, the nucleus tractus solitarius, a critical relay station, receives visceral afferents and has significant input from the paraventricular nucleus of the hypothalamus and the locus ceruleus; both these regions also project to the bladder (Chai and Steers, 1997). Consequently, emotional upheavals, pain and stress are likely to affect bladder function. The frontal cortex and the septum areas of the brain appear to exert inhibitory control over the detrusor, but not the external urethral sphincter. In contrast, activation of specific dorsomedial areas in the pons induces bladder contraction and urethral sphincter relaxation (Chai and Steers, 1997). Adequate bladder emptying in reinforced by a urethrovesical reflex. Dorsomedial descending projections synapse on the Onuf nucleus and inhibit somatic efferents originating there. Other local circuits inhibit sphincter function, while relaxation of the urethra may occur by firing of parasympathetic neurons (Chai and Steers, 1997) or release of nitric oxide (NO).

NEUROIMMUNE INTERACTIONS

A variety of neurotransmitters and neuropeptides (Kinder and Mundy, 1991), along with cytokines modulate the neural regulation of bladder function. A simplified scheme is shown in Figure 16.1. Emotional or physical stress can release corticotropin-releasing hormone (CRH) or its analogue urocortin (Ucn) from the hypothalamus, which then stimulate the nucleus tractus solitarius to release ACh/neuropeptides in the bladder inducing hyperactivity. Other stressful stimuli, perceived by higher cortical centres through bladder afferents or by direct stimulation of the hypothalamus by bladder-generated cytokines, induce release of CRH/Ucn from the hypothalamus or the DRG. Norepinephrine released from sympathetic ganglia will tend to inhibit bladder hyperactivity and immune cell activation (Elenkov et al., 2000).

CRH/Ucn released from DRG can stimulate bladder mast cells, directly or via neuropeptides. For instance, stress-induced bladder mast cell activation is inhibited by neonatal capsaicin treatment, which depletes sensory neurons of their substance P (SP) content, and by neurotensin (NT) receptor antagonists (Theoharides et al., 1998). Mast cells, in turn, release vasodilatory, proinflammatory, tissue damaging and neurosensitizing molecules. Cytokines stimulate recruited leukocytes to release interleukin (IL)-1 which then induces selective release of IL-6 from mast cells (Figure 16.1). IL-1 and IL-6, as well as histamine and leukotrienes, can stimulate further CRH/Ucn release from the hypothalamus, and possibly DRGs (Chrousos, 1995). Leukocytes also release CRH/Ucn locally in the bladder reinforcing this neuroimmune circuit (Karalis et al., 1997) (Figure 16.1). Hypothalamic CRH activates the hypothalamic–pituitary–adrenal (HPA) axis and results in catecholamine and corticosteroid secretion, which eventually downregulate the local immune response in the bladder.

Bladder afferents synapsing at the sacral DRG contain CGRP, cholecystokinin, glutamate, NO, SP, and vasoactive intestinal peptide (VIP) (Chai and Steers, 1997). Depletion of these neuropeptides by the neurotoxin capsaicin transiently reduces bladder pain and irritative voiding symptoms (Barbanti et al., 1993; Fowler, 2000; Kim and Chancellor, 2000; Lazzeri et al., 2000; Silva et al., 2000). Moreover, selective neurokinin (NK)-1 receptor antagonists block SP-induced bladder muscle contraction and plasma extravasation (Montier *et al.*, 1994), while an NK-2 receptor antagonist blocked agonist induced contractions of detrusor muscle strips (Palea *et al.*, 1996; Rizzo and Hey, 2000). Marked depletion of VIP, neuropeptide Y (NPY) and SP by bladder distention may explain the transient pain relief experienced by interstitial cystitis (IC) patients following hydrodistention (Lasanen *et al.*, 1992). Bladder distention may also alter urothelial integrity, and permit penetration of noxious molecules into the bladder wall with associated urothelial leakage and inflammation (Leppilahti *et al.*, 1999). Interestingly, neither NK-1 nor 2 receptor antagonists inhibit volume-induced micturition (Chai and Steers, 1997) and NK-1 receptor antagonists fail to induce adequate analgesia in humans (Hill, 2000).

Considerable evidence confirms the central role of mast cells in bladder neurohormonal interactions (Figure 16.2). Mast cells were first recognized by Friedrich von Recklinghausen in 1863; they were described and named by Paul Erhlich in 1887, because of the metorchromasia (change in colour from blue to violet) when stained with acidified toluidine blue (Figure 16.3). Mast cells were later shown to participate in allergic reactions (Galli, 1993). Mast cells are bone marrow-derived (Galli, 1993) and are essential for Type I hypersensitivity reactions in which immunoglobulin E (IgE) attaches to specific surface binding proteins (FcRI) and triggers mast cell degranulation by compound exocytosis (Theoharides, 1996). In this process, the perigranular membranes fuse with the plasma membrane resulting in secretion of a number of biologically active molecules (Theoharides, 1996). In addition to IgE and allergens, various food dyes and preservatives, drugs, hormones, and toxins trigger mast cell secretion. More importantly, mast cells are also activated by a number of neuroimmune triggers that include kinins, NT (Carraway *et al.*, 1982), somatostatin (SRIF) (Theoharides and Douglas, 1978), SP (Fewtrell *et al.*, 1982), NPY (Grundemar *et al.*, 1994), and ACh (Spanos *et al.*, 1996) (Table 16.1). Bladder mast cells are also activated by acute immobilization stress and this activation can be reduced by SP depletion, NT receptor antagonists, or neutralization of peripheral immune CRH (Theoharides *et al.*, 1998).

Upon activation, mast cell release numerous biologically active mediators (Klein *et al.*, 1989; Plaut *et al.*, 1989); these are grouped in: (a) granule-stored, preformed molecules; and (b) those synthesized *de novo* (Table 16.2). The first group includes heparin, histamine, proteases, phospholipases and chemotactic substances. Newly synthesized

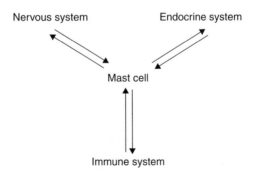

Figure 16.2   Mast cells interacting with the hormonal, immune and nervous systems.

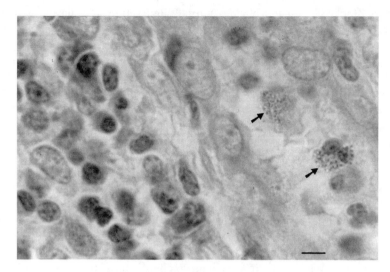

Figure 16.3   Photomicrographs of granular mast cells staining violet with acidified toluidine blue in an area of inflammation from the bladder of a female IC patient. Bar = 10 μm. (See Colour Plate II.)

TABLE 16.1
Neuroimmunoendocrine triggers of mast cell activation.

Anaphylatoxins	*Neuropeptides*
C3a, C5a	Bradykinin
	CGRP
	Endorphins
	NT
	SRIF
	SP
	VIP
Cytokines	
IL-1	
TNF-α	
Growth factors	*Neurotransmitters*
NGF	Acetylcholine
SCF	Purines
Hormones	*Radicals*
ACTH	Free oxygen
CRH	Hydroxy
Estradiol	
PTH	
Urocortin	

molecules include cytokines (Galli, 1993; Kobayashi *et al.*, 2000), especially IL-6 (Galli, 1993), leukotrienes (LTC$_4$), prostaglandins (PGD$_2$), platelet activating factor (PAF) and NO (Serafin and Austen, 1987). Mast cells also release VIP (Cutz *et al.*, 1978) and tumour necrosis factor-α (TNF-α) (Galli, 1993), both of which are potent vasodilatory molecules.

TABLE 16.2
Mast cell derived mediators.

Molecules	Major actions
Prestored	
Enzymes	
Arylsulfatases	Lipid/proteoglycan hydrolysis
Carboxypeptidase A	Peptide processing
Chymase	Angiotensin II synthesis
Kinogenases	Kinins synthesis vasodilation, *pain*
Phospholipases	Arachidonic acid generation
Tryptase	Tissue damage, inflammation, *pain*
Biogenic amines	
Histamine	Vasodilation, angiogenesis, mitogenesis, *pain*
5-HT (serotonin)	Vasoconstriction, *pain*
Chemokines	
IL-8, MCP-1, MCP-3, MCP-4, RANTES	Chemoattraction, leukocyte infiltration
Cytokines	
IL-1,2,3,4,5,6,9,10,13,16	Inflammation, leukocyte migration, *pain*
INF-γ; MIF; TNF-α	Inflammation, leukocyte proliferation/activation
Growth factors	
CSF, GM-CSF, b-FGF, NGF	Endothelial immune and neuronal cell growth
Peptides	
Bradykinin	Vasodilation, *pain*
Endorphins	Analgesia
SP	Anti-inflammatory (?)
SRIF	Inflammation, *pain*
VIP	Vasodilation
Proteoglycans	
Chondroitin sulphate	Cartilage synthesis, anti-inflammatory
Heparin	Angiogenesis, NGF stabilization
Hyaluronic acid	Cell surface recognition
De novo synthesized	
LTB$_4$	Leukocyte chemotaxis
LTC$_4$	Vasoconstriction, *pain*
NO	Vasodilation
PAF	Platelet activation and serotonin release
PGD$_2$	Vasodilation, *pain*

CSF, colony stimulating factor; b-FGF, fibroblast growth factor; 5-HT, 5-Hydroxytryptamine; GM-CSF, granulocyte monocyte-colony stimulating factor; INF-γ, Interferon-γ; MIF, macrophage inflammatory factor; NGF, nerve growth factor; PAF, platelet activating factor; SRIF, somatostatin; TGF-β, transforming growth factor-β; TNF-α, tumour necrosis factor-α.

Mast cells also synthesize and secrete nerve growth factor (NGF) (Xiang and Nilsson, 2000) and SP (Toyoda *et al.*, 2000).

Mast cell-neuronal associations are being increasingly invoked as a possible explanation for the pathophysiology of inflammatory disorders (Goetzl *et al.*, 1990; McKay and Bienenstock, 1994; Marshall and Waserman, 1995) (Figure 16.2). Functional interactions have been proposed between mast cells and neurons both within the CNS (Theoharides, 1996;

Rozniecki *et al.*, 1999) and in the periphery (Foreman, 1987; Church *et al.*, 1989). For instance, bladder mast cells identified immunocytochemically by the content of their unique protease, tryptase, were localized next to SP-positive neuronal processes (Figure 16.4). Electron microscopic studies reveal that mast cells are intimately associated with sensory nerve fibres that contain neuropeptides (Wiesner-Menzel *et al.*, 1981; Newson *et al.*, 1983; Skofitsch *et al.*, 1985; Stead *et al.*, 1989). Direct nerve stimulation activates mast cells (Dimitriadou *et al.*, 1991), and mast-cell derived histamine stimulates peripheral neurons (Christian *et al.*, 1989).

Antidromic stimulation of the lumbrosacral dorsal roots in the rat results in plasma extravasation in pelvic organs, including the urinary bladder (Pinter and Szolcsanyi, 1995). IV administration of SP and a specific NK-1 receptor agonist increase vascular permeability in rat urinary bladder (Abelli *et al.*, 1992). So does IV administration of the mast cell secretagogue, compound 48/80 (Eglezos *et al.*, 1992), an effect that is abolished by capsaicin pretreatment, thought to interrupt mast cell activation by CGRP (Eglezos *et al.*, 1992).

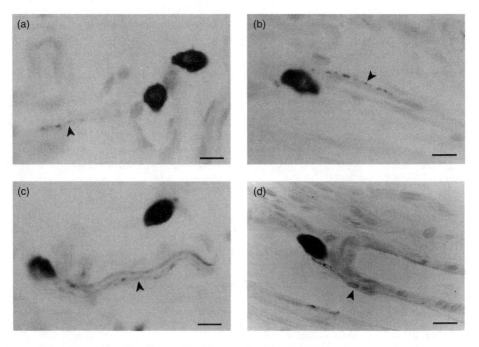

Figure 16.4   Photomicrographs of human bladder mast cells close to neuronal processes examined immunocytochemically for tryptase and SP from a female patient with IC. Note mast cells and SP-positive nerve fibres (solid arrow head). Bladder biopsies (1–2 mm³) were fixed in 2% paraformaldehyde for 2 h at 24°C and were snap-frozen in liquid nitrogen. Frozen 7 μm sections were cut with a cryostat (Jung CM 3000, Leica, Inc., Deerfield, IL, USA), were thaw-mounted and were incubated with 1 : 1000 mouse anti-tryptase monoclonal antibody (Chemicon International, Inc., Tamecula, CA) for 1 h. After three washes in phosphate-buffered saline, 1 : 200 horse anti-mouse IgG-biotin was added to the slides for 30 min, followed by 1 : 200 streptavidin-rhodamine (Pierce,) for 30 min. The slides were then washed in normal saline, were mounted in aqueous mounting medium and this process was repeated for SP. Bar = 10 μm.

Plasma extravasation is inhibited in animals systemically treated with capsaicin (Gronemeyer et al., 1992), a procedure that also blocks dura mast cell activation and vascular permeability induced by antidromic trigeminal nerve stimulation (Dimitriadou et al., 1991), or by acute stress (Pinter and Szolcsanyi, 1995).

MODELS OF BLADDER NEUROIMMUNE INTERACTIONS

Mast cell involvement in bladder pathophysiologic processes is supported by evidence from a number of animal studies. An autoimmune mouse model of bladder inflammation is characterized by mast cell proliferation (Bullock et al., 1992). Type 1 fimbriated E. coli can induce mast cell degranulation and histamine release in experimental animals (Malaviya et al., 1994). Bladder mast cells are increased in feline interstitial cystitis (Buffington et al., 1997), an animal model showing urothelial pathophysiological changes similar to those seen in humans with IC (Elbadawi, 1991); inflamed bladder tissue in this model expresses high affinity binding sites for SP (Fleckestein et al., 1975). SP and bradykinin potentiate the purinergic component of neurogenic excitatory motor innervation of strips of guinea pig urinary bladder (Patra and Westfall, 1996). Intravesical instillation of PGE_2 facilitates micturition, an effect decreased by pretreatment with NK-1 and NK-2 receptor antagonists (Ishizuka et al., 1995). Such inhibition is not seen when PGE_2 is given IV, indicating that the local bladder circuitry may be different (Ishizuka et al., 1995). Acute intravesical administration of ovalbumin in actively sensitized female rats induces plasma protein extravasation (Ahluwalia et al., 1998); prior degranulation of mast cells completely abolishes this response (Wilson et al., 2000). Plasma extravasation is blocked by an NK-1, but not NK-2, receptor antagonist, as well as by a bradykinin-2 receptor antagonist (Ahluwalia et al., 1998).

Intravesical administration of SP or lipopolysaccharide (LPS) does not induce experimental cystitis in mast cell deficient mice (Bjorling et al., 1999). Neurogenic cystitis induced by invasion of rat CNS by the pseudorabies virus is dependent on bladder mast cell activation (Jasmin et al., 2000). Increased micturition frequency due to intravesical acid administration-induced cystitis is reduced by electrical stimulation of the S1 dorsal foramina, but not through inhibition of afferent c-fibre activity (Wang et al., 2000). Intravesical NGF increases the number of bladder contractions and decreases volume, but pre-treatment with capsaicin reduces by 50% only the latter, suggesting that A fibres are also involved (Chuang et al., 2001).

In sensitized mice, intravesical antigen introduction causes mast cell activation, plasma extravasation and inflammation (Saban et al., 2000). However, similar treatment of NK-1R $(-/-)$ double knockout mice leads to mast cell activation, but no inflammation, implying mast cell-to-neuron directional signals (Saban et al., 2000). These studies suggest that bladder mast cells release bradykinin or SP, which induce inflammation through NK-1 receptors. However, SP, NKA, VIP and bradykinin also stimulate release of histamine from isolated guinea pig urinary bladder (Saban et al., 1997). Such bi-directional communication is supported by morphological studies showing nerve-mast cell interactions in humans (Elbadawi and Light, 1996; Letourneau et al., 1996) and guinea pigs (Keith et al., 1995). It was recently shown that communication between murine superior cervical ganglion cells and the mucosal mast cell-like rat basophil leukemia (RBL) cells occurred through SP acting on NK-1 receptors (Cao *et al.*, 1999; Suzuki *et al.*, 1999; Saban *et al.*, 2000). Sympathetic neurons extend axonal projections towards RBL cells when cultured together (Blennerhassett and Bienenstock, 1990; Blennerhassett *et al.*, 1991), and such contacts lower RBL cell resistance, possibly 'priming' them for activation (Blennerhassett *et al.*, 1992); these effects could be mimicked by micromolar concentrations of SP which are normally incapable of inducing secretion (Janiszewski *et al.*, 1992). Sympathetic nerve contacts also induce RBL cell 'maturation', as evidenced by a significant increase in

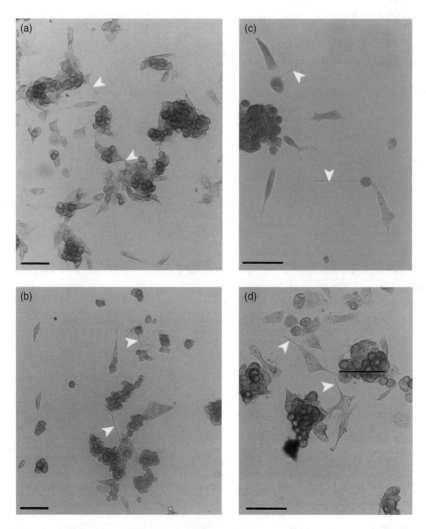

Figure 16.5   Light micrographs of RBL with co-cultured PC-12 cells for 5 days in the presence of NGF and IL-3. (a, b) Note neuronal projections from PC-12 cells (white arrowheads) making contact with RBL cells, (c, d) Note numerous clusters containing PC-12 and RBL cells in close proximity. Bar = 50 μm.

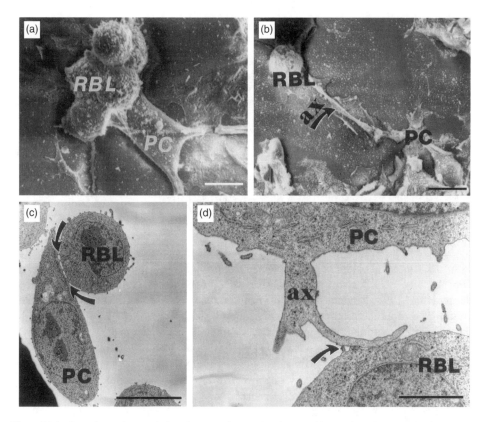

Figure 16.6   Scanning and transmission electron micrographs of RBL with co-cultured PC-12 cells for 5 days in the presence of NGF and IL-3. (a, b) Scanning electron micrographs of RBL cells making direct contact (curved arrow) with PC-12 cells, (c, d) Transmission electron micrographs of (c) one RBL cell with direct contact points (curved solid arrows) with one PC-12 cell (bar = 10 μm) and of (d) one axonal projection (ax) from a PC-12 cell making contact (curved, solid arrow) with an RBL cell (bar = 1 μm).

secretory granules documented by electron microscopy (Blennerhassett and Bienenstock, 1998).

We co-cultured RBL cells with PC-12 cells, (pheochromocytoma cells) (Dichter *et al.*, 1977) in the presence of IL-3 and NGF and demonstrated anatomical contacts with each other (Figure 16.5a–d). In fact, PC-12 cells extended neuronal processes that appeared to touch RBL cells (Figure 16.5a–d). These 'intimate' cell contacts were apparent with both scanning (Figure 16.6a and b) and transmission (Figure 16.6c and d) electron microscopy. RBL cells grown together with PC-12 cells also acquired responsiveness to SP, as shown by the increased release of histamine (Table 16.3). These results indicate that mucosa-like mast cells have the ability to express NK-1 receptors (Cao *et al.*, 1999; Suzuki *et al.*, 1999) or respond to SP through direct activation of G-proteins (Mousli *et al.*, 1990a,b).

Table 16.3
Effect of SP on histamine release from RBL cells
cultured with PC-12 cells.

Conditions	Histamine (% total)
RBL (control)	3.0 ± 0.4
RBL + SP	3.8 ± 0.5
RBL + NGF/IL-3 + SP	5.3 ± 1.2
RBL + PC − 12 (control)	4.1 ± 1.2
RBL + PC − 12 + SP	$25.0 \pm 9.7*$

* $p < 0.05$ (n = 5).

INTERSTITIAL CYSTITIS, AN EXAMPLE OF NEUROIMMUNE PATHOLOGY

Increasing evidence indicates that neuroimmune interactions play a major role in interstitial cystitis (IC), a sterile bladder condition occurring primarily in women (Pontari and Hanno, 1995; Parsons, 1996; Thompson and Christmas, 1996; Sant, 1997). IC also occurs in men (Berger et al., 1998; Novicki et al., 1998) and is often confused with chronic pelvic pain syndromes or non-bacterial prostatitis (Theoharides et al., 1990). IC is characterized by a history of severe urinary urgency, frequency, suprapubic and pelvic pain, as well as dyspareunia, in the absence of bacteriuria (Sant, 1991; Pontari and Hanno, 1995). Clinical diagnosis is usually made by criteria adopted by the NIH/NIDDK for research studies (Hanno et al., 1990): medical history, negative urine cultures, cystoscopic finding of ulcers or petechial mucosal haemorrhages (glomerulations) upon bladder distention under general or spinal anesthesia, and absence of transitional carcinoma or other pathology of the bladder. 'Classic' IC is characterized by Hunner's ulcers and severe inflammation, but is present in less than 10% of IC patients. Typical pathological findings in both 'classic' and the more common non-ulcer IC include submucosal and detrusor edema with suburothelial hemorrhages (Mattila, 1982; Mattila et al., 1983). Neither the severity of symptoms (Johansson and Fall, 1990) nor the cystoscopic findings parallel the degree of bladder inflammation (Messing and Stamey, 1978; Johansson and Fall, 1990). In fact, common IC patients can be divided in two groups: (a) those with inflammation who experience significant relief with bladder distention; and (b) those with mild or no inflammation who do not benefit (Erickson et al., 1994). IC patients with bladder inflammation have higher urine IL-6 levels (Erickson et al., 1997) which correlate with bladder pain (Lotz et al., 1994).

One theory for the cause of IC is that of a defective urothelium, (Parsons et al., 1980) due to reduced bladder glycosaminoglycans (GAG), which allows potential triggers to penetrate the normally impermeable bladder lining (Parsons et al., 2001) and activate local nerve endings (Parsons et al., 1998). This theory is the basis of the KCl sensitivity test (Parsons et al., 2001). However, other studies have demonstrated that the urothelium is intact (Skofitsch et al., 1985) and the levels of urinary sulphfated GAGs are not different from those of controls (Wei et al., 2000). Moreover, a recent assessment of intravesical KCl as a diagnostic test for IC was not validated as the sensitivity was 69.5%, and the specificity was only 50% (Chambers et al., 1999). Another, complementing, theory calls for increased bladder innervation associated with higher numbers of activated mast cells. Even though there is no clinical infection in IC patients, bacteria or bacterial products (Domingue *et al.*, 1995) may contribute to activation of bladder mast cells (Bjorling *et al.*, 1999).

Bladder neuronal proliferation has been demonstrated in IC patients (Christmas *et al.*, 1990; Lundeberg *et al.*, 1993), and an increase in nerve fibres containing NPY and CGRP has also been noted (Christmas *et al.*, 1990; Hohenfellner *et al.*, 1992). Nerve fibres containing SP, adjacent to mast cells, are increased in the bladder submucosa of IC patients (Pang *et al.*, 1995a) (Figure 16.3), as are bladder cells expressing message for the SP receptor (Marchand *et al.*, 1998). SP in the urine of IC patients was not increased compared to controls (Campbell *et al.*, 2001), but this is not surprising since SP acts as a neurotransmitter and its action is limited to synaptic clefts or other junctional areas between nerve termini and target cells. Bladder distention reduces IC symptoms and depletes SP-positive nerves in the rat urinary bladder (Lasanen *et al.*, 1992), while intravesical capsaicin, which acutely stimulates neuronal release of SP, relieves pain in patients with hypersensitive bladder disorders (Barbanti *et al.*, 1993). Rat bladder mast cells can be activated *in vitro* by the neurotransmitter ACh (Theoharides and Sant, 1994; Spanos *et al.*, 1996), as well as by acute stress (Spanos *et al.*, 1997; Alexacos *et al.*, 1999). Urinary pain sensation and urinary bladder hyperreflexia may also be associated with increased bladder expression of pituitary adenylate cyclase-activating polypeptide, as was shown to be the case in cyclophosphamide induced chronic cystitis (Vizzard, 2000).

Mast cell mediators have a variety of pathophysiologic effects that are relevant to the clinical and pathologic findings in IC (Table 16.2). Many studies have reported increased numbers of activated bladder mast cells in IC (for reviews see Theoharides and Sant, 2001). A recent report from the IC database concluded that the only significant pathological finding was increased of tryptase positive bladder mast cells (Tomaszewski *et al.*, 2001). Moreover, we recently showed massive extracellular tryptase from activated bladder mast cells in IC (Theoharides *et al.*, 2001). The factors responsible for the proliferation and/or migration of mast cells in IC are unknown, but damaged urothelial cells produce cytokines, for example interleukin (IL)-3, IL-6 and stem cell factor (SCF) (Galli, 1993). Bladder mast cells in IC may be maximally activated by SCF (Theoharides *et al.*, 1995). The SCF receptor is downregulated in IC suggesting mast cell overstimulation (Peeker *et al.*, 2000). These findings confirm our findings of increased bladder mast cells in non-ulcer IC (Theoharides *et al.*, 1995), and down regulation of the mast cell SCF receptor (Pang *et al.*, 1998). NGF is also increased in the bladders of IC patients (Lowe *et al.*, 1997) and is known to stimulate mast cell proliferation and activation (Marshall *et al.*, 1990; Matsuda *et al.*, 1991). Overgrowth of nerves in the bladder of IC patients may render mast cells responsive to neuropeptides (Shanahan *et al.*, 1985).

Electron microscopy of bladder mast cells from control patients demonstrate that 70% are intact with numerous homogeneous, electron dense secretory granules. In contrast, only 20% of bladder mast cells in IC patients had intact granules; the rest (80%) contained electron dense content at different stages of dissolution (Theoharides *et al.*, 1995) indicating mediator release. This intragranular activation (Letourneau *et al.*, 1996), termed 'piece-meal degranulation' (Dvorak *et al.*, 1992a), is associated with differential mediator release (Theoharides *et al.*, 1982; Kops *et al.*, 1990), especially cytokines (Gagari *et al.*, 1997).

The two main mast cell subtypes differ in histochemical properties, type of granule-associated proteoglycans, neutral proteases and cytokines, as well as morphology, sensitivity to secretagogues and susceptibility to inhibitory drugs (Fox *et al.*, 1988; Bradding *et al.*, 1995; Lutzelschwab *et al.*, 1997). The typical connective tissue mast cell (CTMC) is commonly present in the skin submucosa and stains violet with toluidine blue (Figure 16.3), while the atypical mucosal mast cell (MMC), is found mostly in the bladder and the gastrointestinal tract (Shanahan *et al.*, 1985; Schwartz, 1987; Galli, 1993). Unlike CTMC, MMC secretory granules are susceptible to aldehyde fixation and do not stain with Giemsa or toluidine blue. Consequently, in tissues fixed in 10% formalin (Dundore *et al.*, 1996) MMC are not stained leading to undercounting (Dundore *et al.*, 1996). This may explain conflicting reports of mast cells in IC patients being increased only in the lamina propria or the detrusor (Larsen *et al.*, 1982; Kastrup *et al.*, 1983; Aldenborg *et al.*, 1986; Lynes *et al.*, 1987; Christmas and Rode, 1991), or not at all (Dundore *et al.*, 1996). In addition, the inclusion of transitional carcinoma patients in 'control' groups has led to confusing results (Hanno *et al.*, 1990; Dundore *et al.*, 1996) because of a high number of bladder mast cells associated with the tumour (Theoharides *et al.*, 1995).

Aldenborg and colleagues first compared fixation methods and showed that formaldehyde fixation did not recognize mucosal mast cells stained with toluidine blue. Using iso-osmotic formaldehyde/acetic acid, they reported increased number of mast cells in the mucosa of only ulcerative IC, but increased mast cells in the detrusor of both ulcer and non-ulcer IC (Aldenborg *et al.*, 1986). Lynes *et al.* (1987) also noted a significant increase in detrusor mast cells in ulcer IC, many of which were degranulated. This was the first report showing that mast cells in detrusor and submucosa are degranulated (65 and 50%, respectively) (Lynes *et al.*, 1987). Mast cell density did not correlate with severity of symptoms, but did correlate with degree of inflammation (Lynes *et al.*, 1987). Mast cells are significantly increased in the lamina propria and detrusor in ulcer IC (Johansson and Fall, 1990). In a study of mast cells in IC, bacterial cystitis, and controls, (Christmas and Rode, 1991) there was increase in mast cells in the urothelium/submucosa of non-ulcer IC patients (Christmas and Rode, 1991).

Mast cells are consistently increased in 'classic' IC (Enerbäck *et al.*, 1989) and it was proposed that >30 mast cells/mm^2 in the detrusor muscle may be sufficient for diagnosis (Larsen *et al.*, 1982; Theoharides *et al.*, 1995). Larsen *et al.* (1982) reported no differences in bladder lamina propria mast cells between IC and controls, but found that they were increased and partially degranulated in the mascularis (Larsen *et al.*, 1982). The submucosa of non-ulcer IC had more mast cells than controls and this difference was larger in the muscularis (Feltis *et al.*, 1987). Kastrup *et al.* (1983) and Kruger and Bloom (1974) found increased detrusor mast cells in IC patients.

Recent papers employed immunocytochemistry for tryptase staining (Pang *et al.*, 1995b). Tryptase is a unique protease found in both CTMC and MMC (Figure 16.4). The mast cell proteases, tryptase and chymase, are used to categorize the types of mast cells in IC. Tryptase positive/chymase negative mast cells (MMC) were about 150/mm^2 in the detrusor and 80/mm^2 in the mucosa of IC patients (Yamada *et al.*, 2000), while tryptase positive/chymase positive mast cells (CTMC) were about 125/mm^2; the authors concluded that these numbers were significantly increased over control (Yamada *et al.*, 2000). Ulcer IC was associated with 6–10 fold increase of detrusor tryptase-positive mast cells, while non-ulcer IC had only twice as many mast cells (Peeker *et al.*, 2000).

The presence of potential 'markers' in the urine of IC patients was reviewed recently (Erickson, 2001). Urine prostaglandin E_2 (PGE_2) was not statistically increased in IC (Lynes et al., 1987), nor were PGD_2, PGE_2, PGF_2, thromboxane B_2, or TNF-α (Felsen et al., 1994). Bladder mast cells from IC patients were more responsive to IgE and antigen than those from control patients (Frenz et al., 1994a,b). In vivo bladder mast cell activation was documented by demonstration of urine histamine increase in IC, but not in controls, following bladder hydrodistention (Yun et al., 1992). Elevated histamine has been documented in the bladder walls of IC patients (Kastrup et al., 1983; Lynes et al., 1987; Enerbäck et al., 1989). The histamine content of either spot or 24 urine was not different between control and IC patients (Yun et al., 1992; El-Mansoury et al., 1994), but the histamine metabolites 1,4-methylimidazole acetic acid (1,4-MIAA) (Holm-Bentzen et al., 1987) and methylhistamine (El-Mansoury et al., 1994) were significantly elevated in IC. Tryptase, a specific mast cell marker was also elevated in the urine, but not serum of IC patients (Boucher et al., 1995). IL-6 was elevated in the urine of IC patients, was shown to originate from the bladder and correlated with pain scores (Lotz et al., 1994). Another study of a small number of IC patients also reported elevated urine IL-6 levels (Felsen et al., 1994). In a later study, the severity of inflammation was shown to be associated with urine IL-6 levels (Erickson et al., 1997). IL-6 is a major cytokine secreted from mast cells (Galli, 1993). IL-6 release can be induced from mast cells by bacterial lipopolysaccharide (LPS) (Leal-Berumen et al., 1994) and SCF (Gagari et al., 1997). These biochemical data confirm activation of mast cells in IC, as had been previously demonstrated ultrastructurally (Lynes et al., 1987; Theoharides et al., 1995; Elbadawi and Light, 1996). Activated mast cells also occur in men with IC (Theoharides et al., 1990), suggesting that chronic prostatitis may be a variant of IC (Berger et al., 1998; Novicki et al., 1998). This premise is supported by the finding that pentosanpolysulphfate used in IC (Rovner et al., 2000) is also effective in the treatment of non-bacterial prostatitis (Nickel et al., 2000). The flavonoid quercetin was also shown to reduce symptoms of IC (Katske et al., 2001) and of chronic, abacterial prostatitis (Shoskes et al., 1999). Flavonoids are powerful antioxidants, anti-inflammatory compounds which block secretion from both CTMC and MMC (Middleton et al., 2000). In fact, the structurally related flavonoid myricetin was recently shown to inhibit NF-κB activity in activated endothelial cells and prevented monocyte adhesion (Tsai et al., 1999). Combining the flavonoid quercetin (Middleton et al., 2000), with chondroitin sulphate (Theoharides et al., 2000) which also inhibits mast cell activation, in the food supplement (Algonot-Plus® -see Algonot.com) may provide an alternative therapeutic approach to the treatment of IC, chronic prostatitis and related disorders.

IC AND IRRITABLE BOWEL SYNDROME (IBS)

Abdominal tenderness associated with IC is often mistaken for, or co-exists with, pelvic inflammatory disease, endometriosis or IBS (Sant, 1991; Koziol et al., 1993). Such presentation is not surprising because the innervation of bladder and intestine is very similar (Burnstock, 1990). We first reported that about 40% of IC patients have IBS, while more than 50% have a history of allergies and asthma (Theoharides and Sant, 1994). These findings were later substantiated in a large number of patients (Koziol et al., 1993; Alagiri et al., 1997). Many of these IC patients report worsening of their symptoms by acute emotional stress (Rothrock *et al.*, 2001). IBS is also well known to be precipitated or exacerbated by stress (Mayer, 2000; Mayer *et al.*, 2001).

IBS is similar to IC in that it occurs more often in females, is characterized by abdominal pain, urinary frequency and fatigue (Farthing, 1995). Increased number of mast cells have been documented in the terminal ileum (Weston *et al.*, 1993) colon (Akdis *et al.*, 2000) and caecum (O'Sullivan *et al.*, 2000) of IBS patients. Increased number of mast cells close to SP positive fibres (Figure 16.4) were identified in both the colon and the bladder of a patient with both IBS and IC (Pang *et al.*, 1996). The fact that the underlying pathophysiological process may be similar in the bladder and the intestine is supported by the fact that patients who had undergone cystoplasty developed IC symptoms and the intestine used for bladder augmentation was characterized by increased numbers of mast cells (Kisman *et al.*, 1991; Singh and Thomas, 1996). Mast cell involvement is further suggested by the fact that patients with bronchial asthma also have bladder hypersensitivity (Yamada *et al.*, 1998) and a human anti-IgE antibody blocked passive sensitization of the human bladder (Saban *et al.*, 1997). Stress may 'reactivate' previous inflammation, especially when it occurs together with some trigger (Collins, 2001). Prior inflammation increases the response of rat colon to stress, and this may explain the higher incidence of IBS in patients with inflammatory bowel disease (Collins *et al.*, 1996). In fact, chemical colitis in rats alters the myenteric nerve function (Jacobson *et al.*, 1997). Close anatomical relationship between bladder neurons and mast cells (Elbadawi and Light, 1996; Letourneau *et al.*, 1996) has also been reported in Crohn's disease (Dvorak *et al.*, 1992b).

Intense daily stress correlates significantly with bladder pain, urgency and frequency in IC patients (Rothrock *et al.*, 2001). Patients with stress from orofacial pain have a higher incidence of IC and IBS (Korszun *et al.*, 1998). IC patients also have increased urinary levels of norepinephrine (NE) (Stein *et al.*, 1999). In a rat immobilization stress model, bladder mast cells (Spanos *et al.*, 1997), vascular permeability (Alexacos *et al.*, 1999) and urine IL-6 levels (Boucher *et al.*, 2000) were all increased. Acute psychological stress in rats result in bladder (Spanos *et al.*, 1997; Alexacos *et al.*, 1999) and intestinal (Castagliuolo *et al.*, 1996a; Castagliuolo *et al.*, 1998; Theoharides *et al.*, 1999) mast cell activation, actions mediated via local CRH-induced mast cell activation (Theoharides, 1996), leading to pro-inflammatory effects (Chrousos, 1995). The action of CRH may involve additional neuropeptides, especially NT, in both intestinal (Castagliuolo *et al.*, 1996b; Castagliuolo *et al.*, 1999), and bladder response to acute stress (Theoharides *et al.*, 1998). Combination of cold and immobilization stress increased the number of degranulated bladder mast cells in rats, and induced urothelium degeneration vacuolation and opening of tight junctions (Ercan *et al.*, 1999). Similar findings have been reported in the colon of rats subjected to immobilization stress (Theoharides *et al.*, 1999). These changes were prevented either by capsaicin administration neonatally or by infusion of vagal and celiac ganglia (Peeker *et al.*, 2000). The effect of capsaicin may not be limited to sensory nerves as it also causes thinning of bladder urothelium and submucosa (Byrne *et al.*, 1998). Although these studies implicate c-fibres, stress-induced bladder mast cell activation is partially dependent on NT-containing nerve fibres (Theoharides *et al.*, 1998). Moreover, NGF-induced bladder hyperactivity is not be blocked by capsaicin (Chuang *et al.*, 2001). *In vivo*, NGF may be released from activated mast cells (Xiang and Nilsson, 2000). In line with the possibility that other neuronal pathways, besides c-fibres, are involved in neuropathic bladder pain, it was shown that retrograde infection of rats with pseudorabies virus led to bladder inflammation which was prevented when CNS–bladder circuits were interrupted (Jasmin *et al.*, 1998).

Female sex hormones worsen autoimmune disorders (Ahmed *et al.*, 1985; Schuurs and Verheul, 1990), the prevalence of which switches in adolescence from a male to female predominance. Allergic reactions also may worsen in association with the menstrual cycle (Zondek and Bromberg, 1945; Gibbs *et al.*, 1984). Many pre-menopausal women with IC report worsening of their symptoms premenstrually or during ovulation (Pontari and Hanno, 1995). This effect may be explained by the fact that estradiol augments mast cell histamine secretion *in vivo* (Conrad and Feigen, 1974), secretion triggered by SP *in vitro* (Vliagoftis *et al.*, 1992), and bladder mast cell secretion *in situ* (Spanos *et al.*, 1996). Bladder mast cells in IC exhibit increased expression of high affinity estrogen receptors (Pang *et al.*, 1995b), and estrogens induce proliferation of bladder mast cells in guinea pigs (Patra *et al.*, 1995).

CONCLUSIONS

The recognition of neuroimmune interactions provides new possibilities for the understanding and therapy of bladder diseases, especially those involving pain, such as interstitial cystitis (Theoharides and Sant, 2001). IC is increasingly considered as a condition with chronic neuropathic pain (Wesselmann, 2001) and no available treatment appears to have made a significant impact (Propert *et al.*, 2000). Nevertheless, bladder mast cell activation may contribute to the pathophysiology, as evidenced by the fact that the best studied drug in IC, pentosanpolysulphfate, was recently shown to be a potent inhibitor of mast cell secretion (Chiang *et al.*, 2000). Blockade of mast cell activation may be the best way to interrupt the mast cell–neuron stimulation cycle (Theoharides and Sant, 2001).

ACKNOWLEDGEMENTS

Aspects of this work were supported in part by grants DK42409 and DK44816 from the NIH/NIDDK, from Kos Pharmaceuticals, Inc. (Miami, FL) and Theta Biomedical Consulting and Development Co., Inc. (Brookline, MA), as well as pilot grants from the Interstitial Cystitis Association (New York, NY). We appreciate Dr A. Tishler's kind supply of PC-12 cells. Thanks are due to Ms Sharon Titus and Ms Yvonne Battad for their word-processing skills.

REFERENCES

Abelli, L., Nappi, F., Perretti, F., Maggi, C. A., Manzini, S. and Giachetti, A. (1992) Microvascular leakage induced by substance P in rat urinary bladder: involvement of cyclo-oxygenase metabolites of arachidonic acid. *Journal of Autonomic Pharmacology*, **12**, 269–276.
Ahluwalia, A., Giuliani, S., Scotland, R. and Maggi, C. A. (1998) Ovalbumin-induced neurogenic inflammation in the bladder of sensitized rats. *British Journal of Pharmacology*, **124**, 190–196.

Ahmed, S. A., Penhale, W. J. and Talal, N. (1985) Sex hormones, immune responses and autoimmune diseases. *American Journal of Pathology*, **121**, 531–551.

Akdis, C. A., Akdis, M., Trautmann, A. and Blaser, K. (2000) Immune regulation in atopic dermatitis. *Current Opinions in Immunology*, **12**, 641–646.

Alagiri, M., Chottiner, S., Ratner, V., Slade, D. and Hanno, P. (1997) Interstitial cystitis: unexplained associations with other chronic disease and pain syndromes. *Urology*, **49**, 52–57.

Aldenborg, F., Fall, M. and Enerbäck, L. (1986) Proliferation and transepithelial migration of mucosal mast cells in interstitial cystitis. *Immunology*, **58**, 411–416.

Alexacos, N., Pang, X., Boucher, W., Cochrane, D. E., Sant, G. R. and Theoharides, T. C. (1999) Neurotensin mediates rat bladder mast cell degranulation triggered by acute psychological stress. *Urology*, **53**, 1035–1040.

Barbanti, G., Maggi, C., Beneforti, P., Baroldi, P. and Turini, D. (1993) Relief of pain following intravesical capsaicin in patients with hypersensitive disorders of the lower urinary tract. *Journal of Urology*, **71**, 686–691.

Berger, R. E., Miller, J. E., Rothman, I., Krieger, J. N. and Muller, C. H. (1998) Bladder petechiae after cystoscopy and hydrodistension in men diagnosed with prostate pain. *Journal of Urology*, **159**, 83–85.

Bjorling, D. E., Jerde, T. J., Zine, M. J., Busser, B. W., Saban, M. R. and Saban, R. (1999) Mast cells mediate the severity of experimental cystitis in mice. *Journal of Urology*, **162**, 231–236.

Blennerhassett, M. G. and Bienenstock, J. (1990) Apparent innervation of rat basophilic leukaemia (RBL-2H3) cells by sympathetic neurons *in vitro*. *Neuroscience Letters*, **120**, 50–54.

Blennerhassett, M. G. and Bienenstock, J. (1998) Sympathetic nerve contact causes maturation of mast cells *in vitro*. *Journal of Neurobiology*, **35**, 173–182.

Blennerhassett, M. G., Janiszewski, J. and Bienenstock, J. (1992) Sympathetic nerve contact alters membrane resistance of cells of the RBL-2H3 mucosal mast cell line. *American Journal of Respiratory Cell and Molecular Biology*, **6**, 504–509.

Blennerhassett, M. G., Tomioka, M. and Bienenstock, J. (1991) Formation of contacts between mast cells and sympathetic neurons *in vitro*. *Cell and Tissue Research*, **265**, 121–128.

Boucher, W., El-Mansoury, M., Pang, X., Sant, G. R. and Theoharides, T. C. (1995) Elevated mast cell tryptase in urine of interstitial cystitis patients. *British Journal of Urology*, **76**, 94–100.

Boucher, W. S., Letourneau, R., Huang, M., Kempuraj, D., Green, M., Sant, G. R. and Theoharides, T. C. (2000) Intravesical sodium hyaluronate inhibits rat urine mast cell mediator increase triggered by acute immobilization stress. *Journal of Urology*, **167**, 380–384.

Bradding, P., Okayama, Y., Howarth, P. H., Church, M. K. and Holgate, S. T. (1995) Heterogeneity of human mast cells based on cytokine content. *Journal of Immunology*, **155**, 297–307.

Buffington, C. A. T., Chew, D. J. and Woodworth, B. E. (1997) Animal model of human disease-feline interstitial cystitis. *Comparative Pathology Bulletin*, **29**, 3.

Bullock, A. D., Becich, M. J., Klutke, C. G. and Ratliff, T. L. (1992) Experimental autoimmune cystitis: a potential murine model for ulcerative interstitial cystitis. *Journal of Urology*, **148**, 1951–1956.

Burnstock, G. (1990) Innervation of bladder and bowel. *Ciba Foundation Symposium*, **151**, 2–18.

Byrne, D. S., Das, A., Sedor, J., Huang, B., Rivas, D. A., Flood, H. J., DeGroat, W., Jordan, M. L., Chancellor, M. B. and McCue, P. (1998) Effect of intravesical capsaicin and vehicle on bladder integrity control and spinal cord injured rats. *Journal of Urology*, **159**, 1074–1078.

Campbell, D. J., Tenis, N., Rosamilia, A., Clements, J. A. and Dwyer, P. L. (2001) Urinary levels of substance P and its metabolites are not increased in interstitial cystitis. *British Journal of Urology International*, **87**, 35–38.

Cao, T., Gerard, N. P. and Brain, S. D. (1999) Use of NK1 knockout mice to analyze substance P-induced edema formation. *American Journal of Physiology*, **277**, R476–R481.

Carraway, R., Cochrane, D. E., Lansman, J. B., Leeman, S. E., Paterson, B. M. and Welch, H. J. (1982) Neurotensin stimulates exocytotic histamine secretion from rat mast cells and elevates plasma histamine levels. *Journal of Physiology*, **323**, 403–414.

Castagliuolo, I., LaMont, J. T., Qiu, B., Fleming, S. M., Bhaskar, K. R., Nikulasson, S. T., Kornetsky, C. and Pothoulakis, C. (1996a) Acute stress causes mucin release from rat colon: role of corticotropin releasing factor and mast cells. *American Journal of Physiology*, **271**, 884–892.

Castagliuolo, I., Leeman, S. E., Bartolac-Suki, E., Nikulasson, S., Qiu, B., Carraway, R. E. and Pothoulakis, C. (1996b) A neurotensin antagonist, SR 48692, inhibits colonic responses to immobilization stress in rats. *Proceedings of the National Academy of Sciences USA*, **93**, 12611–12615.

Castagliuolo, I., Wang, C.-C., Valenick, L., Pasha, A., Nikulasson, S., Carraway, R. E. and Pothoulakis, C. (1999) Neurotensin is a proinflammatory neuropeptide in colonic inflammation. *Journal of Clinical Investigation*, **103**, 843–848.

Castagliuolo, I., Wershil, B. K., Karalis, K., Pasha, A., Nikulasson, S. T. and Pothoulakis, C. (1998) Colonic mucin release in response to immobilization stress is mast cell dependent. *American Journal of Physiology*, **274**, 1094–1100.

Chai, T. C. and Steers, W. D. (1997) Neurophysiology of micturition and continence in women. *International Urogynecology Journal*, **8**, 85–97.

Chambers, G. K., Fenster, H. N., Cripps, S., Jens, M. and Taylor, D. (1999) An assessment of the use of intravesical potassium in the diagnosis of interstitial cystitis. *Journal of Urology*, **162**, 699–701.

Chiang, G., Patra, P., Letourneau, R., Jeudy, S., Boucher, W., Green, M., Sant, G. R. and Theoharides, T. C. (2000) Pentosanpolysulfate (Elmiron) inhibits mast cell histamine secretion and intracellular calcium ion levels: an alternative explanation of its beneficial effect in interstitial cystitis. *Journal of Urology*, **164**, 2119–2125.

Christian, E. P., Undem, B. J. and Weinreich, D. (1989) Endogenous histamine excites neurones in the guinea-pig superior cervical ganglion *in vitro*. *Journal of Physiology*, **409**, 297–312.

Christmas, T. J. and Rode, J. (1991) Characteristics of mast cells in normal bladder, bacterial cystitis and interstitial cystitis. *British Journal of Urology*, **68**, 473–478.

Christmas, T. J., Rode, J., Chapple, C. R., Milroy, E. J. and Turner-Warwick, R. T. (1990) Nerve fibre proliferation in interstitial cystitis. *Virchows Arch [A]*, **416**, 447–451.

Chrousos, G. P. (1995) The hypothalamic-pituitary-adrenal axis and immune-mediated inflammation. *New England Journal of Medicine*, **332**, 1351–1362.

Chuang, Y. C., Fraser, M. O., Yu, Y., Chancellor, M. B., deGROAT, W. C. and Yoshimura, N. (2001) The role of bladder afferent pathways in bladder hyperactivity induced by the intravesical administration of nerve growth factor. *Journal of Urology*, **165**, 975–979.

Church, M. K., Lowman, M. A., Rees, P. H. and Benyon, R. C. (1989) Mast cells, neuropeptides and inflammation. *Agents and Actions*, **27**, 8–16.

Collins, S. M. (2001) Stress and the gastrointestinal tract IV. Modulation of intestinal inflammation by stress: basic mechanisms and clinical relevance. *American Journal of Physiology*, **280**, G315–G318.

Collins, S. M., McHugh, K., Jacobson, K., Khan, I., Riddell, R., Murase, K. and Weingarten, H. P. (1996) Previous inflammation alters the response of the rat colon to stress. *Gastroenterology*, **111**, 1509–1515.

Conrad, M. J. and Feigen, G. A. (1974) Sex hormones and kinetics of anaphylactic histamine release. *Physiological Chemistry and Physics*, **6**, 11–16.

Cutz, E., Chan, W., Track, N. S., Goth, A. and Said, S. I. (1978) Release of vasoactive intestinal polypeptide in mast cells by histamine liberators. *Nature*, **275**, 661–662.

Dichter, M. A., Tischler, A. S. and Greene, L. A. (1977) Nerve growth factor-induced increase in electrical excitability and acetylcholine sensitivity of a rat pheochromocytoma cell line. *Nature*, **268**, 501–504.

Dimitriadou, V., Buzzi, M. G., Moskowitz, M. A. and Theoharides, T. C. (1991) Trigeminal sensory fiber stimulation induces morphologic changes reflecting secretion in rat dura mast cells. *Neuroscience*, **44**, 97–112.

Dixon, J. S. and Gilpin, C. J. (1987) Presumptive sensory axons of the human urinary bladder; a fine structural study. *Journal of Anatomy*, **151**, 199–207.

Domingue, G. J., Ghoniem, G. M., Bost, K. L., Fermin, C. and Human, L. G. (1995) Dormant microbes in interstitial cystitis. *Journal of Urology*, **153**, 1321–1326.

Dundore, P. A., Schwartz, A. M. and Semerjian, H. (1996) Mast cell counts are not useful in the diagnosis of nonulcerative interstitial cystitis. *Journal of Urology*, **155**, 885–887.

Dvorak, A. M., McLeod, R. S., Onderdonk, A., Monahan-Earley, R. A., Cullen, J. B., Antonioli, D. A., Morgan, E., Blair, J. E., Estrella, P., Cisneros, R. L., Silen, W. and Cohen, Z. (1992a) Ultrastructural evidence for piecemeal and anaphylactic degranulation of human gut mucosal mast cells *in vivo*. *International Archives of Allergy and Immunology*, **99**, 74–83.

Dvorak, A. M., McLeod, R. S., Onderdonk, A. B., Monahan-Earley, R. A., Cullen, J. B., Antonioli, D. A., Morgan, E., Blair, J. E., Estrella, P., Cisneros, R. L., Cohen, Z. and Silen, W. (1992b) Human gut mucosal mast cells: ultrastructural observations and anatomic variation in mast cell-nerve associations *in vivo*. *International Archives of Allergy and Immunology*, **98**, 158–168.

Eglezos, A., Lecci, A., Santicioli, P., Giuliani, S., Tramontana, M., Del Bianco, E. and Maggi, C. A. (1992) Activation of capsaicin-sensitive primary afferents in the rat urinary bladder by compound 48/80: A direct action on sensory nerves? *Archives International Pharmacodynamics Therapeutics*, **315**, 96–109.

El-Mansoury, M., Boucher, W., Sant, G. R. and Theoharides, T. C. (1994) Increased urine histamine and methylhistamine in interstitial cystitis. *Journal of Urology*, **152**, 350–353.

Elbadawi, A. (1991) Anatomy and innervation of the vesicourethral muscular unit of micturition. In *Clinical Neurology*, edited by R. J. Krane and M. B. Siroky, pp. 5–23. Boston/Toronto/London: Little, Brown and Company.

Elbadawi, A. and Light, J. K. (1996) Distinctive ultrastructural pathology of nonulcerative interstitial cystitis. *Urology International*, **56**, 137–162.

Elenkov, I. J., Wilder, R. L., Chrousos, G. P. and Vizi, E. S. (2000) The sympathetic nerve – an integrative interface between two supersystems: the Brain and the immune system. *Pharmacological Reviews*, **52**, 595–638.

Enerbäck, L., Fall, M. and Aldenborg, F. (1989) Histamine and mucosal mast cells in interstitial cystitis. *Agents and Actions*, **27**, 113–116.

Ercan, F., San, T. and Cavdar, S. (1999) The effects of cold-restraint stress on urinary bladder wall compared with interstitial cystitis morphology. *Urology Research*, **27**, 454–461.

Erickson, D. R. (2001) Urine markers of interstitial cystitis. *Urology*, **57**, 15–21.

Erickson, D. R., Belchis, D. A. and Dabbs, D. J. (1997) Inflammatory cell types and clinical features of interstitial cystitis. *Journal of Urology*, **158**, 790–793.

Erickson, D. R., Simon, L. J. and Belchis, D. A. (1994) Relationships between bladder inflammation and other clinical features in interstitial cystitis. *Urology*, **44**, 655–659.

Farthing, M. J. G. (1995) Irritable bowel, irritable body, or irritable brain? *British Medical Journal*, **310**, 171–175.

Felsen, D., Frye, S., Trimble, L. A., Bavendam, T. G., Parsons, C. L., Sim, Y. and Vaughan, E. D., Jr. (1994) Inflammatory mediator profile in urine and bladder wash fluid of patients with interstitial cystitis. *Journal of Urology*, **152**, 355–361.

Feltis, J. T., Perez-Marrero, R. and Emerson, L. E. (1987) Increased mast cells of the bladder in suspected cases of interstitial cystitis: a possible disease marker. *Journal of Urology*, **138**, 42–43.

Fewtrell, C. M. S., Foreman, J. C., Jordan, C. C., Oehme, P., Renner, H. and Stewart, J. M. (1982) The effects of substance P on histamine and 5-hydroxytryptamine release in the rat. *Journal of Physiology*, **330**, 393–411.

Fleckestein, A., Nakayama, K., Fleckestein-Grun, G. and Byon, Y. K. (1975) Interactions of vasoactive ions and drugs with Ca^{++} dependent excitation-contraction coupling of vascular smooth muscle. In *Calcium Transport in Contraction and Secretion*, edited by E. Carafoli *et al.*, pp. 555–566. Amsterdam: Elsevier.

Foreman, J. C. (1987) Peptides and neurogenic inflammation. *Brain Research Bulletin*, **43**, 386–398.

Fowler, C. J. (2000) Intravesical treatment of overactive bladder. *Urology*, **55**, 60–64.

Fox, C. C., Wolf, E. J., Kagey-Sobotka, A. and Lichtenstein, L. M. (1988) Comparison of human lung and intestinal mast cells. *Journal of Allergy Clinical Immunology*, **81**, 89–94.

Frenz, A. M., Christmas, T. J. and Pearce, F. L. (1994a) Does the mast cell have an intrinsic role in the pathogenesis of interstitial cystitis? *Agents Actions*, **41**, C14–C15.

Frenz, A. M., Christmas, T. J. and Pearce, F. L. (1994b) Functional evaluation of mast cells in interstitial cystitis. *Journal of Urology*, **151**, 219.

Gabella, G. and Davis, C. (1998) Distribution of afferent axons in the bladder of rats. *Journal of Neurocytology*, **27**, 141–155.

Gagari, E., Tsai, M., Lantz, C. S., Fox, L. G. and Galli, S. J. (1997) Differential release of mast cell interleukin-6 via c-kit. *Blood*, **89**, 2654–2663.

Galli, S. J. (1993) New concepts about the mast cell. *New England Journal of Medicine*, **328**, 257–265.

Gibbs, C. J., Coutts, I. I., Lock, R., Finnegan, O. C. and White, R. J. (1984) Premenstrual exacerbation of asthma. *Thorax*, **39**, 833–836.

Goetzl, E. J., Cheng, P. P. J., Hassner, A., Adelman, D. C., Frick, O. L. and Speedharan, S. P. (1990) Neuropeptides, mast cells and allergy: novel mechanisms and therapeutic possibilities. *Clinical and Experimental Allergy*, **20**, 3–7.

Gosling, J. A., Dixon, J. S. and Jen, P. Y. (1999) The distribution of noradrenergic nerves in the human lower urinary tract. *European Urology*, **1**, 23–30.

Gronemeyer, H., Benhamou, B., Berry, M., Bocquel, M. T., Gofflo, D., Garcia, T., Lerouge, T., Metzger, D., Meyer, M. E., Tora, L., Vergezac, A. and Chambon, P. (1992) Mechanisms of antihormone action. *Journal of Steroidal Biochemical Molecular Biology*, **41**, 3–8.

Grundemar, L., Krstenansky, J. L. and Håkanson, R. (1994) Neuropeptide Y and truncated neuropeptide Y analogs evoke histamine release from rat peritoneal mast cells. A direct effect on G proteins? *European Journal of Pharmacology*, **258**, 163–166.

Hanno, P. M., Levin, R. M., Monson, F. C., Teuscher, C., Zhou, Z. Z., Ruggieri, M., Whitmore, K. and Wein, A. J. (1990) Diagnosis of interstitial cystitis. *Journal of Urology*, **143**, 278–281.

Hill, R. (2000) NK_1 (substance P) receptor antagonists – why are they not analgesic in humans? *Trends in Pharmacological Sciences*, **21**, 244–246.

Hohenfellner, M., Nunes, L., Schmidt, R. A., Lampel, A., Thuroff, J. W. and Tanagho, E. A. (1992) Interstitial cystitis: increased sympathetic innervation and related neuropeptide synthesis. *Journal of Urology*, **147**, 587–591.

Holm-Bentzen, M., Sændergaard, I. and Hald, T. (1987) Urinary excretion of a metabolite of histamine (1,4-methyl-imidazole-acetic-acid) in painful bladder disease. *British Journal of Urology*, **59**, 230–233.

Ishizuka, O., Mattiasson, A. and Andersson, K.-E. (1995) Prostaglandin E_2-induced bladder hyperactivity in normal, conscious rats: involvement of tachykinins? *Journal of Urology*, **153**, 2034–2038.

Jacobson, K., McHugh, K. and Collins, S. M. (1997) The mechanism of altered neural function in a rat model of acute colitis. *Gastroenterology*, **112**, 156–162.

Janiszewski, J., Bienenstock, J. and Blennerhassett, M. G. (1992) Substance P induces whole cell current transients in RBL-2H3 cells. *American Journal of Physiology*, **263**, 736–742.

Jasmin, L., Janni, G., Manz, H. J. and Rabkin, S. D. (1998) Activation of CNS circuits producing a neurogenic cystitis: evidence for centrally induced peripheral inflammation. *Journal of Neuroscience*, **18**, 10016–10029.

Jasmin, L., Janni, G., Ohara, P. T. and Rabkin, S. D. (2000) CNS induced neurogenic cystitis is associated with bladder mast cell degranulation in the rat. *Journal of Urology*, **164**, 852–855.

Johansson, S. L. and Fall, M. (1990) Clinical features and spectrum of light microscopic changes in interstitial cystitis. *Journal of Urology*, **143**, 1118–1124.

Karalis, K., Louis, J. M., Bae, D., Hilderbrand, H. and Majzoub, J. A. (1997) CRH and the immune system. *Journal of Neuroimmunology*, **72**, 131–136.

Kastrup, J., Hald, J., Larsen, L. and Nielsen, V. G. (1983) Histamine content and mast cell count of detrusor muscle in patients with interstitial cystitis and other types of chronic cystitis. *British Journal of Urology*, **55**, 495–500.

Katske, F., Shoskes, D. A., Sender, M., Poliakin, R., Gagliano, K. and Rajfer, J. (2001) Treatment of interstitial cystitis with a quercetin supplement. *Techniques in Urology*, **7**, 44–46.

Keith, I. M., Jin, J. and Saban, R. (1995) Nerve–mast cell interaction in normal guinea pig urinary bladder. *Journal of Comparative Neurology*, **363**, 28–36.

Kim, D. Y. and Chancellor, M. B. (2000) Intravesical neuromodulatory drugs: capsaicin and resiniferatoxin to treat the overactive bladder. *Journal of Endourology*, **14**, 97–103.

Kinder, R. B. and Mundy, A. R. (1991) Neurotransmitters. In *Clinical Neuro-Urology*, edited by R. J. Krane and M. B. Siroky, pp. 83–92. Boston/Toronto/London: Little, Brown and Company.

Kisman, O. K., Lycklama a Nijeholt, A. A. B. and van Krieken, J. H. J. M. (1991) Mast cell infiltration in intestine used for bladder augmentation in interstitial cystitis. *Journal of Urology*, **146**, 1113–1114.

Klein, L. M., Lavker, R. M., Matis, W. L. and Murphy, G. F. (1989) Degranulation of human mast cells induces an endothelial antigen central to leukocyte adhesion. *Proceedings of the National Academy of Sciences USA*, **86**, 8972–8976.

Kluck, P. (1980) The autonomic innervation of the human urinary bladder, bladder neck and urethra: a histochemical study. *Anatomical Record*, **198**, 439–447.

Kobayashi, H., Ishizuka, T. and Okayama, Y. (2000) Human mast cells and basophils as sources of cytokines. *Clinical Experimental Allergy*, **30**, 1205–1212.

Kops, S. K., Theoharides, T. C., Cronin, C. T., Kashgarian, M. G. and Askenase, P. W. (1990) Ultrastructural characteristics of rat peritoneal mast cells undergoing differential release of serotonin without histamine and without degranulation. *Cell and Tissue Research*, **262**, 415–424.

Korszun, A., Papadopoulos, E., Demitrack, M., Engleberg, C. and Crofford, L. (1998) The relationship between temporomandibular disorders and stress-associated syndromes. *Oral Surgery, Oral Medicine, Oral Pathology, Oral Radiology, and Endodontics*, **86**, 416–420.

Koziol, J. A., Clark, D. C., Gittes, R. F. and Tan, E. M. (1993) The natural history of interstitial cystitis: a survey of 374 patients. *Journal of Urology*, **149**, 465–469.

Kruger, P. G. and Bloom, G. D. (1974) Structural features of histamine release in rat peritoneal mast cells; a study with toluidine blue. *International Archives of Allergy and Applied Immunology*, **46**, 740–752.

Larsen, S., Thompson, S. A., Hald, T., Barnard, R. J., Gilpin, C. J., Dixon, J. S. and Gosling, J. A. (1982) Mast cells in interstitial cystitis. *British Journal of Urology*, **54**, 283–286.

Lasanen, L. T., Tammela, T. L. J., Liesi, P., Waris, T. and Polak, J. M. (1992) The effect of acute distension on vasoactive intestinal polypeptide (VIP) neuropeptide Y (NPY) and substance P (SP) immunoreactive nerves in the female rat urinary bladder. *Urology Research*, **20**, 259–263.

Lazzeri, M., Beneforti, P., Spinelli, M., Zanollo, A., Barbagli, G. and Turini, D. (2000) Intravesical resiniferatoxin for the treatment of hypersensitive disorder: a randomized placebo controlled study. *Journal of Urology*, **164**, 676–679.

Leal-Berumen, I., Conlon, P. and Marshall, J. S. (1994) IL-6 production by rat peritoneal mast cells is not necessarily preceded by histamine release and can be induced by bacterial lipopolysaccharide. *Journal of Immunology*, **152**, 5468–5476.

Leppilahti, M., Kallioinen, M. and Tammela, T. L. (1999) Duration of increased mucosal permeability of the urinary bladder after acute overdistension: an experimental study in rats. *Urology Research*, **27**, 272–276.

Letourneau, R., Pang, X., Sant, G. R. and Theoharides, T. C. (1996) Intragranular activation of bladder mast cells and their association with nerve processes in interstitial cystitis. *British Journal of Urology*, **77**, 41–54.

Lotz, M., Villiger, P., Hugli, T., Koziol, J. and Zuraw, B. L. (1994) Interleukin-6 and interstitial cystitis. *Journal of Urology*, **152**, 869–873.

Lowe, E. M., Anand, P., Terenghi, G., Williams-Chestnut, R. E., Sinicropi, D. V. and Osborne, J. L. (1997) Increased nerve growth factor levels in the urinary bladder of women with idiopathic sensory urgency and interstitial cystitis. *British Journal of Urology*, **79**, 572–577.

Lundeberg, T., Liedberg, H., Nordling, L., Theodorsson, E., Owzarski, A. and Ekman, P. (1993) Interstitial cystitis: correlation with nerve fibres, mast cells and histamine. *British Journal of Urology*, **71**, 427–429.

Lutzelschwab, C., Pejler, G., Aveskogh, M. and Hellman, L. (1997) Secretory granule proteases in rat mast cells. Cloning of 10 different serine proteases and a carboxypeptidase A from various rat mast cell populations. *Journal of Experimental Medicine*, **185**, 13–29.

Lynes, W. L., Flynn, S. D., Shortliffe, L. D., Lemmers, M., Zipser, R., Roberts, L. J. I. and Stamey, T. A. (1987) Mast cell involvement in interstitial cystitis. *Journal of Urology*, **138**, 746–752.

Malaviya, R., Ross, E., Jakschik, B. A. and Abraham, S. N. (1994) Mast cell degranulation induced by type 1 fimbriated *Escherichia coli* in mice. *Journal of Clinical Investigation*, **93**, 1645–1653.

Marchand, J. E., Sant, G. R. and Kream, R. M. (1998) Increased expression of substance P receptor-encoding mRNA in bladder biopsies from patients with interstitial cystitis. *British Journal of Urology*, **81**, 224–228.

Marshall, J. S., Stead, R. H., McSharry, C., Nielsen, L. and Bienenstock, J. (1990) The role of mast cell degranulation products in mast cell hyperplasia. *Journal of Immunology*, **144**, 1886–1892.

Marshall, J. S. and Waserman, S. (1995) Mast cells and the nerves – potential interactions in the context of chronic disease. *Clinical and Experimental Allergy*, **25**, 102–110.

Matsuda, H., Kannan, Y., Ushio, H., Kiso, Y., Kanemoto, T., Suzuki, H. and Kitamura, Y. (1991) Nerve growth factor induces development of connectivce tissue-type mast cells *in vitro* from murine bone marrow cells. *Journal of Experimental Medicine*, **174**, 7–14.

Mattila, J. (1982) Vascular immunopathology in interstitial cystitis. *Clinical Immunology and Immunopathology*, **23**, 648–655.

Mattila, J., Pitkänen, R., Vaaslasti, T. and Seppänen, J. (1983) Fine-structural evidence for vascular injury in patients with interstitial cystitis. *Virchows Archiv [A]*, **398**, 347–355.

Mayer, E. A. (2000) The neurobiology of stress and gastrointestinal disease. *Gut*, **47**, 861–869.

Mayer, E. A., Naliboff, B. D., Chang, L. and Coutinho, S. V. (2001) V. stress and irritable bowel syndrome. *American Journal of Physiology*, **280**, G519–G524.

McKay, D. M. and Bienenstock, J. (1994) The interaction between mast cells and nerves in the gastrointestinal tract. *Immunology Today*, **15**, 533–538.

Messing, E. M. and Stamey, T. A. (1978) Interstitial cystitis: early diagnosis, pathology and treatment. *Urology*, **12**, 381–392.

Middleton, E., Jr., Kandaswami, C. and Theoharides, T. C. (2000) The effects of plant flavonoids on mammalian cells: Implications for inflammation, heart disease and cancer. *Pharmacological Reviews*, **52**, 673–751.

Montier, F., Carruette, A., Moussaoui, S., Boccio, D. and Garret, C. (1994) Antagonism of substance P and related peptides by RP 67580 and CP-96,345, at tachykinin NK_1 receptor sites, in the rat urinary bladder. *European Journal of Pharmacology*, **251**, 9–14.

Mousli, M., Bronner, C., Landry, Y., Bockaert, J. and Rouot, B. (1990a) Direct activation of GTP-binding regulatory proteins (G- proteins) by substance P and compound 48/80. *Federation of European Biochemical Societies Letters*, **259**, 260–262.

Mousli, M., Bueb, J.-L., Bronner, C., Rouot, B. and Landry, Y. (1990b) G protein activation: a receptor-independent mode of action for cationic amphiphilic neuropeptides and venom peptides. *Trends in Pharmacological Sciences*, **11**, 358–362.

Newson, B., Dahlström, A., Enerbäck, L. and Ahlman, H. (1983) Suggestive evidence for a direct innervation of mucosal mast cells. *Neuroscience*, **10**, 565–570.

Nickel, J. C., Johnston, B., Downey, J., Barkin, J., Pommerville, P., Gregoire, M. and Ramsey, E. (2000) Pentosan polysulfate therapy for chronic nonbacterial prostatitis (chronic pelvic pain syndrome category IIIA): a prospective multicenter clinical trial. *Urology*, **56**, 413–417.

Novicki, D. E., Larson, T. R. and Swanson, S. K. (1998) Interstitial cystitis in men. *Urology*, **52**, 621–624.

O'Sullivan, M., Clayton, N., Breslin, N. P., Harman, I., Bountra, C., McLaren, A. and O'Morain, C. A. (2000) Increased mast cells in the irritable bowel syndrome. *Neurogastroenterology Motility*, **12**, 449–457.

Palea, S., Corsi, M., Artibani, W., Ostardo, E. and Pietra, C. (1996) Pharmacological characterization of Tachykinin NK_2 receptors on isolated human urinary bladder, prostatic urethra and prostate. *Journal of Pharmacology & Experimental Therapeutics*, **277**, 700–705.

Pang, X., Boucher, W., Triadafilopoulos, G., Sant, G. R. and Theoharides, T. C. (1996) Mast cell and substance P-positive nerve involvement in a patient with both irritable bowel syndrome and interstitial cystitis. *Urology*, **47**, 436–438.

Pang, X., Marchand, J., Sant, G. R., Kream, R. M. and Theoharides, T. C. (1995a) Increased number of substance P positive nerve fibers in interstitial cystitis. *British Journal of Urology*, **75**, 744–750.

Pang, X., Cotreau-Bibbo, M. M., Sant, G. R. and Theoharides, T. C. (1995b) Bladder mast cell expression of high affinity estrogen receptors in patients with interstitial cystitis. *British Journal of Urology*, **75**, 154–161.

Pang, X., Sant, G. R. and Theoharides, T. C. (1998) Altered expression of bladder mast cell growth factor receptor (c-kit) expression in interstitial cystitis. *Urology*, **51**, 939–944.

Parsons, C. L. (1996) Interstitial cystitis. *International Journal of Urology*, **3**, 415–420.

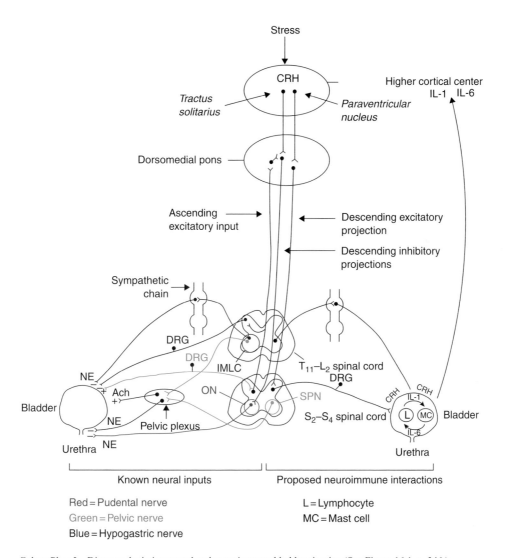

Stress

CRH

Higher cortical center
IL-1   IL-6

Tractus solitarius

Paraventricular nucleus

Dorsomedial pons

Ascending excitatory input

Descending excitatory projection

Descending inhibitory projections

Sympathetic chain

DRG

DRG

IMLC

NE

Ach

ON

T_{11}–L_2 spinal cord

DRG

Bladder

NE

Pelvic plexus

SPN

S_2–S_4 spinal cord

CRH

CRH

IL-1

L   MC   Bladder

IL-6

Urethra   NE

Urethra

Known neural inputs

Proposed neuroimmune interactions

Red = Pudental nerve

Green = Pelvic nerve

Blue = Hypogastric nerve

L = Lymphocyte

MC = Mast cell

Colour Plate I   Diagram depicting neural and neuroimmune bladder circuits. (See Figure 16.1, p. 346.)

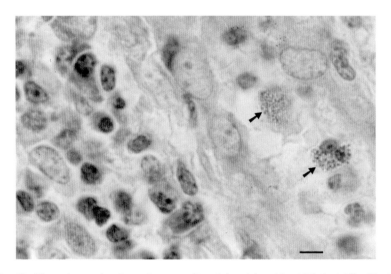

Colour Plate II   Photomicrographs of granular mast cells staining violet with acidified toluidine blue in an area of inflammation from the bladder of a female IC patient. Bar = 10 μm. (See Figure 16.3, p. 350.)

Parsons, C. L., Greenberger, M., Gabal, L., Bidair, M. and Barme, G. (1998) The role of urinary potassium in the pathogenesis and diagnosis of interstitial cystitis. *Journal of Urology*, **259**, 1862–1866.

Parsons, C. L., Stauffer, C. and Schmidt, J. D. (1980) Bladder-surface glycosaminoglycans: an efficient mechanism of environmental adaptation. *Science*, **208**, 605–607.

Parsons, C. L., Zupkas, P. and Parsons, J. K. (2001) Intravesical potassium sensitivity in patients with interstitial cystitis and urethral syndrome. *Urology*, **57**, 428–433.

Patra, P. B., Tibbitts, F. D. and Westfall, D. P. (1995) Endocrine status and urinary mast cells: possible relationship to interstitial cystitis. *NIH IC Symposium* Jan. 9–11, 1995 (abstract).

Patra, P. B. and Westfall, D. P. (1996) Potentiation by bradykinin and substance P of purinergic neurotransmission in urinary bladder. *Journal of Urology*, **156**, 532–535.

Peeker, R., Enerbäck, L., Fall, M. and Aldenborg, F. (2000) Recruitment, distribution and phenotypes of mast cells in interstitial cystitis. *Journal of Urology*, **163**, 1009–1015.

Pinter, E. and Szolcsanyi, J. (1995) Plasma extravasation in the skin and pelvic organs evoked by antidromic stimulation of the lumbosacral dorsal roots of the rat. *Neuroscience*, **68**, 603–614.

Plaut, M., Pierce, J. H., Watson, C. J., Hanley-Hyde, J., Nordan, R. P. and Paul, W. E. (1989) Mast cell lines produce lymphokines in response to cross-linkage of FCϵRI or to calcium ionophores. *Nature*, **339**, 64–67.

Pontari, M. A. and Hanno, P. M. (1995) Interstitial cystitis. In *Campbell's Urology*, edited by P. C. Walsh *et al.*, pp. 1–19. New York: W.B. Saunders Company.

Propert, K. J., Schaeffer, A. J., Brensinger, C. M., Kusek, J. W., Nyberg, L. M. and Landis, J. R. (2000) A Prospective study of interstitial cystitis: results of longitudinal followup of the interstitial cystitis data base cohort. *Journal of Urology*, **163**, 1434–1439.

Rizzo, C. A. and Hey, J. A. (2000) Activity of nonpeptide tachykinin antagonists on neurokinin a induced contractions in dog urinary bladder. *Journal of Urology*, **163**, 1971–1974.

Rothrock, N. E., Lutgendorf, S. K., Kreder, K. J., Ratliff, T. and Zimmerman, B. (2001) Stress and symptoms in patients with interstitial cystitis: a life stress model. *Urology*, **57**, 422–427.

Rovner, E., Propert, K. J., Brensinger, C., Wein, A. J., Foy, M., Kirkemo, A., Landis, J. R., Kusek, J. W., Nyberg, L. M. and Interstitial Cystitis Data Base Group (2000) Treatments used in women with interstitial cystitis:the interstitial cystitis data base (ICDB) study experience. *Urology*, **56**, 940–945.

Rozniecki, J. J., Dimitriadou, V., Lambracht-Hall, M., Pang, X. and Theoharides, T. C. (1999) Morphological and functional demonstration of rat dura mast cell-neuron interactions in vitro and in vivo. *Brain Research*, **849**, 1–15.

Saban, M. R., Saban, R. and Bjorling, D. E. (1997) Kinetics of peptide-induced release of inflammatory mediators by the urinary bladder. *British Journal of Urology*, **80**, 742–747.

Saban, R., Haak-Frendscho, M., Zine, M., Presta, L. G., Bjorling, D. E. and Jardieu, P. (1997) Human anti-IgE monoclonal antibody blocks passive sensitization of human and rhesus monkey bladder. *Journal of Urology*, **157**, 689–693.

Saban, R., Saban, M. R., Nguyen, N. B., Lu, B., Gerard, C., Gerard, N. P. and Hammond, T. G. (2000) Neurokinin-1 (NK-1) receptor is required in antigen-induced cystitis. *American Journal of Pathology*, **156**, 775–780.

Sant, G. R. (1991) Interstitial cystitis. *Monographs in Urology*, **12**, 37–63.

Sant, G. R. (1997) Interstitial cystitis. *Current Opinion Obstetrics and Gynecology*, **9**, 332–336.

Schuurs, A. H. and Verheul, H. A. (1990) Effects of gender and sex steroids on the immune response. *Journal of Steroid Biochemistry*, **35**, 157–172.

Schwartz, L. B. (1987) Mediators of human mast cells and human mast cell subsets. *Annals of Allergy*, **58**, 226–235.

Serafin, W. E. and Austen, K. F. (1987) Mediators of immediate hypersensitivity reactions. *New England Journal of Medicine*, **317**, 30–34.

Shanahan, F., Denburg, J. A., Fox, J., Bienenstock, J. and Befus, D. (1985) Mast cell heterogeneity: effects of neuroenteric peptides on histamine release. *Journal of Immunology*, **135**, 1331–1337.

Shoskes, D. A., Zeitlin, S. I., Shahed, A. and Rajfer, J. (1999) Quercetin in men with category III chronic prostatitis: a preliminary prospective, double-blind, placebo-controlled trial. *Urology*, **54**, 960–963.

Silva, C., Rio, M. and Cruz, F. (2000) Desensitization of bladder sensory fibers by intravesical resiniferatoxin, a capsaicin analog: long-term results for the treatment of detrusor hyperreflexia. *European Urology*, **38**, 444–452.

Singh, G. and Thomas, D. G. (1996) Interstitial cystitis: involvement of the intestine. *European Urology*, **29**, 322–324.

Skofitsch, G., Savitt, J. M. and Jacobowitz, D. M. (1985) Suggestive evidence for a functional unit between mast cells and substance P fibers in the rat diaphragm and mesentery. *Histochemistry*, **82**, 5–8.

Spanos, C., El-Mansoury, M., Letourneau, R. J., Minogiannis, P., Greenwood, J., Siri, P., Sant, G. R. and Theoharides, T. C. (1996) Carbachol-induced activation of bladder mast cells is augmented by estradiol – implications for interstitial cystitis. *Urology*, **48**, 809–816.

Spanos, C. P., Pang, X., Ligris, K., Letourneau, R., Alferes, L., Alexacos, N., Sant, G. R. and Theoharides, T. C. (1997) Stress-induced bladder mast cell activation: implications for interstitial cystitis. *Journal of Urology*, **157**, 669–672.

Stead, R. H., Dixon, M. F., Bramwell, N. H., Riddell, R. H. and Bienenstock, J. (1989) Mast cells are closely apposed to nerves in the human gastrointestinal mucosa. *Gastroenterology*, **97**, 575–585.

Stein, P. C., Torri, A. and Parsons, C. L. (1999) Elevated urinary norepinephrine in interstitial cystitis. *Urology*, **53**, 1140–1143.

Suzuki, R., Furuno, T., McKay, D. M., Wolvers, D., Teshima, R., Nakanishi, M. and Bienenstock, J. (1999) Direct neurite-mast cell communication *in vitro* occurs via the neuropeptide substance P[1]. *Journal of Immunology*, **163**, 2410–2415.

Theoharides, T. C. (1996) Mast cell: a neuroimmunoendocrine master player. *International Journal of Tissue Reactions*, **18**, 1–21.

Theoharides, T. C. and Douglas, W. W. (1978) Somatostatin induces histamine secretion from rat peritoneal mast cells. *Endocrinology*, **102**, 1637–1640.

Theoharides, T. C. and Sant, G. R. (1994) The role of the mast cell in interstitial cystitis. *Urologic Clinics of North America*, **21**, 41–53.

Theoharides, T. C. and Sant, G. R. (2001) New agents for the medical treatment of interstitial cystitis. *Expert Opinion Investigational Drugs*, **10**, 527–552.

Theoharides, T. C., Bondy, P. K., Tsakalos, N. D. and Askenase, P. W. (1982) Differential release of serotonin and histamine from mast cells. *Nature*, **297**, 229–231.

Theoharides, T. C., Flaris, N., Cronin, C. T., Ucci, A. and Meares, E. (1990) Mast cell activation in sterile bladder and prostate inflammation. *International Archives of Allergy and Applied Immunology*, **92**, 281–286.

Theoharides, T. C., Kempuraj, D. and Sant, G. R. (2001) Images in clinical urology: massive extracellular tryptase from activated bladder mast cells in interstitial cystitis. *Urology*, **58**, 605–606.

Theoharides, T. C., Letourneau, R., Patra, P., Hesse, L., Pang, X., Boucher, W., Mompoint, C. and Harrington, B. (1999) Stress-induced rat intestinal mast cell intragranular activation and inhibitory effect of sulfated proteoglycans. *Digestive Diseases and Sciences*, **44**, 87S–93S.

Theoharides, T. C., Pang, X., Letourneau, R. and Sant, G. R. (1998) Interstitial cystitis: a neuroimmunoendocrine disorder. *Annals of the New York Academy of Sciences*, **840**, 619–634.

Theoharides, T. C., Patra, P., Boucher, W., Letourneau, R., Kempuraj, D., Chiang, G., Jeudy, S., Hesse, L. and Athanasiou, A. (2000) Chondroitin sulfate inhibits connective tissue mast cells. *British Journal of Pharmacology*, **131**, 10139–10149.

Theoharides, T. C., Sant, G. R., El-Mansoury, M., Letourneau, R. J., Ucci, A. A., Jr and Meares, E. M., Jr. (1995) Activation of bladder mast cells in interstitial cystitis: a light and electron microscopic study. *Journal of Urology*, **153**, 629–636.

Theoharides, T. C., Kempuraj, D. and Sant, G. R. (2001) Mast cell involvement in interstitial cystitis: a review of human and experimental evidence. *Urology*, **57**, 47–55.

Thompson, A. C. and Christmas, T. J. (1996) Interstitial cystitis – an update. *British Journal of Urology*, **78**, 813–820.

Tomaszewski, J. E., Landis, J. R., Russack, V., Williams, T. M., Wang, L., Hardy, C., Brensinger, C., Matthews, Y. L., Abele, S. T., Kusek, J. W. and Nyberg, L. M. (2001) Biopsy features are associated with primary symptoms in interstitial cystitis: results from the interstitial cystitis database study. *Urology*, **57**, 67–81.

Toyoda, M., Makino, T., Kagoura, M. and Morohashi, M. (2000) Immunolocalization of substance P in human skin mast cells. *Archives of Dermatological Research*, **292**, 418–421.

Tsai, S.-H., Liang, Y.-C., Lin-Shiau, S.-Y. and Lin, J.-K. (1999) Suppression of TNFα- mediated NFκB activity by myricetin and other flavonoids through downregulating the activity of IKK in ECV304 cells. *Journal of Cellular Biochemistry*, **74**, 606–615.

Van de Merwe, J. P. and Arendsen, H. J. (2000) Interstitial cystitis: a review of immunological aspects of the aetiology and pathogenesis, with a hypothesis. *British Journal of Urology International*, **85**, 995–999.

Vizzard, M. A. (2000) Up-regulation of pituitary adenylate cyclase-activating polypeptide in urinary bladder pathways after chronic cystitis. *Journal of Comparative Neurology*, **420**, 335–348.

Vliagoftis, H., Dimitriadou, V., Boucher, W., Rozniecki, J. J., Correia, I., Raam, S. and Theoharides, T. C. (1992) Estradiol augments while tamoxifen inhibits rat mast cell secretion. *International Archives of Allergy and Immunology*, **98**, 398–409.

von Heyden, B., Jordan, U. and Hertle, L. (1998) Neurotransmitters in the human urethral sphincter in the absence of voiding dysfunction. *Urological Research*, **26**, 299–310.

Wang, Y., Zhou, Y., Mourad, M. S. and Hassouna, M. M. (2000) Neuromodulation reduces urinary frequency in rats with hydrochloric acid-induced cystitis. *British Journal of Urology International*, **86**, 726–730.

Wei, D. C., Politano, V. A., Selzer, M. G. and Lokeshwar, V. B. (2000) The association of elevated urinary total to sulfated glycosaminoglycan ratio and high molecular mass hyaluronic acid with interstitial cystitis. *Journal of Urology*, **163**, 1577–1583.

Wesselmann, U. (2001) Interstitial cystitis: a chronic visceral pain syndrome. *Urology*, **57**, 32–39.

Weston, A. P., Biddle, W. L., Bhatia, P. S. and Miner, P. B., Jr. (1993) Terminal ileal mucosal mast cells in irritable bowel syndrome. *Digestive Diseases and Sciences*, **38**, 1590–1595.

Wiesner-Menzel, L., Schulz, B., Vakilzadeh, F. and Czarnetzki, B. M. (1981) Electron microscopical evidence for a direct contact between nerve fibers and mast cells. *Acta Dermatologica Venereologica (Stockholm)*, **61**, 465–469.

Wilson, J. L., Carmody, J. J. and Walker, J. S. (2000) The importance of the hypothalamo-hypophyseal-adrenal axis to the anti-inflammatory actions of the k-opioid agonist PNU-50, 488H in rats with adjuvant arthritis. *Journal of Pharmacology and Experimental Therapeutics*, **294**, 1131–1136.

Xiang, Z. and Nilsson, G. (2000) IgE receptor-mediated release of nerve growth factor by mast cells. *Clinical and Experimental Allergy*, **30**, 1379–1386.

Yamada, T., Murayama, T., Mita, H. and Akiyama, K. (2000) Subtypes of bladder mast cells in interstitial cystitis. *International Journal of Urology*, **7**, 292–297.

Yamada, T., Murayama, T., Mita, H., Akiyama, K. and Taguchi, H. (1998) Alternate occurrence of allergic disease and an unusual form of interstitial cystitis. *International Journal of Urology*, **5**, 329–335.

Yun, S. K., Laub, D. J., Weese, D. L., Lad, P. M., Leach, G. E. and Zimmern, P. E. (1992) Stimulated release of urine histamine in interstitial cystitis. *Journal of Urology*, **148**, 1145–1148.

Zondek, B. & Bromberg, Y. M. (1945) Endocrine allergy, I. allergic sensitivity to endogenous hormones. *Journal of Allergy*, **16**, 1–16.

Index